PRAISE FOR

# HOW TO BREW

"I have always considered *How to Brew* the best complete resource for both new and experienced brewers. In this new edition, John has made a great resource even better and up-to-date with the latest information and techniques. *How to Brew* has all you need to go from complete novice to expert brewer. If you brew, you should own this book."

**—JAMIL ZAINASHEFF,** Chief Heretic, Heretic Brewing Company

"Not only is *How to Brew* one of the most critical and comprehensive DIY homebrewing books available today, I have even seen it on the bookshelves at many great craft breweries."

**—SAM CALAGIONE,** CEO and Founder, Dogfish Head Craft Brewery

"Owning *How to Brew* is like having a brewmaster as your best friend. In the 30-plus years since the American craft beer revolution got its start, countless brewing books have appeared. None, however, has achieved the status of *How to Brew*, which is thorough, comprehensive, and beautifully organized. And now, this new expanded and enhanced edition improves on the original. It's a considerable feat to create a book that is invaluable both to first-time brewers and professional brewmasters, but John has done it with a book that is essential for everyone who is serious about brewing."

**—JIM KOCH,** Founder & Brewer of Samuel Adams

"Whenever I'm asked about what books I'd recommend to a brewer, I always recommend John Palmer's *How to Brew*. It is jam-packed with information that will help beginning brewers get started, and the more advanced technical brewing chemistry and science details that experienced brewers need to become great brewers. It works at all levels."

**—MITCH STEELE,** COO and Brewmaster, New Realm Brewing Company

"John Palmer's *How to Brew* has been a great resource for homebrewers ever since he self-published the first edition in 2000. As the former owner of a homebrew supply store myself, I appreciate John's focus on how to avoid some common pitfalls that many aspiring brewers stumble over with his wise emphasis on "the top five priorities." From the basics (equipment and raw materials), to the critical (cleanliness), to the fun part (making your own beer recipes), this book covers the brewing process from start to finish. This new edition offers more information with an updated layout and expanded table of contents, which make it even easier to use. Anyone contemplating homebrewing, or looking to step up their homebrewing game, should start here."

**—KEN GROSSMAN,** Founder and Brewmaster, Sierra Nevada Brewing Co.

# HOW TO
# BREW

## Everything You Need to Know
## to Brew Great Beer Every Time

Fourth Edition

## By John J. Palmer

BREWERS
PUBLICATIONS™

Brewers Publications™
A Division of the Brewers Association
PO Box 1679, Boulder, Colorado 80306-1679
BrewersAssociation.org
BrewersPublications.com

Proudly printed in the United States of America.

10 9 8 7 6 5 4 3 2 1
ISBN-13: 978-1-938469-35-0
Library of Congress Cataloging-in-Publication Data

Names: Palmer, John J., 1963- author.
Title: How to brew : everything you need to know to brew great beer every
  time / by John J. Palmer.
Description: Boulder, Colorado : Brewers Publications, a division of the
  Brewers Association, [2017] | Includes bibliographical references and
  index.
Identifiers: LCCN 2016057537 (print) | LCCN 2016058117 (ebook) | ISBN
  9781938469350 | ISBN 9781938469367 (E-book) | ISBN 9781938469404 (Enhanced
  E-book)
Subjects: LCSH: Brewing--Amateur's manuals.
Classification: LCC TP570 .P275 2017 (print) | LCC TP570 (ebook) | DDC
  663/.3--dc23
LC record available at https://lccn.loc.gov/2016057537

Publisher: Kristi Switzer
Technical Editor: Randy Mosher, Chris Colby
Copyediting: Iain Cox
Indexing: Doug Easton
Production and Design Management: Stephanie Johnson Martin and Jason Smith
Cover and Interior Design: Kerry Fannon
Production: Justin Petersen
Interior Images: Ed Dorroh of Fermentography.com, Brewers Association and as noted.
Cover Photo: Souders Studios and Square Pixels
Interior Illustrations: Danny Rubin (pages 260, 273, 449-457)
Enhanced e-book videos: Luke Trautwein. Special thanks to The Brew Hut and Dry Dock Brewing Co.

This book is dedicated to the veterans of the Homebrew Digest, the original homebrewing forum, who inspired me to take up this hobby so many years ago. I am always overwhelmed by the generosity of brewers for sharing information, their time, and their passion for beer and brewing, and I always hope to be able to return it in kind.

# Table of Contents

# Section IV Appendices

# Preface to the Fourth Edition

The first edition of *How to Brew* was written in the late 1990s and subsequently published online at HowToBrew.com in June of 2000. It was very well received and immediately generated requests for a hard copy. The second edition was self-published in 2001 and began selling at homebrewing supply shops around the country. Brewers Publications became interested and a thoroughly revised third edition was published in 2006, and it has been a best seller ever since.

However, brewing technology does not stand still. This fourth edition that you are holding incorporates the many advances in both home and craft brewing that have occurred in the past ten years.

- The variety of malts, hops, and yeast strains available has exploded, doubling or tripling in number.
- We have advanced our understanding of the yeast cycle and fermentation, so that warm maturation of lagers is now common practice, producing a clean lager beer in three weeks instead of three months.
- American IPA is now the most popular craft beer style and has changed the way we dry hop beers, from long and cold to short and relatively warm.
- Brew-in-a-bag (BIAB), which I first wrote about in 2007 as part of *Brewing Classic Styles* with Jamil Zainasheff, has become a mainstream brewing technique for all-grain brewing, using less equipment and higher water-to-grist ratios to shorten the brew day.
- In 2013, Brewers Publications released the *Water* book, written by me and Colin Kaminski, as part of the Brewing Elements series, and that knowledge has been incorporated into this edition of *How to Brew*.
- Brewing equipment has undergone a renaissance as well; part of the hobby of homebrewing was innovating and repurposing common kitchen items into brewing equipment that you built yourself, and while it still is part of the hobby, there is a lot of purpose-made stainless steel brewing equipment to choose from these days.

When I first wrote *How to Brew*, most people were brewing beer in order to have beer styles that were otherwise not available. Import beers were available in specialty bottle shops, but were often old and oxidized. Today, craft beer is everywhere, and, despite the preponderance of IPA, many ale and lager styles are readily available. Today's homebrewer is brewing for the pure pleasure of creating a beer, rather than to fill a void in availability. Therefore, my focus in this latest edition is to help you understand how to brew the best beer.

To do this I have listed my "Top Five Brewing Priorities for Brewing Great Beer" in chapter 1, and each chapter thereafter builds on those priorities to help you understand where the different ingredients and processes fit into the overall scheme of things. Some chapters may be more technical than you are expecting, but hopefully I did a good job of laying out the big picture to illustrate what is most important, so that you can skim through or skip over those pages until you have gained more experience and are actually interested in that subject.

I have learned a lot over the past ten years and I hope this edition will help you understand more about your brewing and enable you to brew your best beer ever.

# Acknowledgments

There are many people I must thank again and again for sharing their knowledge, insight, and patience with me as I strive to become a better brewer and write a better book. Those people are: Aaron Hyde, Aaron Justus, A. J. deLange, Amaey Mundkur, Anders Cooper, Dr. Berne Jones, Bob Hansen, Dr. Brad Smith, Dr. Brian Kern, Dr. Charlie Bamforth, Charlie Essers, Dr. Chris Colby, Dr. Chris White, Christoph Neugrodda, Colin Kaminski, Dan Bies, Dana Johnson, Drew Beechum, Dylan Dunn, Ed Dorroh, Ernie Rector, Dr. Evan Evans, Fred Scheer, Gary Glass, Gil Sanchez, Gordon Lane, Gordon Strong, Dr. Graham Stewart, Greg Doss, James Spencer, Jamil Zainasheff, Jeremy and Ting Raub, John Blichmann, John Mallett, Jon Herskovits, Jordon Guerts, Josiah Blomquist, Justin Crossley, Kai Troester, Kara Taylor, Karen Fortmann, Kristi Switzer, Dr. Ludwig Narziss, Malcolm Frazer, Mark Jilg, Mark Sammartino, Dr. Marshall Schott, Martin Boan, Martin Brungard, Matt Brynildson, Mercedes Hemmer, Mike Maag, Mike McDole, Mitch Steele, Neva Parker, Randy Mosher, Ray Daniels, Rick Blankemeier, Ronaldo Dutra Ferreira, Ryan Brooks, Stan Hieronymous, Stephen Mallory, Steve Alexander, Steve Gonzalez, Steve Kinsey, Thomas Weyermann, Todd Peterson, and Dr. Tom Shellhammer.

I would also like to thank the members of the American Homebrewers Association for great fellowship over the years, and, in particular, all of the homebrewing clubs, here in the US and around the world, that have made me feel so welcome at your events. Cheers!

Of course, I could not have done this at all without the support of my wife and family, who put up with the constant travel and kitchen experiments. Thank you.

John Palmer
May 1, 2017

# Abbreviations

| | |
|---|---|
| AA | alpha acids |
| ABV | alcohol by volume |
| DME | dried malt extract |
| DMS | dimethyl sulfide |
| EBC | European Brewery Convention |
| EPA | Environmental Protection Agency |
| FDA | Food and Drug Administration |
| FG | final gravity |
| FWH | first wort hopping |
| HDPE | high-density polyethylene |
| HEPA | high-efficiency particulate arrestance |
| HLT | hot liquor tank |
| IBU | international bitterness unit |
| °L | degrees Lovibond |
| LME | liquid malt extract |
| MSDS | material safety data sheet |
| OG | original gravity |
| PBW | Powder Brewery Wash |
| SRM | Standard Reference Method |

**Units (non-metric units are US unless stated otherwise)**

| | |
|---|---|
| cm | centimeter |
| Eq | equivalent |
| fl. oz. | fluid ounce |
| g | gram |
| gal. | gallon |
| in. | inch |
| " | inch |
| kg | kilogram |
| kPa | kilopascal |
| L | liter |
| lb. | pound |
| mEq | milliequivalent |
| mg | milligram |
| mL | milliliter |
| mol | mole |
| N | normality (Eq/L) |
| oz. | ounce |
| ppm | parts per million |
| psi. | pounds per square inch |
| qt. | quart |

# Introduction

My purpose in writing *How to Brew* is to introduce everyone to the immensely satisfying world of brewing your own beer. You may think that it's complicated and worry that you can't do it, but I want to assure that you can; it's no harder than any other cooking process. If you can heat a can of soup, make scrambled eggs, or bake muffins from a mix, then you can brew beer. Just follow the instructions and understand that the ingredients, time, and temperature all affect the outcome. Let's use the muffins as an example. A typical mix will include flour, sugar, leavening agents, and perhaps some fruit, such as blueberries. You supply the shortening, eggs, and some milk or water. The instructions explain how to mix the ingredients together, place the batter in the specified type of baking pan, and bake it in the oven for a specific time at a specific temperature. It's really quite easy once the process is broken down into a clear set of instructions.

It's the same when brewing your own beer. You follow the recipe instructions to mix and boil the wort, cool it and add the proper amount of yeast, and then ferment the wort at the recommended temperature for the recommended amount of time. With practice you will develop a gut instinct for each part of the process, just as in baking.

Many homebrewing beginners put too much emphasis on the recipe. While a good recipe plays a large part in creating a good beer, the cooking and fermentation of the recipe is much more important to a successful brew. A recipe is simply a collection of ingredients and a plan; it is the execution of that plan that makes all the difference between success and failure. I am going to teach you how to brew. Once you understand how the brewing process works, you will be able to brew a good beer from any reasonable recipe.

In the first section of this book we are going to lay the groundwork for the rest of your brewing education. As with every new skill, it helps to learn to do things the right way the first time, rather than learning via shortcuts that you will have to unlearn later on. On the other hand, you don't need to know how an internal combustion engine works when you are learning how to drive. You just need to know that it needs fuel and oil to work.

To learn to brew beer you don't need to learn how yeast metabolizes the malt sugars, but you do need to understand that eating sugar is what yeast does, and you need to understand what the yeast needs from you to get the job done. Once you understand that, you can do your part, the yeast can do its part, and good beer will happen. As you gain some familiarity with the brewing process you can delve deeper into its inner workings and probably make your beer even better.

So, in section 1, "Brewing Beer Kits," you will learn to drive. Chapter 1, "Brewing Your First Batch of Beer," will provide an overview of the entire process for producing beer from a kit on your kitchen stove. You can use this one chapter to brew a beer right now—today. Chapter 2, "Cleaning and Sanitizing," explains why good preparation, including sanitation, is important, and how to go about it. Chapter 3 gives a brief overview of malt and malt extract, before we go on to chapter 4, "Brewing with Beer Kits and Extracts," which examines the key aspects of do-it-yourself beer kits and how to use them properly. Chapter 5 covers the different kinds of hops, why to use them, how to use them, and how to measure them for consistency in your brewing. Chapter 6, "Yeast and Fermentation," explains what yeast is and what it needs to grow, and examines how the yeast ferments wort into beer so you will understand what you are trying to do, all without going into excruciating detail. Building on this, chapter 7 discusses the different types of yeast strains available, how to prepare them, and how to propagate your favorite yeast cultures.

The last ingredient chapter in section 1, chapter 8, "Water for Extract Brewing," cuts to the chase with a few dos and don'ts about a very complex subject. From there, section 1 moves into the physical processes of brewing. Chapter 9, "Brewing with a Full-Volume Boil," scales up the process with a full-volume boil of six gallons, or 23 L. Chapter 10, "Priming, Bottling, and Kegging," explains each step of how to package your five gallons of new beer into something you can really use.

Everybody wants to brew their favorite beer that they buy at the store, and it's usually a lager. So, chapter 11, "How to Brew Lager Beer," examines the key differences of lager brewing, building on what you have already learned so far. In a similar fashion, chapter 12, "Brewing Strong Beers," addresses another common desire of the homebrew beginner, which is to brew something really strong that is still enjoyable to drink. For those wishing to experiment further, section 1 finishes with chapter 13, "Brewing Fruit Beers," and chapter 14, "Brewing Sour Beers."

Section 1 is a long section, but you will learn to brew, and brew right the first time. In section 2, chapters 15 to 19 delve deeper into the techniques needed for all-grain brewing, including a detailed look at malted barley and adjuncts, so you can take greater control of the ingredients and, thus, your beer. Chapter 20, "Brewing Your First All-Grain Batch," puts all of this into practice. Section 2 finishes with chapters 21 and 22, which give a detailed discussion of water chemistry and how you can manipulate it to suit whatever style of beer you wish to make. Finally, section 3, "Recipes, Experimenting, and Troubleshooting," and the appendices in section 4 will give you the road maps, the tools, and the repair manual you need to always achieve your goals.

It is my sincere hope that this book will help you to derive the same sense of fun and enthusiasm that I have experienced, and that it will enable you to brew some really outstanding beer.

Good Brewing,
John

# Section I
## Brewing Beer Kits

# Brewing Your First Batch of Beer

**1**

## What Do I Do?

If you are like me, you are probably standing in the kitchen wanting to get started. Your beer kit[1] and equipment are on the counter, and you are wondering how long this will take and what to do first. Well, my best advice would be for you to read the first ten chapters, which cover everything from cleaning and sanitation to ingredients to boiling, fermentation, and packaging. These chapters will teach you all the fundamentals of the brewing process, so you won't be misled by incomplete instructions in the kit, and you will have an outstanding first batch.

But, if you are like me, you probably want to do this right now while you have some time. Therefore, this first chapter is designed to walk you through your first batch—start to finish—and give you a complete overview of the brewing process. Brewing can be broken down into three main events: making the wort, fermenting the wort, and bottling, or packaging, the beer. What is wort? That's what we call the sugar solution that is boiled with hops and then left to ferment into beer. Making the wort today is going to take about three hours, and it will be about one month before you can drink your first batch.

---

[1] If you don't have a beer kit, then either assemble the ingredients for the "Cincinnati Pale Ale" recipe in this chapter, or buy one, such as a Palmer's Premium™ Beer Kit, from your local brew shop or online supplier.

Today's brew is going to use what I call the "Palmer Brewing Method," a partial boil method where only half of the wort will actually be boiled with the hops. The rest of the wort (actually malt extract) will be added to the kettle at the end of the boil to pasteurize and this higher gravity wort will then be diluted in the fermentor to the final volume of about 5.5 gallons (gal.), or 21 liters (L) for those readers using metric units.

The instructions in this first chapter are the bare bones; to understand all of the whats and whys of brewing you will need to finish reading the first ten chapters of section 1 of this book, "Brewing Beer Kits." These chapters will discuss all of the brewing ingredients and processes in detail, and explain the purpose behind each step. You will then understand what you are doing, rather than doing it that way just because "that's what it said. . ."

However, you can read all of that tomorrow while your beer is fermenting. Let's get started!

## Before We Get Started: The Top Five Priorities for Brewing Great Beer

Do you want a great beer? Success or failure starts here. This list is prioritized from highest to lowest; meaning that, if you make mistake in a higher priority, it can't be fixed by doing a lower priority correctly. Don't worry, I will walk you through all of this as we go, but I want you to understand the big picture first.

1. *Sanitation.* Good sanitation is the most important factor for brewing great beer. Brewing is all about preparing and fermenting a wort to your specification. Good sanitation ensures that your chosen yeast is the only microorganism in the brew.

2. *Fermentation temperature control.* After good sanitation, a healthy fermentation is the most important factor for brewing great beer, and good temperature control is key. Yeast are living organisms and their activity is controlled by temperature.

3. *Proper yeast management.* Good beer needs well-managed yeast. After temperature, the most important factor for managing the fermentation is pitching the proper quality and quantity of yeast. These topics are discussed in chapters 6 and 7.

4. *The boil.* The ingredients are cooked during the boil. If the wort is not cooked right, the beer will not taste right. Yes, you can undercook or overcook your beer. This will be discussed more in chapter 4.

5. *The recipe.* The definition of a good recipe is that it has the right proportions of ingredients to provide both complexity and balance of the flavors. A typical recipe will consist of a majority of a pale "base" malt, with additional specialty malts for signature flavors or accents, and enough hops to provide a balance of bitterness, flavor, and aroma to the beer. It is important to realize that a great recipe will not overcome poor brewing techniques and a good recipe does not need to be complicated.

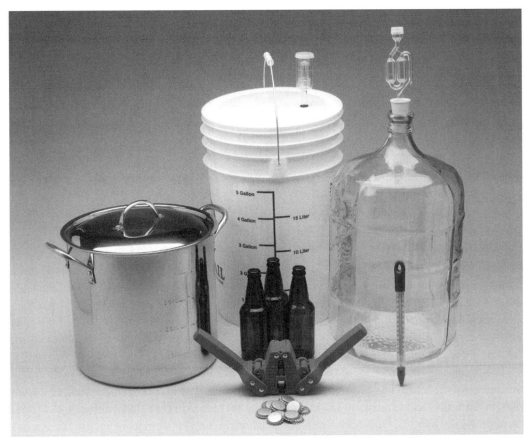

**Figure 1.1.** Here you see the typical equipment a beginning brewer will use. Fermentors, brew kettle, bottle capper, bottle caps, and bottles.

## Brew Day

### Equipment Needed to Brew Today

The following is a list of the minimum equipment you will need for this first batch. Most of these items are available at homebrewing supply shops as part of a beginner's equipment kit. Before we begin, please note that both US units of measure and metric units are used throughout the book. The conversion conventions used in this book are given in appendix H.

*Airlock.* Two basic kinds of airlock are available: single-piece or "bubbler" airlocks, and three-piece airlocks (fig. 1.2). Airlocks are filled with water or sanitizer (do not use bleach!) to prevent contamination from the outside air. Three-piece airlocks have the advantage of disassembly for more thorough cleaning, but can inadvertently allow the liquid within the airlock itself to be sucked back inside the fermentor if the internal pressure drops, which can be caused by either a drop in temperature as the wort cools or by lifting the plastic bucket if the walls of the bucket are not very rigid. Bubbler-type airlocks will not suck liquid back inside the fermentor, but are more easily clogged by fermentation gunk and they cannot be disassembled for cleaning. Both types are inexpensive. If your fermentor is a carboy, you will also need a drilled rubber stopper to hold the airlock.

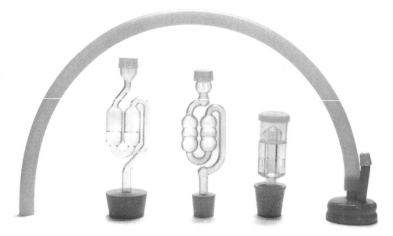

**Figure 1.2.** The basic types of airlocks available for homebrewers, and a blow-off hose.

*Boiling kettle.* For a boiling kettle, a heavy duty 5 gal. (19 L) aluminum or stainless steel stock pot is a good choice. Stainless steel is tougher and easier to maintain, but is usually more expensive than aluminum. The kettle must be able to comfortably boil a minimum of 3 gal. (11.4 L), while allowing for the fact that the wort will foam up as it is boiling. Choose a kettle that has a thick (about 0.1" or 2.5 mm) or aluminum-clad bottom, as this will help prevent scorching.

*Fermentor.* The fermentor should be able to hold at least 5 gal. of wort with about 3 in. (8 cm) of headspace for the fermentation foam. A 6 gal. food-grade plastic bucket is recommended for beginners, because they are very easy to work with and are inexpensive (the actual volume to the rim is closer to 7 gal. [26.5 L]). Carboys are also available, made from either glass or plastic. The carboy shown in figure 1.3 can also use a blowoff hose that ends in a bucket of water, which takes the place of an airlock. Buckets are nice because they typically come with a spigot for easy draining, whereas carboys are nice because you can more easily see the activity of fermentation.

**Figure 1.3.** A variety of bucket and carboy-type fermentors. Blow-off hoses, ending in a small bucket of water, can be used in place of an airlock.

*Grain bag.* A grain bag is typically a medium-sized muslin or nylon mesh bag, which is used for steeping crushed specialty malts to add more flavor to the recipe. In a pinch, a large clean sock will work as a makeshift grain bag.

*Plastic wrap or aluminum foil.* Plastic wrap or aluminum foil are very handy for covering jars or fermentors to keep them clean and sanitized until ready for use. These items are typically sanitary right out of the box.

*Pyrex® measuring cup.* A quart-sized or larger Pyrex measuring cup will quickly become one of your most valuable tools for brewing. It can be used to measure boiling water and is easily sanitized.

*Stirring spoons.* You will need a big, long-handled food-grade plastic or metal spoon for stirring the wort during the boil, and a regular spoon that you can use when rehydrating the yeast.

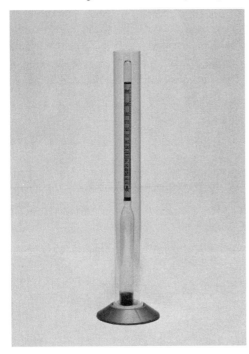

**Figure 1.4.** A hydrometer, which measures the density of a solution relative to water to give its specific gravity. Water has a specific gravity of 1.000.

*Sanitizer.* Chemical sanitizers are necessary to ensure that the yeast is the only microorganism growing in your beer. Popular no-rinse sanitizers are iodophor and Star San. See chapter 2 for more information on sanitizers.

*Thermometer.* Electronic digital thermometers are readily available and generally inexpensive, but always calibrate your thermometer using water that is boiling or otherwise at a known temperature. Having an accurate thermometer is most important for mashing.

*Hydrometer.* A hydrometer (fig. 1.4) is needed to measure specific gravity. Specific gravity is a measure of the density of a solution relative to water, which has a specific gravity of 1.000. In wort, the more sugar dissolved, the higher the specific gravity. The hydrometer measures specific gravity by how high it floats when immersed. Original gravity (OG) is the gravity of the wort before fermentation and final gravity (FG) is the gravity after. As the fermentation progresses, the yeast consumes the sugars and the OG decreases toward the anticipated FG. Advice on using hydrometers can be found in appendix A.

## Preparation (30 Minutes)

### 1. Assemble the Ingredients

You may have purchased a brewing kit at a homebrewing shop that contains the ingredients and instructions to brew a particular style of beer (fig. 1.5). A typical 5 gal. (19 L) beer kit consists of one or two containers of malt extract, steeping grains, hops, and yeast. These preassembled kits are the simplest to use.

If you don't have a kit, then head to a homebrew supply store and buy the ingredients outlined in the "Cincinnati Pale Ale" recipe below. We will be using the Palmer Brewing Method to brew

**Figure 1.5.** A typical homebrewing recipe kit that includes malt extract, hops, and yeast.

the recipe. The Palmer Brewing Method is a partial boil method, meaning that roughly half of the malt extract will be boiled with the hops in 3 gal. (11.4 L) of water. This portion of the total malt extract is listed as one of the ingredients under "Wort A" in the recipe given below. The remaining extract will be added to the boiling kettle after the boil, where it will be pasteurized by the residual heat, and then be diluted in the fermentor with an additional 3 gal. (11.4 L) of water to reach the full recipe volume. The purpose of this method is to reduce the volume of wort that needs to be boiled, saving time and energy while producing the same flavor development that would occur during a full-volume boil using the same ingredients. The reasons for this are explained in chapter 4, "Brewing With Beer Kits and Extracts."

## Cincinnati Pale Ale

*American Pale Ale*

**Original gravity:** 1.042
**Final gravity:** 1.010
**IBU:** 30

**SRM (EBC):** 5 (10)
**ABV:** 4.2%

| Palmer Brewing Method | | |
|---|---|---|
| **Wort A** | **Gravity points** | |
| 2.5 lb. (1.14 kg) pale ale DME | 37.5 | |
| 0.5 lb. (225 g) caramel 80°L malt—steeped | 2.5 | |
| Boil Gravity for 3 Gallons | 1.040 | |
| **Hop schedule*** | **Boil time (min.)** | **IBUs** |
| 0.5 oz. (15 g) Nugget (12% AA) | 60 | 21 |
| 0.5 oz. (15 g) Cascade (7% AA) | 15 | 6 |
| 0.5 oz. (15 g) Amarillo (10% AA) | Steep 15 | 3 |
| **Wort B (add after boil)** | **Gravity points** | |
| 2.5 lbs. (1.14 kg) pale ale DME | 37.5 | |
| **Yeast strain** | **Pitch (billions of cells)** | **Fermentation temp.** |
| American ale | 200 | 65°F (18°C) |

*Note: Different hop varieties may be substituted depending on availability, just be sure to use varieties with similar alpha-acid (AA) percentages (±1%–2%). Chapter 5 contains further details about using hops.*

**2. Clean and Sanitize**

It cannot be overstated: the most important factor for success in brewing is good cleaning and sanitization. Clean first, then sanitize. Clean all equipment that will be used during the brew with a mild, unscented dishwashing detergent, and be sure to rinse well. Some equipment will need to be sanitized for use after the boiling stage, as outlined below in table 1.1.

You can easily sanitize most of your equipment by filling your fermentor bucket with a couple of gallons (7–8 L) of water and adding the recommended amount of no-rinse chemical sanitizer at a typical concentration of 1 fluid ounce per gallon (1 fl. oz./gal.), or 8 milliliters per liter (8 mL/L). Swirl the sanitizer solution to make sure the walls of the fermentor are sanitized as well. Soak all items that need to be sanitized in this fermentor bucket for five minutes (see the manufacturer's instructions for specific minimum sanitization times).

After soaking and sanitizing your equipment, dump the sanitizing solution and cover the fermentor with the sanitized lid. Place the small spoon and the thermometer in the measuring cup and cover completely with plastic wrap to keep them sanitary. Chapter 2 discusses cleaning and sanitizing in more depth.

### TABLE 1.1. CLEANING AND SANITIZATION CHECK LIST

| | | | |
|---|---|---|---|
| Boiling kettle | ❏ Clean | | |
| Big stirring spoon | ❏ Clean | | |
| Regular stirring spoon | ❏ Clean | ❏ Sanitize | |
| Pyrex measuring cup | ❏ Clean | ❏ Sanitize | |
| Fermentor and lid | ❏ Clean | ❏ Sanitize | |
| Airlock | ❏ Clean | ❏ Sanitize | |
| Thermometer | ❏ Clean | ❏ Sanitize | |
| Hydrometer | ❏ Clean | ❏ Sanitize | |

## Making Wort (1 Hour)

Now we begin the fun part of brewing—creating the wort.

**Figure 1.6.** Have your ingredients laid out and ready before you begin your brew.

### 3. Create the Wort

Pour 3 gal. (11.4 L) of clean, low-mineral water in the boiling kettle. Also, pour another 3 gal. (11.4 L) of the same water into your cleaned and sanitized fermentor. It is best to use a low-mineral water source, such as distilled water, when brewing with malt extract, because the extract already contains minerals from the water used in the creation of the extract. You will be boiling the malt extract using the water in the boiling kettle and then diluting the resulting wort in the fermentor to make a total of about 5.5 gal. (21 L). You can expect to lose some water to evaporation during the boil (around 0.5 gal., or about 2 L). More water will be lost to the trub (hop and protein residue), so we start out with about 5.5 gal. in the fermentor to end up with 5 gal. (19 L) of finished beer.

### 4. Mix and Heat the Wort

Add 2.5 lb. (1.14 kg) of pale malt extract to the cold water in the boiling kettle and stir to dissolve. (Hint: dry malt extract dissolves without clumping in cold water.) If you are brewing from a pre-assembled beer kit you purchased, I recommend you follow that kit's instructions (the principles should be the same). At this point, begin heating the wort and stir frequently to prevent scorching of any undissolved malt extract on the bottom of the kettle.

**Figure 1.7.** Stirring the malt extract into cold water in the kettle.

### 5. Steep the Grains

If your purchased kit does not contain crushed grain, proceed to step 6. Put the 0.5 lb (225 g) of crushed grain in your grain bag. Heat the wort to a temperature of 120–170°F (49–77°C). The steep may be started cold, placing the grain bag in the kettle as it heats, but do not exceed 170°F (77°C). Submerge the bag and stir to make sure all of the grain is wetted. The grain bag is steeped in the hot wort just like a teabag for 30 minutes. At the end of this time, the steeped grain is removed and the wort is brought to a boil. Steeping the grains in wort as opposed to plain water improves the wort pH and, along with moderating the steeping temperature, also reduces the risk of bitter tannin extraction from the grain husks. Likewise, do not squeeze the grain bag to get all the wort out after steeping. However, a gentle squeeze to prevent dripping on the stove is fine.

**Figure 1.8.** Steeping the grains in the wort.

### 6. Boil the Wort

If you have not done so already, because you skipped step 5, bring the wort to the boil. As the wort boils, foam will form on the surface. This foam will persist for a few minutes until the wort goes through what is called the "hot break" stage (when it stops foaming). The wort will easily boil over during this foaming stage, especially when hops are first added, so stay close by and stir frequently. Let the wort boil for 5–10 minutes before adding the first hop addition. If it begins to boil over, blow on it, spray it with a little cold water from a spray bottle, turn the heat down, or do a combination of any of those three things.

Putting a few copper pennies[2] into the kettle will help prevent boilovers. Adjust the heat so that the wort in your kettle is boiling moderately, not just simmering. It should be bubbling and visibly churning at the surface, but not to the extent that it splashes out of the kettle. Do not cover your kettle during the boil, because there are volatiles that need to boil off and it is also more likely the kettle will boil over (see chapter 4, "Brewing with Beer Kits and Extract," for more information).

---

2    Yes, I know that US pennies are mostly zinc and just plated with copper, but that is fine. It's the copper plating that is important, because copper doesn't corrode in wort, and other coin metals, such as nickel, can cause haze.

**Figure 1.9.** A good rolling boil is important.

### 7a. First Hop Addition

Add 0.5 oz. (15 g) of Nugget hops to the kettle and start timing the hour of boiling.

Note: Different hop varieties may be substituted depending on availability or preference; just make sure the percentage of alpha acids (% AA) is nearly the same (within 1%) for the variety you are substituting. See chapter 5 for a more detailed discussion of hop alpha acid and quantifying hop bitterness.

### 7b. Second Hop Addition

After 45 minutes has elapsed, add the second hop addition of 0.5 oz. (15 g) of Cascade hops. These will be boiled for 15 minutes before the heat is turned off.

### 7c. Third Hop Addition

At the end of the hour, turn off the heat and add the last hop addition of 0.5 oz. (15 g) of Amarillo hops. This last hop addition will steep for 15 minutes in the hot wort before cooling or chilling.

### 8. Add the Remaining Malt Extract

Immediately after you have finished adding the last hop addition, slowly add the remaining 2.5 lb. (1.14 kg) of dry malt extract—that is, the part of the extract listed under "Wort B" in the Palmer Brewing Method recipe—while stirring gently to help prevent the extract from clumping and forming floating blobs. Crush any blobs against the side of the kettle with the spoon, and stir until all of the extract has dissolved. Let the kettle sit for 15 minutes before cooling. It will only take a couple of minutes for the heat of the wort to pasteurize the added extract, but the rest of the 15

**Figure 1.10.** Adding the first hop addition to the wort.

minute period is to allow the aromatic oils from the last hop addition to diffuse into the wort. Chapter 2 has more information on heat pasteurizing and chapter 5 has more information on hop oils and hop steeping.

### 9. Chill the Wort

After the 15 minute steep, the wort needs to be chilled to the fermentation temperature.

For best results, the wort should be cooled quickly. First, because handling hot wort is dangerous, and second, because it is convenient—quick cooling allows you to get on with your brew day. Only once your wort has cooled to the fermentation temperature can you can pitch your yeast and be done.

Hot wort (generally anything above 120°F [49°C]) is a safety hazard, and wort between 90°F and 140°F (32–60°C) is also easily contaminated by airborne yeast and bacteria. What follows are a few options for chilling your wort.

*Pouring into cold water.* The smaller volume of the partial boil method allows the hot wort to be mostly cooled by pouring it into the cold water in the fermentor. It is important to understand that this method will not fully chill the wort, but only take it down to about 140°F (60°C), which is a risky zone for contamination by bacteria. The fermentor will need to be sealed and allowed to cool overnight to the fermentation temperature before you can proceed. If your sanitation is good, the wait will not a problem.

Keep in mind that handling hot wort is dangerous, but the use of pot holders and a modicum of care are usually all that is needed to use this method successfully (fig. 1.11). But there are other ways to cool your wort more quickly and save you time and worry.

**Figure 1.11.** Pouring hot wort into cold water in fermentor. Notice the hot pads for protection and towel on the floor in case of spillage.

**Figure 1.12.** Place the kettle into an ice bath to chill it quickly. Keep the kettle covered to prevent contamination.

*Cold water bath.* Placing the kettle in the kitchen sink or a tub filled with ice water will chill it down to 70°F (21°C) in about 20–30 minutes. The ice water can be circulated around the kettle to speed up the cooling. You can also stir the wort to improve the cooling, if you are careful. Do not get the cooling water inside the kettle, because this is a contamination risk. If the cooling water gets warm, replace it with colder water. The closer you can get the wort to your fermentation temperature, the better.

*Copper wort chillers.* The best solution for cooling your wort quickly is to use a copper wort chiller. A wort chiller is a coil of copper tubing that is used as a heat exchanger to cool the wort in place. Wort chillers are a necessity for chilling full volume boils, because you can leave the wort on the stove instead of carrying it to a sink or bathtub. Five gallons (19 L) of boiling hot wort weighs almost 45 lb. (20 kg) and is dangerous to carry.

There are two basic types of wort chillers: immersion and counterflow. Immersion chillers are the simplest and work by running cold water through the coil. The chiller is immersed in the wort and the water carries the heat away. Counterflow chillers work in an opposite manner. The hot wort is drained from the kettle through the copper tubing while cold water flows around the outside of the chiller. Immersion chillers are often sold in homebrew supply shops or can be easily made at home. Instructions for building both types of chiller are given in appendix D.

*No chill.* There is also the option of no-chill, if you have the proper equipment. The proper equipment in this case is a 5.3 gal. (20 L) high-density polyethylene (HDPE) jerry can for drinking water. These are common in Australia, where this no-chill technique was invented. The basic procedure is that right at the end of the boil, as soon as the heat is off, you drain the boiling hot wort into the jerry can, squeeze out all the air, and seal the lid. The heat of the wort sanitizes the container. The wort is allowed to cool overnight (and often through the next day) until it has cooled to fermentation temperature. The wort can then be aerated by pouring it to a fermentation bucket, or aerated in the jerry can and fermented with an airlock in that.

Assuming you don't have a copper wort chiller at this point, the simplest method at this stage is pouring your wort into the cold water in the fermentor.

### 10. Pouring the Wort into the Fermentor

Pour the hot (or cooled) wort into the cold water in the fermentor. Pouring the wort through a strainer to remove most of the spent hops and hot break material is optional, but often helpful. This material, called "trub," will not hurt the fermentation; in fact, retaining some trub is nutritionally beneficial for the yeast. But some styles, like IPA, have so much spent hops in the wort that the hops can soak up a lot of beer after fermentation, lowering your yield. Straining the trub for the Cincinnati Pale Ale recipe is not necessary.

Cover the fermentor (if using a bucket) and move it to a cool room. Clean and sanitize the airlock and stopper, if you have not already done so. Fill the airlock to the indicated level with water and insert it into the lid. Allow the wort to cool to the fermentation temperature (65–70°F [18–21°C]) before pitching the yeast. Ideally, the wort should be at fermentation temperature and your yeast pitched within minutes—rather than hours—of finishing the boil to reduce the risk of bacterial contamination before fermentation. However, if your sanitation is good the batch should be fine, even if the wort cools slowly overnight and you don't pitch your yeast until the next day.

### 11. Aerate the Wort

When your wort has cooled to the fermentation temperature and you are ready to pitch the yeast, you should first aerate the wort to provide the oxygen the yeast need to grow big and strong so it can ferment your wort completely. This is the only time during the brewing process where you actually *want* to aerate, or add oxygen, to your wort or beer. The yeast will use this oxygen to synthesize nutrients it needs for growth. See chapters 6 and 7 for a complete discussion of yeast and fermentation.

The best way to aerate wort is with an aeration wand, which is a long tube with an airstone, or carbonation stone, on the end. Using an aquarium air pump and filter, you pump HEPA-filtered air through the wand into the bottom of the fermentor for 5–15 minutes. This will supply about 8 parts per million (ppm) of oxygen for the yeast to use. (If you were wondering, HEPA stands for high-efficiency particulate arrestance.)

Alternatively, you can aerate your wort by pouring it back and forth a few times into the clean and sanitized boiling kettle (see fig. 1.14). However, this method has the risk of airborne contamination, so make sure you do it in a clean room and pitch the yeast immediately after.

### 12. Pitch the Yeast

No, this doesn't mean to throw the yeast away. It means to throw it into your wort. Open two packets of dried ale yeast (a single packet of dried yeast is usually 10 g). Measure 1 cup (250 mL) of warm (77–85°F [25–30°C]), pre-boiled water into your measuring cup and add the yeast. Allow the yeast to sit for 15 minutes before stirring. Stir the yeast gently and allow it to thoroughly rehydrate for 10–30 minutes before pouring (pitching) it into the fermentor containing your (now cooled) wort. You end up with healthier yeast if you rehydrate first in plain water rather than simply sprinkling dry yeast directly onto the wort. Chapter 7 goes into more detail about yeast management.

**Figure 1.13.** An immersion wort chiller placed in the boiling kettle.

**Figure 1.14**. Pouring the chilled wort for aeration.

**Figure 1.15**. Rehydrating the yeast.

**Figure 1.16**. Pitching the yeast.

### 13. Fermentation

Fermentation should start within 12–36 hours. Choosing a location that has a stable temperature in the range 65–70°F (18–21°C) is critically important for beer flavor. A warmer temperature of 75°F (24°C) is okay, but above 80°F (26°C) the beer will exhibit solventlike or phenolic off-flavors. If the temperature falls 5°F (2°C) or more below the recommended range, the yeast will perform sluggishly and may not ferment well, which often leads to raw pumpkin (acetaldehyde) and butter (diacetyl) flavors. For best results, the temperature of the room should be steady within the recommended range and not fluctuate between day and night.

### 14. Clean Up

Now is the time to wash out your boiling kettle and other equipment. Only use the cleaners recommended in chapter 2, and rinse well.

## Fermentation Week(s)

Be prepared to amaze your family and friends with a bubbling airlock! (You laugh now. . .) The science of fermentation is discussed in detail in chapter 6, "Yeast and Fermentation," which will help you understand what is going on in your fermentor. Chapter 7, "Yeast Management," more fully explains how to select, grow, and care for your yeast to achieve the best fermentations.

**Figure 1.17.** Fermenting wort in a glass carboy. Clear glass or plastic carboys allow you to more easily see the activity of fermentation.

### 15. Leave it alone!

The airlock will start bubbling steadily after about 24 hours, the exciting evidence of fermentation. Figure 1.17 shows what it looks like inside the fermentor. The fermentation will proceed like this for two to four days, depending mainly on the temperature and amount of yeast pitched. The yeast creates alcohol, carbon dioxide, and a host of important flavor compounds as it ferments the wort sugars.

The airlock will bubble vigorously for the first few days and then decrease dramatically as the fermentable sugars are consumed by the yeast. Visible activity may cease altogether within one week, but the yeast are still active. For best results, allow the fermentor to sit undisturbed for at least one week after visible activity in the airlock has slowed (although it typically takes about two weeks in total). This will give time for the beer to condition and mature, and improve its clarity for bottling. As it clarifies, the beer will appear to get darker due to less haze scattering the ambient light.

## Bottling Day

The second big day in your career as a homebrewer comes two weeks later, when fermentation is complete. Everything outlined below is more thoroughly discussed in chapter 10, "Priming, Bottling and Kegging."

**Figure 1.18.** Bottling equipment. You will need a bottle capper, caps, and either a siphon with a bottle filler attachment or a bottling bucket with a filler attachment.

To bottle your beer, you will need:

*Bottles.* You will need at least 48 non-twistoff 12 fl. oz. (350 mL) bottles for a typical 5 gal. batch. Alternatively, you could use 30 of the larger 22 fl. oz. (650 mL) bottles to reduce capping time. Twistoff bottles ("twistoffs") do not re-cap well and are more likely to break during capping. Champagne bottles also work well, if you have the right size caps and your capper can accommodate them.

 *Bottling bucket.* You will also need a 6 gal. (23 L) food-grade plastic pail with attached spigot and fill-tube to use as a bottling bucket (fig. 1.19). The finished beer is racked into the bottling bucket for priming prior to bottling. Racking into the bottling bucket makes for clearer beer with less sediment in the bottle. The spigot is used instead of a bottle filler, allowing greater control of the fill level without the hassle of a siphon.

*Bottle capper.* Two styles of bottle capper are available: hand cappers and bench cappers. Bench cappers are mounted to a metal stand and can be operated with one hand, allowing the other hand to hold the bottle, as opposed to a hand capper, which requires both hands to operate.

*Bottle caps.* For bottle caps, either standard or oxygen-absorbing crown caps are available.

*Bottle brush.* For a bottle brush, you will find a long-handled, nylon bristle brush is necessary for the first hardcore cleaning of used bottles.

*Siphon.* Available in several configurations, siphons usually consist of clear plastic tubing with a racking cane and bottle filler. A siphon is a less-recommended alternative to a bottling bucket.

*Racking cane.* A racking cane consists of a rigid plastic tube with a sediment stand-off (a cap) to make sure the trub stays out of the bottle when siphoning.

*Bottle filler.* Finally, you will need a bottle filler, which consists of a rigid plastic tube with a spring-loaded valve at the tip.

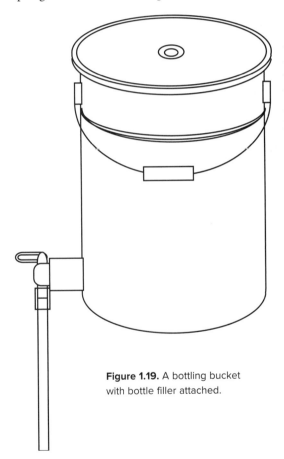

**Figure 1.19.** A bottling bucket with bottle filler attached.

### 16. Prepare Your Bottles

A typical 5 gal. (19 L) batch requires roughly 48 12 fl. oz. (350 mL) bottles for bottling. Thoroughly clean and sanitize the bottles before use. If you are using old bottles, check them for dirt or mold deposits inside. Bottles with deposits may need to be scrubbed with a bottle brush to get them clean. Always clean first, then sanitize.

### 17. Prepare Your Bottle Caps

The bottle caps should be clean (assuming you're using new caps), but it doesn't hurt to sanitize them too. Some homebrewers use flip-top (Grolsch®-style) bottles. The ceramic part of the flip-top lids can be sanitized along with the bottles. The rubber seals can be sanitized separately, like bottle caps.

### 18. Prepare Your Priming Sugar Solution

Adding a priming solution just before bottling gives the yeast a little sugar to re-ferment, which helps carbonate the beer in the bottle. For your priming solution, bring to a boil ¾ cup (4.7 oz. [133 g] by weight) of corn sugar or ⅔ cup (4 oz. [113 g] by weight) of table sugar in two cups (0.5 L) of water. Look to make sure the sugar is thoroughly dissolved. Cover the pan and allow it to cool.

### 19. Combine the Beer and Priming Sugar Solution

The best method is to use a separate container, the same size as your fermentor, as a "bottling bucket". Clean and sanitize the bottling bucket, then gently pour your priming solution into it. Next, siphon the beer from the fermentor into the bottling bucket (fig. 1.20). Do not simply pour the beer into the bottling bucket, and do not let the beer splash as you siphon it in. Instead, put the end of the siphon under the surface of the beer as it fills. The swirling motion of the beer as it enters the bottling bucket will be sufficient to evenly mix the priming solution into the beer with little aeration.

If you don't have a bottling bucket, you can gently pour the priming solution into the fermentor and gently stir it in. Allow the sediment in the fermentor to settle for 15–30 minutes before proceeding to the next step. You can fill the bottles using the bottle filler attachment on your siphon, but it's better to have a dedicated bottling bucket (figs. 1.21 and 1.22).

**Figure 1.20**. Racking from fermentor to bottling bucket.

**Figure 1.21**. Filling bottles from the bottling bucket.

**Figure 1.22**. Filling bottles from a siphon with a bottle filler attachment.

### 20. Bottle and Cap Your Beer
Carefully fill the bottles to within about ¾ to 1 in. (2.0–2.5 cm) below the rim with the primed beer, place a sanitized bottle cap on each bottle, and crimp it using the bottle capper. At this stage it is helpful to have a friend operate the capper while you fill the bottles, or vice versa.

### 21. Leave to Carbonate
Place the filled and capped bottles away from light in a warm environment (room temperature, 70–80°F [21–27°C]). The bottles will take about two weeks to carbonate, depending on the temperature, and will have a thin layer of yeast on the bottom. The priming and bottling processes are discussed in more detail in chapter 10.

## Serving Your Beer
At last, you get to sample the fruit of your labors. It's been about a month since brew day, and you are ready to open your first bottle and see what kind of wonderful beer you have created. During the past two weeks the yeast remaining in the beer has consumed the priming sugar, creating just enough carbon dioxide to carbonate your beer perfectly.

Okay, so maybe you couldn't wait that long and you already opened a bottle. You may have noticed the beer wasn't fully carbonated or that it seemed to have a "green" or yeastlike flavor. It may have an aroma or flavor that is apple-ciderlike or buttery. These flavors are the sign of a young beer. The two-week maturation period not only adds carbonation, but also gives the yeast time to clean up some of the off-flavors it created during fermentation, as well as settle itself out, leaving a clean-tasting—and clean-looking—beer. Further descriptions of off-flavors and what they may signify are given in chapter 25.

### 22. Chill Your Beer

Once carbonated, your bottled beer should be stored cold to help preserve its flavor. It will keep for approximately six months, depending on how well you managed to avoid exposure to oxygen during the last stage of fermentation and during the bottling process. The beer will naturally oxidize as it ages, losing some of its hop character and acquiring stale flavors. Only a few beer styles age well; most should be consumed within six months. The optimal temperature for serving beer depends on the style, varying from 40–55°F (4–12°C). In general, the darker the beer, the warmer it should be served, but that guideline varies as well.

### 23. Pour Your Beer

To pour the beer without getting yeast in your glass, tip the bottle slowly to avoid disturbing the yeast layer on the bottom of the bottle. With practice, you will be able to pour everything but the last quarter-inch of beer without getting any yeast in your glass.

### 24. Savor the Flavor

Finally, enjoy the aroma, then take a deep draught and savor the flavor of the beer you created. Over time, pay attention to the aroma, flavor, bitterness, sweetness, carbonation level, and more. These observations are your first steps to understanding your beer and designing your own recipes.

## But Wait! There's More!

Your first batch is a success and you are on your way to a brighter future. But, as you keep brewing, keep reading. The following chapters will again lead you through extract brewing, but this time with greatly expanded information about the process and the huge variety of hops, yeast strains, and malts that can make each brew deliciously and uniquely your own.

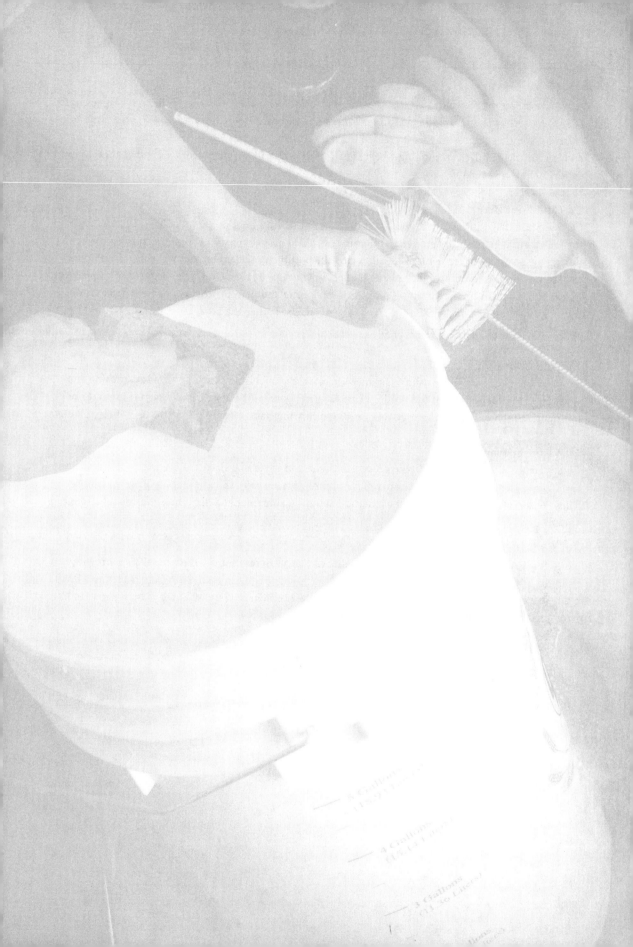

# Cleaning and Sanitizing 2

Cleaning and sanitizing are the most important factors for assuring a successful batch of beer. No matter how great the recipe, how fresh the hops, how much attention you pay to brewing, poor hygiene can wreck a beer like nothing else. However, it's just one part of an overall program of preparation and organization for your brewing. Good preparation prevents nasty surprises. You don't want to be halfway through your brew day and realize that your yeast is old. You don't want to pour good wort into a fermentor that you forgot to clean.

There are two types of brewers—lucky and consistent. The lucky brewer will sometimes produce an outstanding batch of beer, but just as often one that is not. He brews by the seat of his pants, innovating and experimenting with mixed results. The consistent brewer has more outstanding batches than poor ones. He may be an innovator and an experimenter, but the difference is that the consistent brewer takes note of what he did and how much of it he did so that he can always learn from his results. Good organization and record keeping make the difference between luck and skill.

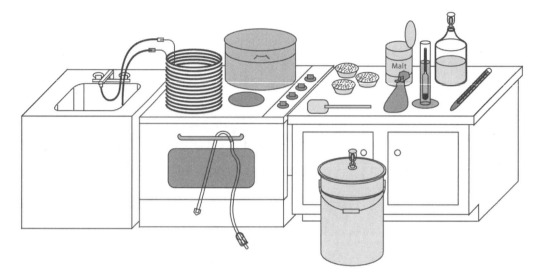

**Figure 2.1.** All the equipment and ingredients for the day's brew are set out on the counter, cleaned, sanitized and ready to go. The crushed specialty malt is tied into a muslin grain bag, and the hops have been weighed and put in three separate bowls.

## Be Organized, Be Prepared.

Organization and advance preparation for each of the brewing processes will make everything work better (fig. 2.1). Here are some examples:

*Check the recipe.* Make a checklist of your ingredients and amounts. Plan ahead for how you are going to measure them. Do you need extra bowls or measuring cups? Do you have good, chlorine-free water out of the tap, or should you buy some?

Create a recipe form to help you be consistent, and keep these forms in a brewing notebook. There are several brewing spreadsheets and software programs available online that can be a big help, or if you want to keep things simple, paper forms work just fine (see an example in fig. 2.2). A brewer needs to be able to repeat good batches and learn from poor ones. If you have a bad batch and want to ask another brewer for their opinion, they are going to want to know all the brewing details: ingredients and quantities, how long you boiled, how you cooled, the type of yeast, how long it fermented and at what temperature, what the fermentation looked like, and possibly more besides. There are so many possible causes for "it tastes funny," that you really need to keep track of everything that you did so you can figure out where it may have gone wrong and fix it the next time. Chapter 25 will help you identify possible causes for most common problems.

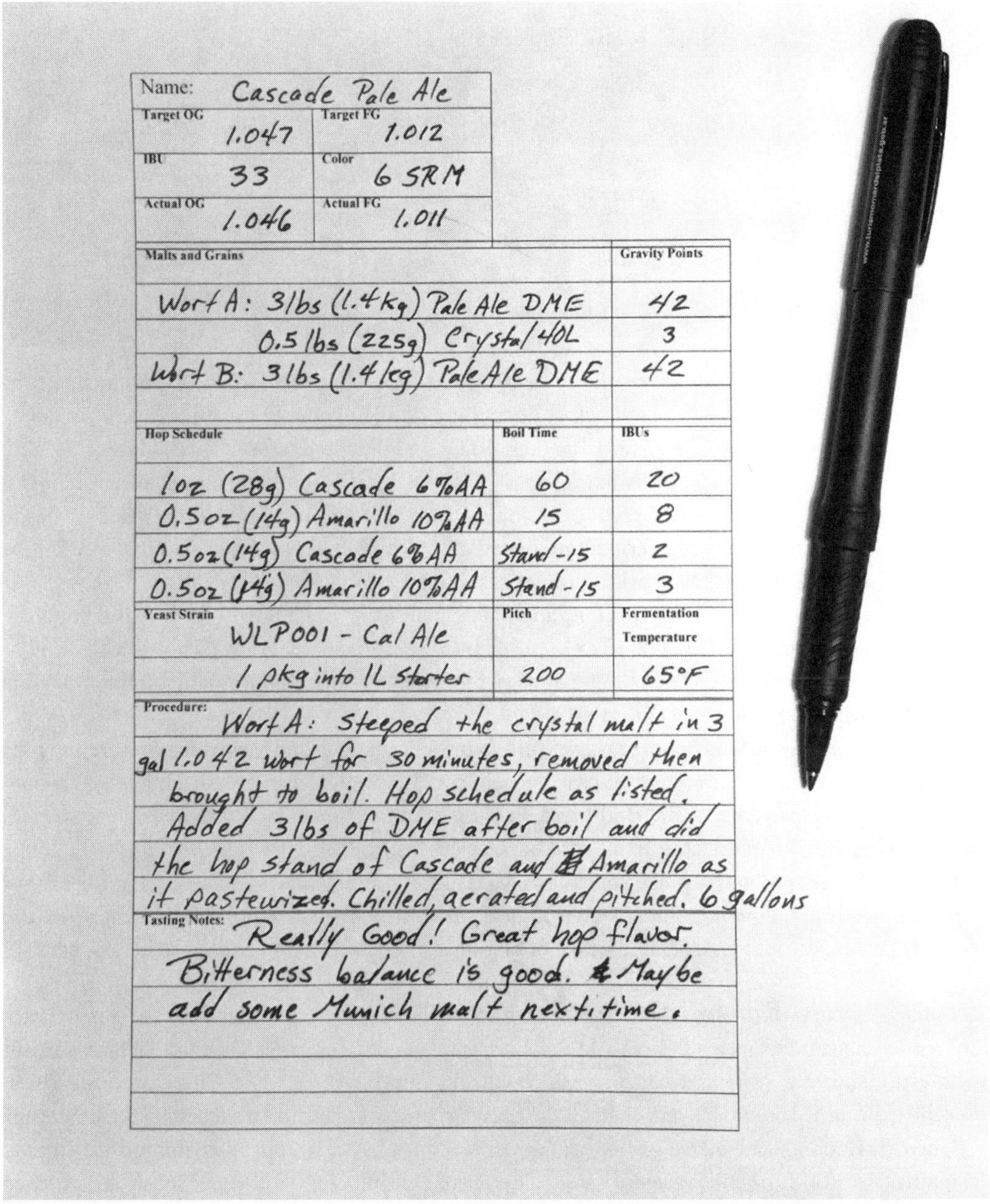

| Name: | Cascade Pale Ale | | |
|---|---|---|---|
| **Target OG** | 1.047 | **Target FG** | 1.012 |
| **IBU** | 33 | **Color** | 6 SRM |
| **Actual OG** | 1.046 | **Actual FG** | 1.011 |

| Malts and Grains | Gravity Points |
|---|---|
| Wort A: 3 lbs (1.4 kg) Pale Ale DME | 42 |
| 0.5 lbs (225g) Crystal 40L | 3 |
| Wort B: 3 lbs (1.4 kg) Pale Ale DME | 42 |
| | |

| Hop Schedule | Boil Time | IBUs |
|---|---|---|
| 1oz (28g) Cascade 6%AA | 60 | 20 |
| 0.5oz (14g) Amarillo 10%AA | 15 | 8 |
| 0.5oz (14g) Cascade 6%AA | Stand -15 | 2 |
| 0.5oz (14g) Amarillo 10%AA | Stand -15 | 3 |

| Yeast Strain | Pitch | Fermentation Temperature |
|---|---|---|
| WLP001 - Cal Ale | | |
| 1 pkg into 1L starter | 200 | 65°F |

**Procedure:** Wort A: Steeped the crystal malt in 3 gal 1.042 wort for 30 minutes, removed then brought to boil. Hop schedule as listed. Added 3 lbs of DME after boil and did the hop stand of Cascade and Amarillo as it pasteurized. Chilled, aerated and pitched. 6 gallons

**Tasting Notes:** Really Good! Great hop flavor. Bitterness balance is good. Maybe add some Munich malt next time.

**Figure 2.2.** An example recipe form useful for record keeping.

*Clean your equipment.* Make a checklist of the equipment you will be using and note whether it needs to be sanitized or only cleaned (table 2.1). Always make sure all your equipment is ready to go when you are. Don't try to clean something at the last minute just as you need it; you are inviting trouble. You may want to purchase utensils specifically for brewing, so you don't stir the wort with the same spatula used for cooking onions. Further instruction on cleaning methods is given later in this chapter.

## TABLE 2.1—CLEANING AND SANITIZATION CHECK LIST

| | | | | |
|---|---|---|---|---|
| Brewpot | ❐ | Clean | | |
| Stirring spoon | ❐ | Clean | | |
| Regular spoon | ❐ | Clean | ❐ | Sanitize |
| Measuring cup | ❐ | Clean | ❐ | Sanitize |
| Yeast Starter jar | ❐ | Clean | ❐ | Sanitize |
| Fermentor and lid | ❐ | Clean | ❐ | Sanitize |
| Airlock | ❐ | Clean | ❐ | Sanitize |
| Thermometer | ❐ | Clean | ❐ | Sanitize |

*Sanitizing.* Anything that contacts the cooled wort and yeast must be sanitized. This includes the fermentor, airlock, and any of the following, depending on your transfer methods: funnel, strainer, stirring spoon, and racking cane. Sanitizing techniques are discussed later in this chapter.

*Preparing the yeast.* This step is paramount. Without yeast, there will be no beer. The yeast should be prepared ahead of the brewing session, especially if you need to make a yeast starter to increase the cell count. A good fermentation makes the difference between success and failure. Chapter 7 has more detailed information on yeast preparation.

*The boil.* Weigh out your hop additions and place them in separate bowls for the different addition times during the boil. If you are going to steep crushed specialty grain in hot—not boiling— wort (see chapter 4), this step needs to be done before you start the boil. Actually boiling the grain causes astringent tastes like those from an old tea bag.

*Cooling after the boil.* The wort must be cooled to the fermentation temperature before pitching the yeast. A quick chill from boiling helps to prevent contamination from beer-spoiling bacteria and wild yeast, and helps generate the cold break (a precipitation of protein and lipids) in the wort. A good cold break reduces the amount of chill haze in the final beer.

Thorough preparation will make your brewing go more smoothly and reduce the chance of disasters like missed steps, forgotten ingredients, or weak yeast. In short, having the equipment ready and the process planned out will make the whole operation simple, keep it fun, and make better beer.

## Brewing Priority Number 1—Good Sanitation

Good growing conditions for the yeast in the beer are also good for other undesirable microorganisms, especially wild yeast and bacteria, which can be a problem (fig. 2.3). Therefore, cleanliness must be maintained through every stage of the brewing process to ensure a successful fermentation.

The definition and objective of sanitization is to reduce bacteria and other microorganisms to insignificant, or at least manageable, levels. The terms "clean," "sanitize," and "sterilize" are often used interchangeably but are definitely not the same thing. Here are the definitions:

**Clean**　　To be free from dirt, stain, and foreign matter.
**Sanitize**　　To reduce undesirable microorganisms to negligible levels.
**Sterilize**　　To kill or eliminate microorganisms, either by chemical or physical means.

Cleaning is the process of removing visible dirt, grime, and other foreign matter from a surface, thereby eliminating the sites that can harbor bacteria. Cleaning is usually done with an alkaline

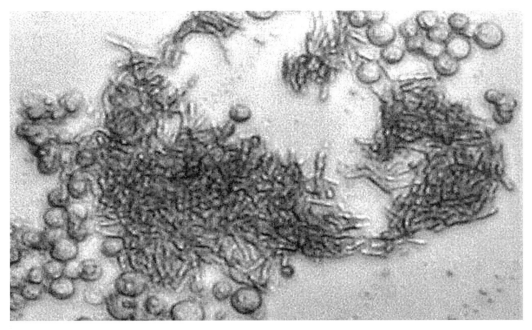

**Figure 2.3.** The round things are yeast cells, and the small rods are bacteria. Magnification 500×. (Image was taken from fermenting wort with spoilage organisms.)

cleaning agent or detergent and hard work. Sanitized is not the same as sterile. Chemical agents used by homebrewers will clean and sanitize, but will not sterilize or eliminate all bacterial spores and viruses. This is okay; sterilization is usually not necessary in brewing, and brewers can be satisfied as long as their sanitization procedure consistently reduces these contaminants to negligible levels. All commercial beer is brewed in sanitized vessels; no production breweries sterilize their equipment, as that would be impractical.

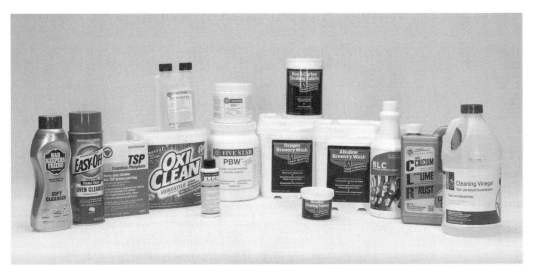

**Figure 2.4.** A range of cleaning products commonly available to homebrewers.

## Cleaning Products

Cleaning requires a certain amount of scrubbing, brushing, and hard work. Dirt and other deposits can shield bacteria from sanitizers that will ultimately contaminate your beer. Several cleaning products available to the homebrewer (fig. 2.4) are discussed below, followed by recommendations for use with each type of brewing equipment.

### Percarbonates

Percarbonate-based cleaners are the best choice for equipment cleaning, in my opinion. Sodium percarbonate (e.g., OxiClean™) is sodium carbonate (familiar as washing soda) reacted with hydrogen peroxide, and it is an effective cleaner for all types of brewing equipment. It rinses easily. Several products containing percarbonates are approved by the Food and Drug Adminstration (FDA) as cleaners in food-manufacturing facilities. Use commercial percarbonate cleaners according to the manufacturer's instructions, but generally use 2 tablespoons per gallon (8 mL/L) and rinse after cleaning.

PBW™ (Powder Brewery Wash) from Five Star Chemicals & Supply, Inc. and Straight A from Logic, Inc. are the best of these products, combining the percarbonate with sodium metasilicate, along with other agents, making the cleaner more effective at removing proteins and preventing the corrosion of copper and aluminum that can occur when using strong alkaline solutions.

Note: If used with hard water, these cleaners may cause calcium carbonate scale (limescale) on surfaces left to soak in the cleaning solution for several days. Limescale from hard water can only be removed with an acidic cleaner, like CLR® from Jelmar. Acidic sanitizers, described in the "Sanitizers" section below, are not intended for acid cleaning or descaling.

### Detergents

Dishwashing detergents and cleansers should be used with caution when cleaning your brewing equipment. These products often contain perfumes that can be absorbed into plastic equipment and released back into the beer. In addition, some products will leave behind a film that might be tasted in the beer and interfere with head retention. Several rinses with hot water may be necessary to remove all traces of the detergent. Detergents containing phosphates generally rinse more easily than those without, but because phosphates are pollutants, they are not ingredients in most household products. A mild unscented dishwashing detergent like Ivory™ brand is a good choice for most of your routine equipment cleaning needs. Only stubborn stains or burned-on deposits will require something stronger.

### Automatic Dishwashers

Using dishwashers to clean equipment and bottles is a popular idea among homebrewers, but there are a few limitations:

- The narrow openings of hoses, racking canes, and bottles usually prevent the water jets and detergent from reaching inside.
- If dishwasher detergent does make it inside items with narrow openings, there is no guarantee that it will be properly rinsed out again.
- Dishwasher drying additives (Jet Dry®, for example) can ruin the head retention of beer. Drying additives work by depositing a chemical film on an item, allowing the item to be fully wetted such that water droplets don't form, so preventing spots. This wetting action destabilizes the proteins that form the bubbles in the head of a beer.

With the exception of spoons, measuring cups and wide mouth jars, it is probably best to only use automatic dishwashers for heat sanitizing, not cleaning. Heat sanitizing is discussed later in this chapter.

### Bleach

Bleach is one of the cheapest cleaners available to the homebrewer. It forms a caustic solution when dissolved in cold water that is good at breaking up organic deposits like food stains and brewing gunk. Bleach is an aqueous solution of chlorine, chlorides, and hypochlorites. These chemical agents all contribute to bleach's bactericidal and cleaning powers, but they are also corrosive to the metals used in brewing equipment. Bleach should not be used for cleaning brass or copper, because it causes both blackening and corrosion. Bleach can be used to clean stainless steel, but with the following caveats:

- Do not leave stainless steel in contact with chlorinated water for more than an hour.
- Rinse the item thoroughly after cleaning, and dry the item completely if it is going to be stored.

### Oven Cleaner

Every once in a while, brewers will scorch the bottom of their brew pot, resulting in a black, burned-on deposit that is difficult to remove. The easiest solution is to apply oven cleaner and allow it to dissolve the deposit. As soon as the burned-on crud has been removed, it is important to thoroughly rinse the area of any oven cleaner residue to prevent subsequent corrosion of the metal. Burned-on deposits are usually the only situation where oven cleaner is needed. Percarbonate cleaners will normally be able to clean other deposits, even the tough ones.

Sodium hydroxide (NaOH), commonly known as lye or caustic soda, is the main ingredient of most heavy-duty caustic cleaners, such as oven and drain cleaner, although potassium hydroxide (KOH) is also used. Spray-on oven cleaner is the safest and most convenient way to use these chemicals.

Even in moderate concentrations, these chemicals are very hazardous, and should only be used in a well-ventilated area while wearing rubber gloves and goggle-type eye protection. If you get caustic cleaner in your eyes it can cause severe burns or blindness. These caustic cleaners can also cause a lot of damage to skin without much pain. . . at first. That initial slippery feeling is the oils and lipids in your skin being dissolved and turned into soap. Vinegar can neutralize caustic cleaner that gets on your skin. If you get caustic cleaner in your eyes, flush with copious amounts of water immediately and seek medical attention. Always read the manufacturer's safety advice on the label.

Sodium hydroxide and potassium hydroxide are very corrosive to aluminum and brass, but copper and stainless steel are generally resistant. Pure sodium hydroxide should not be used to clean aluminum brew pots, because its high pH causes the dissolution of the protective aluminum oxide surface layer, which can result in corrosion and give the beer a metallic taste. Oven cleaner, however, should not be a problem when it is used properly.

## Cleaning Suggestions

### Cleaning Plastic

There are basically three kinds of plastic that you will be cleaning: opaque white high-density polyethylene (HDPE), hard clear polycarbonate, and clear soft vinyl tubing. You will often hear the polyethylene referred to as "food-grade plastic," though all three of these plastics are food grade.

Polyethylene is used for utensils, fermenting buckets, and fittings. Polycarbonate is used for racking canes and measuring cups. The vinyl tubing is used for siphons and similar equipment.

The main thing to keep in mind when cleaning plastics is that they may absorb odors and stains from the cleaning products you use. Dish detergent is your best bet for general cleaning, but scented detergents should be avoided. Bleach is useful for light-duty cleaning, but the odor can remain and bleach tends to cloud vinyl tubing. A percarbonate-based cleaner (described above) has the benefit of cleaning as well as bleaching without the odor and clouding problems. Dishwashers are a convenient way to clean plastic utensils, but the heat might warp polycarbonate items.

## Cleaning Glass

Glass has the advantage of being chemically inert, so it will not react with any cleaning agent you might use, but you will need to use bottle and carboy brushes to effectively clean inside bottles and carboys. Limescale deposits can form on glass when using bleach and percarbonate cleaner in hard water areas, but an acid rinse solves this problem. Be very careful if handling glass carboys when wet, because they can be very slippery and break easily.

### BREWING TIP: CLEANING AND SANITIZING BOTTLES

Dishwashers are great for cleaning the outside of bottles and heat sanitizing, but will not effectively clean the insides.

If your bottles are dirty or moldy, soak them at room temperature in a sodium percarbonate-type cleaner (like PBW) or in a mild bleach solution for an hour or two to soften any residue. You'll still need to scrub them thoroughly with a bottle brush to remove any stuck-on residue.

To eliminate the need to scrub the insides of bottles in the future, rinse them thoroughly immediately after each use.

## Cleaning Copper

For routine cleaning of copper and other metals, percarbonate-based cleaners like PBW are the best choice. Vinegar (dilute acetic acid) is very effective for heavy deposits of blue, green, or black copper oxides, especially when heated. White distilled vinegar is commonly available in grocery stores at a standard concentration of 5% acetic acid by volume. Acid cleaners such as CLR (Calcium, Lime & Rust Remover) are also very effective, but stronger acids such as hydrochloric (muriatic) acid for swimming pools should not be used because they are more likely to corrode the part than clean it.

Brewers who use immersion wort chillers are always surprised how bright and shiny the chiller is the first time it comes out of the wort. If the chiller wasn't bright and shiny when it went into the wort, guess where the grime and oxides ended up? Yep, in the wort. The oxides of copper are more readily dissolved by the mildly acidic wort than is the copper itself. By cleaning copper tubing with acetic acid or CLR once before the first use and rinsing with water immediately after each use, the copper will remain clean with no oxide or wort deposits that can harbor bacteria. Cleaning copper with vinegar should only occasionally be necessary. With time, the copper will achieve a dull finish that does not change, even after emerging from the wort. This is good; it means that the surface oxides are passive and not dissolving into the wort. There is no need to clean copper, brass, and aluminum brewing equipment until it is shiny.

Cleaning and sanitizing copper and brass with bleach solutions is not recommended, because it will cause corrosion of copper and brass. The corrosion products, usually a mix of black, green, and blue, will quickly dissolve in the mildly acidic wort, and potentially harm the yeast during fermentation.

Note: I discuss the cleaning of metals and metal toxicity more thoroughly in appendix G.

### Cleaning Brass

Some brewers who use brass fittings in conjunction with their wort chillers or other brewing equipment are concerned about the lead present in brass alloys. The amount of lead on the surface of brass parts is miniscule and not a health concern. However, a solution of two parts white vinegar (5% acetic acid) to one part hydrogen peroxide (as a common 3% solution) will remove tarnish and surface lead from brass parts when they are soaked for 5–15 minutes at room temperature. The brass will turn a buttery yellow color as it is cleaned. If the solution starts to turn green, then the parts have been soaking too long and the copper in the brass is beginning to dissolve. At this point the solution is contaminated, and the part should be re-cleaned in a fresh solution.

### Cleaning Stainless Steel and Aluminum

For general cleaning, mild detergents or percarbonate-based cleaners are best for steel and aluminum. Bleach should not be used to clean stainless steel and aluminum because it is just too corrosive. Do not clean aluminum shiny bright, because this removes the protective aluminum oxide surface layer and can result in a metallic taste in beer. This detectable level of aluminum is not hazardous. There is more aluminum in one common antacid tablet than would be present in an entire batch of beer made in an aluminum pot.

Common oxalic acid-based kitchen cleansers from the grocery store are very effective for cleaning stubborn stains, deposits, and rust from stainless steel. They also work well for copper. Examples are Bar Keeper's Friend® Cleanser and Polish, Revere® Copper and Stainless Steel Cleaner, and Kleen King® Stainless Steel & Copper Cleaner. Follow the manufacturer's directions and always rinse thoroughly with water afterwards. These cleansers are more effective at removing stains, heat tint, and corrosion from stainless steel and other brewing metals than any other cleaning products.

## STAINLESS STEEL AND PASSIVATION

Stainless steel resists corrosion because of a highly corrosion-resistant layer of chromium and nickel oxides on the surface that protects the iron component of the alloy from rusting. When this layer is intact, the stainless steel is considered to be "passivated." This corrosion-resistant layer will form instantly on clean stainless steel; the key is that the surface has to be free of contaminants, such as oil, dirt, and sugar. There are industrial methods for stronger passivation that involve cleaning with strong acids to remove contaminants, but these are not necessary for day-to-day maintenance of your brewing equipment. Cleaning the stainless steel thoroughly as described in this chapter will ensure that your stainless steel equipment remains passivated and thereby resistant to rust and corrosion. The stainless steel surface doesn't need to be shiny to be passive; in fact, a dull, grayish surface indicates the most passive surface. However, if the surface turns tan, blue, or brown, that surface is not passive and should be cleaned to bare metal. See appendix G, "Brewing Metallurgy," for more information on cleaning metal equipment.

## Beerstone Removal

Beerstone is a composite coating of protein and calcium oxalate that is more difficult to remove than typical hard water limescale. It precipitates from beer over time to deposit a film on your equipment. It is undesirable not only because the coating is rough and can harbor bacteria, but it can also initiate corrosion of the stainless steel around the edges of the deposit. Removal of beerstone requires a two-step process: an acidic cleaner to dissolve the oxalates and carbonates, followed by a caustic detergent or percarbonate-based cleaner to break up the protein component. A blend of nitric acid and phosphoric acid is commonly used in the brewing industry, and Five Star Chemicals introduced a homebrewing product in 2016. An alternative but less effective method is to soak the item with a strong solution of PBW for a couple of hours (or overnight), and follow with an acid soaking to dissolve the exposed salts. Phosphoric acid, vinegar, or CLR can be used for the final step.

## Sanitizing Products

Only when your equipment is clean can you sanitize it.

All items that contact the wort after the boil need to be sanitized: fermentor (including the lid), airlock, rubber stopper, yeast rehydration or starter container, thermometer, funnel, and siphon. Your bottles will need to be sanitized also, but that can wait until bottling day. Chemical sanitizers can be prepared in the fermentor or other bucket, and all your equipment can be soaked directly in that. Heat sanitizing methods work for metal and glass, but cannot be used for most plastics. Chemical sanitizers can be used on plastics as well as metal and glass, but may require rinsing and may cause discoloration or corrosion, depending on the agent. Recommended products are discussed below.

**Figure 2.5.** Chemical sanitizers most commonly used by the homebrewer: Star San, Iodophor, and Peracetic Acid.

## Chemical Sanitizers

### Acidic Anionic Surfactants

Acidic sanitizers, like Star San from Five Star Chemicals, are your best choice. They were developed especially for sanitizing brewing equipment, and work by penetrating the bacterial cell wall to disrupt cell membrane permeability and internal cell function. These sanitizers require only 30 seconds of contact time and do not require rinsing. Unlike bleach and iodophor, acid sanitizers will not cause off-flavors at higher-than-recommended concentrations. The recommended usage is 1 fl. oz. per 5 gal. of water (8 mL/L, but 10 mL/L is fine). They are designed to foam to help coat surfaces, and the foam is just as effective for sanitizing as immersion in the solution. The working solution has a long usage life and an open bucket of it will remain active for several days; the same solution will keep indefinitely in a closed container, such as a spray bottle.

Star San is only effective when the pH of the solution is less than 3.5. At a pH above 3.5 the solution turns cloudy and loses its bactericidal properties. Therefore, the viability of the working solution can be judged by its clarity. For best results, Star San should be mixed with distilled water. Alkaline ground water (i.e., water from a well) will cause the working solution to turn cloudy and lose potency with time. This is why it is a no-rinse sanitizer—when the fermentor or bottle has been drained and filled with wort or beer, the higher pH of the wort and beer neutralizes the sanitizer so that the yeast are unaffected. As a matter of fact, even though there can be a huge amount of foam in vessels like carboys after draining, it will have no effect on fermentation or flavor. I have proven this to myself by intentionally fermenting in a carboy that was full of foam, and the fermentation, flavor, and head retention of the beer were fine.

Star San is my preferred sanitizer for all usages. In my experience, using it in a spray bottle is the most convenient way to sanitize most of my brewing equipment. For kegs and carboys, I find pouring in 0.5 gallon (roughly 2 L) of Star San working solution and shaking to coat the walls works very well. After a minute of shaking, I can drain the sanitizer to a small bucket and use it for soaking small parts as needed.

One last note on Star San: because it is listed as a sanitizer and bactericide by the US FDA and Environmental Protection Agency (EPA), the container must list disposal warnings that are suitable for pesticides. Do not be alarmed, at the recommended dilution for use Star San is not harmful to your skin.

### Iodophor

Iodophor is an iodine solution that is a very effective sanitizer with a short contact time. One tablespoon in 5 gal. of water (or 15 mL per 19 L) is all that is needed to sanitize equipment, if you ensure a two-minute soak time. Soaking equipment longer, for 10 minutes at the same concentration, will disinfect surfaces to hospital standards. This dilution produces a concentration of 12.5 ppm of titratable iodine. At 12.5 ppm the working solution has a faint brown color that you can use to monitor its viability. If the solution loses its color, it no longer contains enough free iodine to work.

At 12.5 ppm the iodophor solution does not need to be rinsed, but that is based on letting the item drip dry to allow the solution to evaporate. Do not mix the solution stronger than 25 ppm (1 fl. oz. per 5 gal., or 30 mL/L), because iodine residue can cause bloodlike off-flavors in the beer. The iodine in iodophor will stain plastic with long exposures, and the iodine in the stain can leach into the beer, causing off-flavors. Replace your plastic if it becomes stained. Iodine is a halogen, like chlorine, but it is less corrosive to stainless steel. Even though the recommended concentration (12.5–25 ppm) is well below the taste threshold, I rinse everything with a little bit of cooled boiled water to avoid any chance of off-flavors, but that's just me.

### Peracetic (Peroxyacetic) Acid

Peracetic acid is a no-rinse sanitizer that should be mixed and used according to the manufacturer's instructions. Peracetic acid is highly effective, but is more commonly used in industry because it is irritating to skin and mucus membranes, can cause asthma attacks in sensitive people, and is corrosive at high concentrations (i.e., 10 times that used in sanitizer solutions). Despite its pungent, vinegar-like odor, peracetic acid will not cause off-flavors in beer.

*Note: The peracetic acid solution used for cleaning brass is not an effective sanitizer.*

### Chlorine Dioxide

Chlorine dioxide is another commercial no-rinse sanitizer that has become available to the home-brewing market outside the US. Chlorine dioxide is not bleach and will not contribute chlorine or chlorophenol flavors to the beer. The solution is made by acidifying sodium chlorite (*not* chloride) with an acid (citric or lactic are recommended) to a pH of 2–3, then diluting the solution to a working strength of about 50 ppm. This working solution has a short contact time of about 30 seconds. The concentrated solution will turn lime green when activated and be a faint yellow green when diluted. The yellow-green color will fade as the working solution decomposes, and the sanitizer has a useful life of about 4 hours. Mix and use according to the manufacturer's instructions. Chlorine dioxide is highly corrosive in concentrated form and the dried residue of the concentrate is combustible.

### Bleach

Bleach is the cheapest and most readily available sanitizer, but I do not recommend it unless nothing else is available. It requires a longer sanitizing time and can impart medicinal off-flavors to the beer. The sanitizing solution is made by adding 1 tablespoon of bleach to 1 gal. of water (4 mL/L). Let the items soak for 20 minutes, then drain. Rinsing is supposedly not necessary at this concentration, but many brewers, myself included, rinse with some boiled water anyway to ensure no off-flavors from the chlorine. Stainless steel should only be soaked in bleach solution for the minimum sanitization time (20 minutes). Do not soak stainless steel overnight in bleach, as it will cause pitting and corrosion.

## Heat Sanitizing

Heat sanitizing is often used to sanitize glass bottles, but is most commonly used to sanitize brewing ingredients that are added to the wort after the boil. Heat sanitizing is commonly known as pasteurization after Louis Pasteur, the father of microbiology. Pasteurization is defined as the 5-log decrease (i.e., a 99.999% reduction) in the number of live microbes in a product. Times and temperatures for pasteurization vary depending on the types of bacteria expected and the nutrient load. In other words, what is the likelihood of microbe "X" reproducing to a significant level following pasteurization?

In brewing, 1 minute at 140°F (60°C) is defined as 1 pasteurization unit (PU), but the recommended number of PU to apply depends on the situation. A typical light lager beer is commonly "flash pasteurized" by heating it to between 160°F and 165°F (71–74°C) for 15–30 seconds, which is the equivalent of 8–12 PU. Stronger beers, specifically beers with more residual nutrients, are pasteurized longer, usually to 12–25 PU.[1] Suggested times and temperatures for pasteurization of unfermented wort are given in table 2.2. The intensity of pasteurization increases exponentially with temperature. For example, 4 minutes at 175°F (79°C) is approximately 1100 PU, so don't go overboard with the time or temperature.

---

[1]    Klimovitz and Ockert (2014).

Note: pasteurization will not kill bacterial spores or mold spores. Pasteurization is intended to reduce beer spoilage bacteria to negligible levels. Therefore, don't be surprised if a jar of wort that you intended to save for a yeast starter develops mold after a couple of months in the refrigerator.

## TABLE 2.2—SUGGESTED PASTEURIZATION TEMPERATURES AND TIMES FOR WORT

| Temperature °F | Temperature °C | Time (minutes) |
|---|---|---|
| 150 | 66 | 37 |
| 155 | 68 | 15 |
| 160 | 71 | 6 |
| 165 | 74 | 2.5 |
| 170 | 77 | 1 |
| 175 | 79 | 0.5 |
| 180 | 82 | 0.1 |

Note: Temperature and time combinations are estimated to give 100 pasteurization units.

## Dishwashers

Dishwashers can be used to heat sanitize (but not sterilize) brewing equipment, but be careful with plastic items that may warp. The steam from the drying cycle will effectively sanitize all surfaces. Bottles and other equipment with narrow openings should be pre-cleaned. Run the equipment through the full wash cycle without using any detergent or rinse agent to avoid having any residue drying inside the items. Dishwasher rinse agents will destroy the head retention on your glassware. If you pour a beer with carbonation and no head, this might be the cause.

## Heat Sterilizing

Heat is one of the few means by which the homebrewer can actually sterilize an item. Brewers that grow and maintain their own yeast cultures need to sterilize their growth media to insure against contamination. Both dry heat (oven) and steam (autoclave, pressure cooker, or dishwasher) can also be used for sanitizing.

## Oven

Dry heat is less effective than steam for sanitizing and sterilizing, but many brewers use it for sanitizing glass bottles and yeast culturing equipment. Temperatures and times required for dry heat sterilization are given in table 2.3. Although the durations seem long, remember this process kills all

## TABLE 2.3—DRY HEAT STERILIZATION

| Temperature | Duration (hours) |
|---|---|
| 338°F (170°C) | 1 |
| 320°F (160°C) | 2 |
| 302°F (150°C) | 2.5 |
| 284°F (140°C) | 3 |
| 250°F (121°C) | 12 |

Note: Times indicated begin when the item has reached the indicated temperature.

microorganisms, not just most, as in sanitizing. To be sterilized, items need to be heat-proof at the given temperature. Glass and metal items are prime candidates for heat sterilization.

Homebrewers can bake their bottles in the oven using this method and have a ready supply of clean sterile bottles. The opening of the bottle can be covered with a piece of aluminum foil prior to heating to prevent contamination after cooling and during storage. They will remain sterile indefinitely if kept wrapped.

*One note of caution:* Ordinary glass bottles are susceptible to thermal shock and breakage and should be heated and cooled slowly (e.g., 5°F [2°C] per minute). Put the bottles in the oven when the oven is cold. You can assume all beer bottles are made of un-tempered soda-lime glass, whereas any glassware that says Pyrex® or Kimax® has been tempered or made from borosilicate glass, which is much stronger and more resistant to thermal shock.

### Autoclaves and Pressure Cookers

Autoclaves and pressure cookers can be used for heat sanitizing via steam. Steam conducts heat more efficiently than dry air, so the cycle time is much shorter than when using dry heat. Steam also conducts heat more uniformly than dry air, and glass is less likely to crack from thermal stress. The typical amount of time it takes to sterilize a piece of equipment in an autoclave or pressure cooker is 20 minutes at 257°F (125°C) at 20 pounds per square inch (psi; 20 psi is equivalent to 138 kilopascals). These devices are only needed for serious yeast culturing and other microbiological work.

### Final Thoughts on Cleaning and Sanitizing

Clean all equipment as soon as possible after use. This means rinsing out the boiling kettle, fermentor, and other items like airlocks, stirring spoons, and tubing as soon as you are finished using them. It is very easy to get distracted and come back to find that the syrup or yeast has dried hard as a rock, and your equipment is stained. If you are pressed for time, keep a large container of water handy and just toss things in to soak until you can clean them later.

You can use different methods of cleaning and sanitizing for different types of equipment. A summary of cleaners and sanitizers is given in table 2.4 and 2.5, respectively. You will need to decide which methods work best for you in your brewery. Good preparation will make each of the brewing processes easier and more successful.

## TABLE 2.4—CLEANING PRODUCTS SUMMARY TABLE

| Cleaner | Amount | Comments |
| --- | --- | --- |
| Percarbonate-based cleaners | 1 fl. oz./gal. 8 mL/L | Best all-purpose cleaners for grungy brewing deposits on all brewing equipment. Most effective in warm water. Will not harm most metals. |
| Detergents | (squirt) | It is important to use unscented detergents that won't leave any perfume odors behind. Be sure to rinse well. |
| Automatic dishwasher | Normal amount of automatic dishwasher detergent. | Convenient for utensils and glassware. Do not use scented detergents or those with rinse agents. |
| Oven cleaner (Spray-on) | Follow product instructions. | Useful for dissolving burned-on sugar from a brew pot. Take care when handling. |
| White distilled vinegar | Full strength as necessary. | Most effective when hot. Useful for cleaning copper wort chillers. |
| Peracetic acid | 2:1 ratio by volume of white distilled vinegar (5% acetic acid) to hydrogen peroxide (3% solution). | Use for removing surface lead and cleaning tarnished brass. See appendix G for more info. |
| Kitchen cleansers (oxalic acid-based) | As-needed with non-abrasive scrubby pad. | Sold as stainless steel and copper cookware cleaner. Use for removing stains and oxides. Chlorine and bleaching cleaners are not recommended. |

Note: 1 tablespoon = 0.5 fl. oz. (15 mL).

## TABLE 2.5—SANITIZERS SUMMARY TABLE

| Sanitizer | Amount | Comments |
|---|---|---|
| Star San | 1 fl. oz. per 5 gal. 30 mL per 19 L (1.6 mL/L) | No-rinse. Can be used via immersion or spraying. Will sanitize clean surfaces with 30 sec. contact time. Allow to drain before use. Mix with distilled water for best results. |
| Iodophor | 12.5—25 ppm 1 tablespoon per 5 gal. = 12.5 ppm (~1mL/L) | No-rinse. Will sanitize with 10 min. contact time at 12.5 ppm. Allow to drip-dry before use. |
| Peracetic acid | Mix and use according to manufacturer's instructions. | No-rinse. Will sanitize with 2 min. contact time at 300 ppm. Can irritate skin and respiratory passages; corrosive at higher concentrations. Allow to drain completely before use. |
| Chlorine dioxide | No-rinse. Mix and use according to manufacturer's instructions. | Will sanitize with 30 sec. contact time at 50 ppm. Higher concentrations can be corrosive. |
| Bleach | 1 tablespoon/gal. 4 mL/L | Bleach will sanitize equipment in 20 minutes. It must be rinsed thoroughly to prevent chlorophenol flavors. |
| Dishwasher | Full wash and heat dry cycle without detergent. | Bottles must be clean before being put in dishwasher for sanitizing. Place bottles upside down on rack. |
| Oven | 338°F (170°C) for 1 hr. | Renders bottles sterile, not just sanitized. Allow bottles to both heat and cool slowly to prevent thermal shock. |

Note: 1 tablespoon = 0.5 fl. oz. (15 mL).

# Malt and Malt Extract 3

## A Brief Discussion of Barley and Malting

Barley (*Hordeum vulgare*) is a cereal grain, similar to oats, rye, and wheat. What makes barley different and best suited for brewing is that it retains its husk after threshing. These insoluble husks create a filter bed that results in better flow and separation of the wort after mashing than other cereal grains. For comparison, imagine trying to drain water through a bowl of oatmeal!

Like all cereals, raw barley is tough—you can break a tooth trying to chew it. This is why cereal grains were historically stone-ground to make flour. Sometime around ten thousand years ago, somebody discovered that if the seed is allowed to sprout, it gets softer. This discovery led to the development of the malting process, in which barley kernels are soaked with water and allowed to germinate for a short time before being dried and stored for later use. Germination unlocks several enzymes in the seed, which begins the breakdown of the protein-carbohydrate matrix surrounding the starch reserves that the seedling needs to grow. This starch reserve is what brewers use to make beer. One way that maltsters carefully evaluate the progress of malting is by squeezing the kernel between their fingers to gauge the degree of modification, that is, how much that stiff matrix has been broken down throughout the kernel.

When the maltster is satisfied with the degree of modification, the malt is dried with warm air under carefully controlled conditions. This preserves the enzymes that will later be used to convert the starch reserves to fermentable sugars. This base malt can then be kilned or roasted at higher temperatures to develop toasted or roasted flavors, such as cookie or biscuit, dark bread crust, and even cocoa and coffeelike flavors.

**Figure 3.1.** A sampling of a range of malt showing how the color affects beer color.

Malt flavor development comes principally from Maillard reactions, which are non-enzymatic browning reactions common to all cooking processes. Maillard reactions involve the chemical bonding of simple sugars with amino acids to create melanoidins (brown-colored compounds) and various heterocyclic compounds that are the source of the smells and flavors from cooked food that we know and love. Caramelization, a different set of chemical reactions that only involve sugars, also occurs during the kilning and roasting of malt, creating the sweet caramel and toffee flavors found in caramel and crystal malts.

In short, maltsters can produce a wide variety of malts with unique flavors and aromas that brewers can use to make interesting beers. Malts and modification will be discussed in much more detail in chapter 15, but I wanted to give you a basis for understanding the use of malt extract and steeping grains in your beer kits.

From a brewer's point of view, there are two kinds of malts: those that need to be mashed and those that only need to be steeped. The difference is whether the malt has uncooked residual starch. Base malts and kilned specialty malts need the precise temperatures of a mash to facilitate the enzymes in their conversion of residual starches to soluble and fermentable sugars. This will also be discussed in more detail in chapter 15.

Other specialty malts, such as caramel and roasted malts, do not need to be mashed. Caramel malts have had their starches converted to sugars by heat and enzyme action right inside the hull during the malting process. These malts contain fermentable and unfermentable sugars, which leave a pleasant caramel-like sweetness in the beer. Caramel malts are available with different ratings in degrees Lovibond (°L), which is a measure of color, each malt having a different degree of fermentability and characteristic sweetness. Roasted malts have had their starches converted to soluble extract by roasting at high temperatures, giving these malts a deep red-brown or black color and bittersweet, cocoa, or coffeelike flavors. These specialty malts can simply be steeped like tea to release their characteristic flavors to the wort.

## Malt Extract Production

Most of the malt extract produced around the world is not made for brewing. It is used in various food products—everything from malted milk to breakfast cereals, baking additives to pet foods. There are two primary grades of barley: malting grade and feed grade. There are also several sub-classes within these grades. The barley that is used to make food extracts is a lower grade of malting barley. These lower-grade malting barleys typically have smaller kernels with higher protein levels, less convertible starch, and a higher proportion of husk material by weight. The barley that is used for brewing beer is universally the highest grade, and today it is relatively easy to find malt extract that is made exclusively for brewing.

**Figure 3.2.** Brewhouse for production of malt extract. (Photograph courtesy of Briess Malt & Ingredients Co.)

Brewer's malt extract is dehydrated brewing wort. It starts out in the brewhouse the same as if you were brewing a beer. The malted barley soaks in hot water to create the mash, which converts the barley's starch reserves into the fermentable sugar solution we call wort. When making beer, the wort is boiled with hops and fermented with yeast. When making malt extract, the wort is typically boiled without hops or yeast and is transferred to evaporators after boiling instead of to a fermentor. Malt extract can consist of a single variety of base malt or a combination that may include specialty malts or adjuncts, depending on the type of extract being made. Adjuncts are fermentables that do not come from malt.

When making beer, a brewer boils the wort to accomplish several things, including the reduction or elimination of volatile compounds that cause off-flavors, to coagulate proteins that contribute to haze (i.e., the hot break), and to isomerize the hop alpha acids for bitterness. Manufacturers of brewing grade malt extract boil their wort for the same reasons, although it is usually unhopped. The wort is boiled long enough to eliminate the undesired volatile compounds, like dimethyl sulfide (DMS), and to coagulate the hot break proteins. The wort is then run into vacuum chambers for dehydration to make a shelf-stable product, at 80% solids (i.e., 20% water), without the use of preservatives. A partial vacuum allows the water to be boiled off at a lower temperature, which means the wort is not thermally stressed and its original flavor and color is preserved. To make a hopped malt extract, either hops can be added to the boil or iso-alpha acid hop extracts can be added at the end of the boil. Using malt extract takes a lot of the work out of homebrewing.

Malt extract is sold in both liquid (syrup) and powdered forms. The syrups are approximately 20% water, so four pounds of dry malt extract (DME) is roughly equal to five pounds of liquid malt extract (LME). Dry malt extract is produced by heating and spraying the liquid extract through an atomizer into a tall heated chamber. The small droplets dry and cool rapidly as they settle to the floor. Because of this extra step, DME powders are only about 3% water; DME also has better shelf stability than LME, and is typically not hopped.

## Summary

In summary, malt extract is not some mysterious substance, but simply concentrated wort ready for brewing and fermentation. Malt extract makes brewing easier by eliminating wort production from the mash; this allows a new brewer to focus on the process of fermentation, which is more important. A good fermentation of a poor wort will make a better beer than the poor fermentation of a good wort.

The next chapter explores beer kits in more detail, and will teach you all about steeping grains, gravity points, and boiling methods.

# Brewing with Beer Kits and Extracts

4

## Choosing a Good Kit

A typical beer recipe kit will include malt extract to serve as the base, specialty grains to add signature flavors, hops to be added at particular times in the boil to create a suitable aroma and bitterness, yeast for fermentation, and the instructions. With more than a hundred different styles of beer, and many recipes for each style, there are lots of kits to choose from. You can even create your own from a mashing recipe kit by substituting malt extract for the base malt and steeping the specialty malts. I will explain how later in this chapter.

Most beer styles can be brewed using extract alone or by combining malt extract and specialty grains. Only a few beer styles require mashing for their signature flavors.

There really is no easy guideline to say that one brand of kit is better than another; they are all pretty good these days. What is often more important is the freshness and quality of packaging of the ingredients. Kits will stay fresh for several months if they are well packaged in oxygen-barrier materials.

Choosing what kind of kit to buy can be difficult. The answer depends on what kind of brewer you are. Are you looking for simplicity? Or are you looking for a challenge?

**Figure 4.1.** A sampling of the wide range of kits available today.

The amount of effort to brew beer from a kit can vary. Some require more space, more equipment, or more attention to detail than others. Some general things to consider when deciding on a kit to use:

- No-boil kits are available, and are convenient if you have limited space and facilities. They are typically pre-hopped, but may contain additional aroma hops for steeping. You simply have to mix the ingredients together in the fermentor and pitch the yeast.
- Is it extract-only, extract with steeping grains, or extract and partial mash? Extract-only is quick and easy, but the other types involve just a few more steps and should only take up to an additional hour of time to brew. No beer kit is actually difficult to brew; some just require more attention to detail than others.
- Is it an ale or lager? Ale styles can be fermented at room temperature; lager styles need cool fermenting and conditioning temperatures, which generally means a cooler or spare refrigerator is essential.
- Is it a low, medium, or high-gravity style? Low-gravity beers are easier to ferment. High-gravity beers require more attention to yeast and fermentation and more can go wrong in a high-gravity beer. Experience with low-gravity brews makes success with high-gravity beers more likely.

Kits may be nationally branded, or assembled by your local homebrew shop. Your homebrew shop may have an edge with regards to freshness, and can give you sage advice on brewing their kits. However, national brands are often tried-and-true recipes, and are also typically well packaged for good shelf life. Whichever way you go, be sure to choose a kit that has well-written instructions and well-packaged ingredients. The fresher, the better.

## Shopping for Extract

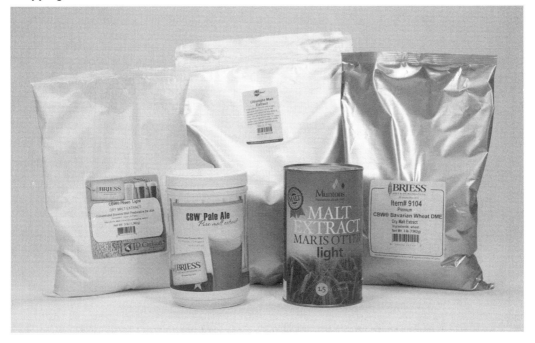

**Figure 4.2.** Malt extract comes in a variety of different brands and types.

The freshness of ingredients is important, particularly for LME. Fresh LME has a bit more fresh malt aroma than DME, but the latter has a better shelf life. Liquid extract typically has a maximum shelf life of about two years, depending on storage conditions; cooler is better. Liquid extract will darken and develop off-flavors like licorice, molasses, and ink aromas from Maillard reactions as it ages. Oxidation of the fatty acid compounds in old LME can cause blunt, stale, or soapy flavors. These oxidation off-flavors from old extract are collectively known as "extract twang." Some home-brewers complain that they can't make good beer with LME, but it's just a matter of freshness. Check the "use by" dates on the cans (within six months is generally best), or buy from a shop that has a high turnover of stock. Dry malt extract has a better shelf life (five years) than the liquid, because the extra dehydration slows the chemical reactions. If you can't get fresh LME, use DME.

Malt extract is commonly available in pale, amber, and dark varieties, and these can be mixed depending on the style of beer desired. Specialized malt extracts, such as Pilsner, pale ale, wheat, rye, Munich, Vienna and other specific styles, are arriving all the time. The quality of malt extracts and beer kits has improved greatly in the last 15 years. An all-extract brewer will be quite satisfied brewing entirely from beer kits as long as they follow the guidelines in this book.

## How Much Extract to Use

Calculating how much malt extract to use to achieve a target recipe original gravity (OG) is very simple. A pound of LME, when dissolved in water to make one gallon of wort, typically yields an OG of 1.034–1.038 (as measured by a hydrometer). A pound of DME dissolved to make one gallon of wort yields an OG of 1.041–1.045. These yield values are often referred to as points per pound per gallon (PPG). This brewing unit can be stated just as easily in kilograms and liters, becoming points per kilogram per liter (PKL). Table 4.1 lists typical malt extract yields in these units.

### TABLE 4.1—TYPICAL MALT EXTRACT YIELDS IN PPG AND PKL

| Type of Malt Extract | Typical Yield in PPG | Typical Yield in PKL |
|:---:|:---:|:---:|
| Liquid | 36 | 300 |
| Dry | 42 | 350 |

PKL, gravity points per kilogram per liter; PPG, gravity points per pound per gallon.

## Mass Gravity Volume Equation

The PPG unit (or PKL unit) is helpful when calculating values for a recipe, because you can put the yield values into what I call the "mass gravity volume equation," which is:

$$\text{mass of extract} \times \text{PPG} = \text{gravity points} \times \text{volume of wort}$$

Which can be rearranged, like so:

$$\text{mass of extract} = (\text{gravity points} \times \text{volume of wort}) / \text{PPG}$$

You can use this rearranged equation to calculate how much malt extract is needed to create a particular volume of wort of a particular gravity. Simply multiply the desired gravity points and wort volume by the PPG (or PKL) number. Remember to multiply by gallons when using PPG, or liters when using PKL. The gravity points are the three digits after the decimal point in a hydrometer reading. For example, a recipe gravity of 1.056 would have 56 gravity points.

Let's look at a couple of examples:

How many pounds of LME are needed (which we call "$X$") to make 6 gal. of 1.056 wort?

$$X \text{ lb.} \times 36 \text{ PPG} = 56 \text{ gravity points} \times 6 \text{ gal.}$$

Rearrange to give

$$X \text{ lb.} = (56 \times 6) / 36$$

$$= 336 / 36$$

$$= 9.3 \text{ lb.}$$

If using PKL, 6 gal. = 22.7 L, and 36 PPG = 300 PKL, so this becomes:

$$X \text{ kg} \times 300 \text{ PKL} = 56 \text{ gravity points} \times 22.7 \text{ L}$$

$$X \text{ kg} = (56 \times 22.7) / 300$$

$$= 1288 / 300$$

$$= 4.2 \text{ kg}$$

The equation can be algebraically manipulated to solve for any of the terms. For example, what is the predicted gravity of 5 lb. (2.27 kg) of DME in 3 gal. (11.4 L) of water?

$$5 \text{ lb.} \times 42 \text{ PPG} = X \text{ gravity points} \times 3 \text{ gal.}$$

$$(5 \times 42)/3 = X$$

$$= 70, \text{ or a hydrometer reading of } 1.070$$

And if you are using metric units:

$$2.27 \text{ kg} \times 350 \text{ PKL} = X \text{ gravity points} \times 11.4 \text{ L}$$

$$(2.27 \times 350)/11.4 = X$$

$$= 69.7, \text{ or a hydrometer reading of } 1.070$$

## Converting All-Grain (Mashing) Recipes to Extract

It is actually quite easy to convert most recipes for all-grain brewing into recipes that use malt extract and specialty grains instead. You use the same concept of gravity points and volume as above to calculate how much malt extract to use in place of the all-grain base malt. For gravity point yields from malts rather than malt extract, the rule of thumb for base malt (i.e., two-row, Pilsner, pale ale, Vienna, and Munich) is 27 PPG, or 225 PKL. Taking the ratio of the respective PPG values for the base malt and the malt extract gives you a conversion factor, which allows you to calculate the amount of extract to use in place of the base malt. One key thing to remember is you need to account for the fact that some malt extract will contain a proportion of base malt. This will be illustrated in the second of the two examples that follow. Table 4.2 lists some common malt extracts and their conversion factors.

For the first example, let's use a simple all-grain stout recipe consisting of 10 lb. (4.5 kg) of pale ale malt, 1 lb. (450 g) of crystal 40°L malt, and 1 lb. (450 g) of roast barley.

1. Start with the base malt grain. We know that the base malt will yield 27 PPG (225 PKL), compared to 36 PPG (300 PKL) from LME. The ratio between the base malt and LME is 27/36 (or 225/300), which gives us a conversion factor of 0.75. In other words, we would need 75% as much LME to make up the same number of gravity points as the base malt grain. Therefore, 10 lbs. (4.5 kg) becomes 7.5 lbs. (3.38 kg) of pale ale LME.

2. The specialty malts don't change, since they will be steeped in a grain bag, so the stout recipe becomes: 7.5 lb. (3.38 kg) of liquid pale malt extract, 1 lb. (450 g) of crystal 40°L malt, and 1 lb. (450 g) of roast barley.

For the second example, let's look at slightly tricker conversion that arises when the malt extract has two components instead of just one. It's really not that hard. In this example, let's convert an all-grain wheat beer recipe to LME. A common wheat beer recipe would be 6 lb. (2.72 kg) of wheat malt and 6 lb. (2.72 kg) of Pilsner malt.

1. Start with the conversion for the wheat malt. A typical wheat LME contains 65% wheat malt and 35% base malt (barley). For wheat malt versus wheat LME, the conversion factor is 1.2 (table 4.2). Therefore, the quantity of wheat LME is 1.2 × 6, which equals 7.2 lb. (if using metric, 1.2 × 2.72 = 3.26 kg).

2. The catch is that every pound or kilogram of wheat LME also contributes 35% by weight of barley malt extract, and this needs to be taken into consideration when calculating the substitution for the Pilsner malt in the recipe. Pilsner malt is your base malt, so 6 lb. (2.7 kg) of Pilsner malt would require 0.75 × 6 = 4.5 lb. of Pilsner LME (if using metric, 0.75 × 2.72 = 2.04 kg). However, the 7.2 lb. total wheat LME is already contributing 2.5 lb. (1.13 kg) of barley malt extract (i.e., 35% of 7.2 lb. [3.26 kg]). We can subtract this existing barley malt extract amount from the Pilsner LME total needed (i.e., 4.5 lb. – 2.5 lb.), so, in the end, we only need 2 lb. (910 g) of additional Pilsner malt extract to complete the recipe.

3. The wheat malt recipe becomes: 7.2 lb. (3.26 kg) of wheat LME and 2 lb. (910 g) of Pilsner LME. You can apply the same method when using other extracts with multiple components, such as Munich or rye.

## TABLE 4.2—WEIGHT-FOR-WEIGHT CONVERSION FACTORS WHEN SUBSTITUTING GRAIN FOR MALT EXTRACT

| Malt extract | Composition of malt extract[a] | Conversion factor[b] |
|---|---|---|
| Pilsner LME | 100% Pilsner | 0.75 |
| Pilsner DME | 100% Pilsner | 0.64 |
| Pale ale LME | 100% pale ale | 0.75 |
| Pale ale DME | 100% pale ale | 0.64 |
| Wheat LME | 65% wheat<br>35% base malt | 1.2 wheat<br>(+ 35% base malt) |
| Wheat DME | 65% wheat<br>35% base malt | 0.88 wheat<br>(+ 35% base malt) |
| Vienna LME | 100% Vienna | 0.75 |
| Munich LME | 50% Munich<br>50% base malt | 1.5 Munich<br>(+ 50% base malt) |
| Munich DME | 50% Munich<br>50% base malt | 1.3 Munich<br>(+ 50% base malt) |
| Rye LME | 20% rye<br>70% base malt<br>10% Caramel 40°L | 1.3 rye<br>(+ 70% base malt,<br>+ 10% C40 malt) |

DME, dried malt extract; LME, liquid malt extract.

[a] Compositions of extracts are examples of Briess CBW® extracts.

[b] Multiply the weight of grain in recipe by the conversion factor to obtain quantity of extract needed.

Note: Base malt is a generic description for malted barley. Pilsner malt and pale ale malt are both base malts made from two-row barley, but pale ale malt is generally kilned to a higher color and flavor than Pilsner malt. The base malt used in wheat or Munich malt extract, for example, could be either a Pilsner or pale ale malt, or something in between, depending on the manufacturer.

## TIPS FOR WORKING WITH EXTRACT

Brewing with malt extract has its quirks and techniques. It is basically sugar. Sugar is easier to mix into cold water without clumping, but it dissolves slowly. Malt extract dissolves best in hot water, but tends to clump. The best advice for both liquid and dry extract? Stir!

- Liquid malt extract is a very dense syrup that will settle at the bottom of the pot where it will burn if the burner is on. The syrup dissolves best in hot water, but take the pot off the burner to avoid scorching (black flakes) until it is fully dissolved.

- Add dry malt extract slowly while stirring constantly. It disperses best in cold water, but is very slow to dissolve. Dry malt extract tends to clump when added to hot water and can form small floating rocks that are difficult to dissolve. But they will, eventually!

### Gravity versus Fermentability

I will discuss the concept of attenuation in chapter 6, but in this chapter I will talk about fermentability. Generally, brewers talk about the fermentability of the extract in the wort, and attenuation of the yeast in the beer (i.e., before versus after fermentation). Fermentability and attenuation are really the same property—the difference is the direction.

The fermentability of most malt extracts today is about 75%, meaning that if the OG of the wort is 1.040, then the final gravity (FG) will be 1.010. However, this can vary by manufacturer. In all-grain brewing, the brewer can manipulate mash conditions to produce a higher or lower proportion of unfermentable sugars, and therefore a correspondingly higher or lower FG. This is also true for the malt extract production process, since it is just a specialized form of brewing. A higher FG can provide a maltier, more full-bodied beer that is appropriate for some styles, such as foreign extra stout. All-grain brewers also have the option of adding dextrin malt to the mash to provide more unfermentable sugars. Extract brewers can achieve the same goal by adding maltodextrin powder, which is a purified form of the same thing. Maltodextrin is not sweet and can be slow to dissolve. It contributes about 40 PPG (340 PKL) and is not fermented by normal brewing yeast.

Generally, the higher the OG, the higher the FG, and the higher the percentage of alcohol by volume (ABV) will be (table 4.3). However, a higher FG does not automatically mean that the beer will be sweeter tasting. The final sugar profile from the mash, the hop bitterness character from the boil, and the fermentation by a particular yeast strain will determine how sweet the beer tastes. Two beers can have the same OG and FG but taste very different.

### TABLE 4.3—PERCENTAGE ABV ASSUMING 75% FERMENTABILITY

| OG | 1.030 | 1.040 | 1.050 | 1.060 | 1.070 | 1.080 | 1.090 |
|-----|-------|-------|-------|-------|-------|-------|-------|
| FG | 1.007 | 1.010 | 1.012 | 1.015 | 1.018 | 1.020 | 1.022 |
| ABV | 2.9% | 4% | 5% | 6.1% | 7.2% | 8.4% | 9.7% |

*ABV, alcohol by volume; FG, final gravity; OG, original gravity.*

*Note: A more complete table is presented in chapter 9.*

**Figure 4.3.** Typical crushed grain with a one-cent coin included for scale.

## Steeping Specialty Grains

Steeping specialty grain is like making tea. Caramel-type malts, such as caramel 20°L and caramel 60°L, have already been mashed in the hull during the malting process, so their sugars are available simply by steeping. Roast malts, like chocolate malt, black malt, and roast barley, have had their starches solubilized by roasting. Although these roast malts release their soluble extract more slowly than base or caramel malt, they contribute a lot of flavor. The grains of roast malts can be crushed ahead of time, placed in a muslin or nylon bag, and steeped to release their flavors. Steeping longer will yield more flavor, similar to steeping black tea.

Steeping time, temperature, and particle size are the factors that influence how well the specialty grain is extracted into the wort. A finer crush will yield more extract and yield it sooner. If you need more of a specialty grain's particular character, you can either crush it finer (or double or triple crush it, i.e., run it through the mill two or three times), or you can just use more of the crushed grain at the regular crush setting. The typical mill gap setting of 0.04 in. (1 mm) provides a good range of particle size; if the mill gap were much smaller, the powder would be difficult to contain within the grain bag. Note, however, that wheat malt may be physically smaller than barley malt and sometimes benefits from double crushing.

Crushed specialty grain is typically steeped in hot wort (140–170°F [60–75°C]) for 30 minutes. Caramel malts release their extract rapidly, while roasted malts yield their extract more slowly. After steeping, the grain bag is removed from the wort before the boil begins.

### Typical Steeping Yield

Typical yields for various specialty malts are given in table 4.4. For best results, the grain should be fairly loose in the grain bag; it should not be swollen tight like a pillow. If necessary, use two or

## TABLE 4.4—TYPICAL SPECIALTY MALT STEEPING YIELDS

| Malt type[a] | Malt color (degrees Lovibond) | PPG | PKL |
|---|---|---|---|
| Carapils® | 2 | 8 | 67 |
| Carafoam® | 2 | 27 | 225 |
| Caramel 10 | 10 | 16 | 134 |
| Caramel 20 | 20 | 17 | 142 |
| CaraVienne | 20 | 17 | 142 |
| Carastan | 35 | 17 | 142 |
| Melanoidin malt | 35 | 28 | 234 |
| Caramel 40 | 40 | 17 | 142 |
| Carawheat® | 55 | 8 | 67 |
| Caramel 60 | 60 | 18 | 150 |
| Caramunich® | 60 | 18 | 150 |
| Caramel 80 | 80 | 16 | 134 |
| Caramel 120 | 120 | 15 | 125 |
| Special "B" | 130 | 14 | 117 |
| Amber malt | 25 | 18 | 150 |
| Brown malt | 60 | 8 | 67 |
| Pale chocolate | 225 | 20 | 167 |
| Chocolate | 350 | 24 | 200 |
| Dark chocolate | 420 | 25 | 209 |
| Carafa Special II® | 450 | 25 | 209 |
| Carafa Special III® | 500 | 26 | 217 |
| Black Prinz® | 500 | 26 | 217 |
| Midnight wheat® | 550 | 27 | 225 |
| Roast barley | 500 | 27 | 225 |
| Black malt | 550 | 27 | 225 |

PKL, (gravity) points per kilogram per liter; PPG, (gravity) points per pound per gallon.

[a] The malts Carapils through Special "B" are stewed, caramel-type specialty malts, and the amber through black malts are dry toasted and roasted specialty malts.

Notes: Steeping data is experimental and was obtained by steeping 120 g in 1 L (equivalent to a concentration of 1 lb./gal.) at 160°F (71°C) for 30 minutes. All malts were crushed in a two-roller mill at the same setting. Your results may differ.

more bags rather than trying to put all of the grain into one bag. The wort should be able to flow through the grain for best extraction.

One other word of advice: don't squeeze the grain bag to get every drop of wort! Simply hold the grain bag over the pot and let it drain for a minute. Pale and caramel steeping grains give up their extract fairly easily, and the grain doesn't retain much extract as its drained. On the other hand, roast malts can have quite a bit of retained extract, but squeezing the grain bag also releases tannins and other astringent compounds into the wort. A brief squeeze to prevent dripping on the stove is fine, but don't attempt to wring more wort out of the bag; you will be sorry you did.

## Steeping Temperature

The temperature of the steep has a big effect on the flavor. Hot steeping obtains a flavor and aroma profile from the grains similar to that obtained if the grains were mashed, because the steeping temperature is very similar to the mash conditions. Hot steeping is the most common technique for adding specialty grain flavors to extract brews. Spices may also be steeped in hot water, or a small amount of hot wort, to create a tea that can be added separately to the wort or beer later in the brewing process. Spice teas prepared this way should be added after the boil to avoid boiling off the very aromas that you wanted to capture.

Cold steeping of specialty grains is very similar to the cold brewing of tea and coffee and has similar effects on flavor and aroma. Cold steeping for several hours or overnight reduces the astringent character of roasted grain and yields more of its fresh aroma, but results in a less intense flavor. Cold steeping is helpful when you have alkaline brewing water (water and extract brewing are discussed further in chapter 8), and is also useful if adding spices, such as cinnamon, allspice, and coriander, among others.

It is important to boil or pasteurize cold-steeped wort (or tea) to prevent possible contamination of the beer, so add it at the end of the boil. Thirty seconds at temperatures of 175°F (79°C) or greater should be adequate for pasteurization. As a quick guide, 1 gal. of water at room temperature (70°F [21°C]) would bring 3 gal. of water at 210°F (99°C) down to about 175°F (79°C). Therefore, any addition less than one volume into three volumes (i.e., 1:3) can be safely added at the end of the boil, but any addition larger than 1:3 should be added a few minutes before the end of the boil.

Spices and roast specialty grains should never be boiled. Boiling brings out the worst flavors from spices and roasted malts, even when boiled in wort. Roasted malt tastes like burnt food when boiled. Boiled cinnamon is very sharp and woody. Likewise, if a specialty grain is hot steeped too long (generally more than 30 minutes), or too hot (near boiling) more astringent tannin compounds (a.k.a. polyphenols) will be extracted from the grain husks. This is particularly true for roast malts. The compounds give the wort a dry, lip-puckering taste, just like black tea that has been steeped too long. Old homebrewing practices for steeping specialty grains had the brewer putting the grain in the pot and bringing it to a boil before removal. That method often resulted in tannin extraction.

Water chemistry plays a role in tannin extraction as well. Steeping roasted malts in low-alkalinity water (i.e., water with low bicarbonate levels) may produce conditions that are too acidic and acrid flavors can result. Likewise, steeping the lightest crystal malts in highly alkaline water may produce conditions that are too alkaline and tannin extraction can result.

The previous edition of this book (and every other homebrewing instruction book) used tap water for steeping, and the grains were steeped before adding the malt extract. In this edition, I am recommending steeping the grains in wort, as I did in chapter 1, to prevent any tannin extraction due to water chemistry. The advantage of steeping in wort is that you reduce the astringent aspect of roast malts with only a small reduction in yield (10%–20% based on my experiments, which equates to only one or two points of the OG). High-gravity worts will decrease the yield more, but a 1.020 gravity wort is more than sufficient to moderate the pH. Steeping spices in wort produces characteristic flavors and aromas without the earthiness or acrid flavors that can result from steeping in water. See chapter 8 for more information on how water chemistry affects extract brewing.

Steeping differs from mashing in that there is no enzyme activity taking place to convert grain or adjunct starches to sugars. Steeping specialty grains is entirely a leaching and dissolution process of existing sugars and flavor compounds into the wort. If diastatic (enzymatic) malts are steeped, that's a mash. You can "steep" the enzymatic malts, but, as will be explained later in Section 2 of this book, if you don't steep them under the right conditions, you won't get good conversion of the starches and the yield will be poor.

---

## BOILING AND FLAVOR DEVELOPMENT

Boiling has always been part of the brewing process going back thousands of years. Many of the flavors we expect from beer are due, at least in part, to the length and intensity of the boil. The boiling of the wort is a crucial stage in the brewing process, because a critical factor in the development of a beer's flavor are the Maillard reactions that occur during the boil. These reactions are affected by time and temperature as well as the initial ingredients. Therefore, here are some guidelines for the boiling of wort:

1  The wort must be boiled. Boiling times vary across styles; historically it could have been up to five hours, but currently 60–90 minutes is considered the norm.

2. The intensity of the boil should be medium; not a flat simmer or a languid boil, nor a leaping violent cauldron, but a rolling boil, where the entire surface of the wort has continuous activity. Medium.

3. The evaporation rate can be a good point of reference in trying to gauge the intensity of the boil. As a general rule of thumb, you are looking for an evaporation rate of about 10%–15% per hour, though it could be as high as 20% depending on your equipment. You are not trying to make steam, you are not trying to concentrate your wort (though that happens)—you are cooking it!

---

## Full Boil versus Partial Boil

When you are brewing with malt extract, you are brewing with concentrated wort. Wort that has, in fact, been previously boiled. As we saw in chapter 3, malt extract manufacturers boil the wort under a partial vacuum in order to make it easier to remove the water. This also allows the water to boil at a lower temperature, which reduces the Maillard reactions so that the malt extract doesn't become overcooked by the time you are done brewing with it. It was standard practice among extract brewers many years ago (circa 1970–1990) to boil all of the extract in a reduced amount of water—boiling only three gallons instead of six—because most electric kitchen stoves cannot put out enough heat to boil six gallons.

For example, a five-gallon recipe for a beer with an OG of 1.050 would use about seven pounds of LME. A full-volume boil of that recipe would call for six gallons being boiled down to five, with the wort's gravity increasing from about 1.042 to 1.050. In other words, for a 1.050 OG beer, the wort should start to cook and flavor development from Maillard reactions start to occur (see sidebar, "Boiling and Flavor Development") at a gravity of 1.042.

If that same seven pounds of LME boiled in only three gallons of water on the kitchen stove, it would create a boil gravity of 1.084. The aroma and flavor compounds produced by Maillard reactions that occur during cooking depend on the concentration of sugars and amino acids in the wort. Therefore, boiling the wort at a higher gravity results in different flavors being created, flavors that are normally not associated with that style of beer. Consequently, extract-brewed beers had a bad reputation for many years for not being as good as all-grain versions of the same recipe and style. While some of the flavor differences between the methods could be due to particular ingredient choices in the recipes, most of it was due to these differences in boil gravity and the associated Maillard reactions.

In the previous (third) edition of *How To Brew* I introduced the "Extract Late" method, whereby roughly half of the extract is boiled in roughly half of the total water with the hop additions included. This achieves a boil gravity and set of Maillard reactions more appropriate to the recipe and style. The remaining extract is added after the boil, after the heat is turned off, and allowed to pasteurize for about 10 to 15 minutes before cooling. This high-gravity wort is then diluted to the full volume in the fermentor.

I doubt I was the first person to come up with the method, but I was one of the first to write about it extensively. "Extract Late" is a rather clumsy title, so in this newest edition we will just call it the "Palmer Brewing Method," since many people call it that already. If you tackled the step-by-step guide in chapter 1, "Brewing Your First Batch of Beer," then you will already have seen an example of the Palmer Brewing Method in action.

A full-volume boil (e.g., boiling seven gallons down to six) is the most traditional method for brewing beer and produces the traditional, or most expected, flavors. The Palmer Brewing Method—utilizing "wort A" and "wort B" components, with the boil and hop additions being done to wort A—comes very close to producing the same flavors and aromas as a full-volume boil using the same recipe. Most recipes in this book will have a Palmer Brewing Method version to make it easier to brew these recipes at home. If you have a large kettle (e.g., a 10 gal. or 40 L kettle) then you can easily conduct a full-volume boil by combining both the wort A and wort B ingredients in the total amount of water. However, you will probably need to purchase a propane or natural gas burner to supply enough heat.

## Summary

Malt extract is simply concentrated wort. It makes brewing easier by reducing much of the time and effort needed to produce the wort, letting a new homebrewer focus on the fermentation process. Most beer kits use malt extract to provide the bulk of the fermentable sugars, using steeping grains to provide the specialty malt flavors appropriate to the style. A good beer kit is simply an extract recipe in prepackaged form, which takes the guesswork out of brewing until you have gained enough experience to create your own recipes.

Steeping specialty grains opens the door to a lot of beer styles that are difficult to brew with extract alone. There are only a few beer styles, such as Belgian *wit*, that require the mashing of adjuncts for their signature flavors and characters. Many homebrewers are perfectly happy using malt extract and steeping grains, because of their ease of use and the great variety of beer that can be brewed. The Palmer method for partial boils makes it easier to brew 5 gal. (19 L) kits on the kitchen stove.

Section 2 of this book will explain the all-grain process, called mashing, which allows you to extract the sugars from the malted grain yourself, giving you total control in producing the wort and tailoring its fermentability.

# Hops

## What Are They?

Hops are the strobili—the cone-like female reproductive structures—that constitute the flower of the hop plant (*Humulus lupulus*). The hop plant is a climbing vine that is native to the temperate regions of North America, Europe, and Asia. The species has separate male and female plants. Only the female vines (known as bines) produce the cones. The vines will climb 30 ft. (9 m) or more up any available support and are commonly trained onto strings or wires when grown commercially. The leaves of the hop plant resemble grape leaves, while the cones of the hop itself vaguely resemble pinecones in shape, but are light green, thin, and papery. At the base of the bracts (the thin, papery outer leaves of the hop) are the yellow lupulin glands that contain the essential oils and resins that are so prized by brewers for their bittering and aroma qualities. The hop oils and resins also inhibit bacterial growth, and this natural preservative property is one reason hops were first used in beer.

Hops have been cultivated for use in brewing for over 1,000 years. Hops are most productive between 35° and 55° latitude in both the northern and southern hemispheres. The earliest known cultivation was in central Europe, and by the early 1500s cultivation had spread to western Europe and Great Britain. At the beginning of the twentieth century, only a couple dozen varieties of hop were being used for brewing worldwide; today, there are over two hundred. The focus of breeding programs has been to increase the alpha acid bittering compounds, while improving yield and disease resistance.

**Figure 5.1**. Hop cones.

**Figure 5.2**. Lupulin glands at the base of the bracts.

## Hop Bitterness

The primary bittering agents in hops are derived from the alpha acids, called humulones, which are present in hop resin. These alpha acids are neither bitter-tasting or soluble in water until they are isomerized by boiling. Isomerization means that the configuration of the molecules has changed, but not the chemical formula. In other words, the molecules are made of the same atoms but in a different arrangement, which alters the properties. When the hops are boiled in the wort, isomerization makes the alpha acids—now called iso-alpha acids (iso-humulones)—intensely bitter and more water-soluble. The longer the boil, the more alpha acids are isomerized, the more iso-alpha acids end up dissolved in the wort, and the more bitter the beer will be. Typically, after an hour of boiling hops, around 25%–30% of the alpha acid content is isomerized.

Hop resin also contains beta acids, called lupulones, but these beta acids do not isomerize in the boil. Instead, beta acids oxidize during storage, with the oxidized beta acids becoming more water soluble and bitter. This is the main source of bitterness in "aged hops." Alpha acids also oxidize if stored at room temperature. While these oxidation products are not the same as isomerized alpha acids, they are bitter. However, oxidized alpha acids are not as bitter as oxidized beta acids, and oxidized beta acids are not as bitter as iso-alpha acids.[1] Although oxidized beta acids are more bitter than oxidized alpha acids, modern hop varieties contain two or three times the level of alpha acids than beta acids. This affects the overall contribution of each type of hop acid to perceived bitterness.

Hop polyphenols (a different type of compound from hop acids) from the hop cones are thought to influence the perceived bitterness of a beer, particularly beers that are dry-hopped.

---

[1] Algazzali and Shellhammer (2016) reported that oxidized beta acids are about 84% (±10%) as bitter as iso-alpha acids, whereas oxidized alpha acids are about 66% (±13%) as bitter as iso-alpha acids.

However, a study that combined chemical analysis with the results from a trained tasting panel showed that the perceived bitterness in selected commercial beers correlated almost entirely with levels of iso-alpha acids and oxidized alpha acids.[2] So, while hop polyphenols may affect perceived bitterness, their effects are variable and not as significant as oxidized alpha acids.

To summarize, here are the key points to remember when considering hop bitterness:

**Compounds that impart bitterness**
- Iso-alpha acids: These compounds are intensely bitter, and are the most prevalent bittering agent in hopped beer.
- Oxidized alpha acids: These are less bitter than oxidized beta acids, but present in greater amounts. Most prevalent in old or aged hops.
- Oxidized beta acids: These are more bitter than oxidized alpha acids, but present in lower amounts. These are the primary bittering agent in aged hops, such as those used in sour (*lambic*-style) beers.
- Hop polyphenols: The bittering effects of this class of compounds are variable and not as significant as hop acids.

**Compounds that do not impart bitterness**
- Raw alpha acids[3]: These are the non-isomerized and non-oxidized, water-insoluble acids that naturally occur in hop resin.
- Raw beta acids: These are the non-oxidized, water-insoluble acids that naturally occur in hop resin.
- The decomposition products of the bittering compounds described above.

## THE HISTORY OF THE IBU TEST

Brewers have been attempting to classify and quantify hop bitterness in beer since the late 1800s. The alpha acids (humulone-type) and beta acids (lupulone-type) from lupulin glands were isolated about this time, and by the early 1900s, scientists realized that those particular components were not present in beer—the hop bitterness and aroma components were transformed by the brewing process. During the 1920s, 30s, and 40s, the molecular structures of these compounds were investigated to determine how alpha and beta acid bitterness was manifested in beer. In 1939, W. Windisch described the bittering potential of the oxidation products of beta acids, and in 1947, isomerized alpha acid was isolated from beer by F. Goveart and M. Verzele. These discoveries lead to efforts to develop a reliable and repeatable test method for measuring bitterness in beer.

In 1953, F.L. Rigby and J.L. Bethune developed a chemical method for isolating iso-alpha acids from beer. In fact, they were able to use this technique on raw lupulin from hops and determine that there were actually three main types of both alpha and beta acids: (normal) humulone, cohumulone, and adhumolone; and lupulone, colupulone, and adlupulone. The cohumulone form is easier to isomerize than the others, but was commonly perceived to give a rougher bitterness to the beer. Even though this view is debatable, selection of low cohumulone character was encouraged as new hop varieties were developed. Many of today's high-alpha-acid hop

2    Hahn, Lafontaine, and Shellhammer, "A holistic examination of beer bitterness" (abstract presentation, World Brewing Congress, Denver, CO, August 17, 2016).
3    Fritsch and Shellhammer (2007).

varieties, like Magnum and Horizon, have lower cohumulone levels than older, lower-alpha-acid varieties of the past, such as Galena and Cluster.

The Rigby-Bethune testing method was sound, but it took an entire day to test one sample. Today, high-performance liquid chromatography (HPLC) can measure alpha acid and beta acid levels in an hour. However, chromatography methods require technical knowledge and experience to accurately interpret the results. A simpler and more robust method was needed for measuring bitterness units (BUs; usually referred to as international bitterness units, or IBUs, in America) in daily beer production.

Two competing spectrophotometric methods (i.e., measuring the light passing through a specimen) for measuring iso-alpha acid content were published in 1955. The first method, by Rigby and Bethune, developed another multiple-step, solvent-extracted sample. Solvent extraction works on the principal that the organic hop compounds (even iso-alpha acids) are more soluble in solvent than they are in water. They reported that the method over-reported the iso-alpha acid levels by 30% compared to their original method for the same beers. The authors concluded this was due to "interfering substances" and that the results for iso-alpha acids were probably not the best indicator for overall beer bitterness.

The second method, proposed by the team of A.B. Moltke and M. Meilgaard (in communication with Rigby and Bethune), took a step back and used a simpler solvent extraction method to isolate a menagerie of hop compounds that were chemically similar to iso-alpha acid. This sample was then measured spectrophotometrically using light with a 275 nanometer wavelength and the number was compared to iso-alpha acid measurements of the same beers derived from the Rigby-Bethune chemical method. This data was analyzed to develop a linear regression equation for perceived bitterness as a function of the spectrophotometrically measured iso-alpha acids. Over the next ten years, the American Society of Brewing Chemists (ASBC) and the European Brewery Convention (EBC) went back and forth with different variations and equations based on the Moltke-Meilgard method. By 1968, the ASBC and EBC had both settled on the following equation that is still in use today (see "Beer-23: Beer Bitterness," in *ASBC Methods of Analysis*):

$$BU = 50 \times Absorbance @275nm,$$

where 50 is a coefficient rounded down from 51.2, based on the slope of the correlation and ratio of solvent used.

## Hop Aroma and Flavor

The aromas associated with hops come from their essential oils (table 5.1), and these oils contain roughly 500 different chemical compounds that have been identified so far. Many exist in only the barest of trace amounts, while others, such as myrcene, can make up 50% of the total essential oil content. The interaction of these compounds is very complex. Brewing scientists have attempted to recreate the aroma of a particular hop by combining the primary individual oils in the appropriate proportions and failed miserably. Of course, there is nothing wrong with adding hop oils to a beer to enhance its hop character, but the whole of hop aroma is greater than the sum of its parts.

Hop flavor is a combination of hop bitterness, resin (mouthfeel), and aroma. Hop aroma from the essential oils is hard enough to describe; flavor is much more difficult. Suffice to say that middling short boil times that contribute a combination of isomerization and residual oil and resins to the wort can be perceived as flavor in the final beer. Generally, these are boil times of less than 20 minutes. Hop steeps, where the hops are soaked in hot wort before cooling, will also contribute hop flavor to the beer.

Table 5.2 lists seven primary characters that are often used to describe hop aroma: floral, fruity, citrus, vegetal, herbal, resinous, and spicy. Of course, this listing is subjective, and if you ask three different experts you will get three different lists. Some characterization plots break the aroma down into 12 or more different characters, but these seven get the point across. It is important to understand that aroma is subjective, and different people can perceive a particular aroma differently. For example, some people might describe mint as herbal and others would describe it as spicy, and they both would be correct.

For more discussion of the essential hop oils, see Stan Hieronymus's book, *For the Love of Hops*, from Brewer's Publications (2012).

## TABLE 5.1—AROMAS OF THE MAIN ESSENTIAL (AROMATIC) OILS FOUND IN HOPS

| Oil | Aroma description |
| --- | --- |
| Myrcene | Sweet carrot, celery, green leaves |
| Humulene | Herbal, woody, spicy clove |
| Caryophyllene | Spicy, cedar, lime, floral |
| Farnesene | Woody, citrus, sweet |
| β-Damascenone | Honey, berry, rose, blackcurrant, Concord grape |
| β-Ionone | Raspberry, violets |
| Linalool | Floral, lavender |
| Geraniol | Floral, rose, marigold, geranium |
| Nerol | Floral, wisteria |
| Citronellol | Citrus, lemon, citronella oil |
| Terpineol | Citrus, fruity |
| Humulenol | Spicy, pineapple, cedar, sagebrush |
| Humulol | Spicy, herbal, hay |
| 4MMP[a] | Muscat grapes, blackcurrants, onion |

[a] 4MMP = 4-mercapto-4-methylpentan-2-one

## TABLE 5.2—CATEGORIZATION OF HOP AROMA

| Floral | Fruity | Citrus | Vegetal | Herbal | Resinous | Spicy |
|--------|--------|--------|---------|--------|----------|-------|
| Geranium | Apple | Grapefruit | Celery | Tarragon | Pine | Fennel |
| Rose | Berries | Orange | Tomato leaves | Marjoram | Juniper | Black pepper |
| Jasmine | Peach | Lemon | Green pepper | Lavender | Heather | Nutmeg |
| Lily of the Valley | Melon | Lime | Cabbage | Dill | Tobacco | Clove |
| Lavender | Passionfruit | Bergamot | Hay | Sage | Woody | Mint |

## Hop Variety Categories

Today, hops are cultivated in many different countries. European, English, American, and Pacific hop varieties are readily available on the open market. In addition, there are Chinese and South African varieties as well, but these are seldom seen outside of their own regions due to high domestic demand. The European varieties were the original brewing hops, and these are often referred to as "landrace" or "noble" hops. These early hops have delicate floral, spicy, and resinous aromas that for centuries were the definition of what hops should smell like. The English varieties developed next and these have traditionally had more of an herbal, earthy, and fruity aroma compared to the European varieties. The American varieties have a predominately citrus character, with undertones of herbal, resinous, and spicy aromas. The Pacific varieties (i.e., New Zealand and Australia) have a strong tropical fruit character with citrus and floral notes.

There was a time when each region could be described by one word: European, spicy; English, herbal; American, citrus; and Pacific, fruity. But, with the craft beer movement accelerating since the early 1990s, hop breeding and development has exploded. Today it is common to find German varieties, such as Mandarina Bavaria and Hüll Melon, that smell of fruit; American varieties, such as Citra and Mosaic, that smell tropical; and English and Pacific varieties that smell like citrus.

There are some hop varieties that help define particular styles, such as the minty flavor of Northern Brewer for California common beer, the floral character of Spalter and Hersbrucker for German lagers, Cascade for American pale ale, and Fuggle and East Kent Goldings for British pale ale. As you are learning to brew, and learning what defines the character of particular beer styles, you will probably attempt to brew classic recipes faithfully by using the prescribed malts, hops, and yeast, and that is a good idea. If you want to get a great foundation in brewing and tasting different beer styles, I heartily recommend brewing all of the recipes in *Brewing Classic Styles* by Jamil Zainasheff and John Palmer (Brewers Publications, 2007).

However, one of the first substitutions any brewer will make is for a hop that is not readily available, or for one that they just don't particularly like. In fact, I would suggest that new brewers consider the hop callouts in any beer recipe to be mere guidelines for a couple of reasons. First, because bitterness is bitterness—there is very little difference in the flavor of that bitterness across different hop varieties. Now, that being said, the flavor and aroma hop selections of a recipe are a bit more important, because the dominant character of those hops may indeed be characteristic of the style. There is still room for substitution though within that hop's variety category. For example, if a recipe for American pale ale calls for Cascade, a good substitute would be Centennial or Amarillo. If you need to substitute for German Tettnang in a German Pilsner recipe, you could readily use Spalt Select or Saaz.

Don't be afraid to substitute hops, just be reasonable about it. You would probably not want to substitute Cascade or Chinook for a German hop if you are trying to faithfully brew a German Pilsner style, because the characters are just too different. But brew what you like; everyone's flavor preferences are different, some prefer fruity, others prefer floral, and still others prefer resinous. Gradually you will find the hop varieties you prefer, and as you fine tune your favorite recipe(s) you will probably decide on particular hops that have to be in there. That's the purpose of a recipe.

Table 5.3 lists some common hop varieties and suggested substitutions. I used to try to keep this book up to date on all varieties, but the hop world is moving far too fast. Get online and find out what's new.

## TABLE 5.3—COMMON HOP VARIETIES AND SUGGESTED SUBSTITUTIONS

| Category | European Hop Varieties | English Hop Varieties | American Hop Varieties | Pacific Hop Varieties |
|---|---|---|---|---|
| **General Character** | Floral, Spicy, Resinous | Resinous, Fruity, Spicy | Citrus, Herbal, Resinous | Fruity, Citrus, Floral |
| **Substitution: European-like** | Hallertauer Mittelfrüh Tettnang Spalt Select Saaz Hersbrucker | Target Challenger Northdown Progress | Crystal Mt. Hood Horizon Cluster Wakatu | Helga Pacifica Sylva Ella |
| **Substitution: English-like** | Magnum Opal Merkur Smaragd | East Kent Goldings Fuggles West Goldings Variety Sovereign | Glacier Columbia Willamette Galena | Green Bullet Fuggle Wye Challenger Pacific Gem Super Pride |
| **Substitution: American-like** | French Triskel Hallertau Blanc | Admiral Pioneer Epic Pilgrim | Amarillo Cascade Centennial Ahtanum | Dr. Rudi Waimea Sicklebract Galaxy |
| **Substitution: Pacific-like** | Hüll Melon Mandarina Bavaria | Archer Olicana Jester | Mosaic Citra Simcoe Amarillo | Nelson Sauvin Riwaka Motueka Topaz |

*Note on using this table: Hops are arranged according to region of origin and principle characters. Hops may be substituted within a subgroup, and across the categories in the same row. Have fun exploring the similarities and differences!*

## Using Hops

Each of the following hopping methods will develop a different hop character in the beer. Some of the methods differ by temperature, some by time, and some by both. Several of the methods will often be used in the same recipe. In general, the longer you boil the hops in the wort the more bitterness, less flavor, and less aroma you will produce. Alpha acid isomerization can occur when the wort temperature reaches 185°F (85°C), but is most active during the boil.

### Mash Hopping

Mash hopping is putting hops in the mash, and it is said to contribute some aroma and flavor to the beer. Other possible benefits are that the alpha acids lower the mash pH by a small amount, and that the hop cones can act to loosen up the grain bed for better lautering. Hop pellets, on the other hand, would in principle have the opposite effect and potentially impede lautering. (All-grain techniques, including mashing and lautering, are covered in section 2 of this book.)

An experiment conducted for the 2014 National Homebrewers Conference by David Curtis and the Kalamazoo Libation Organization of Brewers indicated that the bitterness contribution from mash hopping was about 30% of that from a 60 min. boil using the same amount of hops.[4] In addition, the overall hop character (bitterness, flavor, and aroma) was generally perceived as being less than that of a 60 min. boil.

My opinion is that mash hopping is a waste of hops, but there are many brewers who swear by it for contributing more character to their India pale ales (IPAs). Your mileage may vary.

### First Wort Hopping

First wort hopping (FWH) consists of adding hops to the boil kettle as the wort is received from the lauter tun. The hops steep in the hot wort as the boil kettle is being filled, which may take a half hour or longer. The essential (aromatic) oils are normally insoluble and tend to evaporate to a large degree during the boil. The idea is that, by letting the hops steep in the wort prior to the boil, the oils have more time to oxidize to more soluble compounds, resulting in more flavor and aroma being retained during the boil. However, it is important to understand that the original German study[5] only examined bittering additions, so we are only talking about the difference in hop character between first wort hopping and 60 min. bittering. Late hop addition character would most likely overwhelm any FWH character.

An experiment conducted for the 2014 National Homebrewers Conference[6] indicated that the bitterness contribution from FWH was about 110% of that from a 60 min. boil using the same amount of hops. There wasn't a significant increase in overall hop character as perceived by the participants.

I often use FWH for my beers, because it doesn't seem to hurt anything and I get more utilization from my hops.

### Bittering

The primary use of hops is for bittering. Bittering hop additions are boiled for 45–90 min. to isomerize the alpha acids; 60 min. is most common. Typically, a maximum of 30% of the alpha acids are isomerized during a 90 min. boil. Most of this occurs within the first 45 min., and between

---

4    Curtis (2014).
5    Preis and Mitter (1995).
6    Curtis (2014).

45 and 90 minutes the isomerization only increases by about 5%, and only a little (<1%) at longer boil times. The aromatic oils of the bittering hops tend to boil away if boiled for too long, leaving little hop flavor and no aroma.

If you consider the cost of bittering a beer in terms of the amount of alpha acid per unit weight of hop used, it is more economical to use a half ounce of a high-alpha acid hop rather than one or two ounces of a low-alpha acid hop. You can save your more expensive (or scarce) aroma hops for flavoring and finishing. A more detailed discussion of hop utilization follows later in this chapter, where you can refer to table 5.5 for the percent utilization of hops as a function of boil time and boil gravity.

## Flavoring

Adding flavoring hops midway or later in the boil allows both isomerization of alpha acids and evaporation of light aromatics. This yields moderate bitterness and retains residual oils that are perceived as hop flavor. These flavoring hop additions are added at any time within the final 30 min. of the boil. Any hop variety may be used, but usually the lower alpha acid or mid-range alpha acid varieties are chosen. However, high alpha acid varieties that have pleasant flavors, such as Galena and Challenger, can be used. Often small amounts—0.25–0.5 oz. (7–15 g)—of several varieties will be combined at this stage to create a more complex character.

## Finishing, Hop Bursting, and Hop Steeping

When hops are added during the final minutes of the boil, less of the aromatic oils are lost to evaporation and more hop aroma is retained. One or more hop varieties may be used, in amounts varying from 0.25–4.0 oz. (7–120 g), depending on the character desired. A total of 1–2 oz. (30–60 g) is typical. These finishing or aroma hops are usually added 15 min. or less before the end of the boil, or are added at "knockout" (when the heat is turned off) and allowed to steep 10–30 min. before the wort is cooled.

Hop bursting is a practice that is common with American IPA brewers, where most of the bitterness of the beer comes from finishing hop additions added in the last 15 minutes before knockout. This allows some isomerization and more retention of oils. This technique uses much more hops to achieve the desired bitterness, which also produces a lot more hop flavor in the beer than traditional hopping schedules.

Hop steeping, or whirlpool hopping, comes from commercial brewing practice where the wort is directed to a whirlpool after the boil to separate spent hops and trub before going through the plate heat exchanger for chilling. The hot wort will often spend 30–60 min. in the whirlpool before being chilled, and brewers would use this opportunity to add more hop oil to the wort. The temperature in the whirlpool is hot (>185°F [>85°C]) but not boiling. This causes some isomerization to occur, but also preserves more of the essential oils than finish hopping or hop bursting. It is important to understand that whirlpool hopping is a necessity that grew out of commercial brewers having to separate the trub. Hop steeping is a good way to improve hop flavor and aroma in a beer, but the whirlpool action is not necessary.

If you are attempting to clone a commercial beer that uses whirlpool hopping, you need to find out how long the hops are in the whirlpool and at what temperature. You can estimate the bitterness contribution by taking 40% of the calculated bitterness for boiling for that amount of time. While perceived bitterness is more than just the level of isomerized alpha acids, 40% of calculated bitterness is a good starting point. This is based on a study by Malowicki and Shellhammer (2005) that showed the amount of isomerization at 194°F (90°C) is roughly 40% of that at 212°F (100°C); this fell to about 15% at 176°F (80°C). If maximum hop aroma and flavor is your goal, you may be better off boiling some of your hop-standing hops for an equivalent time to get all of the bitterness contribution, and then adding the rest (and maybe some extra) at knockout, chilling them quickly to capture all of the oils.

In some setups, a "hop back" is used—the hot wort is run through a small chamber full of fresh hops before the wort enters a heat exchanger or chiller. This is essentially the same as hop steeping, or whirlpool hopping; in effect, you are using the kettle or whirlpool as your hop back.

A word of caution when adding hops at knockout or using a hop back—depending on several factors relating to the hops you are using (e.g., the amount, variety, and freshness), the beer may take on a grassy taste, due to hop polyphenols that would normally be neutralized by the boil. If the hop character ends up too grassy, then I would suggest boiling your hop addition for a little more time. If there is not enough fresh hop character, then I would suggest using dry hopping.

### Dry Hopping

Dry hopping, where hops are added to the beer at the end of fermentation, is probably the best way to get fresh hop aroma into a beer. As the name implies, the hops are added dry. Many varieties of hop are appropriate for dry hopping, and several varieties can be combined to give the beer a more complex character. While you may be tempted to use a large amount of low-alpha acid aroma hops, you need to consider that this will also add a lot of plant material, which can contribute tannic or grassy flavors. Even though the tannic quality will probably subside after a few weeks, many craft brewers use higher-alpha acid varieties instead, such as Centennial, Galaxy, and Citra, because these hops generally have higher oil content by weight, which means less vegetative matter in the tank. Choose your variety with care because not every hop is appropriate for every beer style.

Dry hops for IPAs are generally added to give 3–5 days of contact time at temperatures of 50–70°F (10–21°C). Allow less contact time at warmer temperatures. The hops should be removed after this period to reduce the grassiness that can come from the cones. Historically, dry hopping rates for British pale ales were 0.5–1 lb. per barrel (note: these are imperial units). By contrast, British IPAs were historically hopped at 5–9 lb. per barrel (imperial units), but you have to remember that the typical hop then was half the strength of a typical hop today, and that the beers were often held in the casks for a year or more before serving, which greatly decreased the bitterness and aroma of the hops. Today's IPAs are generally dry hopped at 1–2 lb. per US beer barrel (31 gal.), which works out to 0.5–1 oz./gal. (7.5–15 g/L). And remember, not every beer is an IPA; just because you can dry hop any style doesn't necessarily mean you should.

Dry hopping actually does add bitterness to the beer (usually 1–5 IBU), but that bitterness comes from oxidized alpha and beta acids, and hop polyphenols, not from isomerized alpha acids.

Note: Dry hopping means what it says, that is, you add them dry. You don't have to pre-boil or sanitize them. Contamination and beer spoilage from hops just doesn't happen. However, dry hopping re-introduces oxygen to the beer, and this can cause oxidation and staling. Therefore, many brewers add dry hops toward the end of fermentation while there is still active yeast able to scavenge the oxygen, rather than afterward when the beer is off the yeast. Another method for reducing the potential for oxidation is to boil some water to drive out the oxygen, refrigerate it to cool, and then carefully disperse the hops into that water and pour the resulting slurry into the beer. A third method, when dry hopping in kegs, is to put the hops into the keg first, pressurize and purge the keg with carbon dioxide, and then rack the beer into the purged keg.

**Figure 5.3**. Clockwise from left, dried whole cones, fresh whole cones, and pelletized hops.

## Hop Forms—Pellets, Plugs, and Whole Hops

It's rare for any group of brewers to agree on the best form of hops. Each of the common forms—whole, plug, or pellet—has its own advantages and disadvantages. What form works best for you will depend on where in the brewing process the hops are being used, and will probably change as your brewing methods change.

Whichever form of hops you choose to use, freshness is important. Fresh hops smell fresh, herbal, and spicy, like resinous needles, and have a light green color like freshly-mown hay. Old hops, or hops that have been improperly stored, are often oxidized and smell like pungent cheese and may have turned brown. Ideally, hop suppliers pack hops in oxygen-barrier bags and keep them cold to preserve the freshness and potency. Hops that have been stored warm or in non-barrier (thin) plastic bags can easily lose 50% of their bitterness potential in just a few months. Most plastics are permeable to oxygen; so when buying hops at a homebrew supply store, check to see if the hops are kept in a cooler or freezer and if they are stored in oxygen-barrier containers. If you can smell the hops when you open the cooler door, then the hop aroma is leaking out through the packaging and they are not well protected from oxygen. If the stock turnover in the brewshop is high, non-optimum storage conditions may not be a problem. Ask the shop owner if you have any concerns.

## TABLE 5.4—HOP FORMS AND THEIR RELATIVE MERITS

| Hop form | Advantages | Disadvantages |
|---|---|---|
| Whole | Easy to strain from wort.<br><br>Best aroma, if fresh.<br><br>Good for dry hopping. | Will oxidize faster than pellets or plugs.<br><br>They soak up wort, resulting in some wort loss after the boil.<br><br>Bulk makes them more difficult to weigh. |
| Plugs | Retain freshness longer than whole hops.<br><br>Convenient 0.5 oz. (15 g) units.<br><br>Plugs behave like whole hops in the wort. | Can be difficult to break apart into smaller amounts<br><br>Soak up wort just like whole hops. |
| Pellets | Easy to weigh<br><br>Small increase in utilization due to shredding.<br><br>Best storability. | Turns into hop sludge in bottom of kettle that is difficult to strain.<br><br>Aroma content tends to be less than other forms due to amount of processing.<br><br>Hard to contain when dry hopping—creates floaters. |

**Figure 5.4.** Cascade hops on the vine.

**Figure 5.5:** Measuring hops on a digital scale

## How to Measure Hops

Hops used by homebrewers are typically measured by weight in either ounces or grams (pounds and kilograms for commercial brewers). Hop callouts in recipes specify the weight of the addition, the percent of alpha acids (% AA), and the time of the addition. The hop addition times are measured from the end of the boil, that is, the amount of time that they are boiled before the heat is turned off.

The % AA value is usually listed on the bag. If the hops have been stored cold, that number is usually fairly accurate. However, if the hops have been stored warm, the actual % AA may be significantly lower. Hops can lose as much as 50% of their bitterness potential in six months if stored poorly. Oxidized alpha acids are still bitter, but they are only about 66% as bitter as iso-alpha acids, and they will not isomerize in the boil.

## Hop Utilization and (International) Bitterness Units

Hop resins containing the alpha acids act like oil in water. Heating the hops causes isomerization of the alpha acids, and this change in their molecular geometry makes the alpha acids more soluble in the wort (which is, after all, mostly water). The percentage of the total alpha acids from the hops that are isomerized and survive into the finished beer as iso-alpha acids is termed the hop "utilization."

## ALPHA-ACID UNIT (AAU)

Alpha-acid units (AAUs) or home bittering units (HBUs), are the weight of hops (in ounces) multiplied by the percentage of alpha acids (% AA). This unit is convenient for describing hop additions in a recipe, because it indicates the total bittering potential from a particular hop variety while allowing for year-to-year variation in the % AA.

AAUs as a unit are not used as much as they used to be, partly because hops have higher levels of alpha acids today, and partly because specifying hops in grams is much more common as homebrewing grows around the world.

Calculating an AAU is simple.

AAU = weight of hop addition in oz. × % AA

For example, a recipe callout could read:

1.5 oz. of Cascade (5% AA) at 60 minutes.

AAU = 1.5 × 5
= 7.5

So, instead, the callout could read:

7.5 AAU of Cascade at 60 minutes.

If next year the % AA for Cascade is 7.5%,

7.5 AAU = $X$ oz. × 7.5% AA
7.5 / 7.5 = $X$
= 1

Therefore, you can work out that you would only need 1 oz. rather than 1.5 oz. of Cascade to arrive at the same 7.5 AAU contribution.

The isomerization rate for alpha acids is solely dependent on temperature. Isomerization starts at about 175°F (80°C), and the rate reaches a maximum at the temperature of the boil. Therefore, the total amount of isomerization for any hop addition depends mainly on boiling time. Higher elevations lower the boiling temperature of the wort, and therefore lower the isomerization rate and total isomerization as well.[7] Under homebrewing conditions, hop utilization generally tops out at about 30%. Many factors, like boil temperature, affect hop utilization, but the poor solubility of the hop resins is always a problem (the resins consist of the bittering compounds). Most of the losses that affect hop utilization are due to the iso-alpha acids and other bittering compounds being carried out of solution. This is due to these bittering compounds sticking to the trub and the walls of the boil kettle. Think of the oil and water analogy, the hop resins are going to stick in a thin layer to every surface they contact in the wort, which, in addition to the kettle and trub, includes the fermentor, the chiller, the tubing, and the yeast.

Therefore, one of the more significant factors for utilization is the batch size. The reason is simply the surface area-to-volume ratio of the kettle and fermentor: the larger the batch size, the less surface area per unit volume for the resins to stick to. This is probably the main reason why commercial brewers get better utilization than homebrewers, because the batch sizes are 10 to 100 times larger.

---

[7] However, using a pressure-cooker to increase the isomerization rate is a bad idea, because it will change the Maillard reactions and subsequently affect the flavor or the beer.

The other important factors are the amount of trub generated in the wort and the yeast mass in the fermentor. Both of these effects can be estimated by looking at the OG of the beer. The wort gravity doesn't affect the solubility of the iso-alpha acids and other bittering compounds directly. Instead, the higher the wort gravity, the more protein in the wort, and the more solid material will be generated during the hot and cold break. A higher OG also means that more yeast has to be pitched to ferment it, leading to more losses. Of course, the wort composition also affects utilization; for example, a high-adjunct, low-protein wort will generate less break material than an all-malt wort.

The last factor affecting utilization is the hop form, that is, whether you use pellets, plugs, or whole hops. Hop pellets seem to give better utilization because the processing has crushed and squeezed the lupulin glands and made the compounds inside more accessible. Pellets seem to have about 10%–15% better utilization than the rate for whole cones or plugs, but all this means is in relation to what you would achieve anyway for a particular boiling time, all else being equal. For example, if boiling whole hops for 50 min. gives you 20% utilization, then pellets may give you 22%–23%, that is, a 10%–15% increase on what you could achieve with whole hops. Pellets are not going to increase your total hop utilization from, say, 20% to 30% over whole hops, all else being equal.

There are several different models for calculating bitterness units (BUs) currently in use among homebrewers. The difference between them all is how the utilization is calculated. The Tinseth model[8] is probably the most commonly used and is presented in the section below under "Hop Utilization Equation Details." This utilization model consists of a boil time function and wort gravity function. These two functions are multiplied together to give the combined utilizations found in table 5.5.

*A final note on bittering and utilization.* It is nearly impossible to model every factor that will affect the utilization across everyone's equipment and brewing process. The key to using the following equations is to understand that the results are a benchmark, a number you can use to measure how much hop bitterness you intend the beer to have. Calculate your recipes using the model, taste your beers, then adjust your hop additions based on the numbers and your perception of those numbers as realized in the beer. Experienced brewers understand that different brewers brewing the same recipe will make different beers. The beers may be very similar or they may be very different—there are a lot of variables at play. Also, don't get obsessed trying to calculate utilization or BUs to three decimal places. People have a bitterness resolution of about five bitterness units, which means that you can readily taste the difference between a 20 and a 25 BU beer, but not between a 28 and 31.

### Calculating Hop Bitterness Units

Alpha acid units simply tell you how much alpha acid is going into the wort. The BUs estimate how much of that alpha acid will be isomerized and make it into the final beer. The equation for calculating BUs is:

For oz./gal.,

$$BU = \text{weight of hops} \times \% \text{ AA} \times \% \text{ utilization} \times (75/\text{final volume}).$$

For g/L,

$$BU = \text{weight of hops} \times \% \text{ AA} \times \% \text{ utilization} \times (10/\text{final volume}).$$

---

[8]    Glenn Tinseth, "Glenn's Hop Utilization Numbers," 1995, accessed November 15, 2016, http://www.realbeer.com/hops/research.html.

The proper units for BUs are milligrams per liter (mg/L), so to convert from oz./gal. a conversion factor of 75 (actually 74.89) is needed. For the metric world, to convert from g/L, the conversion factor is 10. (For those of you paying attention to the units, the missing factor of 100 was taken up by the % utilization, i.e., multiplying as 0.28 as opposed to 28%.)

Let's do an example with the following recipe for Joe Ale.

## Joe Ale

*To make 5 gal. (19 L).*

| Extract | Gravity Points |
|---|---|
| 5.5 lb. (2.5 kg) amber DME | |
| Boil gravity for 6 gal. (23 L) | 1.038 |
| **Hop schedule** | **Boil time (min.)** |
| 1.0 oz. (30 g) Mt. Hood 8% AA | 60 |
| 1.5 oz. (45 g) Hersbrucker 4% AA | 15 |

There are three steps to calculating the BUs for this recipe.
1.  Calculate the boil gravity based on the initial boil volume.
2.  Calculate the utilization for each addition based on the boil gravity and boil time.
3.  Calculate the BUs for each addition using the final volume after the boil.

Note that there are two gravity and volume factors in the utilization and BU calculations. The utilization factor depends on the gravity of the wort before the boil, because that is what predicts the amount of trub in the kettle. The final volume factor for the BU equation is the volume of wort after the boil, because now we are calculating the final concentration of the isomerized alpha acid. The hop additions added $X$ amount of alpha acids to the wort. The utilization is the amount of alpha acid isomerized divided by what was lost to the trub and environment. The final bitterness is the concentration of what's left in the volume of wort after the boil. Any dilutions you do in the fermentor will change the concentration and the BUs.

*Calculate the boil gravity.* As we saw in chapter 4, the boil gravity of your wort can be calculated by rearranging the mass gravity volume equation. This allows the boil gravity points to be calculated by multiplying the weight (mass) of extract by the extract potential (in PPG or PKL) and dividing this value by the wort volume before the boil.

$$\text{gravity points} = (\text{mass of extract} \times \text{PPG}) / \text{volume of wort}$$

In this case, the boil volume is 6 gal. (23 L). From table 4.1, we know dry malt extract typically yields 42 PPG (350 PKL), and the Joe Ale recipe above calls for 5.5 lb. (2.5 kg) of extract. So, for US standard units:

$$\text{gravity points} = (5.5 \times 42) / 6$$

$$= 38.5, \text{ or } 1.038$$

And for metric:

$$\text{gravity points} = (2.5 \times 350)/23$$

$$= 38, \text{ or } 1.038$$

Now we can use the known gravity of the boil (i.e., 1.038) to figure out the utilization.

If you have a smaller pot and wanted to use the Palmer method to carry out a 3 gal. (11.4 L) boil, you can still follow the same procedure. Calculate the gravity of the boil based on how much extract you are adding to the water, for instance, 3 lb. into 3 gal. Your boil gravity would be $(3 \times 42)/3 = 42$, or 1.042, but remember that the final volume you use for the BU equation would be total volume of wort in the fermentor after you have added the extra extract and water.

*Utilization.* The utilization is a combination of two factors: one is the isomerization rate and the other is the loss of isomerized alpha acids to the environment as a function of wort gravity. The utilization numbers that Tinseth published are shown in table 5.5. To find the utilizations for the "in-between" boil gravity values, simply interpolate the value based on the numbers for the bounding gravity values at the given time.

For this example, to calculate the utilization for a boil gravity of 1.038 at 60 min., look at the 60 min. utilization values for 1.030 and 1.040, which are 0.276 and 0.252, respectively (table 5.5). There is a difference of 24 between the two, and 8/10ths of the difference is about 19, so the adjusted utilization for 1.038 would be 0.276 − 0.019 = 0.257. The utilization for the 15 min. hop addition is calculated the same way and equals 0.127.

Going back to our Joe Ale recipe and calculating BUs for both additions:

$$BU_{60} = 1 \times 8\% \times 0.257 \times (75/5)$$

$$= 30.8, \text{ or } 31 \text{ BU}$$

and

$$BU_{15} = 1.5 \times 4\% \times 0.127 \times (75/5)$$

$$= 11.4, \text{ or } 11 \text{ BU}$$

In grams and liters, the calculations would be:

$$BU_{60} = 28 \times 8\% \times 0.257 \times (10/19)$$

$$= 30.3, \text{ or } 30 \text{ BU}$$

and

$$BU_{15} = 42 \times 4\% \times 0.127 \times (10/19)$$

$$= 11.2, \text{ or } 11 \text{ BU}$$

Giving a grand total of 41 or 42 BUs depending on how you round it, but, in all honesty, I would probably just call this 40 because you can't taste a difference less than 5 BU and these calculations tend to overestimate the actual BUs anyway.

Please understand that these equations are just an estimate. Each brew is unique; the equipment, temperature of the boil, pH, and losses during fermentation; all of these factors make it nearly impossible for two people to arrive at the same amount of bitterness in their beer for any given hop addition. However, these equations allow us to consistently plan our hop additions, to be consistent within our own setup, and make calculated changes from batch to batch to improve our beers. And that is a good reason to use them.

## Hop Utilization Equation Details

For those of you who are comfortable with the math, the following equations were determined by Tinseth (see table 5.5) from curve fitting a lot of test data while working on his PhD at the University of Oregon. The degree of utilization is composed of a gravity factor and a boil time factor multiplied together. The gravity factor accounts for reduced utilization due to higher wort gravities. The boil time factor accounts for the change in utilization due to boil time:

$$\text{Utilization} = f(G) \times f(t)$$

where

$$f(G) = 1.65 \times 0.000125^{(GB - 1)},$$

$$f(t) = [1 - e^{(-0.04 \times t)}] / 4.15,$$

and

GB is the boil gravity; t is the time in minutes.

The numbers 1.65 and 0.000125 in $f(G)$ were empirically derived to fit the boil gravity (GB) analysis data. In $f(t)$, the number −0.04 controls the shape of the curve for utilization versus time ($t$). The factor 4.15 controls the maximum utilization value. This number may be adjusted to customize the curves to your own system. For example, if you feel that you are having a very vigorous boil or generally get more utilization out of a given boil time for whatever reason, you can reduce the number a small amount to 4.0 or 3.9. Likewise, if you think that you are getting less out of your boil, then you can increase it to 4.25 or 4.35. These adjustments will alter the utilization value for each time and gravity in table 5.5.

## Bitterness Units Nomograph for Hop Additions

To use a nomograph for deriving BU (fig. 5.6 and 5.7), start on the right and draw a straight line from the %AA of your hop through the Weight of the addition, to arrive at the AAUs for that addition. Next, draw a line from the AAUs through the Recipe Volume to arrive at the AAUs per gallon. Now move to the left hand side of the chart and draw a line from your Boil Gravity, through your Boil Time, to determine the Utilization. Finally, draw a line through the points from the Utilization and AAUs per gallon lines to determine the BUs of that hop addition.

## TABLE 5.5—UTILIZATION AS A FUNCTION OF TIME VERSUS BOIL GRAVITY

| Boil time (min.) | Boil gravity | | | | | | | | | |
|---|---|---|---|---|---|---|---|---|---|---|
| | 1.030 | 1.040 | 1.050 | 1.060 | 1.070 | 1.080 | 1.090 | 1.100 | 1.110 | 1.120 |
| 0 | 0.000 | 0.000 | 0.000 | 0.000 | 0.000 | 0.000 | 0.000 | 0.000 | 0.000 | 0.000 |
| 5 | 0.055 | 0.050 | 0.046 | 0.042 | 0.038 | 0.035 | 0.032 | 0.029 | 0.027 | 0.025 |
| 10 | 0.100 | 0.091 | 0.084 | 0.076 | 0.070 | 0.064 | 0.058 | 0.053 | 0.049 | 0.045 |
| 15 | 0.137 | 0.125 | 0.114 | 0.105 | 0.096 | 0.087 | 0.080 | 0.073 | 0.067 | 0.061 |
| 20 | 0.167 | 0.153 | 0.140 | 0.128 | 0.117 | 0.107 | 0.098 | 0.089 | 0.081 | 0.074 |
| 25 | 0.192 | 0.175 | 0.160 | 0.147 | 0.134 | 0.122 | 0.112 | 0.102 | 0.094 | 0.085 |
| 30 | 0.212 | 0.194 | 0.177 | 0.162 | 0.148 | 0.135 | 0.124 | 0.113 | 0.103 | 0.094 |
| 35 | 0.229 | 0.209 | 0.191 | 0.175 | 0.160 | 0.146 | 0.133 | 0.122 | 0.111 | 0.102 |
| 40 | 0.242 | 0.221 | 0.202 | 0.185 | 0.169 | 0.155 | 0.141 | 0.129 | 0.118 | 0.108 |
| 45 | 0.253 | 0.232 | 0.212 | 0.194 | 0.177 | 0.162 | 0.148 | 0.135 | 0.123 | 0.113 |
| 50 | 0.263 | 0.240 | 0.219 | 0.200 | 0.183 | 0.168 | 0.153 | 0.140 | 0.128 | 0.117 |
| 55 | 0.270 | 0.247 | 0.226 | 0.206 | 0.188 | 0.172 | 0.157 | 0.144 | 0.132 | 0.120 |
| 60 | 0.276 | 0.252 | 0.231 | 0.211 | 0.193 | 0.176 | 0.161 | 0.147 | 0.135 | 0.123 |
| 70 | 0.285 | 0.261 | 0.238 | 0.218 | 0.199 | 0.182 | 0.166 | 0.152 | 0.139 | 0.127 |
| 80 | 0.291 | 0.266 | 0.243 | 0.222 | 0.203 | 0.186 | 0.170 | 0.155 | 0.142 | 0.130 |
| 90 | 0.295 | 0.270 | 0.247 | 0.226 | 0.206 | 0.188 | 0.172 | 0.157 | 0.144 | 0.132 |

Source: Glenn Tinseth, "Glenn's Hop Utilization Numbers," 1995, accessed November 15, 2016, http://www.realbeer.com/hops/research.html.

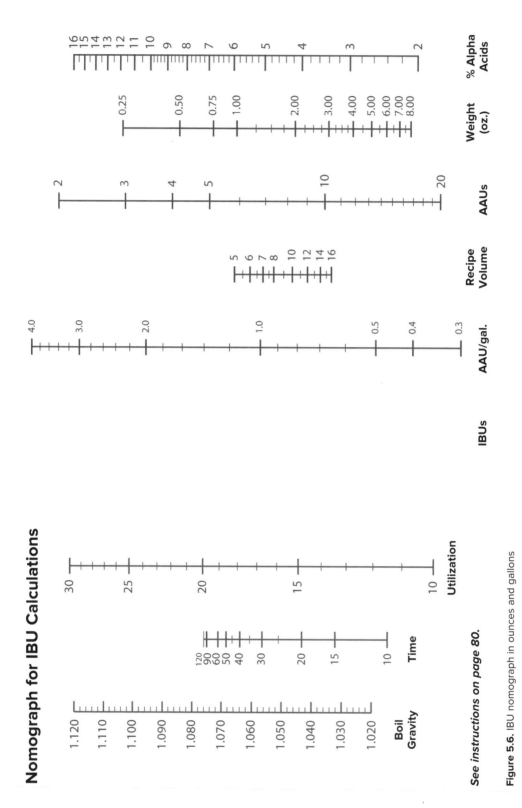

**Nomograph for IBU Calculations**

*See instructions on page 80.*

**Figure 5.6.** IBU nomograph in ounces and gallons

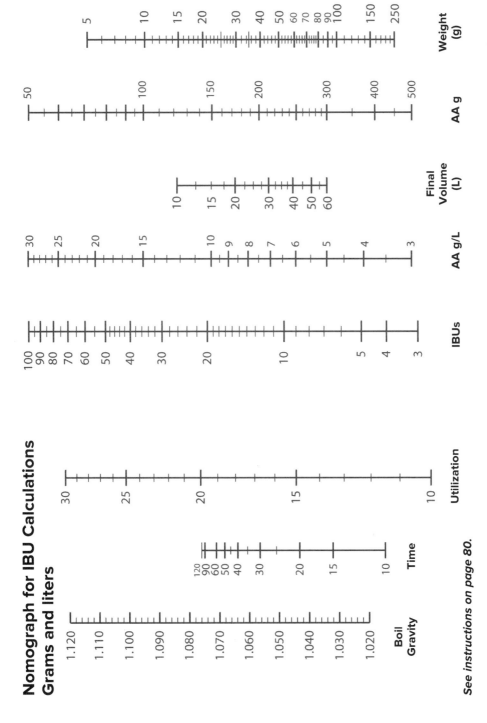

See instructions on page 80.

**Figure 5.7.** IBU nomograph in grams and liters.

# Yeast and Fermentation 6

*A good fermentation of a bad recipe is going to make a*
*better beer than the bad fermentation of a good recipe.*

There was a time when the role of yeast in brewing was unknown. The German Beer Purity Law of 1516—the *Reinheitsgebot*—listed the only permissible materials for brewing as malt, hops, and water. Everyone talks about how Louis Pasteur discovered yeast in the 1850s, but he didn't. What he did was the conduct the experiments and publish the papers that determined yeast were responsible for the fermentation process and defined how it worked. Everyone knew about yeast, in fact, yeast slurry was packaged and sold by the Dutch in 1780, with a compressed cake form entering the market in 1825. Brewers and bakers knew the importance of transferring yeast from batch to batch long before Pasteur; they just didn't understand what it was. They referred to it as the "mother" or "God is Good," among others. The word "yeast" is derived from words that mean "foam" or "to rise." Other scientists at the time ascribed fermentation to a purely chemical reaction catalyzed by air. The idea that a living organism would be responsible was too biologic—it was too old-fashioned for modern thinking. Scientists considered yeast to be a byproduct of fermentation rather than the cause.

Beer would not be beer without yeast and fermentation. If you recall my "Top Five Priorities for Brewing Great Beer" from chapter 1, fermentation temperature and yeast management are second and third, respectively. Why is fermentation temperature placed ahead of yeast itself? All else being equal, the yeast quality, quantity, and activity level (a function of temperature) are all important; but, when you ask expert brewers their opinion, fermentation temperature control is always their first priority. However, before we get into that, let's discuss what yeast is and how it works.

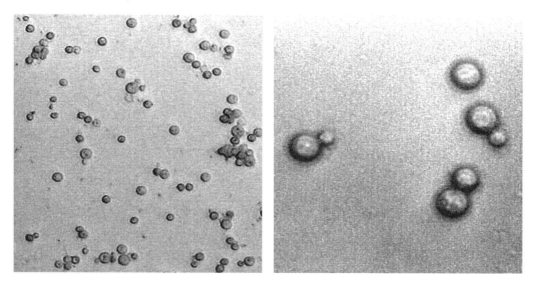

**Figure 6.1.** Aerial view of yeast ranch. Magnification 300×.     **Figure 6.2.** Budding yeast cells. Magnification 1000×.

## How Yeast Work

Brewer's yeast (*Saccharomyces cerevisiae*) is a type of fungus, and the species name literally translates to "sugar fungus beer." It reproduces asexually by budding, that is, splitting off little daughter cells (fig. 6.1 and 6.2). Yeast is unusual in that it can live and grow both with or without oxygen. Most microorganisms can only do one or the other. Yeast can live without oxygen by a process that we refer to as fermentation. The yeast cells take in simple sugars, like glucose and maltose, and produce carbon dioxide and alcohol as waste products. Yeast cells reproduce during fermentation by cloning themselves, budding off little daughter cells that can grow and create daughter cells of their own.

Table 6.1 lists normal ranges of the different sugars (technically known as saccharides) found in a typical beer wort. The main constituent is maltose, followed by assorted dextrins, maltotriose, glucose, sucrose, and fructose. Maltose is a glucose disaccharide, which means that it is made up of two glucose molecules. Maltotriose is a trisaccharide consisting of three glucose molecules. Sucrose (commonly known as table sugar) is a disaccharide that is made of one glucose and one fructose, and occurs naturally in plants (e.g., in sugar cane, beets, maple sap, and nectar). Dextrins (a.k.a. oligosaccharides) are larger sugars consisting of more than three monosaccharide groups.

Yeast consumes sugars in the wort methodically, starting with the monosaccharides (simple sugars) and the disaccharide, sucrose; then moving on to the main wort constituent, maltose; and finally finishing up with the trisaccharide, maltotriose. Interestingly, yeast seems to work on sucrose

## TABLE 6.1—TYPICAL SUGAR SPECTRUM OF WORT

| Sugar | Typical range |
|---|---|
| Glucose | 10%–15% |
| Fructose | 1%–2% |
| Sucrose | 1%–2% |
| Maltose | 50%–60% |
| Maltotriose | 15%–20% |
| Dextrins | 20%–30% |

*Note: Dextrins are not fermented by brewer's yeast.*

first, breaking down this disaccharide into glucose and fructose, then consuming all of the glucose and fructose before moving on to maltose. The last fermentable sugar, maltotriose, typically accounts for 15%–20% of the total sugars in wort, but is usually not fermented completely, although this depends on the yeast strain. For example, lager yeast ferments maltotriose better than strains of ale yeast.

The sugar composition and yeast strain are what determine the fermentability of the wort. If the wort was mashed to have high fermentability (methods of mashing are described in chapter 17), the percentage of unfermentable dextrin would be about 20%, meaning that 80% of the wort was fermentable. As the yeast consumes the malt sugars there is a change in the specific gravity of the wort, or beer, which is the degree of attenuation. In a wort that is 80% fermentable, you might expect the attenuation to be 80%. However, attenuation will only get near 80%. This is because the yeast strain also determines the fermentability of the wort, and most yeast strains attenuate in the range of 67%–77%. For example, a yeast strain having an attenuation of 75% would ferment a beer from 1.040 to 1.010, or from 1.060 to 1.015. If the yeast strain is a low attenuator, then it may not ferment very much of the maltotriose, resulting in an overall attenuation of about 67%. If the yeast strain was a high attenuator, it will ferment most of the maltotriose, resulting in an overall attenuation of about 77%. In addition, wort with a high percentage of monosaccharides (e.g., 30% rather than the normal 15%) can inhibit the production of the enzymes that yeast uses to ferment maltose, resulting in a "stuck" fermentation. This aspect is discussed more in chapter 24, "Developing Your Own Recipes."

Brewer's yeast doesn't actually respire oxygen during any stage of the brewing process. The yeast cells use oxygen, but not in the same "breathe in, breathe out, burn food" sort of respiration that other cells do. Instead, yeast cells use oxygen to chemically synthesize unsaturated fatty acids and sterols they need to build and maintain their cellular membranes.

## WHAT IS A CELL MEMBRANE?

If you compare a yeast cell to the human body, the cell wall is your skin (the epidermis), and the cell membrane is the living tissue underneath. Of course, this analogy is a gross mangling of the facts, but it serves the purpose. The cell wall is the structural container of the yeast cell and the cell membrane directly beneath it is the semi-permeable boundary layer through which the yeast cell interacts with its environment. The cell membrane allows sugars to be taken into the cell and waste products (alcohol and carbon dioxide) to be excreted. A fresh, healthy yeast cell has a pliable membrane; an old yeast cell has a tougher, less permeable membrane that makes it harder for the cell to take in food and excrete waste.

Brewer's yeast produces many other compounds in addition to ethanol (ethyl alcohol) and carbon dioxide. Some are desirable, such as esters and phenols, and some are not, such as the precursors for diacetyl and the production of fusel alcohols. Esters are responsible for the fruity notes in beer, and phenols cause the spicy notes. Both are generally desirable, depending on style. But, you can easily have too much of a good thing. A good example of this principle is diacetyl, a vicinal diketone compound that can be beneficial in very small amounts. Diacetyl gives a buttery note to the flavor of a beer and helps round out the malt flavor in some ale styles, but it is very easy to end up with too much diacetyl in your finished beer. Another undesirable product are fusel alcohols, which are heavier molecular weight alcohols that are readily apparent as "solventlike" notes.

After fermentation, when all the food is used up and their cell membranes are too old to function well, the yeast cells go into hibernation mode. They build up their glycogen and trehalose reserves (i.e., storage carbohydrates), clump together (i.e., flocculate) and settle to the bottom of the fermentor. Different yeast strains flocculate differently and will settle faster or slower as a result. Some yeast practically paint themselves to the bottom of the fermentor, while others are ready to swirl up if you so much as sneeze. Highly flocculent yeasts can sometimes settle out before the fermentation is finished, leaving residual sugars and unacceptable levels of acetaldehyde and diacetyl, which would otherwise be cleaned up by the yeast during the maturation phase of fermentation.

## Defining Fermentation

### The Three Phases of Fermentation

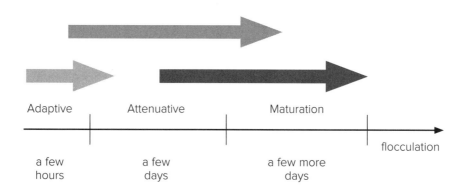

**Figure 6.3.** Three phases of fermentation

The fermentation of malt sugars into beer is a complicated biochemical process. From a brewer's point of view, there are three phases of fermentation after pitching the yeast (fig. 6.3):
1. Adaptation, or lag, phase before activity is seen in the fermentor.
2. High-growth, or high-kräusen, phase when most of the attenuation occurs.
3. Maturation, or conditioning, phase when the yeast becomes quiescent, the kräusen drops, the beer clears, and the flavor improves.

We will discuss each of these phases in more detail in a minute. However, it is worth noting that, from the yeast's point of view, there are also three phases, but a slightly different three:

1.  Adaptation phase, a relatively short period before growth.
2.  High-growth phase, when eating and reproduction take place in earnest.
3.  Stationary phase, when the yeast essentially hibernates.

You may notice that the maturation phase seems to be missing from the yeast's list of priorities. Yes, it is, and it is up to the brewer to understand this and create the proper conditions for it to occur.

Yeast cells do not behave like a synchronized school of fish. They are more like a large group of people—while many are doing the same general activity, some will be more active than others, and some will be less active. In other words, the yeast cells don't transition from one phase to another as a group; instead, the phase is determined by what the majority of them seem to be doing at the time. In the wild, any particular yeast cell within a group could randomly be at any point in its life cycle (fig. 6.4). As brewers, we can get the yeast cells to become more organized by pitching them to yeast starters and controlling the temperature. That's how we can get them all to march in step and ferment more efficiently.

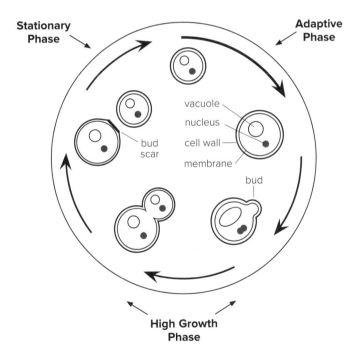

**Stationary Phase**

**Adaptive Phase**

vacuole
nucleus
cell wall
membrane
bud scar
bud

**High Growth Phase**

**Figure 6.4. Diagram of the life cycle of a yeast cell.** The yeast cell starts at the top center of the diagram in the stationary phase. When pitched into wort the cell enters the adaptation phase, where it takes in essential nutrients and physically increases in size. When it has sufficient nutrients and is an appropriate size, the yeast cell begins the reproduction process by producing a bud on its cell wall. The nucleus of the cell, containing the DNA, replicates and divides as well. The vacuole is like a combination storage shed and workshop for the cell, containing nutrients, enzymes, and waste products. The vacuole resources are shared with the budding cell and it is these resources that ultimately limit how many times the cell can bud. The budding cell separates, leaving a bud scar on the cell wall of the mother cell. The new cell—the daughter cell—is smaller than the mother cell and will need to grow before it too can start budding daughter cells of its own. At the end of the growth phase, when nutrients have become limiting, the yeast cell enters the stationary phase and lives off its reserves until pitched to new wort.

### Adaptation (Lag) Phase

Immediately after pitching into wort, the yeast cells start adjusting to the wort conditions, changing their metabolism in response to the sugars and other nutrients present, and regulating enzyme production and other attributes according to their needs in the new environment. At this stage, yeast cells use their own glycogen reserves together with lipids and dissolved oxygen in the wort to synthesize sterols and fatty acids. These substances are critical to maintain cell membrane integrity and function, enabling the cell membrane to be permeable to wort sugars and other wort nutrients. Long-chain fatty acids are used to build structural lipids, while short-chain fatty acids are esterified and excreted. Under oxygen-poor conditions, yeast can produce these sterols and fatty acids using lipids found in wort trub, but that metabolic pathway is much less efficient.

Once the cell membranes are permeable, the yeast cells can start metabolizing the sugars and free amino nitrogen (FAN) in the wort for food and growth. An oxygen-rich wort shortens the adaptation phase, and allows the yeast to quickly revitalize and move on to the next phase of fermentation, the high-growth and attenuation phase.

Under normal conditions, the yeast should work through the adaptation phase and begin the high-growth phase within 12–24 hours. In other words, most, if not all, of the first day after pitching will be taken up with the adaptation phase, with the next phase beginning on day 2, or thereabouts. If 48 hours pass without apparent activity (foamy, yeasty mass on top of the wort and rapid bubbling in the airlock), then a new batch of yeast should probably be pitched.

In the past, a short lag phase has been overemphasized as a benchmark indicator of a good fermentation. It was regarded as "the shorter, the better" because it meant that the yeast were strong and ready to work. While it is a notable indicator, a short lag phase does not guarantee an exemplary fermentation and outstanding beer. A short lag phase only means that the adaptation time was short; for instance, it could be because there was not much oxygen to take up and it was time for the yeast to eat or die. However, these cautions aside, a short lag phase (6–12 hours) and rapid bubbling in the airlock generally means that the yeast are healthy, happy, and able, though 12 hours is probably more the average.

Of course, wort temperature will have a large effect on yeast activity. A common mistake that many homebrewers make is pitching the yeast when the wort has not been chilled enough and is still warmer than 80°F (>27°C), thinking that it will cool down by the time the high-growth phase starts. This is risky—it is very likely that more metabolites (byproducts) will be produced at this early stage than the yeast will reabsorb during the maturation stage, which will affect how clean-tasting the final beer will be.

Managing temperature is a critical part of the fermentation process, and it starts as soon as your yeast is pitched and the adaptation phase begins. If you want a cleaner-tasting beer (i.e., less esters, less fermentation character), the yeast should be pitched into the wort when the wort temperature is at or slightly below the intended fermentation temperature. For example, if the recommended temperature range for California ale yeast is 68–73°F (20–23°C), then pitching the yeast at a wort temperature of 65–68°F (18–20°C) will reduce the amount of byproducts produced early in the high-growth phase, and so moderate the overall amount through your fermentation. If you want more fermentation character in your beer, then raise the wort pitching temperature to the middle of the recommended range. I would not recommend pitching at the top of the range unless you have experience with the yeast and the style, and know that you like the resulting flavors. Also be aware that pitching too cold can stress the yeast, resulting in lots of fermentation character.

Bottom line: a warmer wort temperature at pitching will shorten the adaptation (lag) phase, but assuming your sanitation up to this point has been good, there is no benefit to speeding up this phase to outcompete bacteria or other microbes. The pitching temperature sets the stage for the activity in the high-growth phase. The warmer the wort, the faster the yeast will grow and the more byproducts they will produce.

## High-Growth Phase

The visible churning of the wort and rapid bubbling in the airlock heralds the high-growth phase of fermentation. Once a yeast cell has adapted to its surroundings, it starts to eat, and once it has started to eat, it starts reproducing. The high-growth phase is a period of vigorous fermentation and can last anywhere from 1–3 days for ales, or 2–5 days for lagers, depending on the original gravity and pitching rate. The majority of the attenuation (e.g., 98% of the attenuation) should occur during the high-growth phase.

A head of foamy yeast kräusen will form on top of the beer (fig. 6.5). The kräusen consists of yeast cells and wort proteins, and is a light cream color with islands of green-brown gunk. The gunk is a residue composed of extraneous wort protein, hop resins, polyphenols, and dead yeast. Many brewers skim this residue off the kräusen, especially if they plan on harvesting some of the yeast foam for another batch. (Harvesting from your own fermentation is described in chapter 7.) Fortunately, the residue is relatively insoluble and typically sticks to the sides of the fermentor as the kräusen subsides, so removal is not strictly necessary.

High-attenuation yeast strains tend to be low or poorly flocculent, whereas low-attenuation strains tend to be highly flocculent. Flocculation is the propensity of yeast cells to clump together when they go dormant, which causes them to settle to the bottom of the fermentor at the end of fermentation. These traits will be discussed in more detail in chapter 7.

Most of the esters and other flavor compounds are produced during the high-growth phase. The best way to control these flavors in your beer is to control the growth, and you do that by controlling the temperature. Yeast activity is highly dependent on temperature—too cold and they go dormant, too hot (more than 10°F [5°C] above the recommended range) and they will indulge in an orgy of fermentation that often ruins the flavor of the beer.

Furthermore, the fermentation process during the high-growth phase produces heat. The internal temperature of the fermentor can be as much as 10°F (5°C) above ambient conditions, just due to yeast activity. This is a good reason to keep the fermentation environment well within the recommended temperature range, so that you will get a normal vigorous fermentation where the beer turns out as intended, even if it was warmer than the surroundings.

Many off-flavors, such as acetaldehyde and diacetyl, can be cleaned up by the yeast, but others cannot. High temperatures promote the production of fusel alcohols, leading to solventlike flavors. High temperatures can also promote an excess of esters, leading to banana or bubblegum-flavored beers. Once created, these flavors cannot be reduced by maturation. Temperature control is the key to controlling fermentation and fermentation character.

The high-growth phase will last as long as the yeast's sterol and lipid reserves hold out. When those reserves are exhausted, the yeast cells no longer have the resources to share and they stop producing daughter cells. The yeast cell membranes will become old and less able to transfer food and waste, so general activity slows down as well. This decrease in activity means less carbon dioxide is created, which causes the foamy kräusen to settle back into the beer.

## WORT TEMPERATURE OR ROOM TEMPERATURE CONTROL?

There are two schools of thought about using electronic temperature controllers to manage fermentation temperature. Some advocate controlling the temperature of the wort, or beer, while others advocate controlling the temperature of the room or refrigerator. Both methods can be misused depending on how and when the control is applied. The key is understanding the phases of fermentation and knowing when the temperature of the wort should be controlled, that is:

- Fermentation byproducts are generated during the high-growth phase.

- The temperature of the wort at the beginning of the high-growth phase should be at the lower end of the recommended range to moderate the rate of growth in order to limit the amount of byproducts produced.

- The temperature of the wort or beer should be raised, or allowed to rise, after the high-growth phase to increase the activity of the yeast during the maturation phase.

- The temperature of the fermentation should never decrease during fermentation. The beer temperature should only decrease after maturation is finished. This applies to all beers, even lagers.

Therefore, when you pitch your yeast to the fermentor, the temperature of the wort must not be higher than the set point of whatever controller you are using. In other words, don't pitch until your wort temperature reaches the set point. If the wort is warmer, the yeast will experience high growth (and produce a lot of byproducts) for a short period of time and then lose activity due to the decreasing temperature. You don't want the yeast to slow down halfway through.

If you are controlling the temperature of the wort or beer, verify that the controller does not over-chill by more than 2°F (1°C) below the set point. You will need to monitor the gravity or fermentation activity (i.e., bubbling in the airlock) in order to raise the set point for the conditioning phase.

If you are controlling the temperature of the refrigerator or room instead of the beer, the temperature of the fermentation will rise several degrees above the room temperature. This is okay, provided that the fermentation temperature does not exceed the recommended range. What you will find, in fact, is that the fermentation often meets the higher temperature guideline for good maturation.

This type of temperature control has worked for thousands of years. But be aware that, as yeast activity slows, less heat will be generated—if the room is cold, the yeast may flocculate and enter the stationary phase too early. Be prepared to raise the room temperature set point accordingly.

### Maturation Phase

The start of the maturation phase doesn't mean that fermentation is over. There are now a lot of fermentation byproducts in the beer, including compounds that contribute off-flavors, such as acetaldehyde (green apple or raw pumpkin flavors) and diacetyl (stale butter or milk flavor). The beer is considered to be "green" at this point. Before it can be considered ready for consumption, the beer needs a maturation period to allow the yeast time to clean up these compounds.

The key to maturation is active yeast. At the end of the high-growth phase the yeast cells are worn out and ready to call it a day and go to sleep. The beer isn't finished yet, but the yeast cells are too old and tired to do anything about it. The cells flocculate and settle to the bottom (fig. 6.6).

Physically rousing the yeast (swirling the fermentor) will sometimes help to keep the cells in suspension and working on the beer. However, this doesn't change the fact that the yeast cells

**Figure 6.5.** A healthy, creamy, kräusen of yeast floats on top during the high-growth stage of fermentation. This is a good time to skim off some yeast to save for a future batch.

**Figure 6.6.** As the high-growth stage winds down and the maturation phase takes over, the yeast starts to flocculate and the kräusen settles back into the beer.

are old and tired, and it can take a long time for them to eat the remaining sugars and consume undesirable byproducts like acetaldehyde and diacetyl. How do we keep the yeast from becoming old and tired?

As the brewer, you must balance the pitching rate and wort resources so that the majority of the yeast cells will still be relatively fresh and have not reached their reproduction limit by the time the majority of the fermentable sugars have been consumed. This will leave lots of active yeast cells in suspension that will then look for alternative food sources and clean up the beer.

Maturation consists of the reduction of acetaldehyde and other aldehydes into alcohols, the enzymatic breakdown of diacetyl, and the adsorption of other off-flavors onto the surface of the yeast cells as they settle to the bottom of the fermentor. The mechanisms of maturation were poorly understood for many years, but research over the last fifty or sixty years has improved our understanding of yeast metabolism and the maturation process. We now know how to achieve full maturation in a shorter time. The key is managing the fermentation using pitching rate and temperature.

The details of pitching rate will be discussed more in chapter 7, but for now think of it as managing the number of yeast relative to the amount of fermentables, so that the fermentables run out before the yeast gets tired. Yeast activity is directly affected by temperature, so raising the temperature toward the end of the high-growth phase will greatly increase activity and allow the yeast to clean up the byproducts faster. This technique for raising the temperature towards the end of fermentation is generally known as a "diacetyl rest." A diacetyl rest is most typically used for lager beers, although it is applicable to all yeast strains and will help with the cleanup of other byproducts, such as acetaldehyde, as well.

Therefore, as you see the bubbling in the airlock start to slow down toward the end of the high-growth phase, approximately day 3 for ales and day 4 for lagers, raise the fermentation temperature by 9°F (5°C) for a diacetyl rest. A more precise indicator is when the beer is 2–5 specific gravity points from the final gravity. This rest is more typically used for lager fermentations, but the same principle can apply to ales as well (see table 6.2 for both lager and ale guidelines).

A higher pitching rate will mean a greater proportion of active yeast cells remain toward the end of attenuation, so a smaller temperature increase can achieve sufficient maturation activity. As you might expect, a lower pitching rate will mean fewer active yeast cells remain, so a larger temperature increase may be necessary to fully clean up the beer. Of course, temperature adjustments can only be taken so far. The key is having the right amount of yeast.

A good rule of thumb is that the maturation phase should be as long as the high-growth phase. If you choose not to raise the temperature for maturation then you should probably allow twice as long for the maturation phase, just to be on the safe side. For example, looking at the guidelines for lagers in table 6.2, the suggestion is to raise the temperature of the fermentation by 14–18°F (8–10°C) around day 4, assuming that the high-growth phase (steady bubbling in the airlock) is going to last until day 6. You would then raise the temperature for the diacetyl rest (maturation phase) and leave it there for at least six days. Same idea for ales. The point is to give the yeast all the time and help it needs to clean up the beer. Your yeast may be done sooner, but the extra time doesn't hurt the quality of the beer. Patience is a virtue.

## TABLE 6.2—MATURATION REST GUIDELINES

| Schedule | Lager | Ale |
|---|---|---|
| When (approx. time) | day 4 of 6 | ~day 3 of 4 |
| When (specific gravity) | 2–5 points from FG | 2–5 points from FG |
| Raise temperature by | 14–18°F (8–10°C) | 5–10°F (3–6°C) |
| Leave for | minimum 6 days (up to 12) | minimum 4 days (up to 8) |

### Cold Conditioning (Lagering)

Traditionally, lagering was viewed as a long, slow maturation process, and this view has persisted in the literature right through the latter half of the twentieth century. But it is important to separate the two types of maturation that need to occur in "green" beer. The first type is the reduction of fermentation byproducts by the yeast, as described in the section above. The second type is physical clarification of the beer, that is, clearing it of excess yeast and haze, and this is what we call cold conditioning.

New beer is hazy beer, and hazy beer doesn't taste the same as clarified beer. Haze due to suspended yeast can give the beer a yeastlike, brothy flavor, but these flavors should decrease with time as the yeast flocculates and settles to the bottom. Haze due to protein and polyphenols are harder to get rid of. The protein-polyphenols that cause haze are chemically the same compounds that create astringency in the mouth, and they can promote staling reactions in the beer. A hazy beer can be a good beer, but a clearer beer is often a better tasting beer. There are, however, some styles where haze is an important part of the beer's appearance. Haze and clarifiers are discussed in more detail in appendix C.

Cold conditioning is a process of slowly cooling the beer down by 2°F (1°C) per day to about 9–15°F (5–8°C) below the fermentation temperature to promote the flocculation of the yeast and the coalescence of the protein-polyphenol complexes that cause haze. The protein-polyphenol complexes in beer are held together by hydrogen bonds, which are stronger at colder temperatures. Hence, cold temperatures help these complexes to become large enough to settle out. The cooling rate for settling protein-polyphenol haze does not matter as much as the final temperature, colder being better. The point of slow cooling is to prevent thermal shock of the yeast cells and subsequent excretion of fatty acids and other lipids. These lipids can interfere with head retention and will readily oxidize, creating stale flavors. Thermal shock at any time can cause the yeast cells to release protein signals that cause other yeast cells to shut down to protect against the cold, potentially leading to premature flocculation and underattenuation.

Cold conditioning and lagering are useful tools for clarifying the beer, but it must be understood that it needs to happen after the yeast have completely fermented the beer and cleaned up the byproducts. Otherwise, you risk high levels of diacetyl and aldehydes, and underattenuation. You should only have to cold condition for a week or two to settle most of the haze. Fining agents, such as silica gel, gelatin, and isinglass, can help greatly and will usually clear the beer in a couple of days instead of weeks. Usually, the beer can be left on the yeast for several weeks without causing any flavor problems, but you should be aware that there are risks to leaving your beer on the yeast and trub for too long after fermentation.

As they lie on the bottom of the fermentor, dormant yeast cells can excrete undesirable amino acids, short-chain fatty acids, lipids, and enzymes. As a result, leaving the beer on the trub and yeast cake for too long (more than a month, for example) can lead to soapy, waxy, fatty, and other oxidized flavors. Even worse, after very long times the yeast cells begin to die and break down (i.e., undergo autolysis), which produces meaty, brothy, and soy sauce tastes and odors.

## Building a Better Fermentation
*Step 1.*    Cool the wort to the pitching temperature.
*Step 2.*    Aerate to add oxygen to the wort.
*Step 3.*    Pitch the yeast.
*Step 4.*    Add yeast nutrients if necessary.

Yeast cannot live on sugar alone. Yeast also need minerals, nitrogen, amino acids, and fatty acids to enable them to live and grow. The primary source for these building blocks is the malted barley. Refined sugars like table sugar, corn sugar, and honey do not contain any of these nutrients. However, let's talk about oxygen and aeration first.

## Oxygen and Aeration

The role of oxygen in yeast fermentation has already been discussed at several points in this chapter, but not how to get it into your wort. Commercial breweries generally do it by inline injection of pure oxygen into the wort as it is pumped to the fermentor. At the homebrewing scale it is easier and less complicated to cap and shake the fermentor, or to use an aquarium airstone with either an air pump or small oxygen tank. Another option, one that has worked for thousands of years, is open fermentation, which will be discussed at the end of this chapter.

Depending on the strain, yeast typically needs 8–12 ppm of oxygen for a good fermentation. Without aeration, fermentations tend to be underattenuated. Higher-gravity worts need more yeast cells (i.e., higher pitching rates), and thus need more oxygen, but the higher gravity makes it more difficult to dissolve oxygen into the wort in the first place. Boiling the wort drives out the dissolved oxygen normally present, so aeration of some sort is needed prior to fermentation. For the home-brewer, proper aeration of the wort can be accomplished in the following ways:

- pouring the wort with splashing or spraying (gives about 4 ppm);
- shaking the container (only applicable to small volumes, gives about 8 ppm);
- using a stainless steel airstone with an aquarium air pump to bubble air into the fermentor for 5–10 min. (gives about 8 ppm);
- or using an airstone with an oxygen tank to bubble oxygen into the fermentor for about 1 min. (gives about 10 ppm).

Supporting data is given in table 6.3, but it is apparent that there is a lot of variation in results with pouring and shaking. Airstones are generally more effective and consistent, but cause the wort to foam, which will often ooze out of the fermentor as a result. Anti-foaming agents can be used to minimize this problem.

### TABLE 6.3—OXYGENATION DATA FOR 5 GAL. (19 L) 1.040–1.080 WORT

| Method | Time | Dissolved oxygen level |
|---|---|---|
| Siphon with sprayer attachment | (siphoning time) | 4 ppm[a] |
| Shaking with air (small volume) | 1 min. | 8 ppm[a] |
| Shaking with air (large volume) | 5 min. | 2.7 ppm[b] |
| Airstone with aquarium air pump | 5 min. | 8 ppm[a] |
| Airstone with oxygen, 1 L/min. | 30 sec. 1 min. 2 min. | 5.1 ppm[b] 9.2 ppm[b] 14.1 ppm[b] |
| Airstone with oxygen, rate not specified | 1 min. | 12 ppm[a] |

[a] Greg Doss, David Logsdon, and Company, "The Meaning of Life According to Yeast," Wyeast Labs, http://www.bjcp.org/cep/WyeastYeastLife.pdf.

[b] White and Zainasheff (2010, p. 79).

For the beginning homebrewer using rehydrated dry yeast, I recommend the simplest aeration methods of shaking the starter and pouring the wort. Pouring is also effective if you are doing a partial boil and adding water to the fermentor to make up the total volume. Instead of pouring the wort, you can just pour the water back and forth several times to aerate the water. Dry yeast typically requires less oxygen, because the yeast is grown with high glycogen and lipid reserves before being dehydrated and therefore generally needs less oxygen in the adaptation phase.

If you are using liquid yeast cultures, including yeast harvested from yeast starters or previous batches, I recommend aerating with an airstone using an aquarium pump or oxygen tank. Otherwise, be prepared to form a vortex and stir your arms off. Using a high-speed electric mixer is not recommended, because the shearing action can damage the yeast and denature some of the foaming proteins for head retention.

Using an air pump and airstone to bubble air into the fermentor is effective and saves you from lifting a heavy fermentor to pour or shake the wort. The saturation point of oxygen from the air in chilled wort is about 9 ppm. Most yeast strains require 8–12 ppm of oxygen for adequate growth and activity, but that requirement also depends on the pitching rate and the volume of the wort. The yeast will take up the oxygen quickly, generally in less than an hour. An air pump and airstone should reach an oxygen concentration of 8 ppm for 5 gal. after about 5 min., but I recommend aerating for longer, about 10–15 min., to make sure that your yeast has the time to get all it needs.

The only precaution you need to take, other than sanitizing the airstone and hose, is to be sure that the air going into the fermentor is not carrying any mold spores or dust-borne bacteria. An in-line filter is recommended to prevent airborne contaminants from reaching the wort. One type is a sterile medical syringe filter (fig. 6.7), which can be purchased at hospital pharmacies or at your local brew-shop. An alternative, build-it-yourself bacterial filter can be made by filling a tube with moist cotton balls. The cotton should be changed after each use.

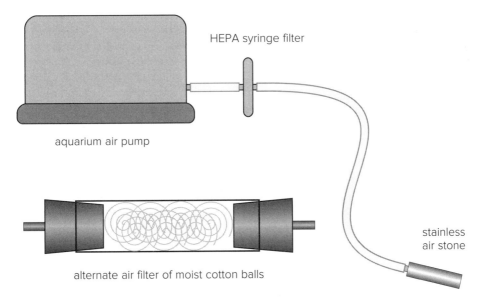

**Figure 6.7.** Here is an example of an aquarium air pump with an airstone and microbial filter. The filter is a HEPA medical syringe filter. An alternative is to make an inline filter from a plastic tube, cotton, and rubber stoppers. The pre-moistened cotton provides the filtering action and should be discarded after each use.

You can also buy small oxygen tanks used for soldering and brazing at the hardware store or welding shop. Pure oxygen has a wort saturation point of 40 ppm, so you only need a relatively short time compared to pumping ordinary air. However, using pure oxygen can result in overoxygenation and off-flavors if done excessively. In addition, oxygenation should only be done before pitching, because pure oxygen is toxic to the yeast. Frankly, these reasons are why I tend to recommend using an air pump rather than an oxygen tank for homebrewing, because it is nearly impossible to over-oxygenate with plain air. Pure oxygen is necessary and practical at the commercial level, where brewers deal with very large worts and have the equipment to monitor oxygen concentrations.

### Free Amino Nitrogen

The second nutrient the yeast need from the wort is nitrogen. Nitrogen is available in the form of amino acids and small peptides, specifically referred to as free amino nitrogen (FAN). Nitrogen is a universal component of amino acids, peptides, and proteins, and is therefore vital for all metabolic processes in yeast (and all living things, for that matter). Malted barley normally supplies all of the FAN the yeast need to grow, sometimes too much, in fact. This is one reason that beers brewed using adjuncts can have cleaner flavors than all-malt beers. However, if the recipe uses too much low-protein adjunct (e.g., corn, rice, honey, or refined sugars), then the wort may not have enough FAN. Both of these conditions can affect beer flavor. High FAN levels, consistent with a highly-modified all-malt wort, would tend to produce more fruity esters. Low FAN levels, consistent with a high-adjunct wort, would tend to have a more floral character. Most worts probably fall between these two extremes.

## FAN, FERMENTATION, AND DIACETYL

The level of FAN in the wort can affect the amount of diacetyl in the beer.

Diacetyl is a vicinal diketone, as is 2,3-pentanedione. While "vicinal diketone" sounds very ominous, it simply means that the molecule consists of two ketone groups side by side, formed from neighboring carbon atoms. Diacetyl can also be called 2,3-butanedione, so you can see how these two compounds are very similar, butane having four carbon atoms in its carbon chain and pentane having five. Diacetyl has a buttery flavor and aroma commonly associated with microwave popcorn (because diacetyl is used as the principle artificial flavor ingredient), while 2,3-pentanedione has a sweeter buttery flavor and aroma that is more like honey-butter. Together, these compounds are the main off-flavors in new beer.

Brewer's yeast is very effective at removing diacetyl and 2,3-pentanedione during the maturation phase, but diacetyl will often appear again later after packaging. Why? The reason is the presence of acetohydroxy acids in the fermentation. Acetohydroxy acids are a metabolic byproduct of the yeast, but the yeast cannot break them down further. Instead, acetohydroxy acids have to undergo oxidation to diacetyl (and 2,3-pentanedione) before the yeast can do their work. Therefore, this oxidation step is the limiting factor.

A strong vigorous fermentation at cool temperatures can seem clean after conditioning, but still have lots of acetohydroxy acid in solution that is simply waiting for a little bit more oxygen or higher temperatures to oxidize. This is another reason that the conditioning phase should take place at a warmer temperature than the high-growth phase, to promote conversion of all the acetohydroxy acid to diacetyl and 2,3-pentanedione so that it can be cleaned up by the yeast.

## Essential Minerals

Minerals present in the brewing water and malt are vital for the yeast. These include both macro-nutrients, such as phosphorus, and other nutrients used in smaller quantities as enzyme cofactors and catalysts for yeast metabolism, such as magnesium and zinc. Magnesium in particular plays a vital role in cellular metabolism, and its function can be inhibited by excessive calcium additions to the water. Brewers adding calcium salts to adjust their water chemistry should include magnesium salts as part of the addition if they experience fermentation problems.

Calcium stimulates yeast growth but is not growth limiting like magnesium. Calcium plays an important role in membrane structure and function and is necessary for yeast flocculation. Only small amounts are required for metabolic function and it is usually completely supplied by the malt. However, calcium reacts with malt phosphates and other compounds, such as oxalates, throughout the brewing process, resulting in insufficient calcium for flocculation and hazy beer.

Sulfur is used by yeast for synthesizing certain amino acids (methionine and cysteine), enzymes, and vitamins. The preferred source for sulfur is the catabolism of the amino acid methionine, although the yeast can use inorganic sulfate from the wort and water when organic sources (i.e., amino acids) are exhausted. Excess sulfur can be stored in the yeast cell vacuole in the form of glutathione, a peptide that can serve as an anti-oxidant.

Additions of zinc can greatly improve the cell count and vigor of the starter. Unlike the other essential minerals, zinc is often deficient in the wort, or trapped in a form that the yeast cannot assimilate. However, adding too much zinc will cause the yeast to produce excessive byproducts and cause off-flavors. Zinc acts as a catalyst and tends to carry over into succeeding generations, therefore, it is probably better to add zinc to either the starter or the main wort, but not both. For best performance, zinc levels should be between 0.1–0.3 mg/L, with 0.5 mg/L being the maximum. If you experience stuck fermentations or low attenuation, and you have eliminated obvious factors, such as temperature, low pitching rate, poor aeration, or low FAN, then low zinc may be a cause.

## Nutritional Supplements

Finally we come to nutritional supplements: vitamins, minerals, and energizers for the yeast. Nutritionally speaking, the wort should supply everything the yeast need, but some high-gravity styles or styles that have a high proportion of refined sugar may need supplements to get the job done. There are many different products but only two main types: fertilizer and dead yeast. Fertilizer-type supplements are simply refined chemicals, such as diammonium phosphate, which can supply much needed nitrogen and phosphorus. In any case, yeast seem to prefer more organic sources for their nutrients, and that's where dead yeast cells come in. Yeast extract (a nicer way of saying dead yeast) is easy for yeast to assimilate. Modern yeast nutrient supplements are usually a combination of essential vitamins, minerals (including zinc), amino acids, and yeast extract, all of which yeast can readily absorb.

All of these aspects of the wort—oxygen, nitrogen, minerals, and other nutrients, are vital for ensuring good yeast health and a good fermentation. These are the internal forces that determine the quality of the beer. But what about the external forces? We have already discussed temperature, so it is time to discuss the fermentors themselves.

**Figure 6.8.** Open fermentation is still fairly common, with each batch protected by its layer of yeast kräusen. Skimming and harvesting yeast is as easy as sanitizing a clean shovel and transferring the yeast to a waiting bucket. The fermentor is usually loosely covered with a lid. (Photo courtesy of Austmann Bryggeri, Trondheim, Norway.)

## Open versus Closed Fermentation

There are two types of fermentation: open and closed. An open fermentor is open to the air, and a closed fermentor is not. An open fermentor can be covered with a lid, but as long as the lid is not sealed to the outside air it is still considered to be open. A closed fermentor usually has an airlock that allows carbon dioxide to vent from the fermentation but doesn't allow outside air in. Closed fermentation using airlocks helps prevent contamination and oxidation of the batch.

The difference between open and closed fermentation is essentially the use of an airlock, but the geometry of the fermentor is usually different as well. Closed fermentors, such as a tank, carboy, or a bucket with a lid, are typically taller than they are wide. Open fermentors are generally broader and shallow, with a 1:1 height to width ratio or less (for the homebrewer, this usually just means using a bucket or bin). This geometry has advantages and disadvantages. The broader shape means open fermentors shed heat more effectively than closed fermentors, and they provide more oxygen to the yeast. Yeast harvesting is easier from open fermentors (fig. 6.8) and the yeast health is usually better. But, open fermentors typically hold less volume and take up more floor space than taller, narrower closed fermentors. In addition, open fermentors usually require racking the beer to a secondary fermentor for maturation, while closed fermentors do not.

The transition from open to closed fermentation happened relatively recently during the twentieth century. It is safe to say that before 1900 most fermentations (both commercial and home) were open, whereas after 2000 most fermentations were closed, although open fermentation is not uncommon even today. The reason for the change? Stainless steel tanks, which didn't come into general use until the 1950s. Prior to this, fermentation tanks were made from wood, brick, or stone and lined with pitch, wax, fiberglass, or epoxy.

Many of the standard guidelines and procedures that we take for granted as being the best practices for fermentation are dictated by the equipment we use, and it is important to understand the reasons why. Let's look at some examples.

*Primary and secondary fermentation.* The distinction between attenuation and maturation, and the racking from a primary to a secondary fermentor comes directly from open fermentation systems where the beer was casked away from oxygen to mature before serving.

*Dropping the beer and rousing the yeast.* It used to be common practice to aerate the beer a couple of times during fermentation by pumping it between open fermentors, or agitating the yeast to get it back in suspension. This was because the role of oxygen was not well understood. Now we have airstones and inline oxygenation to fully aerate the wort before fermentation.

*Racking the beer from the yeast.* It is recommended not to leave the beer on the yeast for too long, because tall, cylindro-conical fermentors place a high hydrostatic pressure on the yeast that increases the likelihood of autolysis after flocculation. This pressure doesn't occur in most shallow open fermentors, or in a small homebrewing cylindro-conical fermentor.

*Long cold lagering.* Historically, packaging the beer in wooden casks to age the beer tended to result in souring, especially if the beer was not kept cold. But cooling the beer slows down the maturation process by the yeast, and thus long cold lagering became the standard practice out of necessity. With modern equipment and controls, maturation can be accomplished much faster, and you don't have to move the beer as much. It can all be done in a single vessel.

## Basic Procedure for a Closed Fermentation

Most homebrewing is done with a closed fermentor, usually in the form of a food-grade plastic bucket with a tight-sealing lid, or a food-grade plastic or glass carboy. Either way, an airlock or blowoff hose is used. Closed fermentation is generally recommended for new brewers because it is simpler—pitch the yeast, affix the airlock, and walk away for two weeks. Open fermentation requires the brewer to pay more attention to the fermentation activity.

The basic procedure for a closed fermentation is the same as described in chapter 1:

1. Pour the chilled wort into the fermentor.
2. Aerate the wort with an airstone, or by rocking or shaking the fermentor.
3. Pitch the yeast, seal the lid, and affix an airlock.
4. The fermentor should be placed in a cool room with a consistent temperature within the fermentation temperature range for the yeast.
5. The fermentation should start in 12–24 hours and will bubble in the airlock for several days. Leave it alone for at least two weeks (after pitching) before bottling.

## Basic Procedure for Open Fermentation

There are many short videos on the Internet that can show you the open fermentation process and give you a good idea of what to expect. Good sanitation is always important. Here is a basic procedure:

1. Chill the wort to the intended fermentation temperature and pour it into the bin with turbulence to get some aeration. For best results, the wort should be aerated with an airstone for at least 5 min.
2. Pitch the yeast into the wort. Drape a clean towel over the fermentor.

3. A kräusen should appear within 24 hours. Skim off the brown crud that floats on top. The yeast is creamy white. Once the kräusen is clean, you can scoop some of the fresh yeast from the kräusen into a closed container and store it in the refrigerator for your next batch.

4. The best time to harvest yeast for your next batch is when the attenuation is three-quarters of the way to final gravity. For example, if the OG is 1.050, and the expected FG is 1.010, you would harvest at about 1.020 specific gravity. For a typical ale fermentation, this is around day 3 after pitching. This is also a good point during the fermentation to raise the temperature 5–10°F (3–5°C). Top cropping of yeast from open fermentations helps adapt the yeast to the method. Every professional brewer will tell you that batch-to-batch repitching of the yeast for the same recipe definitely improves the fermentation performance and the beer flavor after three generations or so.

5. On day 4 (assuming a typical ale fermentation) the beer will need to be covered with a lid and an airlock affixed, or transferred to a secondary fermentor, to allow it to mature away from oxygen. This timing assumes the beer will be fully attenuated by the end of day 5. Your fermentation may vary. Maturation can take anywhere from two days to two weeks depending on the temperature and health of the yeast.

6. If you are transferring the beer to a secondary fermentor for maturation, use a sanitized siphon. Fill the siphon with sanitizing solution, and cover the outlet with your finger until you place the racking cane in the primary fermentor. Allow the sanitizer to drain into another bucket, and then move the outlet to your secondary fermentor as the beer starts to flow. Place the outlet hose on the bottom so that it is quickly covered and doesn't splash. Watch out for any leaks in the siphon that will bubble air into the green beer. The beer should still be fairly cloudy with suspended yeast. The maturation temperature should be the same, or a few degrees higher, as previously discussed.

You would think that open fermentation would have a problem with contamination from the air, but the dense mat of yeast on top of the beer protects it pretty well. Any debris that falls into the fermentor is usually carried to the sides by the yeast mass. An easy way to prevent debris falling in is to drape a clean towel over the bucket or tub. Air can still easily pass through the towel, but it keeps the dust and pet hair out. A lid that comes down over the sides of the fermentor but leaves an air gap is also a good solution. Airborne bacteria and mold spores cannot fall up to get inside, and there is usually so much carbon dioxide coming off the fermentation that insects won't come near it.

An open fermentor for home brewing can be a standard fermentation bucket, a standard brewing kettle, or a large food-grade plastic storage bin. The greater air exposure is part of what gives these beers their greater fermentation character, but it is not necessary to go overboard with it. Before refrigeration, wort would be run into wide shallow coolships to help dissipate the heat of the boil, and fermentors were often wider and shallower than what I have described here in order to dissipate the heat of fermentation, but those are solutions to problems we don't usually have nowadays. A roughly cubic volume is a good starting point. A larger surface area will result in more esters, perhaps too much, depending on your taste. A common HDPE plastic storage bin (fig. 6.9), such as those used for storing clothes and blankets, works well.

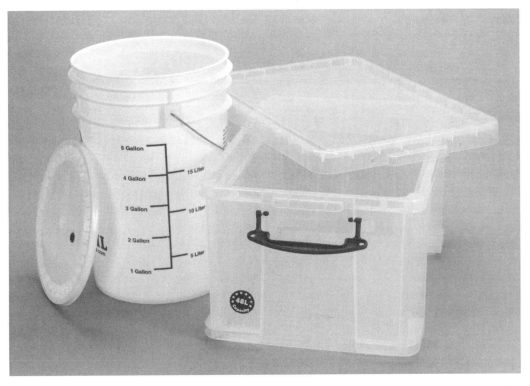

**Figure 6.9.** Open fermentors used to be the norm; any food-grade bucket or bin will work. Cover it loosely with an overhanging lid or drape a clean towel over it to keep debris out. If you are using the lid, there should be an air gap for venting.

Open fermentation usually requires the use of a secondary fermentor for maturation, because oxygen exposure after the high growth phase will oxidize the beer and shorten its shelf life. If the open fermentor is a standard bucket, then sealing with the lid and airlock is sufficient and the beer can mature without being transferred.

Racking or transferring the beer to another fermentor is always an oxidation and contamination risk. Glass or plastic carboys make good secondary fermentors because of the smaller head space. A smaller headspace is better to minimize the oxygen exposure and make it easier for the headspace to be purged by the still-fermenting beer. Do not dilute your beer to increase the volume; even boiled water contains about 1 ppm oxygen, which can cause staling. The head space will be purged fairly quickly anyway.

The maturation time should be 1–2 weeks and then the beer can be bottled or kegged. If the beer was cold conditioned for a long period of time between maturation and packaging, additional yeast may be necessary for priming and carbonating.

Hopefully this chapter has helped you understand the big picture of what fermentation is and how it works. The next chapter will get into the nuts and bolts of yeast management.

# Yeast Management 7

In this chapter, we will discuss yeast management, that is, optimizing the quantity and quality of your yeast to optimize the fermentation. If you manage the yeast, then you manage the beer. Yeast does not want to make beer, it just wants to eat and reproduce; beer is a waste product as far as it's concerned. So to control the beer we need to control the yeast growth—more specifically, the rate of growth and the total amount of growth. Managing these two factors consistently will produce high quality fermentations and good beer consistently. If you manage these two factors inconsistently, although you may get lucky with the odd batch, you will only produce good beer inconsistently.

Fermentation is the most critical part of the brewing process, but it is often taken for granted by new brewers. A lot of thought will be given to the recipe with regards to malts and hops, but the yeast choice will be whatever was readily available. Even if some consideration is given to the yeast strain and fermentation temperature, the pitching conditions are often not planned or controlled. The brewer cools the wort, aerates it a bit, and then sprinkles the yeast on the wort and waits for it to do its thing. There's a saying that is particularly apt here: faith is good, but planning is better.

Imagine you are a small business owner and you are making a product. You need to hire workers to make the product. You want those workers to be reliable, coming to work on time, working consistently with no accidents that would halt production, and able to output a consistently high quality product. You are going to invest in training those workers to do their job the best they can. You are going to organize and manage their resources so that they can work most efficiently. I trust this is becoming obvious by now—manage the yeast and you manage the beer.

Good planning is proper yeast selection, pitching rate, and managing the work environment for best results. We discussed fermentation temperature control in the previous chapter. Here we will focus on selecting the yeast and planning the pitch.

## Yeast Types

There are two main types of brewer's yeast: ale yeast (*Saccharomyces cerevisiae*) and lager yeast (*Saccharomyces pastorianus*[1]). The Latin names are binomials—*Saccharomyces* is the genus, and *cerevisiae* and *pastorianus* are the species names. These yeast species are closely related, and strains within each species are closer still, but they can act quite differently in fermentations.

To put this in perspective, coyote, jackals, gray wolves, and dogs all belong to the genus *Canis*, and gray wolves and dogs are all considered to be the same species, separate from coyote and jackals. Think of brewer's yeast strains being as different from each other as different dog breeds are from each other and from wolves. Every yeast strain produces a different beer. Sometimes the differences are small, such as a bit more ester in the aroma, or a little bit lower finishing gravity. Sometimes the differences are large, such as between ale and lager yeasts—the same wort can produce completely different beers solely due to the type of yeast used. This is especially true when you step outside the *Saccharomyces* genus and conduct the fermentation using "wild yeasts" from the genus *Brettanomyces*, or using bacteria species. (These types of "wild" fermentations are covered in chapter 14.) To stretch the above analogy further, think of foxes, which are "dog-like" but in a different genus (*Vulpes*) from *Canis* altogether.

Ale yeasts are referred to as "top-fermenting" because historically the fermentation activity seemed to be taking place on the top of the wort, while lager yeasts are referred to as "bottom-fermenting" because the activity seemed to be more submerged. While this generalization is not actually true, there is one important difference, and that is temperature. Ale yeasts prefer warmer temperatures, 65–75°F (18–24°C), although they can work 5°F (3°C) either side of that range. Lager yeasts prefer 50–55°F (10–13°C), although they can work 5°F (3°C) either side of that range as well. Some yeast strains can be successfully used across both ranges. These are often referred to as "hybrid-style" strains and may, in fact, be actual hybrids between the two species.

## Characteristic Yeast Strains

There are hundreds of different strains of brewer's yeast available nowadays and each strain produces a different flavor profile. Some Belgian strains produce fruity esters that smell like bananas and cherries, some German strains produce phenols that smell strongly of cloves, although these two examples are rather special. Most yeasts are not that dominating, but it illustrates how much the choice of yeast strain can determine the taste of the beer. One of the main differences between different beer styles is the strain of yeast that is used.

Many major breweries have their own strain of yeast. The yeast strain will have evolved to be unique to the style(s) of beer being made. In fact, yeast readily adapts and evolves to specific brewery conditions, so two breweries producing the same style of beer with the (ostensibly) same yeast

---

[1]    Used to be called *Saccharomyces carlsbergensis.*

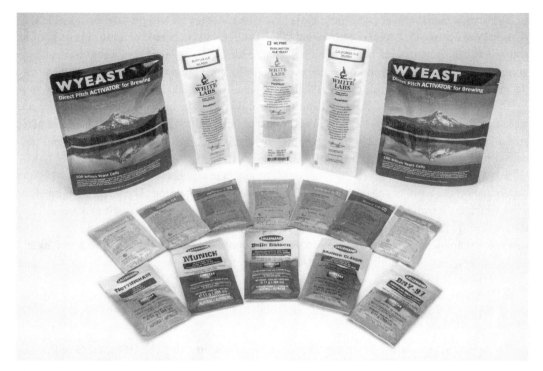

**Figure 7.1.** There are many strains, types (dry or liquid), and brands of homebrewing yeast.

strain will actually have, over time, cultured different yeast strains that produce unique beers. Yeast companies have collected many of these different yeasts from around the world and packaged them for homebrewers (fig. 7.1).

You will find that each company's cultured strain will be subtly different due to the conditions under which it was sampled, stored, and grown. You may find that you prefer one company's strain over another's. Detailed descriptions of each company's cultured strains will be available at your brewshop or on the company's website. Any list will likely soon be rendered incomplete, because new strains are being added to the market all the time.

The earlier analogy to dogs and dog breeds is useful for understanding the diversity of yeast strains. Some strains are very different from one another and others are quite similar. There are families of strains; for example, British ale yeast includes Scottish ale, Irish ale, and English ale. This is much as you have hound dogs and all the varieties of those. Even within a particular strain, such as English ale, there will be many sub-strains, such as London and Yorkshire ale. Unfortunately, there is no standard convention for naming yeast strains; one company's British ale strain may be another company's English or London ale strain. The main point is that the strains are related and similar.

The descriptions below are general, and you should review the manufacturer's information for specific fermentation temperature ranges and flavor character. The descriptions also list whether the strain is generally available in dry or liquid form, or both.

Dry yeast comes from select, hardy strains that have been dehydrated for long-term storability. Dry yeast needs to be rehydrated before use. Liquid yeast is yeast cells in water and it needs to be refrigerated to keep fresh, but it's generally ready to pitch right from the package.

Dry yeast is convenient because the packets have a long storage life and provide many viable yeast cells that can be prepared quickly on brewing day. Dry yeast can be stored for up to two years (preferably in the refrigerator), but the packets do degrade with time. The number of viable yeast cells per gram of dried yeast after rehydration varies with the strain, but a good rule of thumb is that dry yeast has about 10 billion cells per gram, so a 10 g packet contains about 100 billion viable yeast cells.

Dry ale yeasts make good beer, but the rigors of dehydration limits the number of different strains that are available. There are perhaps a couple of dozen different dry yeast strains, compared with several hundred strains in liquid form. Liquid yeast is packaged in pouches or tubes and typically contain about 100 billion cells per package.

*Notes on characteristics for yeast descriptions.* The apparent attenuation of a yeast strain will vary depending on the types of sugar in the wort that the yeast is fermenting. The fermentability of the wort, which is determined by the brewer, sets the limits to which a particular yeast may work. Thus, the number quoted for a particular yeast is an average. Generally, less flocculent yeast strains attenuate more than highly flocculent strains do. For purposes of discussion, the apparent attenuation is ranked as low, medium, and high with the following percentages:

- Low = 65%–70%
- Medium = 70%–75%
- High = 75%–80%

The apparent attenuation percentage is determined by subtracting the OG and FG of the beer and dividing by the OG. A 1.040 OG that ferments to a 1.010 FG would have an apparent attenuation of 75%.

$$\text{Attenuation} = ([OG - FG] - 1)/(OG - 1)$$

$$= ([1.040 - 1.010] - 1)/(1.040 - 1)$$

$$= 30/40, \text{ or } 0.75$$

The "real" attenuation is less. Pure ethanol has a specific gravity of about 0.800. If you had a 1.040 OG beer and got 100% real attenuation, the resulting final gravity would be about 0.991 (corresponding to about 5% alcohol by weight). The apparent attenuation of this beer would be 122%.

The flocculation characteristics of a strain are very general. There are several wort and fermentation factors that will affect flocculation, so the general strain characteristic is only ranked as low, medium, or high. High flocculation means that most of the yeast settles to the bottom after fermentation is complete. Low flocculation means that most of the yeast doesn't settle to the bottom afterwards. Medium flocculation is anything in between, although they tend to be more flocculent than not. Highly flocculent yeasts tend to be low-attenuating and condition poorly, and often require rousing or higher temperatures later in fermentation to keep them active. Less flocculant strains tend to be more attenuative and condition better, but often leave a hazier beer.

## Ale Yeasts

*American ale.* A low ester, neutral-tasting ale yeast. Good for any type of ale. Medium attenuation, medium flocculation. Suggested fermentation temperature range is 65–72°F (18–22°C). Available in dry and liquid form.

*American wheat ale.* This yeast is used to brew American-style *hefeweizen*, which have the haze of Bavarian hefeweizen, but a comparatively moderate banana and clove character. Medium-high attenuation, low flocculation. Suggested fermentation temperature range is 65–70°F (18–21°C). Available in liquid form.

*California ale.* A very clean-tasting ale yeast, producing less esters than other types of ale yeast, and good for accentuating the hop character. Produces a drier beer than American ale. High attenuation, medium flocculation. Suggested fermentation temperature range is 65–73°F (18–23°C). Available in dry and liquid form.

*Australian ale* This all-purpose strain produces a complex malty and estery beer. Great for pale ales, brown ales and porters. Medium attenuation, high flocculation. Suggested fermentation temperature range is 65–70°F (18–21°C). Available in dry and liquid form.

*English ale.* This strain (or family of strains) is known for top-cropping and producing English milds, bitters, and porters. More maltiness and residual sweetness is evident than with American ale yeast. Medium attenuation, medium flocculation. Suggested fermentation temperature range is 65–70°F (18–21°C). Available in dry and liquid form.

*European ale.* This full-bodied strain finishes very malty with low esters and low sulfur. Top-cropping yeast that produces a dense rocky head during fermentation. Especially well-suited to *altbier*. High flocculation, low attenuation. Suggested fermentation temperature range is 65–70°F (18–21°C). Available in dry and liquid form.

*Irish ale.* The slight residual diacetyl is great for stouts. It is clean, smooth, soft and full-bodied. Very nice for any cold-weather ale, at its best in stouts, brown, and red ales. Medium flocculation, medium attenuation. Suggested fermentation temperature range is 65–70°F (18–21°C). Available in liquid form.

*Scottish ale.* This strain is more similar to the American strains than it is to the English strains. It is clean and brings out the best of the malt and hops. Medium flocculation, medium attenuation. Suggested fermentation temperature range is 65–70°F (18–21°C). Available in liquid form.

*Yorkshire ale.* This strain is the quintessential top-cropping English ale yeast that produces a malty complex beer with moderate esters. Specific character varies with the mother brewery it was sourced from. Medium attenuation, medium-high flocculation. Suggested fermentation temperature range is 65–70°F (18–21°C). Available in liquid form.

*Belgian abbey ale.* Lots of fruity esters (banana, spice), and can be tart and dry. Very good for Belgian ales, *dubbel* and *tripel*. This is actually a group of strains, and each particular strain will have its own personality. Medium flocculation, high attenuation (in general). Suggested fermentation temperature range is 65–72°F (18–22°C). Available in dry and liquid form.

*Belgian ale.* Less esters than Belgian abbey yeasts, with a malty bread and biscuit character that complements the spicy phenolics that are the hallmarks of Belgian yeast strains. Medium flocculation, high attenuation. Suggested fermentation temperature range is 68–78°F (20–26°C). Available in dry and liquid form.

*Belgian wit (white).* Mild phenolic character for the classic Belgian *wit* beer style. Slightly tart and fruity. Low flocculation, medium attenuation. Suggested fermentation temperature range is 65–75°F (18–24°C). Available in dry and liquid form.

*German weizen.* Produces the distinctive clove and spice character of wheat beers. The low flocculation of this yeast leaves the beer cloudy ("hefeweizen" means "yeast wheat"), but its smooth flavor makes it an integral part of a true unfiltered wheat beer. Low flocculation, high attenuation. Suggested fermentation temperature range is 65–75°F (18–24°C). Available in dry and liquid form.

*Saison.* There are many strains of *saison* yeast, but they all tend to be highly attenuative. Some can be quite dry and tart, although not actually sour. Character can vary from strong esters to mild phenolics (fruity to spicy). Medium flocculation, high attenuation. Suggested fermentation temperature range is 68–75°F (20–24°C). Available in dry and liquid form.

## Lager Yeasts

*American lager.* Very versatile for most lager styles. Gives a clean malt flavor. Some strains have more residual acetaldehyde than others. Low sulfur and low diacetyl. Medium flocculation, high attenuation. Suggested fermentation temperature range is 50–56°F (9–12°C). Available in liquid form.

*Bavarian lager.* This is the lager yeast strain used by many German breweries. Rich flavor, full-bodied, malty and slight esters. This is an excellent general-purpose yeast for lager brewing. Medium flocculation, high attenuation. Suggested fermentation temperature range is 50–55°F (10–13°C). Available in dry and liquid form.

*Bohemian pils.* Ferments clean and malty, leaving residual maltiness in high-gravity Pilseners. Sulfur produced during fermentation dissipates with conditioning. Very suitable for Vienna and Oktoberfest styles. Medium flocculation, medium attenuation. Suggested fermentation temperature range is 50–55°F (10–13°C). Available in dry and liquid form.

*Czech lager.* Classic dry finish with good maltiness. Moderate esters. Good choice for Pilseners and *bock* beers. Medium flocculation, high attenuation. Suggested fermentation temperature range is 50–55°F (11–13°C). Available in liquid form.

*Danish lager.* Clean, crisp and dry. Soft, light profile that accentuates hop characteristics. Medium flocculation, medium-high attenuation. Suggested fermentation temperature range is 48–56°F (9–13°C). Available in dry and liquid form.

*German lager.* Generally dry, this family of lager strains is noted for their clean, malty character and low esters. Medium flocculation, medium attenuation. Suggested fermentation temperature range is 50–56°F (10–13°C). Available in dry and liquid form.

*Mexican lager.* This lager strain produces a very clean, dry lager. One of my favorite yeasts because of it clean character. Medium flocculation, medium-high attenuation. Suggested fermentation temperature range is 50–56°F (10–13°C). Available in liquid form.

*Munich lager.* A classic lager yeast that is smooth, malty, well rounded, and accentuates hop flavor. It is reported to be prone to producing diacetyl, so use a diacetyl rest. Medium flocculation, medium attenuation. Suggested fermentation temperature range is 48–54°F (9–12°C). Available in dry and liquid form.

## Hybrid Yeasts

*San Francisco lager.* Warm-fermenting bottom-cropping strain. Ferments well to 62°F (17°C), having some of the fruitiness of an ale while keeping lager characteristics. Malty profile, high flocculation, low attenuation. This is the yeast that is used for California common beers. Suggested fermentation temperature range is 58–65°F (14–18°C). Available in dry and liquid form.

*German altbier.* Ferments dry and crisp, leaving a good balance of malt and hops. Produces an extremely rocky head and ferments well down to 55°F (13°C). A good choice for *alt*-style beers. Low flocculation, medium-high attenuation. Suggested fermentation temperature range is 55–64°F (13–18°C). Available in liquid form.

*Kölsch-style ale.* An old German style of beer that is more lager-like in character. Nice maltiness without as much fruit character as other ales. Some sulfur notes that disappear with aging. Low flocculation, high attenuation. Suggested fermentation temperature range is 56–68°F (13–20°C). Available in liquid form.

## What Is the Pitching Rate and Why Does It Matter?

As a brewer you are a yeast rancher, and your job is to provide the right number of healthy yeast to best ferment the wort. You can do this by pitching multiple yeast packages to match the strength of the wort (its OG), or you can pitch a single package into a starter and use that mini-fermentation to grow enough yeast to pitch into the full batch. But how do we know how much yeast to pitch?

The most commonly recommended pitching rate is 1 million cells per milliliter of wort per degree Plato[2] (i.e., 4 gravity points). For the mathematically challenged, I will point out that 1 million per milliliter is the same as saying 1 billion per liter, and that there are typically 100 billion cells in a yeast package. (*Note:* 1 billion = $1 \times 10^9$, or 1,000,000,000.) So, one yeast package could ferment 10 L of 1.040 OG wort. There are roughly 4 L in a gallon (actually 3.78 L), so we can say that this rate roughly equates to one package being capable of fermenting 2.5 gal. of 1.040 wort.

What is usually overlooked is that this recommended rate is for "repitched" yeast—tired yeast like you would get from the bottom of the fermentor from a previous batch. That yeast is not at peak vitality (i.e., health) or viability (i.e., percentage of living cells). Vitality and viability depend on the conditions of the previous fermentation, as discussed in chapter 6. Fresh yeast from a well-prepared starter is at the peak of vitality and viability and only half as much fresh yeast is needed to do the same job as repitched yeast. Going back to our calculation above, this means that one fresh package of liquid yeast, or a rehydrated package of dry yeast, is capable of fermenting 5 gal. (19 L) of 1.040 wort, and this is the basis for the common claim that one package of 100 billion cells can ferment a five-gallon batch. If the wort gravity is greater than 1.040 or the package is a couple of months old, then you would most likely need to use an additional package or a starter to ensure the vitality and viability of the yeast before pitching it into the main batch.

The pitching rate range for a good fermentation using fresh yeast is 0.5–1.0 billion cells per liter per 4 gravity points for ales, and roughly 1.0–1.5 billion per liter per 4 gravity points for lagers. See tables 7.1–7.3 for recommended pitching rates as a function of wort gravity. (For those worried about units, note that table 7.2 gives pitching rates per gallon; and all three tables give gravity readings in degrees Plato as well as specific gravity.) It is a lot easier to underpitch than to overpitch; more is generally better. In fact, I will go so far as to say that it is difficult to hurt a fermentation by pitching too much yeast (unless the yeast is half dead and full of trub from a previous fermentation).

Viewing yeast management as akin to ranching (see sidebar, "The Sheep Analogy for Pitching Rate") illustrates two important aspects of yeast starters and fermentations. The first aspect is that

---

[2]   Degrees Plato is an alternative measurement system for wort gravity that is more commonly used by professional brewers and is based on the weight percentage of sugar in the wort. It is measured with a refractometer. See appendix A for more information on this measurement system.

## THE SHEEP ANALOGY FOR PITCHING RATE

If a rancher puts 100 sheep on 1 acre of land for a day, the sheep will quickly eat all of the grass and no new sheep will be born during that time. If the rancher puts 10 sheep on 100 acres, lots of reproduction will occur but not all of the grass will be eaten in the same amount of time. But, if you put 100 sheep on 10 acres, then all of the grass will be eaten and a predictable amount of reproduction will occur. Of course, a sheep rancher is raising sheep to make wool, while a brewer is raising yeast to make beer, but the point is that both are trying to optimize the number and health of their animals to get the best results.

small yeast starters (such as 1 pint or 500 ml) do not generate much growth (i.e., increased yeast mass). There is simply not enough nutrients in a small starter to allow much growth to occur by the time those nutrients are consumed. Typically, pitching one yeast package to a well-aerated 1 L starter of 1.040 specific gravity will roughly double the yeast count, and pitching to a 2 L starter of same gravity will generate about two-and-a-half times the yeast. Using a stir plate to maintain oxygen levels will help with the total growth, increasing the growth factor by about 25% (e.g., from 2 to 2.5).

The second aspect is that the amount of growth dictates the amount of aroma and flavor compounds generated. Lower pitching rates within the recommended range will result in more growth before the sugars are consumed compared to higher pitching rates. In other words, lower pitching rates tend to result in more fermentation character in the beer (more esters and the like), while higher pitching rates tend to result in less fermentation character, that is, a "cleaner" fermentation. There is a limit of course, and extreme overpitching tends to result in the production of unused metabolic byproducts, greater yeast autolysis, and more off-flavors.

The *rate* of yeast growth is most dependent on temperature. The *total amount* of growth is limited by the aeration and available wort nutrients. You control the flavor of the beer by matching the initial number of yeast (the pitch) to the available nutrients (the size and strength of the wort), and then controlling growth rate with temperature. If you want the next batch of a recipe to be the same as the previous batch, then you need to match the previous fermentation. If you want to clone an award-winning beer, then you need to clone its award-winning fermentation. Match the fermentation to match the beer.

### Pitching Rates and Beer Styles

As we have seen, the pitching rate has a large effect on the character of the beer. In the first few days of fermentation, when the yeast have entered the high-growth phase and are rapidly reproducing, more diacetyl precursors (acetohydroxy acids), acetaldehyde, and fusel alcohols are being produced than at any other time. A low pitching rate means more total cell growth, more amino acid synthesis, and therefore more byproducts. A high pitching rate means less total cell growth and less byproduct.

Ale styles that are noted for their fermentation character, like Belgian ales, should be pitched with fresh yeast toward the low end of the recommended range, 0.5 billion cells per liter per 4 gravity points. Ale styles that are regarded as having a more balanced or light fermentation character, such as southern English brown ale, dry stout, and English pale ale, should be pitched more toward the middle at 0.75 billion cells per liter per 4 gravity points. Ales that are considered to have a very clean character, like American pale ale, blonde ale, and northern English brown ale, should be pitched at 1.0 billion cells per liter per 4 gravity points. Of course, some yeast strains have a very assertive character all on their own, and therefore the pitching rates for those yeasts need to be on the upper end in order to prevent the fermentation going hog-wild and funky. Examples of this kind are saison, Bavarian weizen, and *witbier* yeasts.

## PITCHING RATE AND ESTER FORMATION

The pitching rate affects the ester character of the beer. Lower pitching rates tend to produce more aromatics and esters than higher pitching rates. A lower pitching rate will encourage more yeast reproduction due to the perceived abundance of resources. The yeast will reproduce until the total yeast mass reaches the limit for what the wort resources can support. Esters are metabolic byproducts that are formed by the yeast during the growth phase when they need to eliminate excess waste. The yeast utilize the oxygen from aeration to synthesize sterols and other essential lipids. Short-chain fatty acids are created as intermediate compounds during the synthesis of long-chain fatty acids, which are incorporated into the various types of sterols and lipids that the yeast need to live and reproduce. It is the synthesis of these sterols and lipids that are actually the limiting factor for yeast growth. Each time the yeast cells reproduce by budding, they share their lipid reserves with the daughter cell.

The leftover short-chain fatty acids are essentially industrial waste that is toxic to the yeast. The yeast combines these fatty acids with other waste products (the alcohols) to produce a non-toxic ester that it can excrete into the environment. There are two main stages that facilitate esterification. The reaction starts between an alcohol (ethanol or fusel) and a fatty acid and uses acetyl coenzyme A, which conveys the carbon atoms in the fatty acids into the citric acid cycle, or Krebs cycle, in yeast metabolism (this produces energy for the yeast cell). The esterification reaction is catalyzed by the enzyme alcohol acetyltransferase, which transfers an acetyl group from the alcohol to the ester chain. Different combinations of fatty acids and alcohols produce different esters, and the spectrum of esters produced tends to be specific to the yeast strain.

Yeast tends to produce more esters during stressful fermentations; conditions that amount to either a famine or a feast (orgy). Famine conditions are low nutrients (i.e., low levels of FAN, minerals, or lipids), low oxygen, and low temperatures. Feast conditions are high nutrients, high oxygen, and high temperatures. Both of these sets of conditions are stressful for the yeast, leading to non-optimum metabolism and the formation of more short-chain fatty acids that need to be esterified. Low pitching rates lead to famine conditions. Sufficiently high pitching rates with normal fermentation temperature, oxygen levels, and nutrient levels, are the least stressful conditions for fermentation and tend to produce the least esters. Very high pitching rates (all else being nominal) would push the system toward famine conditions again.

Pitching rates are typically doubled for lager styles, so the corresponding rates for estery lagers like Kölsch and California common beer are 1.25 billion cells per liter per 4 gravity points. Lagers that have a moderate fermentation character, like Dortmunder, Munich *dunkel*, and American lagers, have rates of 1.5 billion cells per liter per 4 gravity points. Very clean styles, like Munich *helles*, Vienna lager, and German Pilsner can use even higher rates of 1.5 billion cells per liter per 4 gravity points to produce the best results.

## TABLE 7.1—PITCHING RATES (BILLIONS OF YEAST CELLS) PER LITER AS A FUNCTION OF WORT GRAVITY

| | | Recommended pitch rate for style (billion cells/L/°P) | | | | | |
| | | Ale | | | Lager | | |
| SG | °P | 0.5 | 0.75 | 1.0 | 1.25 | 1.5 | 1.75 |
|---|---|---|---|---|---|---|---|
| 1.020 | 5.1 | 3 | 4 | 5 | 6 | 8 | 9 |
| 1.025 | 6.3 | 3 | 5 | 6 | 8 | 9 | 11 |
| 1.030 | 7.6 | 4 | 6 | 8 | 9 | 11 | 13 |
| 1.035 | 8.8 | 4 | 7 | 9 | 11 | 13 | 15 |
| 1.040 | 10.0 | 5 | 7 | 10 | 12 | 15 | 17 |
| 1.045 | 11.2 | 6 | 8 | 11 | 14 | 17 | 20 |
| 1.050 | 12.4 | 6 | 9 | 12 | 15 | 19 | 22 |
| 1.055 | 13.6 | 7 | 10 | 14 | 17 | 20 | 24 |
| 1.060 | 14.7 | 7 | 11 | 15 | 18 | 22 | 26 |
| 1.065 | 15.9 | 8 | 12 | 16 | 20 | 24 | 28 |
| 1.070 | 17.1 | 9 | 13 | 17 | 21 | 26 | 30 |
| 1.075 | 18.2 | 9 | 14 | 18 | 23 | 27 | 32 |
| 1.080 | 19.3 | 10 | 14 | 19 | 24 | 29 | 34 |
| 1.085 | 20.5 | 10 | 15 | 20 | 26 | 31 | 36 |
| 1.090 | 21.6 | 11 | 16 | 22 | 27 | 32 | 38 |
| 1.095 | 22.7 | 11 | 17 | 23 | 28 | 34 | 40 |
| 1.100 | 23.8 | 12 | 18 | 24 | 30 | 36 | 42 |
| 1.105 | 24.9 | 12 | 19 | 25 | 31 | 37 | 43 |
| 1.110 | 25.9 | 13 | 19 | 26 | 32 | 39 | 45 |
| 1.115 | 27.0 | 14 | 20 | 27 | 34 | 41 | 47 |
| 1.120 | 28.1 | 14 | 21 | 28 | 35 | 42 | 49 |

°P, degrees Plato; SG, specific gravity.

For details of relationship between °P and SG, refer to appendix A.

## TABLE 7.2—PITCHING RATES (BILLIONS OF YEAST CELLS) PER GALLON AS A FUNCTION OF WORT GRAVITY

| | | Recommended pitch rate for style (billion cells/L/°P) | | | | | |
|---|---|---|---|---|---|---|---|
| | | Ale | | | Lager | | |
| SG | °P | 0.5 | 0.75 | 1.0 | 1.25 | 1.5 | 1.75 |
| 1.020 | 5.1 | 10 | 14 | 19 | 24 | 29 | 34 |
| 1.025 | 6.3 | 12 | 18 | 24 | 30 | 36 | 42 |
| 1.030 | 7.6 | 14 | 21 | 29 | 36 | 43 | 50 |
| 1.035 | 8.8 | 17 | 25 | 33 | 42 | 50 | 58 |
| 1.040 | 10.0 | 19 | 28 | 38 | 47 | 57 | 66 |
| 1.045 | 11.2 | 21 | 32 | 42 | 53 | 64 | 74 |
| 1.050 | 12.4 | 23 | 35 | 47 | 59 | 70 | 82 |
| 1.055 | 13.6 | 26 | 39 | 51 | 64 | 77 | 90 |
| 1.060 | 14.7 | 28 | 42 | 56 | 70 | 84 | 98 |
| 1.065 | 15.9 | 30 | 45 | 60 | 75 | 90 | 105 |
| 1.070 | 17.1 | 32 | 48 | 65 | 81 | 97 | 113 |
| 1.075 | 18.2 | 34 | 52 | 69 | 86 | 103 | 121 |
| 1.080 | 19.3 | 37 | 55 | 73 | 91 | 110 | 128 |
| 1.085 | 20.5 | 39 | 58 | 77 | 97 | 116 | 135 |
| 1.090 | 21.6 | 41 | 61 | 82 | 102 | 122 | 143 |
| 1.095 | 22.7 | 43 | 64 | 86 | 107 | 129 | 150 |
| 1.100 | 23.8 | 45 | 67 | 90 | 112 | 135 | 157 |
| 1.105 | 24.9 | 47 | 71 | 94 | 118 | 141 | 165 |
| 1.110 | 25.9 | 49 | 74 | 98 | 123 | 147 | 172 |
| 1.115 | 27.0 | 51 | 77 | 102 | 128 | 153 | 179 |
| 1.120 | 28.1 | 53 | 80 | 106 | 133 | 159 | 186 |

°P, degrees Plato; SG, specific gravity.

For details of relationship between °P and SG, refer to appendix A.

## TABLE 7.3—PITCHING RATES (BILLIONS OF YEAST CELLS) PER 6 GAL. (23 L) BATCH AS A FUNCTION OF WORT GRAVITY

| | | Recommended pitch rate for style (billion cells/L/°P) | | | | | |
| | | Ale | | | Lager | | |
| SG | °P | 0.5 | 0.75 | 1.0 | 1.25 | 1.5 | 1.75 |
|---|---|---|---|---|---|---|---|
| 1.020 | 5.1 | 58 | 88 | 117 | 146 | 175 | 204 |
| 1.025 | 6.3 | 73 | 109 | 145 | 182 | 218 | 254 |
| 1.030 | 7.6 | 87 | 130 | 174 | 217 | 260 | 304 |
| 1.035 | 8.8 | 101 | 151 | 201 | 252 | 302 | 353 |
| 1.040 | 10.0 | 115 | 172 | 229 | 286 | 344 | 401 |
| 1.045 | 11.2 | 128 | 192 | 257 | 321 | 385 | 449 |
| 1.050 | 12.4 | 142 | 213 | 284 | 355 | 425 | 496 |
| 1.055 | 13.6 | 155 | 233 | 311 | 388 | 466 | 543 |
| 1.060 | 14.7 | 169 | 253 | 337 | 421 | 506 | 590 |
| 1.065 | 15.9 | 182 | 273 | 364 | 454 | 545 | 636 |
| 1.070 | 17.1 | 195 | 292 | 390 | 487 | 585 | 682 |
| 1.075 | 18.2 | 208 | 312 | 416 | 520 | 623 | 727 |
| 1.080 | 19.3 | 221 | 331 | 441 | 552 | 662 | 772 |
| 1.085 | 20.5 | 233 | 350 | 467 | 583 | 700 | 817 |
| 1.090 | 21.6 | 246 | 369 | 492 | 615 | 738 | 861 |
| 1.095 | 22.7 | 258 | 388 | 517 | 646 | 775 | 904 |
| 1.100 | 23.8 | 271 | 406 | 542 | 677 | 812 | 948 |
| 1.105 | 24.9 | 283 | 425 | 566 | 708 | 849 | 991 |
| 1.110 | 25.9 | 295 | 443 | 590 | 738 | 885 | 1033 |
| 1.115 | 27.0 | 307 | 461 | 614 | 768 | 922 | 1075 |
| 1.120 | 28.1 | 319 | 479 | 638 | 798 | 957 | 1117 |

°P, degrees Plato; SG, specific gravity.

*For details of relationship between °P and SG, refer to appendix A.*

It may be surprising to note that strong beers, including old ale, Russian imperial stout, Baltic porter, and barleywine, use high pitching rates. You may think that these beers have a lot of flavor, and they do, but the flavors are typically clean, estery flavors, rather than funky off-flavors. Strong beers need clean fermentations to prevent the off-flavors from overwhelming the beer.

To consistently brew good beer of any style, the brewer needs to understand yeast behavior, and manage that behavior to his or her purpose. No single pitching rate is right for every style of beer. The pitching rates discussed above and listed in tables 7.1, 7.2, and 7.3 are guidelines, not rules. Use a rate that makes sense to you and adjust it up or down as you see fit for your recipe.

## Preparing Yeast and Yeast Starters

**Figure 7.2.** Measuring cup and packets.

## Rehydrating Dry Yeast

For best results, rehydrate dry yeast in warm pre-boiled water before pitching. The rehydration temperature of the water should be 77–86°F (25–30°C) for ale yeast, and 69–77°F (21–25°C) for lager yeast.

Don't just sprinkle dry yeast onto the wort like the manufacturer's instructions often suggest. The manufacturer says that to make it more convenient to use, but often half the yeast cells will die, because they can't draw enough water across their cell membranes due to the high sugar concentration.

*Rehydration Procedure.*
1. Put 1 cup (250 mL) of warm[3] pre-boiled water into a sanitized jar and sprinkle the dry yeast on top. Do not stir it in. Cover with plastic wrap or aluminum foil and wait 15 min. for the yeast to rehydrate.

---

[3]   Remember, 77–86°F (25–30°C) for ale yeast; 69–77°F (21–25°C) for lager yeast.

**Figures 7.3.** Dry yeast that has been rehydrated.

2. Gently stir the rehydrated yeast into the water to ensure it is fully wetted. You want to limit the exposure to oxygen until you are ready to pitch the yeast into your wort. Otherwise, the yeast cells will start burning up their nutrient reserves too soon—all dressed up with nowhere to go.

3. Re-cover the jar and let it sit for another 15 min. to complete rehydration. The yeast will form a creamy layer on the bottom of the jar. It should be pitched within 30 min. for best results. Swirl the jar to re-suspend the yeast immediately before pitching.

## Propagating Yeast with a Starter

Liquid yeast is generally regarded as being better than dry yeast because there are so many more strains available, but it has a shorter shelf life and is more expensive. Most people will buy a single package of liquid yeast to save money and then propagate it with a starter, which usually takes about a day. Dry yeast can be rehydrated and ready to go in an hour, so it is often more convenient to rehydrate more packages than take the time and effort to grow it up with a starter. However, the dehydration process is hard on the yeast, and even rehydrated yeast will sometimes exhibit small amounts of off-flavors, such as phenolics, due to the yeast being stressed. Therefore, revitalizing dry yeast with a starter can be beneficial too, although it is usually not necessary. Beers brewed with dry yeast regularly win medals at the Great American Beer Festival® (GABF).

Remember to size your starter to the recommended total pitch for your wort strength and volume. You can refer to table 7.4 to see estimated growth factors when using packaged yeast. For example, if you are going to brew 6 gal. (23 L) of a 1.065 OG lager, you will need about 545 billion cells, assuming a typical pitching rate (table 7.3). You can propagate sufficient yeast by pitching two packages of yeast (100 billion cells per package) to 4–5 L of 1.040 SG wort (giving 520–560 billion cells), or pitching three packages to 3 L (giving 600 billion cells). The latter would give you more than you need, but that amount of overpitching is not going to hurt the beer. It may be a little too "clean" for your taste, but you will just have to try it and see. The optimum pitching rate will depend on the yeast strain as well; some need higher pitching rates, some lower, to give the best character.

## TABLE 7.4—ESTIMATED GROWTH FACTOR FOR 100 BILLION CELLS (1 PACKAGE) PITCHED INTO 1.040 SG WORT

| Yeast packages | Starter volume in liters | | | | | | | | | |
|:---:|:---:|:---:|:---:|:---:|:---:|:---:|:---:|:---:|:---:|:---:|
| | 1 | 2 | 3 | 4 | 5 | 6 | 7 | 8 | 9 | 10 |
| 1 | 2.0 | 2.6 | 3.0 | 3.4 | | | Long time | | | |
| 2 | | 2.0 | 2.3 | 2.6 | 2.8 | 3.0 | 3.2 | 3.4 | | |
| 3 | | | 2.0 | 2.2 | 2.4 | 2.6 | 2.7 | 2.7 | | |
| 4 | Low growth | | | 2.0 | 2.2 | 2.3 | 2.5 | 2.6 | 2.7 | 2.8 |
| 5 | | | | | 2.0 | 2.1 | 2.3 | 2.4 | 2.5 | 2.6 |

Notes: This table gives estimated propagation factors based on starter volume. For example, 2 packages of yeast pitched to 6 L would have a growth factor of 3, yielding 600 billion cells. The starter wort is assumed have 8 ppm of dissolved oxygen. Empty table cells are not recommended, i.e., low growth factor or excessive time. A growth factor of 2 may take 24–36 hours to finish; a growth factor of 3 may take 36–48 hours.

## TABLE 7.5—GRAMS OF MALT EXTRACT FOR MAKING 1.040 SG YEAST STARTERS

| Starter volume (L) | DME (g) | LME (g) |
|:---:|:---:|:---:|
| 1 | 115 | 134 |
| 2 | 230 | 268 |
| 3 | 345 | 403 |
| 4 | 460 | 537 |

Note: Only liters and grams are given because it makes the propagation estimates simpler, and it doesn't matter whether you are pitching 500 billion cells into 5 gal. or 20 L, it's still 500 billion cells.

## Making a Yeast Starter
You will need:
- a pot for boiling wort
- a flask for fermentation
- an airlock or some aluminum foil
- malt extract or wort
- set of scales that measures in grams (if measuring extract)

*Notes for choosing and sizing the flask:*

**Material.** Glass Erlenmeyer conical flasks (2–5 L in size) are the gold standard because you can actually boil the wort in them, but they are expensive. Plastic juice bottles (2–3 L) work well, but they have to be thoroughly cleaned and sanitized before use, and they aren't heat resistant. The other problem with plastic juice bottles is they often have a dimpled bottom, which means that you can't use them on a stir plate if you are so inclined. Flat-bottomed glass jugs work better on a stir plate, but you still need to cool the wort before filling them.

**Size.** Bigger is always better. You can make a small starter in a big container, but not the other way around.

### Step 1.
Choose the size of the yeast starter based on the combinations given in table 7.4. Mix the malt extract and water in the boil pot according to the weights and volumes given in Table 7.5. Bring the wort to a boil for 10 min. Put the lid on the pot for the last couple of minutes, turn off the stove and let it sit while you prepare for the next step.

You can add a little bit of hops if you want to, but it really is not necessary. Adding a pinch of yeast nutrient (preferably containing zinc) to the starter wort is good to ensure good growth. I prefer to add it to the starter rather than to the main batch.

### Step 2.
Cool the starter wort. Fill the kitchen sink with a couple of inches of cold water. Take the covered pot and set it in the water, moving it around to speed the cooling. When the pot has cooled to room temperature (68–77°F [20–25°C]) pour the wort into the sanitized fermentor flask. Pour all of the wort in, even the sediment. This sediment consists of proteins and lipids, which are actually beneficial for yeast growth at this stage.

Ideally, the starter wort temperature should be the same as what you plan the fermentation temperature to be. This allows the yeast to get acclimated to working at that temperature. If the yeast is started warmer and then pitched to a cooler fermentation environment, it may be shocked or stunned by the change in temperature and may take a couple days to regain normal activity.

### Step 3.
Cap the fermentor flask or bottle with the lid. Shake the starter vigorously to aerate it (see sidebar, "Notes on Aeration of Starters" for more information). Shake the starter BEFORE adding the yeast.

**Figure 7.4**. Making starter wort with dry malt extract.

**Figure 7.5**. Pouring wort into a sanitized fermentor flask.

**Step 4.**

Sanitize the outside of the yeast package and pour the yeast into the flask. Swirl the flask to mix it well. Affix an airlock to the flask and put it somewhere with a steady temperature out of direct sunlight. If you don't have an airlock that will fit, don't worry. Instead, put a clean piece of aluminum foil over the top and fold it over the sides. This way the carbon dioxide will be able to vent without exposing the starter to airborne bacteria. The foil doesn't have to be tight, if fact it's better if it's loose, because that allows extra oxygen to diffuse in and support yeast growth.

The starter will turn cloudy with suspended yeast and may form a kräusen. Depending on the size and pitch ratio of the starter (table 7.4), the total propagation time should be 24–48 hours at 70°F (21°C). You may want to give the starter another day to settle before pitching it into your wort.

**Figure 7.6**. Adding yeast to starter.

One recommendation when pitching a large starter is to chill the starter overnight in the refrigerator to settle all the yeast. Then the unpleasant tasting starter beer can be poured off so only the yeast slurry will be pitched. This helps prevent the taste of the starter from influencing the taste of the final beer if the starter volume is large, but, on the other hand, properly managing the maturation phase of the main fermentation should take care of that.

You don't need to warm up the starter before pitching to your fermentor. Pitching from colder to warmer helps wake the yeast up; it's pitching from warmer to colder that shocks them and puts them to sleep.

## NOTES ON AERATION OF STARTERS

It is actually fairly easy get 8 ppm of dissolved oxygen into a starter by shaking the wort—the key is adequately mixing the air with the wort.

One liter of air (at sea level) contains about 285 milligrams (mg) of oxygen gas. If this amount were entirely dissolved in 1 L of water, you would have a concentration of 285 mg/L, which is equivalent to 285 ppm. This means that 35 mL of air should be able to supply 8 ppm. However, that assumption requires that all of the oxygen molecules in that 35 mL of air come into contact with the water and then dissolve, which doesn't happen. Therefore, more headspace is better. My experiments demonstrated that leaving 20% headspace in a 2 L soda bottle filled with wort will saturate to 8 ppm with 1 min. of vigorous shaking. I recommend shaking in two batches for larger volumes.

The saturation concentration of oxygen in fresh water exposed to the air varies between 7 and 10 ppm within the temperature range of 60–75°F (15–24°C) and elevation from sea level to 6000 ft. (1830 m). The maximum concentration of 10 ppm occurs with colder temperatures at lower elevation, and the lesser concentration with higher temperatures at higher elevations. A typical saturation concentration at room temperature near sea level is between 8 and 9 ppm. Aerating with pure oxygen increases the partial pressure (i.e., the percentage) of oxygen and therefore raises the equilibrium concentration. That is why you can achieve higher dissolved concentrations using a pure oxygen tank.

## NOTES ON USING A STIR PLATE

A stir plate uses a spinning magnet to turn a magnetic stir bar in the wort. The stir bar is itself a plastic-coated magnet. A smaller stir bar (1–2 in. [2.5–5 cm]) is more effective than a larger stir bar (3 in. [8 cm]) at creating a small vortex, which helps pull air into the wort and keeps the wort and yeast moving. The benefit of a stir plate for yeast propagation is that more oxygen is supplied to the yeast for a longer period of time and this allows for more growth, about 25% more compared to a shaken starter. The wort does not need to be shaken before being placed on the stirplate. The stirring action also minimizes the foamy kräusen.

One important consideration is that the stir plate should only be used for the first half of the propagation, for example, 12–18 hours when assuming 24–36 hours of total propagation time. The reason is that the yeast needs an active period of low oxygen to build up its glycogen and trehalose reserves prior to cooling, decanting, and pitching to the main wort. These reserves help the yeast to adapt to its new environment. Taking the yeast directly off the stir plate and immediately cooling it in a refrigerator prior to pitching is more stressful.

## When Is My Yeast Starter Ready to Pitch?

A yeast starter is ready be pitched when it has attained high kräusen (full activity), or after it has settled out. The starter will remain ready for about two days after, depending on the temperature. There is an approximately 18-hour window, between peak activity and flocculation, when the yeast is tired and you shouldn't pitch it.

The composition of the starter wort and the main wort must be very similar if the starter is pitched at or near peak activity. Why? Because the yeast cells in the starter wort have produced a specific set of enzymes for that wort's sugar profile. If those yeast are then pitched to a different wort with a different proportion of sugars, the yeast will be impaired and the fermentation may be affected. Yeast that has been eating a lot of sucrose, glucose, and fructose will quit making the enzyme that allows it to eat maltose—the main sugar of brewer's wort.

**Figure 7.7.** Finished starter with yeast layer on the bottom.

At the end of fermentation the yeast cells build up their glycogen and trehalose reserves; kind of like a bear storing fat for the winter. Glycogen and trehalose are two carbohydrates that act as food reserves for the yeast cell. Yeast cells slowly feed off these reserves when other food is not present, and use this food extensively to fuel the synthesis of essential lipids, sterols, and unsaturated fatty acids upon being pitched into an oxygenated wort. Yeast cells rapidly deplete their glycogen reserves when exposed to oxygen. If glycogen can be compared to the fat that a bear stores for winter, the other component, trehalose, acts more like the bear's heavy fur coat. Trehalose seems to get built up on both the inside and outside of the yeast cell membrane, and is generally believed to make the membrane structure more robust and more resistant to environmental stress.

By allowing the yeast starter fermentation to go to completion, these reserves are built up so that, upon pitching, the yeast starts out with a ready fuel supply and a clean slate to better adapt to the new wort. However, these same reserves are used by the yeast while in hibernation, so if the starter is left too long before pitching (e.g., a week at room temperature or a month in the refrigerator), these reserves may be depleted. The reserves should be replenished with a fresh starter wort fermentation before use.

## Using Yeast from Commercial Beers

There are many quality microbrew beers on the market that are bottle conditioned, that is, naturally carbonated and unfiltered, much the same as homebrewed beers are. The yeast layer from a bottle conditioned beer can be harvested and grown just like other yeast, but there will be a lot fewer viable cells, often less than 1% of what is in a yeast package. Harvesting is common practice among homebrewers to capture special yeast strains or bacteria that would not otherwise be available. This method can be used for cloning some of the specialty styles, such as Belgian wit, Trappist ale, or saison. The only caveat is that some bottle-conditioned beers are packaged with a different yeast than the primary yeast used for fermentation, because the packaging yeast provides for more consistent conditioning in the bottle. In addition, high-alcohol beers are not good candidates, because the yeast is severely weakened and has probably mutated by the time you try to culture it. Harvesting yeast from a bottle-conditioned beer is quite simple.

1. After opening the bottle, thoroughly clean the bottleneck and opening with sanitizer to prevent bacterial contamination.
2. Simply pour the beer into a glass as you normally would, leaving the yeast layer intact on the bottom of the bottle.
3. Swirl up the sediment with the beer remaining in the bottle and pour the yeast sediment into a prepared starter solution as described in the previous section, "Making a Yeast Starter."

For best results, add the sediment from two or three bottles and be sure to use the freshest beer you can find. The starter should behave the same as any other liquid yeast pack starter, though it may take longer to build due to the smaller amount of yeast that you start out with. In fact, you may not notice any activity in the starter for the first couple of wort additions until the amount of yeast builds to higher levels. Add more starter wort as necessary to build the yeast slurry to pitching level. Be sure to taste, or at least sniff, the starter beer to check for contamination. It should be beer-like and not funky.

### Support Your Local Craft Brewery

If you have a quality brewpub or microbrewery nearby, the brewers are often happy to provide yeast to homebrewers. A good brewery produces a lot more yeast than they can use and it is usually free of contamination. I keep a spare, sanitized, plastic pint container in case an opportunity presents itself.

The advantage to obtaining yeast this way is that you usually get very healthy yeast. You are virtually guaranteed of a vigorous, healthy fermentation, without the fuss of preparing a yeast starter a few days beforehand. The yeast will stay viable for a couple of weeks if kept in the refrigerator. But remember, you may want to revitalize the yeast with a small starter if the yeast is stored for a long time.

### Simple Yeast Ranching

Each batch of beer you brew, indeed, every yeast starter you make, is a good source of yeast for a future batch. Simply make a larger starter than the current batch requires and put some in a separate jar in the refrigerator for future use.

If you prefer to obtain yeast from your active fermentation, the best way is to skim it from the kräusen. To do this, you will need to be using a bucket type or open fermenter. First, skim off the green-brown hop and protein residue with a sanitized spoon early in the high-growth phase. As the creamy white kräusen builds up, you can skim this fresh yeast off with a (freshly) sanitized spoon and transfer it to a sanitized jar. Fill the jar with cooled boiled water and place it in the refrigerator. You want to use boiled water for two reasons: one, for sanitation, and two, to avoid exposing the yeast to dissolved oxygen, which would cause the yeast to deplete their glycogen reserves before storage. The lack of oxygen and nutrients in the water will cause the yeast to enter the stationary phase and it will keep for a couple of months. You should pitch this yeast to a starter after storage to revitalize it.

Table 7.6 is a tool to help the budding yeast rancher keep track of which yeast floc is which. The name of the strain and lot # can be entered in the upper left corner (eg., WLP023, Lot #1023025). The matrix starts in cell A1 with an original manufacturer yeast package or a pitch from a brewery. Line A in the matrix is the "mother" line; lines B and C are the daughter lines. The harvest date (H) and the pitching date (P) should be recorded for each generation. (See inset table for data field descriptions.) The harvest date is the birthday of a particular floc and marks the first day of storage of the yeast until it is pitched. Sometimes, two or even three harvests will be made from a fermentation, and these are recorded below it, becoming lines B and C. The prefix "A3" indicates the origin of the yeast line, and the column number indicates the generation number. For example, in cell C3, the harvested yeast is labeled A3C3, identifying it as being the 6th generation overall (3 + 3), but the 3rd generation of line AC3, which was harvested from line A3.

It is important to keep track of how many times a yeast culture has been propagated and to go back to a younger propagation line if you notice that the beer character is changing for the worse with time. The propagation limits varies widely depending on who you talk to; I would suggest seven as a limit. This example is not the only way to keep track of yeast propagation, but it works for me.

If you harvest yeast from the bottom of the fermentor, you will need to separate the yeast from all the trub that is mixed in. Professional brewers most often do this by "acid washing" the yeast—using acid to lower the pH to about 2.5 so that bacteria is inhibited and then using a whirlpool to separate the heavier trub from the lighter yeast. But acid washing tends to inhibit the yeast too, and is not strictly necessary. You can use chilled boiled water and two sanitized jars to separate the healthy yeast (white) away from the majority of the trub, like so:

1. After racking the beer, swirl up the yeast layer on the bottom of the fermentor and pour some into a large sanitized jar (such as a mayonnaise jar).
2. Gently pour some cold, boiled water into the jar and swirl it up to get all the yeast and trub in suspension.
3. Let the jar sit for a minute or three to allow most of the trub to settle to the bottom. Gently pour the cloudy water, which contains suspended yeast, into another sanitized jar. Discard the dark trub.
4. Add some more water to the second jar and repeat this procedure until you have eliminated most of the dead yeast and trub on the bottom of the jar. When the yeast finally settles, you should have a white yeast layer on top of a thin trub layer. Try to eliminate most of the trub, but removing it entirely is not necessary.
5. Store the jar in the refrigerator for up to a couple of months. The yeast will turn brown as it ages. Discard it when it turns the color of peanut butter. Eventually, the yeast will die and autolyze as their nutritional reserves are used up.

Always pitch your harvested yeast to a small starter before using to restore its vitality; you want a small starter (i.e., 100 billion cells/L) because you don't want a lot of growth. Pitching weak yeast to a big starter will most likely tire them out all over again. If the starter smells wrong—phenolic, buttery, or otherwise—the yeast may be contaminated. The dominant smell of a starter should be a bready, yeasty smell, although sulfur smells are common with lager yeast strains.

## Summary

Yeast management is all about the care and feeding of your yeast before fermentation. Some would argue that this chapter should come before the "Yeast and Fermentation" chapter, but I wanted you to understand the bigger picture of the importance of fermentation before we got into the details of pitching rate and propagation.

| (Yeast info.) | 1 | 2 |
|---|---|---|
| A | A1 | A2 |
| | H date | H date |
| | P date | P date |
| | Batch # | Batch # |

## TABLE 7.6—EXAMPLE YEAST RANCHING MATRIX

| WLP023 #1023025 | 1 | 2 | 3 | 4 | 5 | 6 | 7 |
|---|---|---|---|---|---|---|---|
| A | A1 030114 030114 #442 | A2 030414 041214 #443 | A3 041614 042214 #444 | A4 042514 042614 #445 | A5 043014 050314 #448 | A6 050714 051014 #452 | A7 051514 052614 #456 |
| B | A3B1 042514 042514 #446 | A3B2 043014 050314 #449 | A3B3 050614 050914 #451 | A3B4 051214 052014 #454 | | | |
| C | A3C1 042514 042714 #447 | A3C2 043014 050314 #450 | A3C3 050714 051014 #453 | A3C4 051314 052014 #455 | | | |

# Water for Extract Brewing

**8**

The two main things you need to know about brewing water at this stage of your brewing education are:

1. The water should taste clean, without any off-flavors. Bad tasting water will usually make bad tasting beer.
2. The water for brewing with malt extract should be low in dissolved minerals, with all ion concentrations less than 50 ppm. This will be explained.

## Understanding Your Source Water

Our water usually comes from one of two sources: surface water or groundwater. Surface water, such as lakes, rivers, and streams, usually comes from precipitation. Groundwater comes from underground aquifers. Surface water is usually high in organic matter but low in minerals, the latter typically less than 50 ppm for each constituent. Groundwater is usually low in organic matter but high in dissolved minerals, the latter typically greater than 50 ppm and sometimes greater than 100 ppm.[1]

Depending on your location, your water source may change throughout the year due to availability or treatment costs. Surface water usually requires filtration to remove silt, algae, and other organic matter, and chlorine disinfection to eliminate dangerous parasites and bacteria. Groundwater often doesn't require filtration and needs less disinfection, but may require demineralization treatments to prevent mineral deposits in pipes.

---

[1] Parts per million (ppm) is a common measure of concentration. For most substances dissolved in water, ppm is equivalent to milligrams per liter (mg/L), hence, 1 ppm = 1 mg/L.

The first step in determining if your tap water is suitable for brewing is to taste it. Does your water taste good, or does it taste like a swimming pool? A strong chlorine flavor or odor is the most common problem with tap water. The chlorine or chloramine that is added by law to public drinking water for residual disinfection can cause medicinal off-flavors in beer. It is important to remove residual chlorine and chloramine from your brewing water. Bottled water typically does not contain these residual disinfectants.

In the USA, most of the water quality characteristics and issues that affect your tap water are discussed in the annual water quality report that is available (by federal law) from your local water provider. If you are on an independent well, you need to send out samples for water quality testing. Water quality reports will be discussed later in this chapter.

Other problems can be pond-like odors or a heavy mineral taste. These flavors can often be fixed by home treatment, but in most cases it may be easier to just buy bottled water from the store.

Therefore, the next step for using your tap water for brewing is to get rid of the chlorine and chloramine.

## Brewing Water Dechlorination Treatments

The following methods can be used to remove chlorine and other common off-flavors and odors in water.

*Boiling.* If your tap water smells like a swimming pool, it's either chlorine or chloramine (see sidebar, "Chlorine, Chloramine, and Chloride"). Some of these odors can be removed by boiling. Chlorine can generally be removed by pre-boiling, but chloramine will not. Pre-boiling the water will often eliminate bad odors but it's generally less effective than carbon filtration.

*Activated carbon filtration.* Activated carbon filters are a good way to remove many organic odors and flavors from drinking water and they are readily available at most home improvement and hardware stores. However, even though it's how most breweries do it, the average homebrewer may find carbon filtration is less effective for removing chlorine and chloramine because the required flow rate is very slow. How slow? The guidelines that are used by the commercial brewing industry specify an empty bed contact time (EBCT) of 2 min. for chlorine and 8 min. for chloramine.[2]

The EBCT is calculated by dividing the volume of the carbon filter (in gallons or liters) by the flow rate (in gallons or liters per minute). For example, if a home carbon filter had a cartridge volume of 1 L, then the flow rate would have to be 0.5 L/min. to achieve an EBCT of 2 min. It would take 40 min. to filter 20 L (i.e., roughly a 5 gal. batch) of water. Eliminating chloramine would take over two-and-a-half hours to filter the same volume.

Large commercial filtration systems can process a lot of water at the same time, but home units, in all practicality, take too long to provide enough water for the batch. Chemical reduction by metabisulfite is much quicker.

*Metabisulfite.* Sodium metabisulfite and potassium metabisulfite are commonly used in the wine industry to inhibit mold and wild yeast growth on the crushed fruit, but they can also be used to oxidize and remove chlorine and chloramine from brewing water. Metabisulfite can be added by using the chemical itself (typically potassium metabisulfite) or by adding one Campden tablet to the water. One Campden tablet will usually treat 20 gal. (76 L), although using one tablet for only 5 gal. (19 L) is fine. Crush up the tablet and stir to help it dissolve. Chlorine and chloramine are chemically reduced to insignificant levels of chloride (<10 ppm) within a couple of minutes at room temperature. The sulfite ions are oxidized to sulfate ions by the same reaction. Typical dosage rate

---

[2]    Palmer and Kaminski (2013, p. 197–8).

for potassium metabisulfite powder is 10 mg/L, assuming the maximum chloramine or chlorine levels permissible in treated tap water.

Bottom line: metabisulfite powder or Campden tablets are the best way to eliminate chlorine and chloramine from your brewing water.

## CHLORINE, CHLORAMINE, AND CHLORIDE

Chlorine (Cl) is a gaseous chemical element in the periodic table with atomic number 17. It is highly reactive and commonly used as a disinfectant. However, because chlorine is highly reactive it doesn't last long when used as a residual disinfectant, so the water utilities have to use a lot of it. Chlorine has a strong smell, but more importantly it will react with polyphenol compounds in the beer to produce chlorophenols, which have a strong medicinal or plastic smell and flavor that is impossible to remove. The only good thing about chlorine is that it tends to boil away if you pre-boil the water before brewing.

Chloramine ($NH_2Cl$) is produced by reacting chlorine with ammonia and is more stable (lasts longer) than chlorine as a residual disinfectant in water, so the water utilities don't have to use as much. Today, almost all water utilities use chloramine rather than straight chlorine to treat their water. The problem is that "more stable" means the chloramine doesn't boil away and requires extra steps for brewers to eliminate it.

Chloride ($Cl^-$) is the ion of chlorine and doesn't produce chlorophenols. Chloride is actually beneficial for beer flavor in small amounts in brewing water, so don't confuse it with elemental chlorine.

### Brewing Water and the Water Quality Report

Your annual water quality report should provide the information you need to determine whether you have low-mineral water or not. Low-mineral water is defined for our purposes as having less than 50 ppm for each of the important brewing ions. These ions, which come from dissolved minerals (e.g., limestone), are calcium ($Ca^{2+}$), magnesium ($Mg^{2+}$), bicarbonate ($HCO_3^-$), sodium ($Na^+$), chloride ($Cl^-$), and sulfate ($SO_4^{2-}$).

However, water quality reports are mainly concerned with governmental requirements for chemical contaminants, such as heavy metals or pesticides, and the water treatment plant will have treated the water for those issues. The "Secondary" or "Aesthetic Standards" section of the report (i.e., information that is not required by federal law) will usually contain the mineral information that you need for brewing. If your water quality report does not have this information, then a phone call to the water utility can usually provide it. If not, you may have to send a water sample out to a lab for further testing, or test it yourself using a brewing water test kit.

Low-mineral water for extract brewing is really just the start. The following general guidelines will help you better understand what your water quality report can tell you about the mineral content of your brewing water. All of these will be discussed in more detail below.

**General guidelines for brewing water:**
1. Hardness is good.
2. Alkalinity is bad.
3. Water pH can be used as a general indicator of the balance between hardness and alkalinity in the water.

4.  Hardness and alkalinity affect the water and mash pH.
5.  Sodium, chloride, and sulfate do not affect pH, but can affect the beer flavor the same way that salt and pepper season your food.

*Water hardness.* Calcium and magnesium ions in water are responsible for water hardness. It is called hardness for a couple of reasons, the first being that calcium and magnesium ions will bind to part of the soap molecule and cause the formation of soap scum instead of lather; this results in more soap needed to do the cleaning job. The second reason is that hard water can cause hard calcium carbonate scale (limescale) to form on your plumbing, although the alkaline carbonate ions are just as much to blame as the calcium and magnesium ions. Both water hardness and water alkalinity are usually quantified by the unit, "ppm as calcium carbonate ($CaCO_3$)." The reasons for this will be discussed further in chapter 21, where water chemistry is treated in more depth.

Water hardness can also be classified as either temporary or permanent. Temporary hardness precipitates out of the water as white scale when boiled, while permanent hardness stays in solution. Temporary hardness comes from calcium and magnesium ions that are dissolved from carbonates and bicarbonates, like limestone, while permanent hardness comes from calcium and magnesium ions that are dissolved from highly soluble salts, such as chloride and sulfate salts.

Why is hardness good? Because minimum levels of calcium and magnesium dissolved in the wort are important for several aspects of the brewing process, including yeast health, fermentability, and clarity. In extract brewing, however, these minimum levels are already present in the malt extract, and additional calcium and magnesium are generally not needed. Therefore, using low-mineral water such as distilled, reverse osmosis-treated, or bottled drinking water is recommended. Too much water hardness (i.e., more than 150 ppm of $Ca^{2+}$) will make the beer taste minerally.

*Water alkalinity.* The sum of the dissolved carbonate ($CO_3^{2-}$) and bicarbonate ($HCO_3^-$) in tap water constitute the total alkalinity of the water. Total alkalinity is measured by acid titration, and is equal to the quantity of acid of known concentration necessary to titrate the water sample from the sample's original pH (typically 7–9) to a pH of 4.3, in milliequivalents per liter (mEq/L).[3] By convention, this quantity is usually converted to "total alkalinity, ppm as $CaCO_3$," (calcium carbonate).

The bicarbonate ion typically accounts for 97% of total alkalinity in tap water, so the bicarbonate concentration can be used to estimate the total alkalinity of the water according to the equation:[4]

Total alkalinity ppm as $CaCO_3$ = (50/61) × [$HCO_3^-$] in ppm

In addition, it is important to realize that total alkalinity as $CaCO_3$ is roughly equal to the temporary hardness as $CaCO_3$. This is why brewing textbooks used to say to remove the temporary hardness from water before brewing—it's not the hardness that we want to remove, it's the associated alkalinity!

Why is alkalinity bad? Because alkalinity raises the pH of wort and beer. This issue is more important in all-grain brewing (the effects of which are discussed in chapter 21), but it can affect the flavor of malt extract beers as well. High alkalinity will raise the wort and beer pH, causing the malt flavor of the beer to be dull and the hop bitterness to be harsh and lingering. This is the primary

---

[3]  Equivalent (Eq) is a unit for describing equivalency in acid–base reactions. It can be defined as the amount of a dissolved substance that will either supply or react with 1 mole of hydrogen ions ($H^+$), or supply or react with 1 mole of electrons ($e^-$). A milliequivalent (mEq) is one-thousandth of an equivalent (1 mEq = 0.001 Eq).
[4]  This conversion equation will be explained in chapter 21.

reason why I recommend using distilled or reverse osmosis water for extract brewing, which will decrease the total alkalinity coming from both the malt extract and brewing water.

Alkalinity is also the reason that softened water is not recommended for brewing. Most home water softeners only remove calcium and magnesium ions (which you actually want), add sodium ions (which you don't want), and don't remove alkalinity (which you would like to get rid of). This is why softened water—that is, water filtered through a home water softener—should never be used for brewing.

Here are some general guidelines for total alkalinity as $CaCO_3$:

- If your tap water has a total alkalinity of less than 50 ppm, then you have low-mineral water and can use it to brew any malt extract recipe or kit.
- If the total alkalinity is between 50 and 100 ppm, this is moderate. You can use it, but you may notice some flavor issues.
- If your total alkalinity is greater than 100 ppm, then you shouldn't use it. I recommend that you use bottled water instead.
- However, it is important to understand that in the grand scheme of things, brewing with alkaline water is always a minor problem compared to fermentation issues, such as temperature control.

*Water pH.* Pure water has a pH of 7 meaning that it is neither acidic nor alkaline, and is thus neutral. Water that has a pH less than 7 is acidic, while a pH greater than 7 is alkaline. Most potable water has a pH of 7.5–8.5, and generally has more alkalinity than hardness. Potable water with a pH of 9 will generally have more alkalinity and/or less hardness than potable water with a pH of 7. To know by how much, you have to look at the individual ion concentrations.

That's really about all there is to it. Water pH is never very important in evaluating brewing water. Why? Because two completely different mineral profiles (e.g., groundwater vs. surface water) can have the same water pH. It's the particular ion concentrations in the water that affect our beer. pH is a result—that is, pH is a measure of chemical activity and equilibrium—and we don't really care about the chemical equilibrium in the water. What we care about is the pH of the wort and beer, and that equilibrium is a result of the ion constituents in the water reacting with the ions from the malt in the mash. All of this will be explained later in chapters 21 and 22. For now, please understand that you do not have to worry about or adjust water pH.

*Calcium, magnesium, and bicarbonate ions.* It is the levels of calcium, magnesium, and bicarbonate ions that constitute water hardness and alkalinity. Malt extract likely has all the calcium and magnesium the beer needs, and all the alkalinity that the brewer would ever want. In general, good brewing water has a calcium ion concentration of 50–150 ppm, magnesium ion concentration of 5–40 ppm, and a total alkalinity less than 100 ppm as $CaCO_3$.

*Chloride, sulfate, and sodium ions.* Chloride, sulfate, and sodium ions act to season the beer and help bring out certain flavors. Chloride accentuates the maltiness of the beer, making the beer seem fuller and sweeter. Sulfate accentuates the hop character of the beer, making the beer taste drier and crisper. Sodium also accentuates the malt, but at high concentrations (>100 ppm) it tends to give an overall minerally character to the beer. With the exception of the increasingly popular Gose style, sodium is never intentionally added to brewing water, although sometimes it is necessary to add small amounts as part of another salt, such as sodium bicarbonate or sodium metabisulfite.

The general recommendations for sulfate and chloride concentrations are given in table 8.1 below. Brewers use the sulfate-to-chloride ratio of the brewing water as a benchmark for how the water profile will affect the balance of hops and malt in the beer. Hoppy beers, like IPA, will have a sulfate-to-chloride ratio of 5:1, malty beers, like *bock* and Oktoberfest, will have a ratio of 1:1 or even 0.5:1. This is not a rule, however, it is a guideline.

## TABLE 8.1—THE SIX IMPORTANT IONS IN BREWING WATER

| | Suggested minimum | Suggested range | Purpose |
|---|---|---|---|
| Calcium ($Ca^{2+}$) | 50 ppm | 50–150 ppm | $Ca^{2+}$ determines water hardness along with $Mg^{2+}$, and is vital to many of the biochemical reactions in fermentation. It should be adequately supplied by the malt extract. |
| Magnesium ($Mg^{2+}$) | 5 ppm | 0–30 ppm | $Mg^{2+}$ determines water hardness along with $Ca^{2+}$, and is a vital yeast nutrient. It is usually fully supplied by the malt extract. High levels taste sour/bitter. |
| Total Alkalinity as $CaCO_3$ (bicarbonate, $HCO_3$) | N/A | 0–100 ppm (0–120 ppm) | Alkalinity raises wort and beer pH and can cause astringency. Alkalinity is more tolerated in redder/darker beers where it balances the acidity of dark malts. |
| Sodium ($Na^+$) | N/A | 0–100 ppm | $Na^+$ levels can be very high (>300 ppm) due to use of water softeners. High levels can cause mineral or metallic flavors. |
| Chloride ($Cl^-$) | N/A | 50–150 ppm | $Cl^-$ accentuates maltiness, sweetness and fullness. Too much can be cloying and corrosive to equipment. |
| Sulfate ($SO_4^{2-}$) | N/A | 50–400 ppm | $SO_4^{2-}$ accentuates hop bitterness, making it seem drier and crisper. At high concentrations (>400 ppm) it is harsh and unpleasant. |

Notes: The concentrations given are general recommendations for all styles. Specific recommendations for particular styles are given in chapter 22. Water chemistry is explored in more depth in chapter 21.

## Adding Brewing Salts to Season Your Beer

I strongly recommend that you first brew a malt extract recipe or beer kit using low-mineral, or distilled, water without using any salt additions, and see how it tastes. If you decide that you want to enhance the maltiness or hoppiness, then you can add some salts to the water, but be careful that you don't overdo it. It is very easy to oversalt a beer and have it taste minerally.

Many people try to enhance both the maltiness and the hoppiness of the beer by adding both sulfate and chloride. This usually results in a minerally tasting beer. As a general rule of thumb, the sum of the sulfate and chloride ions should not exceed 500 ppm. Table 8.2 lists the common brewing salts and how much each salt contributes to the ion profile of the water.

**Seasoning Guidelines:**
- The maximum recommended concentration for sulfate is 400 ppm.
- The maximum recommended concentration for chloride is 150 ppm.
- Do not try to add both sulfate and chloride at the maximum level, it will taste minerally.
- Generally, you don't have to worry about adding too much calcium, but exceeding 200 ppm can taste minerally.

Calculating the final concentrations for your salt additions is easy. You simply multiply the weight of the salt addition in grams by the ion contribution(s) in ppm per gram for that salt (see table 8.2), divided by the volume of water you are adding it to. In equation form it looks like this:

(weight × ion contribution) / volume = final concentration

For example, 2 g of calcium sulfate added to 5 gal. of water would be:

For $Ca^{2+}$: $(2 \times 61.5)/5 = 24.6$ ppm $Ca^{2+}$ added.

For $SO_4^{2-}$: $(2 \times 147.4)/5 = 59$ ppm $SO_4^{2-}$ added.

Concentrations are additive. Let's say that we had started out with water that already had a $Ca^{2+}$ concentration of 40 ppm, and we added the same addition as above (2 g into 5 gal.) to this water. The final concentration of $Ca^{2+}$ would then be $40 + 24.6 = 64.6$ ppm. It works the same way for additions from multiple salts, such as calcium sulfate and calcium chloride—you would add the calcium concentrations from both contributions together to get the final concentration in the water.

## IPA Example for Adding Salts

Let's look at an example for adding salts to an IPA. This is a hoppy beer that benefits from a crisp, dry finish, and brewers frequently add calcium sulfate (gypsum) to the water to achieve it.

Let's say we decide we want to add 300 ppm of sulfate to the water, based on our experience of brewing that kit with bottled water. Looking at table 8.2, we can see that adding 1 gram of calcium sulfate per gallon contributes 147 ppm of sulfate. To achieve 300 ppm of $SO_4^{2-}$, we therefore want to add about 2 g/gal. of calcium sulfate. This will also contribute about 120 ppm of calcium ions to the water. This means your beer might end up tasting minerally, depending on the concentration of calcium ions already present in the malt extract.

The calcium sulfate salt would be added to the brewing water. Let's assume we are brewing with 6 gal. of water total, with 3 gal. being used to boil the extract with the hops, and 3 gal. being added to the fermentor after the boil (the Palmer Brewing Method, as explained in chapter 4). In this example, 6 g of calcium sulfate would be added to the boil kettle at the beginning of the boil, and another 6 g would be mixed in with the water in the fermentor. If you're using liters, this example requires a calcium sulfate addition of about 0.5 g/L (check table 8.2), which you can see is still an addition of 6 g added to 11.4 L in the boil kettle and 6 g added to 11.4 L in the fermentor.

## TABLE 8.2—ION CONTRIBUTIONS BY SALT ADDITIONS

| Brewing salt, formula, molecular weight (MW) | Contributions at 1 gram of salt per liter | Contributions at 1 gram of salt per gallon | Comments |
|---|---|---|---|
| **Calcium sulfate** $CaSO_4 \cdot 2H_2O$ MW = 172.2 | 232.8 ppm $Ca^{2+}$ 557.7 ppm $SO_4^{2-}$ | 61.5 ppm $Ca^{2+}$ 147.4 ppm $SO_4^{2-}$ | Solubility limit at room temperature is about 2 g/L. Dissolves better in cold water. Stir vigorously. |
| **Magnesium sulfate** $MgSO_4 \cdot 7H_2O$ MW = 246.5 | 98.6 ppm $Mg^{2+}$ 389.6 ppm $SO_4^{2-}$ | 26.0 ppm $Mg^{2+}$ 102.9 ppm $SO_4^{2-}$ | 70 ppm is the maximum recommended concentration for $Mg^{2+}$. |
| **Calcium chloride** $CaCl_2 \cdot 2H_2O$ MW = 147.0 | 272.6 ppm $Ca^{2+}$ 482.3 ppm $Cl^-$ | 72.0 ppm $Ca^{2+}$ 127.4 ppm $Cl^-$ | Dissolves readily. Lowers mash pH. Food-grade salt may not be high purity. |
| **Magnesium chloride** $MgCl_2 \cdot 6H_2O$ MW = 203.3 | 119.5 ppm $Mg^{2+}$ 348.7 ppm $Cl^-$ | 31.6 ppm $Mg^{2+}$ 92.1 ppm $Cl^-$ | Dissolves readily. Lowers mash pH. Food-grade salt may not be high purity. |
| **Sodium chloride** NaCl MW = 58.4 | 393.4 ppm $Na^+$ 606.6 ppm $Cl^-$ | 103.9 ppm $Na^+$ 160.3 ppm $Cl^-$ | Dissolves readily. Avoid iodized salt and anti-caking agents. |

Notes: Adding NaCl is not recommended but the numbers are provided for your information. The other brewing salts are naturally hydrated with a few water molecules ($H_2O$) and this weight is taken into account in determining the ion contributions listed.

Let's think about the suggested sulfate-to-chloride ratio of 5:1 for an IPA. If we are adding 300 ppm of sulfate to the water, this suggests that the maximum concentration of chloride that we would want in the water would be 60 ppm. It is highly likely that there is already at least 60 ppm of chloride in the extract. Therefore, we shouldn't consider using any chloride salts when making additions to this beer.

### Oktoberfest Example for Adding Salts

To season an Oktoberfest beer, we want to enhance the maltiness by adding chloride. Looking at table 8.2, adding 1 g/gal. of calcium chloride adds 127 ppm of chloride. Assuming that there is already some chloride in the extract, 127 ppm is close to the recommended maximum of 150 ppm, so we will decrease the addition from 1 g/gal. to 0.67 g/gal., which equates to 2 g in 3 gal. This equates to $0.67 \times 127 = 85$ ppm of chloride. In using grams and liters, the addition of calcium chloride works out to 0.17 g/L.

## Summary

As we know from chapter 3, malt extract is concentrated wort. As such, most likely all of the minerals that the beer needs for good fermentation and flavor are right there in the extract. Therefore, you should use a low-mineral water, such as bottled spring water, or water that has been de-mineralized by distillation or reverse osmosis, to reconstitute the malt extract. If you know for a fact that your tap water is low in minerals and it tastes good, then you can use it. Chlorine flavors and aromas in brewing water are best removed by carbon filtration or metabisulfite tablets—especially if you're dealing with chloramine. The mineral profile of brewing water can be adjusted with brewing salts, if you know what's in your water to start with. In the USA at least, it should be possible to obtain a water quality report from your water utility company. If not, then you can test it yourself with a home testing kit or send a sample to a water testing laboratory. Adjusting water using brewing salts requires prior experience brewing with a specific malt extract or recipe to know how much you want to add.

# Brewing with a Full-Volume Boil

This chapter is going to walk through the full-volume boil brewing process, using a large kettle and outdoor propane burner. The Palmer Brewing Method for partial boils, as described in chapter 1, works very well on the kitchen stove, but boiling the full 5–6 gal. (19–23 L) requires bigger equipment and we will discuss that shortly (fig. 9.1).

Generally speaking, a full-volume boil is going to give you the best flavor in your beer, and the reason for that is simply because that's the way it has always been done. A full-volume boil develops the flavors that we have come to expect from our beers.

Is there any advantage to full-volume boils over the Palmer method, other than it being the "normal" way to do things? No, not really. The Palmer method matches the boil gravity of a full-volume boil, thereby avoiding the wort darkening and the development of off-flavors that can result from high-gravity, concentrated partial boils. Full-volume boils will generally have a little better hop utilization than partial boils due to boil gravity and kettle size, but the Palmer method mostly avoids underutilization by manipulating the boil gravity.

However, partial boils are really only applicable to brewing with malt extract. All-grain brewing requires that all of the wort be boiled, and you will usually be collecting 1–2 gal. (3–7 L) more wort than your batch size[1] to account for evaporation and trub loss. You will eventually want to try mashing and full-volume boils, so you may as well get started now by boiling full volume with all the extract.

---

[1]   A batch, in case it is not clear, is a brew session. The batch size is the nominal volume of beer after fermentation. Thus, we talk about 5 and 10 gal. (19 and 38 L) even though we may be brewing more wort to make that batch size.

Any recipe in this book can be converted to a full boil by starting with 6 gal. (23 L) of low-mineral water in the boil kettle and combining all of the malt extract from wort A and wort B. The steeping grains are steeped as usual.

The all-grain brewing process will be covered by a walk-through in chapter 20, "Your First All-Grain Batch," but the wort boiling process is essentially what is covered here.

## The Recipe

We are going to brew a porter, a clone recipe of Sierra Nevada's Porter, which has stood the test of time and continues to be one of my favorite recipes.

### "Port O'Palmer" Porter

**Original gravity:** 1.053  
**Final gravity:** 1.014  
**IBU:** 30

**SRM (EBC):** 24 (48)  
**ABV:** 5.2%

| Wort | Gravity points | |
|---|---|---|
| 6 lb. (2.7 kg) pale ale DME | 39 | |
| 0.55 lb. (250 g) caramel 80°L malt – Steep | 1 | |
| 0.55 lb. (250 g) aromatic (Munich 20°L) malt – Steep | 1.5 | |
| 0.55 lb. (250 g) chocolate (350°L) malt – Steep | 2 | |
| 0.33 lb. (150 g) Briess Black Prinz (500°L) malt – Steep | 1.5 | |
| Boil gravity for 6.5 gal. | 1.045 | |
| **Hop schedule** | **Boil time (min.)** | **IBU** |
| 0.5 oz. (15 g) Nugget 13% AA | 60 | 21 |
| 0.5 oz. (15 g) Cascade 6% AA | 30 | 7.5 |
| 0.5 oz. (15 g) East Kent Goldings 5% AA | Steep – 15 | 1.5 |
| **Yeast strain** | **Pitch (billions of cells)** | **Fermentation temperature** |
| English ale | 225 | 65°F (18°C) |

## Equipment Needed

### Kettles

Let's talk about boil kettles. Stainless steel kettles are the most popular choice because they are shiny, durable, and often have convenient accessories like thermometers, volume markings or sight glasses, and draining valves so you don't have to siphon your wort. A modern stainless steel homebrewing kettle is a bit of an investment, but it will last forever. A heavy-duty aluminum stockpot from a restaurant supply store won't have the accessories, but will work just as well as a boil kettle for about half the cost.

Let's talk about kettle sizes. Bigger is generally better, but there are limits. You are typically going to be boiling 6–7 gal. (23–26.5 L) of wort for a 5 gal. (19 L) batch of beer. A 7.5–8 gal. (28–30 L)

**Figure 9.1.** Full-volume boils require more heat and chilling. You will need to invest in a larger kettle, an outdoor propane or natural gas burner, and a wort chiller.

kettle works for malt extract brews. Malt extract has been pre-boiled and doesn't foam as much as an all-grain wort. If you are going to be boiling 7 gal. (26.5 L) of an all-grain wort, I suggest you get a 10 gal. (38 L) kettle to have the extra head space and reduce the likelihood of a boilover.

If you think that at some point you will want to brew 10 gal. (38 L) of beer instead of 5 gal. (19 L), then I suggest you get a 15 gal. (57 L) kettle. You *can* make a 5 gal. (19 L) batch in a 15 gal. (57 L) kettle, although it is a bit too large. In general, I recommend buying a kettle that is 5 gal. (19 L) larger than the batch size.

## Burners

Boiling large volumes of wort on the kitchen stove is just not practical. A standard large electric element on a kitchen stove is typically 2.6 kilowatts (kW), which equates to about 8,900 British thermal units (Btu) from burning propane or natural gas (methane). However, electric elements are more efficient at transferring heat to the kettle because of direct contact with the element. Much of the heat from a gas burner flows past the kettle and is lost to the air. A conservative estimate is that roughly half of the power of a gas burner is lost. However, gas burners still put out twice the heat of an electric stove, because most are at least 40,000 Btu (11.7 kW).

There are two basic shapes of gas burner heads. One is cone-shaped, looking like a showerhead with an insert that spreads the flame to the outside of the cone. The other burner head is a flattened ring, or disc, with lots of individual flame ports arranged in a spoke or concentric ring pattern.

Generally, the ring burners are better at distributing the flame is to the bottom of the kettle and more efficient for the amount of gas burned. The cone type works well, but they tend to use more propane to boil the same amount of wort than the disc type. A gas burner in the range of 50,000 to 100,000 Btu (14.7–29.3 kW) will work well for up to 20 gal. (76 L) batch sizes.

The final consideration is the size and strength of the burner stand, which must be sturdy enough to support a full boil without tipping over. Burners that were specifically designed for boiling large volumes of water, like the kind used for crawfish boils, should be sturdy enough; it's the cheap ones you have to worry about. See the sidebar, "Brew Beer, Not Turkeys." A high quality burner will last forever.

## BREW BEER, NOT TURKEYS

Don't be tempted to buy an aluminum turkey fryer combo from a big discount store. The kettle is really not big enough and the propane burners are often inefficient and underpowered. In addition, a full wort boil weighs twice as much as a turkey in oil. A turkey burner may not be strong enough to support a full wort boil without tipping over or collapsing.

**Figure 9.2.** Several examples of wort chillers.

## Chillers

Cooling large volumes of wort is the other side of the coin (fig. 9.2). The most common solution for homebrewers is the copper coil immersion chiller, where the coil is immersed in the wort at the end of the boil to sanitize it, and then cooling water is run through the coil to chill the wort. These immersion chillers can be made in several configurations, from single coil to multiple coils. They can be purchased at your homebrewing supply shop, or you can make them yourself with a few trips to the hardware store.

Copper wort chillers can also be made in the counterflow configuration, where the hot wort is run through the copper tubing and cooling water is run outside it. Typically, this is done by inserting the copper tubing inside of a garden hose. Complete plans for making these types of chillers are given in appendix D, "Building Wort Chillers."

The third type of chiller is the plate chiller, which are usually the most efficient in terms of the amount of water used to cool the wort. A plate chiller consists of thin plates brazed together in a stack so there is more surface area contact and less mass between the wort and water. The only issue with using plate chillers is to make sure you are adequately screening your hops and trub from the boil kettle to prevent clogging. They also require thorough internal cleaning and sanitizing after each use to prevent contamination of the next batch from anything trapped inside. Plate chillers are a little high-maintenance, but worth it for the efficiency.

One important thing I almost forgot to mention. . .

Immersion chillers chill the wort in your boil kettle. Counterflow and plate chillers chill the wort while it is being transferred to your fermentor. You can use counterflow and plate chillers with gravity, that is, the height difference between the kettle and fermentor, but they generally work better if you have an electric wort pump. These pumps are pretty common; they are centrifugal, impellor-driven pumps that are not self-priming. The wort has to drain from the kettle to the pump inlet, and the outlet has to go up from the pump to the fermentor to help clear air bubbles in the pump chamber, or it won't work well. Wort pumps are usually pumping hot wort so they don't need to be sanitized, but they do need to be disassembled and cleaned between uses.

## Fermentors

I am always annoyed to find out that my children have used my fermenting bucket to wash the dog. Plastic buckets make good fermentors, because they are big and cheap to replace when they are scratched or too stained to continue using. It used to be that plastic buckets and glass carboys (demijohns) were your only real options for fermentors, but today there is a lot more choice (fig. 9.3). There are plastic carboys and wide-mouth plastic jugs in 5, 6.5, and 8 gal. sizes (or 20, 25, and 30 L sizes). There are plastic or stainless steel conical fermentors; there are even stainless steel buckets! All of these options will work, the advantages of one over another is often a matter of how easy they are clean. Narrow-necked glass or plastic carboys are a real pain in that regard, often requiring soaking and a long carboy brush to scrub the inside. Having an easy to clean and sanitize drain valve or racking port to avoid having to siphon the wort to the bottling bucket or keg is very convenient as well.

Conical fermentors have grown in popularity with homebrewers for several reasons. The cone-shaped bottom directs all the sediment (trub and flocculated yeast) to the bottom of the cone where it can easily be discharged through the bottom dump valve. Trub will settle first during fermentation and so can be drawn off first, allowing clean yeast to be gathered at the end of fermentation. If you're planning on extended maturation (more than a month), or a secondary fermentation with fruit, the ability to remove the trub and yeast through the bottom dump valve eliminates the need to rack to a secondary fermentor, which reduces the possibility of oxidation or contamination. After fermentation, the beer itself is removed via a racking port on the side of the cone above the yeast, allowing sediment-free beer to be transferred into a bottling bucket or keg. The racking arm can also be used

**Figure 9.3.** A wide variety of fermentors are available these days.

to take wort samples during fermentation for measuring specific gravity. Dry hops or fruit can be added through the lid on top. Conical fermenters are available in either plastic or stainless steel and in a range of sizes. They can be quite expensive, but offer some advantages for yeast handling and repitching to subsequent batches.

## Brew Day

### Preparation

*1. Preparation.* Clean and sanitize your equipment (chapter 2). Assemble your ingredients. Weigh out and your bag of steeping grain. Is it crushed? Weigh out your hop additions. Do you have yeast? Think about how you are going to chill your wort after the boil. Do you have a chiller? Is it clean? Clean it now before you actually need it! Don't forget to clean and sanitize your fermentor, lid, and airlock too!

*2. Prepare the brewing water.* You will need 6 gal. (23 L) of fresh, clean, low-mineral water in the boil kettle. If the water has chlorine or chloramine, then add half of a crushed Campden tablet to covert the chlorine to chloride (chapter 8).

*3. Add malt extract.* Dissolve the 6 lb. (2.7 kg) of DME into the water in the boil kettle. Stir thoroughly to make sure all of it dissolves. If you are using 7.6 lbs. (3.5 kg) of LME instead, then take care because that tends to settle to the bottom of the kettle and scorch.

*4. Heat and steep.* Heat the wort to 150–170°F (65–75°C) for steeping. It's good to get in the habit of heating to the mashing temperature range even though specialty malts can be steeped at lower temperatures, especially when steeping kilned base malts like Munich and amber specialty malts that have residual starch. See chapter 15 for descriptions of the various types of malts.

Place the crushed specialty malts in the grain bag. There is nearly 2 lb. (900 g) of crushed grain in this recipe and you may need to use two bags. You want the grain to be loose in the bag for maximum extraction.

Steep the crushed grain for 30 min., occasionally stirring or moving the bags around to ensure good extraction. Remove and discard the grain afterwards.

## The Hot Break

5. *Watch the kettle for boilovers.* A foam will rise as the wort comes to a boil. This is good. If the foam suddenly billows over the side, this is a boilover. This is bad. If it looks like the wort is going to boil over, either lower the heat or spray the surface with water from a spray bottle (fig. 9.4). Putting a few copper pennies into the kettle to act as boil initiators will also help prevent boilovers. Anti-foaming additives are also available and they can help tame foaming in both the boil and in the fermentor.

The foam is caused by the proteins in the wort during the boil. The wort will continue to foam until the excess protein coagulates and sinks back into the kettle. You will see clumps of protein floating around in the wort, and it may look like egg-drop soup. This transition from foam to clumps is called the hot break, and usually takes 5–20 min. to occur,

**Figure 9.4**. Watch out for boilovers! Spray the surface with water from a spray bottle to beat down the foam.

depending on the amount of protein in your extract. The malt extract was already boiled once when it was made, so usually there is less foaming and hot break material in extract boils. In this particular boil, however, you are using more malt extract, so there will be more foam. The first hop addition often triggers a great deal of foaming, especially if hop pellets are used. I recommend waiting until the hot break occurs before you do the first hop addition and then start timing the hour (fig. 9.5). The extra boiling time won't hurt.

Covering the kettle with the lid can help with heat retention and help you achieve your boil, but it can also lead to trouble. Murphy's Law has its own brewing corollary: if it can boil over, it will boil over. Covering the kettle and turning your back on it is the quickest way to achieve a boilover. If you cover the kettle, watch it like a hawk. . . a hawk and ten buzzards even.

Once you achieve a boil, only partially cover the kettle, if at all. Why? Because in wort there are sulfur compounds that form and then boil off. If these compounds aren't removed during the boil, they will later form dimethyl sulfide, which contributes a cooked cabbage or cornlike flavor to the beer. If the cover is left on the kettle, or left on such that the condensate from the lid can drip back in, then these flavors will have a much greater chance of showing up in the beer.

**Figure 9.5.** Add your first hop addition after the hot break has occurred and the wort has stopped foaming.

## Hop Additions

6. *First hop addition (t = −60 min.).* Once the hot break has occurred, add the 0.5 oz. (15 g) of Nugget hops. Start the 60 min. countdown. The wort may foam up again for a short time. Maintain a rolling boil; a rolling boil is one that has visible terrain at the surface with high and low areas. It is more than a simmer, but the wort is not spitting and splashing out of the pot either.

7. *Second hop addition (t = −30 min.).* Add the 0.5 oz. (15 g) of Cascade hops 30 min. before the end of the boil. Continue the rolling boil and stir occasionally. There will be clumps of stuff floating in the wort. This is not a concern, it's just the hot break material (i.e., coagulated and precipitated protein). There may be foam or protein floating around on top or clinging to the sides of the kettle. This may be skimmed away if you wish. It won't hurt anything, but it will contribute to the trub later on.

8. *Have you prepared your yeast?* I thought this would be a good time to remind you, since you are just standing watching the kettle boil. Did you prepare a liquid yeast starter a couple of days ago, and is it ready to pitch? Or are you going to use dry yeast? If so, you will want to rehydrate it soon. Chapter 7 has more information on preparing your yeast and pitching rates.

## Kettle Finings

9. *Add Irish moss (t = −5 min.).* Add the kettle finings. Irish moss is a seaweed containing carrageenans, a group of long-chain polysaccharides primarily composed of galactose. All kettle finings are based on carrageenans, and they have long been used to increase the amount of cold break material in wort and improve beer clarity. A common commercial example is Whirlfloc®. Stir the wort for a minute to help dissolve and disperse the finings. Do not boil longer than 5 minutes.

Fining agents containing carrageenans should be added at the end of the boil, because boiling carrageenans for more than a couple of minutes will denature the molecular structure and make them less effective at binding haze-active proteins (i.e., haze-forming proteins) in the wort. Note that adding too little or too much Irish moss (or Whirlfloc) will decrease its effectiveness and result in a condition known as "fluffy bottoms," where sediment tends to swirl up easily in the beer bottle. Follow the manufacturer's instructions for dosage rates. Irish moss helps beer clarity, but it is not essential. Don't worry if you don't have any. Malt extract generally has less protein and therefore less break material than fresh wort, because the malt was boiled during manufacture of the extract.

## Hop Steeping (a.k.a. Whirlpool) Addition

*10. Third hop addition (t= 0 min.).* Turn off the heat and add the third and final hop addition of 0.5 oz. (15 g) of East Kent Goldings at the end of the boil. Stir to break up and disperse the hops. Let the hops steep for 15 min. to absorb the oils from this last addition before chilling.

**Figure 9.6.** Chilling is active, cooling is passive. Get involved—chill!

## Chilling the Wort

*11. Chill the wort.* At the end of the boil, it is best to cool the wort quickly (fig. 9.6). First, because handling hot wort is dangerous, and second, because it is convenient—quick cooling allows you to get on with your brew day. Only once your wort has cooled to your fermentation temperature can you can pitch your yeast and be done.

My tap water is not cold enough for my wort chiller to chill the wort to pitching temperature—I can usually get down to about 85°F (30°C). I could chill lower by using an immersion pre-chiller in a bucket of ice water before that water goes to the chiller in the wort, which is what many people do. What I usually do though, is seal the fermentor and put it in the refrigerator overnight to finish chilling. I will aerate and pitch the next day.

But, let's assume you have colder tap water than I do and have successfully chilled to your fermentation temperature.

## MURPHY'S LAWS OF BREWING

If it can boil over, it will boil over.

The fermentor will be either too hot or too cold.

The weather will change.

Nature always sides with the hidden flaw.

If you don't have time to do it right, you will probably end up doing it over.

The race is not always to the swift, or the battle to the strong, but that's the way to bet. (In other words, the most conscientious brewer may not win all the time, but he or she will probably win most of the time.)

If you keep messing with it, you will probably screw it up.

### Transferring from Kettle to Fermentor

You may have been wondering what to do about all the hops and trub in the bottom of the kettle before you transfer the wort to the fermentor. The hot and cold break material consists of various proteins and fatty acids that can serve as yeast nutrients in moderate amounts. Too much break material can promote off-flavors and staling after fermentation; too little and the yeast will be stressed. Generally, it is good to strain most of the hops and at least some trub from getting into the fermentor using a common wire mesh kitchen strainer. Strain, do not filter. The strained wort should be fairly cloudy or turbid with break material, but you don't need all of it in the fermentor.

*12. Separate the wort from the trub.* In large boils, the most common method for separating the wort from the break material is to whirlpool it to the center and let it settle, allowing clear wort to be drawn off from the side. Rapidly stir the wort in a circle. Continue stirring until all the wort and trub is moving and a small vortex forms. Stop stirring and let the whirlpool slow down and settle for 10 min. or so. The hops and trub will form a pile in the center of the kettle, leaving the edge relatively clear. The wort can now be drawn off with a siphon or side spigot and will leave most of the trub behind. A mesh strainer on the pickup tube or siphon will also help prevent hops from clogging your counterflow or plate chiller.

### Aerate the Chilled Wort and Pitch the Yeast

*13. Aerate the wort.* After the wort has cooled to pitching temperature, it is time to aerate it. The best way to aerate your wort is to use an airstone and an air pump with a HEPA filter. See chapter 7, "Yeast Management," for more information.

*14. Pitch the yeast.* Pitch (pour) the yeast into the fermentor, making sure to add it all. Always pitch from cooler to warmer; the yeast should be cooler than the wort, and the wort should be at the intended fermentation temperature. You never want to pitch from warmer to cooler, because this may shock the yeast and cause it to stop working. See chapter 6, "Yeast and Fermentation," for a thorough discussion of fermentation temperature.

The airlock should be filled to the line with fresh clean water. Many people use vodka or sanitizer solution as alternatives, but it is not necessary. You just want something that will not grow mold or contaminate the batch in case it is inadvertently sucked into the fermentor.

### Conducting Your Fermentation—Quick Review

Active fermentation should start within 12 hours, but it can take up to 24 hours due to lower pitching rates or cooler temperatures. The airlock will bubble regularly. The fermentation activity can be vigorous or slow, either of which is fine. The three important factors for a successful fermentation are pitching enough yeast, good wort nutrients, and maintaining a consistent temperature in the correct range. If you do these right, it is not unheard of for the attenuation phase of an ale fermentation to be done in 48 hours. Three days at 65–70°F (18–21°C) for the attenuation phase is more typical for the porter recipe being described here.

If you are conducting an open fermentation, look for when the kräusen starts to diminish. This would be the time to seal the lid and insert the airlock for the maturation phase. If the fermentation was closed from the beginning, then just sit back and wait. It will help maturation if you can raise the fermentation temperature 5°F (3°C). Allow the beer to mature and condition in the fermentor for two weeks before bottling (fig. 9.7).

Next, in chapter 10, we will discuss how the brewing and fermenting of lager differs from ale. Then, in chapter 11, we will prepare to prime, bottle, and ultimately consume our beer.

**Figure 9.7.** Here is an example of a fermentor with a blow-off tube in a temperature controlled refrigerator.

## ESTIMATING THE ALCOHOL CONTENT

How much alcohol will there be? This is a common question. While there are various laboratory techniques that can be employed to determine it precisely, there is a simple way to estimate it. The easiest is to use a "triple scale" hydrometer, which has a percent alcohol by volume (ABV) scale right on it. You subtract the respective percentages that correspond to your OG and FG, and there you have it.

If you don't have this type of hydrometer, the following table (table 9.1), based on the work of Balling, should satisfy your curiosity. Find the intersection of your OG and FG to read your estimated percent ABV.

## TABLE 9.1—PERCENT ALCOHOL BY VOLUME FROM SPECIFIC GRAVITY HYDROMETER READINGS

| FG | OG | | | | | | | | | |
|---|---|---|---|---|---|---|---|---|---|---|
| | 1.030 | 1.035 | 1.040 | 1.045 | 1.050 | 1.055 | 1.060 | 1.065 | 1.070 | 1.075 |
| 0.998 | 4.1 | 4.8 | 5.4 | 6.1 | 6.8 | 7.4 | 8.1 | 8.7 | 9.4 | 10.1 |
| 1.000 | 3.9 | 4.5 | 5.2 | 5.8 | 6.5 | 7.1 | 7.8 | 8.5 | 9.1 | 9.8 |
| 1.002 | 3.6 | 4.2 | 4.9 | 5.6 | 6.2 | 6.9 | 7.5 | 8.2 | 8.9 | 9.5 |
| 1.004 | 3.3 | 4.0 | 4.6 | 5.3 | 5.9 | 6.6 | 7.3 | 7.9 | 8.6 | 9.3 |
| 1.006 | 3.1 | 3.7 | 4.4 | 5.0 | 5.7 | 6.3 | 7.0 | 7.7 | 8.3 | 9.0 |
| 1.008 | 2.8 | 3.5 | 4.1 | 4.8 | 5.4 | 6.1 | 6.7 | 7.4 | 8.0 | 8.7 |
| 1.010 | 2.6 | 3.2 | 3.8 | 4.5 | 5.1 | 5.8 | 6.5 | 7.1 | 7.8 | 8.4 |
| 1.012 | 2.3 | 2.9 | 3.6 | 4.2 | 4.9 | 5.5 | 6.2 | 6.8 | 7.5 | 8.2 |
| 1.014 | 2.0 | 2.7 | 3.3 | 4.0 | 4.6 | 5.3 | 5.9 | 6.6 | 7.2 | 7.9 |
| 1.016 | 1.8 | 2.4 | 3.1 | 3.7 | 4.4 | 5.0 | 5.7 | 6.3 | 7.0 | 7.6 |
| 1.018 | 1.5 | 2.2 | 2.8 | 3.4 | 4.1 | 4.7 | 5.4 | 6.0 | 6.7 | 7.3 |
| 1.020 | 1.3 | 1.9 | 2.5 | 3.2 | 3.8 | 4.5 | 5.1 | 5.8 | 6.4 | 7.1 |
| 1.022 | 1.0 | 1.6 | 2.3 | 2.9 | 3.6 | 4.2 | 4.9 | 5.5 | 6.2 | 6.8 |
| 1.024 | 0.8 | 1.4 | 2.0 | 2.7 | 3.3 | 4.0 | 4.6 | 5.2 | 5.9 | 6.5 |

FG, final gravity; OG, original gravity.

# Priming, Bottling, and Kegging 10

In this chapter we will focus on getting your hard-won beer into a bottle and ready for drinking. To bottle your beer you will need clean bottles, bottle caps, a bottle capper, and a bottling bucket or a siphon with a bottle filler (fig. 10.1). You will also need some sugar to use for priming—that extra bit of fermentable sugar that is added to the beer at bottling time to provide the carbonation.

## When to Bottle

Beer should be bottled when it has finished fermenting, which should be after it has fully attenuated and matured as described in chapter 6, "Yeast and Fermentation."

Ales of moderate strength are usually ready to bottle two to three weeks after pitching, when fermentation has completely finished. Lagers should only take a little bit longer if they were done right, that is, three to four weeks after pitching. There should be few, if any, bubbles coming through the airlock. If your fermentor is glass, you will see the beer darken and clear as the yeast flocculates. Although two or three weeks may seem like a long time to wait, the flavor won't improve by bottling any earlier. Some older books recommend bottling after the bubbling stops, or in about one week,

**Figure 10.1.** Bottling equipment.

but this is usually bad advice. It is not uncommon for fermentation to stop after 3–4 days and then begin again a few days later due to a temperature change. If the beer is bottled before fermentation is complete, the beer will become overcarbonated and the pressure may exceed the bottle strength. Exploding bottles are a disaster (and messy to boot).

## Bottle Cleaning

Many homebrewers get used bottles from restaurants and bars, or buy them new from homebrew shops. You can also use flip-top bottles (the "Grolsch-style" bottles, also called swing-tops), which are nice to have for their easy-close, easy-open reusability.

If you are going to re-use old bottles, they need to be cleaned thoroughly before sanitizing. Soak them in a cleaning solution and scrub inside and out with a nylon bottle brush. Heavy duty cleaning is needed to ensure there are no deposits in which bacteria or mold spores can hide. After the bottles have been cleaned with a brush, soak them in sanitizing solution or use the dishwasher with the heat cycle to sanitize them. The sanitizing solution needs to reach all surfaces for it to work. If you use an iodophor solution to sanitize, allow the bottles to drain upside down on a rack, or rinse them. Some no-rinse sanitizers are no-rinse because the residue doesn't have residual flavors; others like iodophor are no-rinse if they fully evaporate.

If you are diligent in rinsing your bottles promptly and thoroughly after each use, only the sanitizing treatment will be necessary before each use in the future. By maintaining clean equipment you will save yourself a lot of work.

Also sanitize the priming container, siphon unit, stirring spoon, and bottle caps. But do not boil or bake the bottle caps, as this may ruin the gaskets.

## Priming

I generally don't plan precise carbonation levels for my beers; medium carbonation seems to work pretty well for everything. However, as a beer judge, I know that the carbonation level of a beer can have a profound effect on flavor perception. Low carbonation levels can make a beer seem sweeter and smoother, whereas too low can make it seem flabby and underattenuated. High carbonation levels can make a beer crisper or more assertive, and add clarity between the foretaste and finish, but too much carbonation can make hop bitterness overly aggressive and harsh. This chapter will give you the tools you need to plan the carbonation for your beers, but you need to understand that there are several variables at play when trying to calculate the precise amount of priming sugar to get (for example) 3.0 volumes of carbon dioxide ($CO_2$). See the sidebar, "Volumes of $CO_2$ per Gram of Sugar."

Any sugar—white cane sugar, beet sugar, palm sugar, brown sugar, honey, molasses, even maple syrup—can be used for priming. The darker sugars can contribute a subtle aftertaste (sometimes desired) and are more appropriate for heavier, darker beers. Simple sugars, like corn sugar (glucose) or table sugar (sucrose, refined from either sugar cane or beets), are used most often, although many brewers use dry malt extract (DME) too. Weight for weight, cane sugar generates a bit more $CO_2$ than corn sugar, and both cane and corn sugar carbonate more than DME, so you will need to take that into account. For all priming in general, you want to add 2–3 gravity points of sugar to the beer, although the exact amount depends on several factors that will be discussed later in this chapter.

The priming sugar is only a small percent of the total fermentables in the beer, and has a correspondingly small effect on the flavor of the beer. Most people use corn sugar or ordinary white table sugar (i.e., cane or beet sugar), because they are cheap and easy, and don't change the flavor.

You can prime your beer with any fermentable that you want. Sucrose-based sugars, such as invert sugar syrups, honey, or maple syrup, may add a small accent to the beer, but if you are really looking for a particular flavor, then it probably should be included as one of the main wort ingredients. Brewing purists may insist on using wort or malt extract to avoid any possible flavor change, but I don't think that is necessary.

The big question is, "How much do I use?" The most commonly quoted answer to this question is ¾ cup (roughly 4.7 oz., or 130 g) of corn sugar for a 5 gal. (19 L) batch. This will produce about 2.5 volumes of $CO_2$ in the beer, which is pretty typical of most American and European pale ales. Three ounces (85 g) in the same batch will produce a lower carbonation level of about 2.0 volumes, and 6 oz. (170 g) will produce about 3.0 volumes.

Equivalent amounts of different priming sugars are listed in table 10.1. These amounts take into account the $CO_2$ contribution for each type of saccharide in the priming sugar, the relative percentages of each type of saccharide in the priming sugar, and the total weight percent of fermentable saccharides in the priming sugar. See the "Volumes of $CO_2$ per Gram of Sugar" sidebar for an explanation of the amount of $CO_2$ produced by each type of sugar.

### Residual $CO_2$, Temperature, and Pressure

Use tables 10.2 and 10.3 to determine how much priming sugar to use in ounces per gallon or grams per liter, respectively, but be prepared for a small degree of variability in the results. You may want to use a little more or a little less than the tables indicate due to your altitude.

Keep in mind these numbers are based on a model that estimates how much $CO_2$ is still in solution when primed (called the residual $CO_2$), and how much $CO_2$ the yeast actually produces from the added sugar, all as a function of temperature and pressure. You should understand that the residual amount of $CO_2$ in the beer is the equilibrium volume for the temperature (and atmospheric pressure) at the end of fermentation; all the rest has vented through the airlock. The end of fermentation means exactly that—no more sugar is being fermented and no $CO_2$ is being generated by the yeast. Raising the beer temperature after fermentation will cause more $CO_2$ to vent through the airlock, and lower the residual. Lowering the temperature after fermentation will not generate more $CO_2$ in solution nor will it pull more $CO_2$ into solution, because the amount in the head space is already at equilibrium with the atmospheric pressure from the higher temperature. Lowering the temperature after that point will only pull outside air in through the airlock, which has a much lower partial pressure of $CO_2$.

Therefore, the amount of residual $CO_2$ that is in the beer before priming is given by the highest temperature after fermentation, and that amount is what you are adding to when you prime the beer to reach your target carbonation level. If you live at high elevations, you can assume the residual is lower due to the lower atmospheric pressure. The residual $CO_2$ and gas constants across a typical temperature range for priming beer are given in tables 10.2 and 10.3. These tables list the calculated amounts of table sugar (sucrose) to add per volume beer in ounces per gallon (table 10.2) and grams per liter (table 10.3) to achieve the desired volumes of $CO_2$.

## VOLUMES OF $CO_2$ PER GRAM OF SUGAR

The ideal gas law states that

$PV = nRT$, where:

$P$ is the absolute pressure of a gas, $V$ is the volume the gas occupies, $n$ is the amount of gas (in moles), $R$ is the gas constant, and $T$ is the absolute temperature.

At standard temperature and pressure, 1 mole of gas occupies 22.4 L of volume, hence 22.4 L/mole. However, this value changes with temperature; it's 22.4 L/mole at 32°F (0°C; the defined temperature for standard temperature and pressure). At 70°F (21°C), the gas constant is 24.5 L/mole. Using this number and the molar mass in grams of each of the saccharides, we can calculate the weight of a saccharide needed to generate one volume of $CO_2$ (in liters) in one liter of beer (hence, liters $CO_2$/liter beer. A typical carbonation level of 2.5 volumes would have 2.5 L of $CO_2$ dissolved into 1 L of beer. The residual $CO_2$ in beer at 70°F (21°C) is 0.8 volumes.

Brewer's yeast produces approximately 2 moles $CO_2$ per mole of corn sugar (glucose) it ferments. Glucose is a monosaccharide (i.e., one sugar unit), as is fructose, and yeast ferments them the same. Sucrose (table sugar) and maltose (brewing sugar) are disaccharides, and these are hydrolyzed (split into their constituent monosaccharides) by yeast into glucose and fructose before fermentation. Therefore, yeast produces 4 moles of $CO_2$ per mole of disaccharide sugar. Maltotriose, a trisaccharide and the largest sugar that yeast ferments, produces 6 moles of $CO_2$ per mole when fermented. Knowing this, we can now calculate how much of each type of saccharide is needed to produce one volume of $CO_2$.

For 70°F (21°C) where gas volume is 24.5 L/mol:

$$\left(\frac{\text{molar mass saccharide}}{\text{moles } CO_2 \text{ per mole saccharide}}\right)/24.5 = X \text{ g sugar per volume.}$$

Monosaccharide (e.g., glucose):

$(180.16/2)/24.5 = 3.68$ g per volume.

Disaccharide (e.g., sucrose):

$(342.3/4)/24.5 = 3.49$ g per volume.

Trisaccharide (e.g., maltotriose):

$(504.4/6)/24.5 = 3.43$ g per volume.

However, corn sugar (i.e., glucose, commonly sold as dextrose monohydrate in homebrew stores) has 9% moisture by weight, so its contribution must be divided by the proportion of solids, in other words $3.68/91\% = 4.04$ g per volume.

The equation for calculating the required weight (in grams) of a priming sugar that is a mixture of different types of sugar is:

$$\text{weight (g)} = (\text{target } CO_2 - \text{residual } CO_2) \times (\text{liters of beer}) \times (1/\% \text{ solids}) \times (1/\% \text{ fermentability})$$
$$\times [(\% \text{ monosacch.} \times 3.68) + (\% \text{ disacch.} \times 3.49) + (\% \text{ trisacch.} \times 3.43)],$$

where "% solids" is the opposite of the percentage moisture by weight, and "% fermentability" is the percent solids that is actually fermentable.

For example, liquid malt extract (LME) is about 80% solids and 75% of those solids are fermentable sugar; the rest of the solids are unfermentable carbohydrates, proteins, and lipids. The proportions of the fermentable sugars in LME are roughly 13.5% glucose and fructose, 68% maltose and sucrose, and 18.5% maltotriose. To calculate the amount of LME needed to produce 2.5 volumes $CO_2$ in a 19 L batch of beer, we plug these values into the equation above, so:

$$\text{weight LME} = (2.5 - 0.8) \times 19 \times (1/80\%) \times (1/75\%) \times [(13.5\% \times 3.68) + (68\% \times 3.49)$$
$$+ (18.5\% \times 3.43)]$$

weight LME = 189 g (6.75 oz.)

The sugar proportions for DME are the same, only the percent solids (% solids) value changes.

## TABLE 10.1—PRIMING SUGAR INFORMATION AND EQUIVALENT PRIMING QUANTITIES[a]

| Priming sugar | Percent solids[b] | Extract yield PPG (PKL) | Fermentability | Weight oz. per 5 gal. (g per 19 L) |
|---|---|---|---|---|
| Cane sugar | 100% | 46 (384) | 100% | 4.0 (113) |
| Corn sugar | 91% | 42 (350) | 100% | 4.7 (131) |
| Brown sugar | 95%[c] | 44 (367) | 97%[c] | 4.35 (122) |
| Molasses/treacle | 80% | 36 (300) | 50% | 10.25 (287) |
| Lyle's Golden Syrup® | 82% | 38 (317) | 100% | 5.1 (144) |
| Maple syrup | 67% | 31 (259) | 100% | 6.65 (186) |
| Honey | 80% | 38 (317) | 95% | 5.4 (152) |
| LME | 80% | 36 (300) | 75%[c] | 6.75 (189) |
| DME | 90% | 42 (350) | 75%[c] | 6.0 (168) |

PKL, gravity points per kilogram per liter; PPG, gravity points per pound per gallon.

[a] Priming quantities calculated for 2.5 volumes $CO_2$ in 5 gal. (19 L) batch at 1 atm. (101.3 kPa) and 70°F (21°C).

[b] This is the inverse to percent moisture, it is not the same as percent moisture.

[c] Values are approximate.

## Making the Priming Solution

The best way to prime your beer is to mix your priming sugar into the whole batch prior to bottling. This ensures that all the bottles will be carbonated the same. Older books used to recommend adding 1 teaspoon of sugar directly to the bottle for priming. This is time consuming and imprecise, and bottles may carbonate unevenly and explode.

Bulk priming by adding a solution of dissolved priming sugar to the entire batch has the advantage of achieving a consistent carbonation across all the bottles compared to pouring a spoonful of sugar into each one. However, there are several brands of off-the-shelf priming agents available that can take some of the preparation out of the priming and bottling process. Some of these are canned syrups that you pour into the bottling bucket, others are tablets or drops that you add to the bottle. Whichever you use, be sure to follow the directions carefully to avoid exploding bottles.

Here's how to make and add priming solutions for 5 gal. (19 L) of beer:

1. Select your priming sugar (table 10.1 is a useful reference). Boil that amount of sugar in about 2 cups (500 mL) of water and let it cool. You can add the priming solution in either of two ways, depending on your equipment; I prefer the first option (2a) of the two described below.

2a. If you have a bottling bucket gently pour the priming solution into it. Using a sanitized siphon, transfer the beer into the sanitized bottling bucket (fig. 10.2). Place the outlet beneath the surface of the priming solution. Do not allow the beer to splash into the bottling bucket, because you don't want to add oxygen to your beer at this point. Keep the intake end of the racking tube an inch off the bottom of the fermentor (most racking canes have an end cap that does this for you) to leave the yeast and sediment behind. Stir the primed beer thoroughly after filling the bottling bucket; you want it to be thoroughly mixed but remember not to splash or aerate the beer.

2b.  If you don't have a bottling bucket, open the fermentor and gently pour the priming solution into the beer. Stir the beer gently but thoroughly with a sanitized spoon, trying to mix it in evenly while being careful not to stir up the sediment too much. Wait a half hour for the sediment to settle back down and to allow more diffusion of the priming solution to take place. Use a bottle filler attachment with the siphon to make the filling easier.

**Figure 10.2.** Racking to the bottling bucket.

The numbers presented in tables 10.2 and 10.3 will help you determine more precise amounts of table sugar for priming, based on the final fermentation temperature or the priming temperature of the beer, whichever is higher, because this temperature quantifies how much $CO_2$ is left in the beer before priming (i.e., the residual $CO_2$). To use the tables, find the row for the highest temperature of the beer, and follow it across to the column for the volumes of $CO_2$ you want in the beer after priming. The number listed is the priming weight in ounces per gallon (table 10.2) or grams per liter (table 10.3); multiply that number by the actual final volume of your beer in gallons or liters, respectively, to get the total weight of priming sugar.

The values in tables 10.2 and 10.3 are calculated based on an elevation of 330 ft. (100 m) above sea level, which is where the majority of the world's population lives. If you live at higher elevations, the residual $CO_2$ will be less. For example, if you live in Denver, Colorado, which has an elevation of around 5,200 ft. (1585 m), the residual will be about 0.15 volumes less than the amount given.

Tables 10.2 and 10.3 are for weights of sucrose (table sugar) rather than dextrose (corn sugar) because table sugar is more readily available and it doesn't absorb significant amounts of water. The conversion factor from table sugar to corn sugar is 1.16. Multiply the sucrose weight in table 10.2 by 1.16 to get the equivalent weight of corn sugar.

Here is a list of typical volumes of $CO_2$ for various beer styles:

- British ales                1.5–2.0
- Porter and stout            1.7–2.3
- Belgian ales                1.9–2.4
- American ales               2.2–2.7
- European lagers             2.2–2.7
- Belgian lambic              2.4–2.8
- American wheat beer         2.7–3.3
- German wheat beer           3.3–4.5

## TABLE 10.2—OUNCES OF TABLE SUGAR (SUCROSE) PER GALLON OF BEER TO REACH THE INDICATED TOTAL VOLUMES OF CARBON DIOXIDE, BASED ON THE FINAL FERMENTATION TEMPERATURE.

| Temperature | | Gas volume constant | Volumes residual $CO_2$ | Target carbonation in volumes of $CO_2$ | | | | | | |
|---|---|---|---|---|---|---|---|---|---|---|
| °F | °C | (L/mol) | | 1.5 | 2.0 | 2.5 | 3.0 | 3.5 | 4.0 | 4.5 |
| 41 | 5 | 23.1 | 1.32 | 0.09 | 0.34 | 0.59 | 0.83 | 1.08 | 1.33 | 1.58 |
| 43 | 6 | 23.2 | 1.27 | 0.11 | 0.36 | 0.60 | 0.85 | 1.10 | 1.34 | 1.59 |
| 45 | 7 | 23.3 | 1.23 | 0.13 | 0.38 | 0.62 | 0.87 | 1.11 | 1.36 | 1.61 |
| 46 | 8 | 23.4 | 1.19 | 0.15 | 0.39 | 0.64 | 0.88 | 1.13 | 1.37 | 1.62 |
| 48 | 9 | 23.5 | 1.16 | 0.17 | 0.41 | 0.66 | 0.90 | 1.14 | 1.39 | 1.63 |
| 50 | 10 | 23.5 | 1.12 | 0.18 | 0.43 | 0.67 | 0.91 | 1.16 | 1.40 | 1.64 |
| 52 | 11 | 23.6 | 1.08 | 0.20 | 0.44 | 0.69 | 0.93 | 1.17 | 1.41 | 1.65 |
| 54 | 12 | 23.7 | 1.05 | 0.22 | 0.46 | 0.70 | 0.94 | 1.18 | 1.42 | 1.67 |
| 55 | 13 | 23.8 | 1.02 | 0.23 | 0.47 | 0.71 | 0.95 | 1.19 | 1.44 | 1.68 |
| 57 | 14 | 23.9 | 0.99 | 0.25 | 0.49 | 0.73 | 0.97 | 1.21 | 1.45 | 1.69 |
| 59 | 15 | 24.0 | 0.96 | 0.26 | 0.50 | 0.74 | 0.98 | 1.22 | 1.45 | 1.69 |
| 61 | 16 | 24.0 | 0.93 | 0.27 | 0.51 | 0.75 | 0.99 | 1.23 | 1.46 | 1.70 |
| 63 | 17 | 24.1 | 0.90 | 0.29 | 0.52 | 0.76 | 1.00 | 1.23 | 1.47 | 1.71 |
| 64 | 18 | 24.2 | 0.87 | 0.30 | 0.53 | 0.77 | 1.01 | 1.24 | 1.48 | 1.72 |
| 66 | 19 | 24.3 | 0.84 | 0.31 | 0.54 | 0.78 | 1.02 | 1.25 | 1.49 | 1.72 |
| 68 | 20 | 24.4 | 0.82 | 0.32 | 0.55 | 0.79 | 1.02 | 1.26 | 1.49 | 1.73 |
| 70 | 21 | 24.5 | 0.80 | 0.33 | 0.56 | 0.80 | 1.03 | 1.27 | 1.50 | 1.73 |
| 72 | 22 | 24.5 | 0.77 | 0.34 | 0.57 | 0.80 | 1.04 | 1.27 | 1.50 | 1.74 |
| 73 | 23 | 24.6 | 0.75 | 0.35 | 0.58 | 0.81 | 1.04 | 1.28 | 1.51 | 1.74 |
| 75 | 24 | 24.7 | 0.73 | 0.36 | 0.59 | 0.82 | 1.05 | 1.28 | 1.51 | 1.75 |
| 77 | 25 | 24.8 | 0.72 | 0.36 | 0.59 | 0.82 | 1.06 | 1.29 | 1.52 | 1.75 |
| 79 | 26 | 24.9 | 0.70 | 0.37 | 0.60 | 0.83 | 1.06 | 1.29 | 1.52 | 1.75 |
| 81 | 27 | 25.0 | 0.68 | 0.38 | 0.60 | 0.83 | 1.06 | 1.29 | 1.52 | 1.75 |
| 82 | 28 | 25.0 | 0.67 | 0.38 | 0.61 | 0.84 | 1.07 | 1.30 | 1.52 | 1.75 |
| 84 | 29 | 25.1 | 0.65 | 0.39 | 0.61 | 0.84 | 1.07 | 1.30 | 1.53 | 1.75 |
| 86 | 30 | 25.2 | 0.64 | 0.39 | 0.62 | 0.84 | 1.07 | 1.30 | 1.53 | 1.75 |

Notes: "Volumes of $CO_2$" are liters of $CO_2$ gas dissolved in 1 L of beer. Multiply the indicated ounces by 28.3 to calculate the weight in grams per gallon. Values for gas volume constant, residual $CO_2$, and priming amounts based on 330 ft. (100 m) elevation above sea level.

## TABLE 10.3—GRAMS OF TABLE SUGAR (SUCROSE) PER LITER OF BEER TO REACH THE INDICATED TOTAL VOLUMES OF CARBON DIOXIDE, BASED ON THE FINAL FERMENTATION TEMPERATURE.

| Temperature | | Gas volume constant | Volumes residual $CO_2$ | Target carbonation in volumes of $CO_2$ | | | | | | |
|---|---|---|---|---|---|---|---|---|---|---|
| °F | °C | (L/mol) | | 1.5 | 2.0 | 2.5 | 3.0 | 3.5 | 4.0 | 4.5 |
| 41 | 5 | 23.1 | 1.32 | 0.7 | 2.5 | 4.4 | 6.2 | 8.1 | 9.9 | 11.8 |
| 43 | 6 | 23.2 | 1.27 | 0.8 | 2.7 | 4.5 | 6.4 | 8.2 | 10.1 | 11.9 |
| 45 | 7 | 23.3 | 1.23 | 1.0 | 2.8 | 4.7 | 6.5 | 8.3 | 10.2 | 12.0 |
| 46 | 8 | 23.4 | 1.19 | 1.1 | 2.9 | 4.8 | 6.6 | 8.4 | 10.3 | 12.1 |
| 48 | 9 | 23.5 | 1.16 | 1.3 | 3.1 | 4.9 | 6.7 | 8.5 | 10.4 | 12.2 |
| 50 | 10 | 23.5 | 1.12 | 1.4 | 3.2 | 5.0 | 6.8 | 8.7 | 10.5 | 12.3 |
| 52 | 11 | 23.6 | 1.08 | 1.5 | 3.3 | 5.1 | 6.9 | 8.7 | 10.6 | 12.4 |
| 54 | 12 | 23.7 | 1.05 | 1.6 | 3.4 | 5.2 | 7.0 | 8.8 | 10.6 | 12.5 |
| 55 | 13 | 23.8 | 1.02 | 1.7 | 3.5 | 5.3 | 7.1 | 8.9 | 10.7 | 12.5 |
| 57 | 14 | 23.9 | 0.99 | 1.8 | 3.6 | 5.4 | 7.2 | 9.0 | 10.8 | 12.6 |
| 59 | 15 | 24.0 | 0.96 | 1.9 | 3.7 | 5.5 | 7.3 | 9.1 | 10.9 | 12.7 |
| 61 | 16 | 24.0 | 0.93 | 2.0 | 3.8 | 5.6 | 7.4 | 9.2 | 10.9 | 12.7 |
| 63 | 17 | 24.1 | 0.90 | 2.1 | 3.9 | 5.7 | 7.5 | 9.2 | 11.0 | 12.8 |
| 64 | 18 | 24.2 | 0.87 | 2.2 | 4.0 | 5.8 | 7.5 | 9.3 | 11.1 | 12.8 |
| 66 | 19 | 24.3 | 0.84 | 2.3 | 4.1 | 5.8 | 7.6 | 9.4 | 11.1 | 12.9 |
| 68 | 20 | 24.4 | 0.82 | 2.4 | 4.1 | 5.9 | 7.7 | 9.4 | 11.2 | 12.9 |
| 70 | 21 | 24.5 | 0.80 | 2.5 | 4.2 | 6.0 | 7.7 | 9.5 | 11.2 | 13.0 |
| 72 | 22 | 24.5 | 0.77 | 2.5 | 4.3 | 6.0 | 7.8 | 9.5 | 11.2 | 13.0 |
| 73 | 23 | 24.6 | 0.75 | 2.6 | 4.3 | 6.1 | 7.8 | 9.5 | 11.3 | 13.0 |
| 75 | 24 | 24.7 | 0.73 | 2.7 | 4.4 | 6.1 | 7.9 | 9.6 | 11.3 | 13.0 |
| 77 | 25 | 24.8 | 0.72 | 2.7 | 4.4 | 6.2 | 7.9 | 9.6 | 11.3 | 13.1 |
| 79 | 26 | 24.9 | 0.70 | 2.8 | 4.5 | 6.2 | 7.9 | 9.6 | 11.4 | 13.1 |
| 81 | 27 | 25.0 | 0.68 | 2.8 | 4.5 | 6.2 | 8.0 | 9.7 | 11.4 | 13.1 |
| 82 | 28 | 25.0 | 0.67 | 2.8 | 4.6 | 6.3 | 8.0 | 9.7 | 11.4 | 13.1 |
| 84 | 29 | 25.1 | 0.65 | 2.9 | 4.6 | 6.3 | 8.0 | 9.7 | 11.4 | 13.1 |
| 86 | 30 | 25.2 | 0.64 | 2.9 | 4.6 | 6.3 | 8.0 | 9.7 | 11.4 | 13.1 |

Notes: "Volumes of $CO_2$" are liters of $CO_2$ gas dissolved in 1 L of beer. Values for gas volume constant, residual $CO_2$, and priming amounts based on 330 ft. (100 m) elevation above sea level.

## Bottling Your Beer

### Bottle Filling

The next step is filling the bottles (figs. 10.3 and 10.4). Place the fill tube of your bottling bucket or bottle filler at the bottom of the bottle. Fill slowly at first to minimize aeration and keep the fill tube below the waterline. Fill to about ¾"–1" (2.0–2.5 cm) from the top of the bottles. If the head space fills with foam that is actually a good thing—it minimizes the air in the head space and helps reduce staling. Place a sanitized cap on the bottle and cap. Many people place the caps on the bottles and then wait to cap several at the same time. After capping, inspect every bottle to make sure the cap is secure.

Store the capped bottles at 68–85°F (20–30°C) for two weeks so they fully carbonate. After carbonation, store the bottles cold if you can. The majority of the maturation process, where the fermentation byproducts are cleaned up and haze settles out, should have happened in the fermentor when the yeast was most active. Warm temperatures increase yeast activity and speed up oxidation reactions that cause the beer to go stale. The carbonation temperature for bottle conditioning is a compromise between getting enough yeast activity to carbonate the beer, versus promoting the oxidation reactions that stale the beer. Bottle conditioning warmer for a shorter time may help the yeast take up the oxygen in the head space better, thereby reducing the oxidation of the beer better than carbonating at a cooler temperature over a longer period of time. However, that will depend on the overall health and quantity of yeast in the bottle.

**Figure 10.3.** Bottling using a bottling bucket with a filling tube.

**Figure 10.4.** Bottling using a siphon with a bottle filler attachment.

## Storage

Two common questions are, "How long will a homebrewed beer keep?" and "Will it spoil?" The answer to both these questions depends on you.

The biggest cause of flavor instability in beer, outside of contamination, is oxidation. The total amount of oxidation depends on how careful you were in transferring the beer to the bottling bucket or the keg. The rate of oxidation depends solely on temperature and is roughly reduced by half for every 18°F (10°C) decrease in temperature. Therefore, if you chill the carbonated beer from 77°F (25°C) to 41°F (5°C), the oxidation rate will be reduced to one-fourth, in other words, it will stay fresh four times longer.

The spoilage question obviously depends on how diligent you were with your cleaning and sanitation throughout the brewing process. The bottling equipment is a very common cause of infection, so be sure to thoroughly clean and sanitize your siphoning and bottling equipment after every use, and clean and sanitize it again before use. Think prevention! Many people think it should only be cleaned and sanitized thoroughly before use, and are lazy about cleaning it up after the previous use, just giving it a rinse. This is not a good idea—you don't want to give bugs a chance to gain a foothold between batches.

Finally, it is important to keep the beer out of direct sunlight, especially if you use clear or green bottles. Exposure to sunlight or fluorescent light will cause beer to develop a skunky character. It is the result of a photochemical reaction with hop compounds and sulfur compounds. Contrary to popular belief, this is not a character that Heineken®, Grolsch®, and Molson® strive for in their beer. It is simply the result of a combination of green glass and poor handling by retailers, who store them under fluorescent lighting. Other beers, like Miller High Life®, don't boil hops with the wort but instead use a specially processed hop extract for bittering that lacks the compounds that cause skunking (and flavor). Brown bottles are best unless you make a point of keeping your beer in the dark.

## Drinking Your First Homebrew

One final item that nobody ever remembers to tell new brewers until it's too late: don't drink the yeast layer on the bottom of the bottle.

People will say, "My first homebrew was pretty good, but that last swallow was terrible!" or "His homebrew really gave me gas," or "It must have been spoiled, I had to go to the bathroom right away after I drank it."

Welcome to the laxative effects of live yeast! It is a very effective and healthful probiotic. However, if you do not need increased ease of movement, you can avoid it by pouring carefully.

Pour the beer slowly so you don't disturb the yeast layer. With a little practice, you will be able to pour out all but the last quarter inch of beer. The yeast layer is also good at harboring many bitter flavors. It's where the word "dregs" came from. I remember one time my homebrew club was at a popular watering hole for a Belgian beer tasting. The proprietor prided himself on being a connoisseur of all the different beers he sold there. But our entire club just cringed when he poured for us. The whole evening was a battle for the bottle so we could pour our own. Chimay Grande Reserve, Orval, Duvel—all were poured glugging from the bottle, the last glass-worth inevitably being swirled to get all the yeast from the bottom. It was a real crime—not every beer is a hefeweizen. At least I know what their yeast strains taste like now.

## Kegging Your Beer

Somewhere between the first batch and the seventy-third, a homebrewer may look at the boxes of bottles collecting in the garage and say, "What the hell am I doing?!"

Let's face it, bottling is tedious, bordering on the onerous. It may even suck. Fortunately, there is an alternative, although an expensive one, and that is kegging. Many years ago, during the golden years of the twentieth century, soda pop was dispensed from 5 gal. (19 L) stainless steel kegs. These kegs were phased out towards the end of the century and replaced with plastic bags of syrup in cardboard boxes. These kegs were snapped up by enterprising homebrewers and have become the gold standard for storing and serving home brew. They are well worth the cost in terms of time and effort saved.

A kegging system consists of a keg, a $CO_2$ regulator, and a $CO_2$ tank. A plastic picnic tap (cobra tap) and hose are used to dispense the beer from the keg. The keg itself has several parts, including the lid, the in and out connectors, two stainless steel tubes, and several gaskets. The connectors contain spring-loaded poppets that keep the keg from leaking, though they occasionally need replacing. The connectors are specific to either the gas or liquid side. Beneath each connector is a gasket and dip tube—a short one for the gas in, and a long one for the beer out.

There are two main types of connectors: pin-lock and ball-lock. The pin-lock connectors have pins that engage the hose connections, while ball-lock connectors use balls that lock into a groove. In the case of pin-lock kegs, the poppets snap into the connector body, on a ball-lock keg they do not. There is no advantage to either system, the main difference may be what is most readily available in your area. Just pick one and stick with it.

In the US, a new 5 gal. (19 L) stainless steel keg will typically cost from $150 to $200; used ones are typically half that amount. The $CO_2$ regulator will cost from $75 to $100. A small $CO_2$ gas bottle can cost $100 up to $150, although I find you get a better deal by renting the standard (25 lb., or 11.3 kg) welding $CO_2$ bottles from a welding shop. This way you don't actually buy the tank, you just buy the gas and exchange the tanks each time. The miscellaneous hoses, tap, and fittings may cost another $50 by the time you get out of the store. Then there is the spare refrigerator or freezer to keep it in.

### Reconditioning a Used Keg

Actually, it may be hard to find a used soda keg these days, but if you do, it will need to be thoroughly cleaned. See the recommendations for cleaning stainless steel in chapter 2. The first thing to do is to completely disassemble it and replace the gaskets. A keg uses five gaskets: a large one for the lid, two medium ones for the connectors, and two small ones for the dip tubes (fig. 10.5).

1.  Remove the gasket from the lid and discard. It may be in good shape but the gaskets absorb the odors from soda and will give off-flavors to the beer. Place the lid in a small bucket of cleaning solution to soak.
2.  Unscrew the connectors and remove the poppets. Check the poppet gaskets for nicks that can cause leaks. If they look good, place them in the bucket of cleaning solution.
3.  Remove the gaskets from the connectors and throw them away. Replace them with new ones. Place the connectors in the bucket with cleaning solution.
4.  Remove the dip tubes and clean them inside and out with cleaning solution and a nylon tubing brush. The short dip tube is for the gas side, the long tube is for the

liquid side. Place new gaskets on the dip tubes. Rinse and place the clean tubes in a small bucket of sanitizing solution.

5.    Scrub the lid, poppets and connectors with a nylon brush, rinse, and place in a small bucket of sanitizing solution. Sanitize the lid gasket as well.

6.    Check the inside of the keg for residue. A sponge or nonmetallic scrubby (nylon scouring pad) should be used to clean around the lid area inside and remove any dirt or residue. If the bottom of the keg is dirty, you will probably need to find someone with skinny arms, like a spouse or child, that can reach inside to scrub it. A note from experience: your children are less likely to hold a grudge than your spouse. In addition, you can switch children when you have multiple kegs. Rinse and drain the keg after cleaning.

7.    Sanitize the inside of the clean keg. Some sanitizers can be sprayed to wet the walls, and other cannot. See chapter 2 for more information on how to use sanitizers. Drain the keg after sanitizing; do not rinse. Reassemble the keg with the sanitized parts and seal the lid to keep it sanitary until ready to use.

## Keg Carbonation

You could prime the keg with extra sugar the same way you would when bottling, but almost nobody does. It is much easier to force carbonate with overpressure from the $CO_2$ tank. Residual $CO_2$ doesn't really matter when you are force carbonating. Table 10.4 shows the volumes of $CO_2$ achieved for a range of regulator pressures as a function of beer temperature. For example, carbonating at 15 psi. (103 kPa) at a beer temperature of 45°F (7°C) would produce 2.5 volumes of $CO_2$, which is a typical level for many ales. However, it will take about a week to achieve that equilibrium, and would require the tank to be attached to the keg the whole time.

**Figure 10.5**. Typical ball-lock or pin-lock parts and gaskets. Replace all of the gaskets if you are buying a used keg. There is one gasket for the main lid, one for each of the connection posts, and one for each of the dip tubes.

It is much faster, and more practical, to crank up the pressure to 35–40 psi. (240–275 kPa) for a few hours or overnight to carbonate the beer. Rocking the keg while the gas line is attached will speed the dissolution of the gas into the beer.

Here is a common procedure that many homebrewers use:

1. Rack the beer into a clean and sanitized keg.
2. Install the lid but do not seal it yet. Set the $CO_2$ regulator to a low pressure, such as 5 psi (34 kPa). Attach the gas line to the gas-in connector and allow the flowing gas to purge the head space for a minute before sealing the lid.
3. Increase the regulator pressure to 35 psi. (240 kPa). Listen to the flow of gas into the keg. When the flow decreases, rock the keg gently back and forth a few times. You will hear the gas flow again.
4. Do this several times and then disconnect the gas and place the keg in your refrigerator or cooler to chill to serving temperature.
5. Allow the keg to sit for several hours (or overnight) and then repeat step 3, but with the regulator set to the desired serving pressure (e.g., 15 psi [103 kPa]). Rock the keg to listen for more gas flow. If gas flows, then repeat step 4. If no gas is flowing then the keg should be carbonated and ready to serve. Reattach the gas line from time to time to maintain the serving pressure and equilibrium carbonation.

Don't put the tank and regulator in the refrigerator because the condensation can ruin the regulator. Drill a hole in the side of the refrigerator to run the gas line through if you want to keep pressure on it all the time.

## GETTING A SPARE REFRIGERATOR OR FREEZER

There are several reasons to get a spare refrigerator or freezer when you are homebrewing regularly. You can get a temperature controller for fermenting in them, you can lager in them, you can store beer in them, and you can use them as a kegerator or keezer (a fancy way of saying a converted refrigerator or freezer for serving keg beer).

Refrigerators have the advantage of being front loading, which makes it easier to get fermentors and kegs in and out, but they lose more cold every time you open them. Freezers are top loading, which means they are much more energy efficient, but are harder on your back when lifting things in and out. But new freezers are a lot cheaper than new refrigerators, which tend to have lots of bells and whistles you don't need. The choice is up to you.

Also, be aware that a curbside or Craigslist deal for an old fridge or freezer may cost you more money in electricity than buying a new, more energy efficient one. Old refrigerators are invariably moldy and need lots of cleaning as well.

## TABLE 10.4—RESULTING VOLUMES OF CO$_2$ FROM FORCE CARBONATING AT THE INDICATED PRESSURE AND TEMPERATURE

| Temperature | | Regulator pressure in psi (kPa) | | | | | | | |
|:---:|:---:|:---:|:---:|:---:|:---:|:---:|:---:|:---:|:---:|
| °F | °C | 5 (34) | 10 (69) | 15 (103) | 20 (138) | 25 (172) | 30 (207) | 35 (240) | 40 (275) |
| 35 | 2 | 2.0 | 2.5 | 3.0 | 3.6 | 4.1 | 4.6 | 5.1 | 5.6 |
| 40 | 4 | 1.8 | 2.3 | 2.7 | 3.2 | 3.7 | 4.1 | 4.6 | 5.1 |
| 45 | 7 | 1.7 | 2.1 | 2.5 | 2.9 | 3.4 | 3.8 | 4.2 | 4.6 |
| 50 | 10 | 1.5 | 1.9 | 2.3 | 2.7 | 3.1 | 3.5 | 3.9 | 4.3 |
| 55 | 13 | 1.4 | 1.8 | 2.1 | 2.5 | 2.9 | 3.2 | 3.6 | 3.9 |
| 60 | 16 | 1.3 | 1.7 | 2.0 | 2.3 | 2.7 | 3.0 | 3.3 | 3.7 |
| 65 | 18 | 1.2 | 1.5 | 1.9 | 2.2 | 2.5 | 2.8 | 3.1 | 3.4 |
| 70 | 21 | 1.2 | 1.5 | 1.7 | 2.0 | 2.3 | 2.6 | 2.9 | 3.2 |
| 75 | 24 | 1.1 | 1.4 | 1.6 | 1.9 | 2.2 | 2.5 | 2.8 | 3.0 |
| 80 | 27 | 1.0 | 1.3 | 1.6 | 1.8 | 2.1 | 2.3 | 2.6 | 2.9 |

Notes: For example, force carbonating with the regulator set at 20 psi (138 kPa) at 50°F (10°C) would generate 2.7 volumes of CO$_2$ in the beer.

Values based on Henry's law and give the equilibrium volumes of CO$_2$ achieved when force-carbonating a beer at a specified temperature in a keg. The pressure is the regulator pressure in psi (kPa), and the temperature is the temperature of the beer. These values are based on atmospheric pressure at sea level; add 1 psi per 2000 ft. elevation (5.6 kPa per 500 m).

## Serving from the Keg

If you were to attach the serving tap directly to the liquid-out connector and attempt to pour a beer carbonated with 2 volumes of CO$_2$, you would get a glass of foam; very quickly, in fact. The serving pressure of the keg needs to be balanced by the length of beer line, which consists of vinyl tubing, to the tap. The smaller the tubing's inner diameter, the more resistant it is to flow. Opinions vary, but the recommended residual pressure from the keg for most situations is a positive 0–5 psi. (0–34 kPa) balance. For a typical serving pressure of 10–15 psi. (69–103 kPa) at 40–50°F (4–10°C), you would want about 5–7 ft. (1.5–2 m) of vinyl tubing with a 3/16 in. (5 mm) inner diameter. This will balance most of the pressure and allow you to pour a beer without excessive foaming. The more resistance, the slower it will pour with less foaming. See table 10.5 for a listing of typical beer line specifications.

## Counter-Pressure Filling

But what if you want to fill a couple of bottles or a growler to take to a friend's house? The quickest way to do so is to add a short length of vinyl tubing to the end of your tap and insert that all the way to the bottom of the bottle and begin filling. Having the beer very cold while you do this helps reduce foaming.

A more sophisticated method is to make a counter-pressure bottle filler, or buy one like the Blichmann BeerGun™. These devices use back-pressure to prevent the beer from foaming while filling the bottle. Do-it-yourself plans are readily available on the Internet.

## Kegging versus Cask and Bottle Conditioning

Carbonation can be accomplished by force carbonating in a keg or by priming with extra sugar. The practice of adding priming sugar to a keg is called cask conditioning, and bottle conditioning is adding priming sugar to the bottle. Unfortunately, the word "conditioning" has two meanings. The first meaning is that the beer has matured, that the yeast has done its job in the fermentor of cleaning up the off-flavors of a young beer. The second meaning is that the beer has been carbonated by further yeast activity. Bottle conditioning is primarily carbonation of the beer in the bottle and is not maturation. Carbonation can be done in either the bottle or in the keg, but the two methods do produce different results.

Studies have shown that priming and conditioning is a unique form of fermentation due to the oxygen present in the head space of the bottle or keg, only about 30% of which is used. The other 70% of the oxygen can contribute to staling reactions. Fermentation conditions in the bottle are often very different from those present during the initial fermentation, and this can produce different fermentation characters as the beer ages. In some styles, like Belgian strong ale, bottle conditioning and the resultant flavors are the hallmark of the style. These styles cannot be produced with the same flavors via force carbonation. The difference in character is not good or bad, it is different and needs to be evaluated on its own merits.

Cask conditioned ales are really in a class by themselves because the serving method, dispensing with a beer engine and sparkler at low carbonation levels, gives the beer a very soft mouthfeel. Priming rates for cask ales are quite low compared to other beer styles. The priming rate for cask ales tends to be in the range of 1–2 g/L, or 1.1–1.3 volumes of $CO_2$; roughly half of what a bottled ale of the same style would be. The result is a beer that is more malt forward with a softer and more lingering bitterness. Really big IPAs generally don't do well as cask conditioned ales, because the high bitterness needs more carbonation to help cleanse the palate between sips.

There are lots of resources on the Internet that can help you brew cask conditioned ales. Look for a booklet titled, *Hosting Cask Ale Events,* written and self-published by Randy Baril (2015), which has a lot of great information for priming and serving cask ales.

### TABLE 10.5—VINYL BEER LINE RESISTANCE

| Inner diameter inches (mm) | Resistance per foot | Resistance per meter |
| --- | --- | --- |
| 3/16" (5 mm) | 2.0 psi | 45 kPa |
| 1/4" (6 mm) | 0.75 psi | 17 kPa |
| 5/16" (8 mm) | 0.4 psi | 9 kPa |

## Final Thoughts

In a perfect world, beer would be (and should be) fully matured by the yeast in the fermentor before being packaged. Once carbonated, it should be ready for drinking. In general, I think the idea that strong beers should be aged for a period of months to develop peak flavor is a myth. There is a difference between aging well and improving with age. Ideally, beer should be at the peak of flavor after sufficient maturation in the primary fermentor. Any improvement in flavor after that, in my opinion, tends to indicate that the fermentation (including maturation) could have been better.

Some high-gravity styles, such as old ale, Russian imperial stout, Baltic porter, *doppelbock*, and barleywine, may be the exception to this rule, where some oxidation character would seem to add complexity. But does it really? Or are we just standing around and saying, "Hey, Grandpa is looking really good these days!"? I suppose beauty is in the eye of the beholder.

However, in a conversation with my friend Gordon Strong, a Grandmaster beer judge, he pointed out that, in reality, many beers (not styles, but individual brews) need additional time after fermentation, either in the keg or in the bottle, for the flavors to meld or mellow. While I would argue that this is the result of a poor recipe or poor fermentation, Gordon made a very valid point that my stating beer flavors are, or should be, at their peak when fresh is like saying that yeast is most active at 90°F (32°C); it is true but it is often more complicated than that. In other words, Gordon thinks that saying 'beer is best when freshest' could easily be misapplied, and he's right. Give your beer the time it needs in the fermentor to mature; give your beer the time it needs in the bottle or keg to mature; but, all the while, be cognizant of where any rough edges may be coming from, and try to address them in the brewing process so that your packaged beer has the best flavor for the longest period of time.

# How To Brew Lager Beer

**11**

It is easier to brew a bad lager than a bad ale because the expectations are higher for lagers. Brewers and beer drinkers are more forgiving about off-flavors in ale, calling it "complexity," but lagers should be "clean," with minimal off-flavors from esters, dimethyl sulfide (DMS), diacetyl, acetaldehyde, or fusel alcohols.

## Lager Fermentation

As discussed in chapter 7, "Yeast Management," lager yeast likes lower fermentation temperatures between 50°F and 55°F (10–13°C). Lager yeast is considered a different species than ale yeast, possessing an enzyme that allows it to ferment a sugar called melibiose that ale yeast strains can't ferment. Melibiose doesn't actually occur in beer wort, so don't expect an increase in attenuation from that; the ability of lager yeast to ferment melibiose is purely academic. However, lager yeast strains are also generally better than ale yeast strains at fermenting the trisaccharide maltotriose, and this can improve the attenuation by a point or two.

Lager yeast also produces more sulfur compounds during fermentation than ale yeast. New lager homebrewers are often astonished by the rotten egg smell coming from their fermentors, causing them to think that the batch is contaminated and dump it. That would be a mistake, because these highly volatile sulfur compounds will continue to vent during the maturation phase of fermentation. A previously rank smelling beer that is properly fermented will be sulfur-free and delicious at bottling time.

Historically, all beer was fermented in large open tubs and then kegged (or barreled) for maturation. Lager comes from the German word *lagern*, which means "to store," and this is done in a *lager*, or storehouse. The beer was stored cold to reduce the risk of it going sour and preserve its flavor for later in the year. This practice of long, cold maturation became known as "lagering." The main effect of the near-freezing temperatures when lagering is brilliant beer clarity, which also improves flavor. Lagering, or cold conditioning, promotes the settling of haze-forming protein-polyphenol complexes, which are held together by weak hydrogen bonds, with the result that these complexes only coalesce and settle out at colder temperatures. Beer haze is astringent. The polyphenols that bind to proteins to form haze will also bind to proteins in your saliva and coat your tongue to give a characteristic dry mouthfeel, which is actually the definition of astringency. Lagering improves beer flavor by settling out more haze than a comparable ale fermentation, all else being equal.

## Lower Temperatures Mean Longer Times

The cooler temperatures of lager fermentation (50–55°F [10–13°C]) slow the yeast's growth rate and reduce ester formation by the yeast. However, the cooler temperatures also slow the breakdown of diacetyl and acetaldehyde by yeast. Yeast cells do not work when it is cold, even if they are from a lager yeast strain. Therefore, lager fermentations typically use a diacetyl rest (i.e., raising the fermentation temperature by 5°F [3°C]) toward the end of fermentation to help invigorate the yeast and clean up these byproducts before lagering. The idea that the cells of a lager yeast strain downshift and actively clean up earlier fermentation byproducts at near freezing temperatures is a fallacy.

The lower initial fermentation temperature slows the rate at which yeast works and lengthens both the attenuation and maturation fermentation times, although this can be offset by higher pitching rates as discussed in chapters 6 and 7. The high-growth phase is a period of vigorous fermentation and can last anywhere from two to six days for lagers (compared with 1–3 days for ales), depending on the original gravity and pitching rate. The majority of the attenuation (i.e., 98% of attenuation) should occur during the high-growth phase. The rule of thumb is that the maturation phase may take as long as the attenuation phase, if not longer, depending on temperature. Raising the temperature for a diacetyl, or maturation, rest will shorten the time. Diacetyl rests will be discussed shortly.

## AUTOLYSIS

When a yeast cell dies it ruptures, releasing cell contents that can lead to meaty, brothy, or sulfury off-flavors in the beer. Autolysis is a risk whenever you have a large yeast mass sitting on the bottom of the fermentor for a long time (generally more than a month), but it depends on the overall health of the yeast. Healthy yeast cells can sit quietly for two months without autolysing. A lightly autolyzed beer will have a yeasty or brothy aroma or flavor. A moderately autolyzed beer will have a meaty aroma and flavor that is very similar to the smell of a jar of bouillon cubes. A heavily autolyzed beer will have a smokier, rubbery aroma and flavor, rather like smoked ham, or have more sulfury smells like garbage or sewage.

Autolysis is more common in very tall cylindroconical fermentors because of the high hydrostatic pressure on the yeast. But, hydrostatic pressure is not an issue on the homebrewing scale. The main cause of autolysis in homebrewing is poor initial yeast health and stressful fermentation conditions (e.g., temperature extremes, poor nutrients, and low pitching rates).

## Pitching and Fermenting with Lager Yeast

The key to brewing good lagers is understanding that all the fermentation processes occur more slowly due to the lower temperatures. It takes more time for the same amount of yeast to ferment a wort at a lower temperature than it does to ferment the same wort at a higher temperature. However, we can mitigate this by pitching more yeast.

*The right way.* The only difference between lager fermentation and ale fermentation is a few degrees cooler temperature and a little longer time; everything else should be the same. The wort should be chilled toward the bottom end of your yeast's fermentation temperature range before pitching. The fermentation temperature should be raised toward the end of the attenuation (high-growth) phase to keep the yeast active. The temperature should not be lowered for lagering until the yeast has been given sufficient time to fully maturate the beer and clean up all the byproducts during the diacetyl rest. This best practice can be summarized by: pitch cold, ferment warm.

*The wrong way.* In the past, homebrewers would do it backward. They would cool their wort to room temperature, pitch the yeast, and place the fermentor in a cool cellar or refrigerator to chill it down to the lager yeast fermentation temperature. The fermentation would start quickly, with a short lag phase, but the activity would slowly decrease over the next few days as the beer cooled. The fermentation looked good visually, but tended to have off-flavors such as diacetyl and acetaldehyde, and, in some cases, poor attenuation due to lack of yeast activity during the maturation phase.

*Pitching rate and starters.* The best way to ensure a strong, healthy lager fermentation is to pitch twice as much yeast as you would for an ale. Recommended pitching rates for lagers are given in table 11.1. I recommend that the yeast starter should be grown at no more than 5–9°F (3–5°C) above the intended fermentation temperature. After the starter fermentation finishes (usually 1–2 days), I recommend chilling the starter in the refrigerator overnight to settle all the yeast.

On brewing day, when you are ready to pitch your yeast, take the starter out of the fridge and pour off most of the starter beer. Swirl up the remaining beer and yeast and pitch this into your wort while the yeast is still cold. Experience among homebrewers has shown that pitching yeast from cooler to warmer shortens the acclimation period of the yeast. Also, pitching only the slurry avoids some off-flavors from the starter beer and saves room in the fermentor for your wort.

If you are using dry lager yeast, rehydrate it in pre-boiled water at a temperature of 69–77°F (21–25°C). The rest of the rehydration procedure is the same as previously described in chapter 7.

*Wort temperature at pitching.* The wort should be chilled to your intended fermentation temperature (50–55°F [10–13°C]) before pitching your yeast, or maybe even a degree or two lower. This may require that you place your filled fermentor in a temperature-controlled refrigerator overnight to cool and then pitch your yeast the next day. If your sanitation is good, this delay will not be a problem.

*Lag time.* The lag time for a lager fermentation should be very similar to an ale fermentation, because you pitched twice as much yeast. You should see activity within 24 hours. The fermentation should also display similar vigor (bubbling rate in the airlock) seen with an ale fermentation, but it may be less. The lager high-growth phase should take about the same amount of time as well, but it may be slower.

*Diacetyl rest.* I like to ferment and lager in glass or plastic carboys, because I can see the activity in the beer. During fermentation there are clumps of yeast and trub rising and falling in the beer—it literally looks like it is being stirred with an invisible stick. When you see that kind of activity slow down, and the bubbling in the airlock slows down, you know it's time for the diacetyl rest.

To conduct the diacetyl rest, raise the fermentation temperature by 10–15°F (5–10°C) at a rate of no more than 10°F (6°C) per day towards the end of fermentation when it is about three-quarters complete. This will be 3-5 days after pitching the yeast, when the airlock activity has slowed from rapid (~30 bubbles per minute) to a more sedate pace, such as 4-8 bubbles per minute. It can also be done at the end of fermentation, after airlock activity stops, but works better when the yeast is still active. Either way, give the yeast enough time to consume diacetyl and acetaldehyde before lagering; at least four days, if not a week. It will not hurt the beer to wait a week or even two at warmer temperatures before lagering (assuming you have a good airlock).

*Lagering.* Fermentation, including maturation, should be finished before the beer is cooled to facilitate clarification. To cold condition, or lager, the beer, slowly reduce the temperature by not more than 10°F (6°C) a day down to about 35°F (2°C). Gradual cooling ensures the yeast cells don't go into thermal shock and excrete lipids that can cause off-flavors and hurt head retention. This temperature is usually sufficient to flocculate the yeast and clear the beer of haze, although some noted brewing textbooks suggest going colder, down to 28–30°F (−1°C to −2°C) to get every last bit of haze. However, in my opinion, going colder runs the risk of freezing the beer for no real gain. The lagering time can be 2–4 weeks, although I recommend shorter times (1–2 weeks) if you plan on priming and bottle conditioning to carbonate the beer. There is no difference between lagering times for different strength beers. The yeast should have cleaned up the byproducts during the diacetyl rest, so all that is left to do is allow the yeast and haze to settle, which is purely physical, not biological.

To summarize, to manage your lager fermentation:
1. Pitch more yeast.
2. Pitch at your target fermentation temperature.
3. Conduct a diacetyl rest before you lager.

## Controlling the Fermentation Temperature

There are several options for controlling fermentation temperature for lager, but they essentially boil down to adding a temperature controller, such as those from Johnson Controls or Ranco, to your spare refrigerator or chest freezer. Yes, you can build insulated boxes and use blocks of ice, but after many trips to the hardware store and the time spent transferring ice in and out, it is easier and less expensive in the long run to buy a spare freezer and a controller. They work. By themselves. Enough said.

With a temperature controller in place, the refrigerator or freezer is plugged into the controller and the controller is plugged into the wall. A temperature probe is run inside the fridge and it governs the on/off cycling of the compressor to maintain a narrow temperature range. Here in southern California, I use one to maintain 65°F (18°C) in the summertime for brewing ales. Temperature controllers are readily available at most homebrew supply shops.

Single-stage controllers only control one heating or cooling device; two-stage controllers can control both heating and cooling. A two-stage controller is convenient for lager brewing, because

## TABLE 11.1—RECOMMENDED LAGER YEAST PITCHING RATES (BILLIONS OF YEAST CELLS PER LITER) AS A FUNCTION OF ORIGINAL GRAVITY

| OG | °Plato | Recommended relative pitching rate (billion cells/°P/L) | | |
|---|---|---|---|---|
| | | Lower (1.25/°P/L) | Typical (1.5/°P/L) | Higher (1.75/°P/L) |
| 1.030 | 7.5 | 9 | 11 | 13 |
| 1.035 | 8.8 | 11 | 13 | 15 |
| 1.040 | 10.0 | 12 | 15 | 17 |
| 1.045 | 11.2 | 14 | 17 | 20 |
| 1.050 | 12.4 | 15 | 19 | 22 |
| 1.055 | 13.6 | 17 | 20 | 24 |
| 1.060 | 14.7 | 18 | 22 | 26 |
| 1.065 | 15.9 | 20 | 24 | 28 |
| 1.070 | 17.1 | 21 | 26 | 30 |
| 1.075 | 18.2 | 23 | 27 | 32 |
| 1.080 | 19.3 | 24 | 29 | 34 |
| 1.085 | 20.5 | 26 | 31 | 36 |
| 1.090 | 21.6 | 27 | 32 | 38 |
| 1.095 | 22.7 | 28 | 34 | 40 |
| 1.100 | 23.8 | 30 | 36 | 42 |
| 1.105 | 24.9 | 31 | 37 | 43 |
| 1.110 | 25.9 | 32 | 39 | 45 |

°P, degrees Plato; OG, original gravity.

Notes: For details of relationship between °P and specific gravity, refer to appendix A. Pitching rates are per liter, assuming pitching rates of 1.25 (lower), 1.5 (typical), or 1.75 (higher) billion cells per degree Plato per liter (billion cells/°P/L). For example, for 20 L of 1.050 wort the typical pitching rate would be 20 × 19 billion = 380 billion cells, which is equivalent to about four liquid yeast packages. Many brewers recommend using the higher pitching rate when brewing beers with OG >1.070. Lower rates can be used when trying to promote ester formation.

it allows you to place a small heater inside the refrigerator or freezer that can raise the temperature for the diacetyl rest, as well as prevent the beer from freezing during the wintertime if there is a cold snap. A two-stage controller works between the two devices to maintain the set temperature range—if the probe gets too cold, the controller shuts off the freezer and turns on the heater, if the probe gets too hot, the controller shuts off the heater and turns on the freezer. The controller doesn't work against itself by having both heater and freezer on at the same time.

## OH NO! IT FROZE!

By the way, what if your beer freezes during lagering? Horrors! Well, it happened to me. Let me tell you about my first lager. . .

'Twas a few weeks before Christmas and all around the house, not an airlock was bubbling, in spite of myself. My Vienna was lagering in the refrigerator out there, with hopes that a truly fine beer, I soon could share.

My controller* was useless, 32°F couldn't be set, so I turned the fridge to Low, to see what I would get. On Monday it was 40°, on Tuesday lower yet, on Wednesday morning I tweaked it; it seemed like a good bet.

Later that day when I walked out to the shed, my nose gave me pause, it filled me with dread. In through the door I hurried and dashed, but I tripped on the stoop and fell with a crash. Everything looked ordinary, well what do you know, but just in case, I opened the fridge slow.

When what to my wondering eyes should appear, My carboy was FROZEN, I had made ice beer! My first thought was tragic, I was worried a bit, I sat there and pondered, then muttered, "Aw, sh—!"

More rapid than eagles, my curses they came, and I gestured and shouted and called the fridge bad names. "You bastard! How could you! You are surely to blame! You're worthless, you're scrap metal, not worth the electric bills I'm paying! To the end of the driveway, with one little call, they will haul you away, haul away all!"

Unlike dry leaves that before the hurricane fly, when brewers meet adversity, they'll give it another try. So back to the house, wondering just what to do, five gallons of frozen beer, a frozen airlock too. And then in a twinkling, I felt like a goof, the carboy wasn't broken, the beer would probably pull through.

I returned to the shed, after hurrying 'round, gathering cleaning supplies, towels, whatever could be found. I'd changed my clothes, having come home from work, I knew if I stained them, my wife would go berserk. I was loaded with paper towels, I knew just what to do, I had iodophor-ed water and a heating pad too.

The carboy, how it twinkled! I knew to be wary, the bottom wasn't frozen but the ice on top was scary! That darned fridge, it had laid me low, trying to kill my beer under a layer of snow. I cleaned off the top and washed off the sides, picked up a block of ice and threw it outside. I couldn't find the airlock, it was under the shelf, and I laughed when I saw it, in spite of myself.

The work of a half hour out there in the shed, soon gave me to know, I had nothing to dread. The heating pad was working, the ice fell back in, I re-sanitized the airlock, I knew where it had been. Not an eisbock, but a Vienna I chose; it was the end of the crisis of the lager that froze.

I sprang to my feet, to my wife gave a whistle, and we went off to bed under the down comforter to wrestle. But the fridge heard me exclaim as I walked out of sight, "Try that again, you bastard, and you'll be recycled all right!"

* The now discontinued Hunter Airstat™ was a temperature controller better suited to controlling window-type air conditioners.

## Priming and Bottling Lager

Most of the time there is no difference between priming for lager and priming for ale. There will still be some yeast in the beer even after lagering. But sometimes, such as when the beer has had a long, cold lagering for more than a month, you may need to add fresh yeast; there may not be enough left to carbonate the beer in the bottle in a timely manner. If your lager freezes, chances are likely that the yeast has been impaired, and you should probably add new yeast.

The yeast you add to the fermentor should be the same strain as the yeast you originally pitched. The number of cells in a typical yeast package (dry or liquid) is more than sufficient, and you can mix it with your priming sugar in the bottling bucket. If it is dry yeast, rehydrate it first for best results.

Carbonation (bottle conditioning) should be conducted at room temperature, 65–75°F (18–24°C), not at the lager fermentation temperature. The comparatively high pitching rate to the priming sugar means that almost no yeast growth will occur, meaning few byproducts will be produced. You don't need to lager the beer after carbonation, but you should store it cool or cold for better flavor stability. Bottle conditioning is discussed in more detail in chapter 10.

## Fermentation Off-Flavors in Lager

### Diacetyl and 2,3-Pentanedione

The vicinal diketones, diacetyl and 2,3-pentanedione, are not actually produced by yeast; instead, yeast produces the acetohydroxy acid precursors. Vicinal diketones are created when the aceto-hydroxy acids undergo oxidative decarboxylation (i.e., removal of hydrogen and carbon dioxide). Warm temperatures and oxygen promote this oxidation reaction, and more diacetyl will form in the bottle if the acetohydroxy acid precursors are still present under these conditions. Yeast can remove diacetyl about ten times faster than oxidation can create it, but this is entirely dependent on yeast activity and temperature.

Diacetyl causes a buttery flavor, which is considered an off-flavor in most lagers. A small amount of diacetyl is considered good in other styles of beer, such as dark ales and stouts, but is considered a flaw in most lager styles. A diacetyl rest should be used to remove diacetyl before lagering.

2,3-pentanedione is a very similar compound to diacetyl and is produced in the same manner. It has a sweet, honey-butter or toffee flavor and aroma. A small amount of 2,3-pentadione adds to the perceived sweetness of light lager styles, but as the beer warms it can be excessive. A strong fermentation and diacetyl rest will minimize 2,3-pentanedione in the beer.

### Dimethyl Sulfide

Dimethyl sulfide (DMS) is an off-flavor in most styles. However, DMS is common in many light lagers and is considered to be part of the character in small amounts, rather like diacetyl is in ales. It can have a creamed-corn or cooked cabbage aroma and flavor in pale beers, or a more tomatolike character in dark beers. The production of DMS occurs in the wort during the boil due to the chemical reduction of another compound, S-methylmethionine, which is itself produced during malting. The heat during kilning of base malts reduces the amount of S-methylmethionine in the final malt, so darker base malts, such as pale ale malt and Munich malt, have much less S-methylmethionine than the more lightly kilned Pilsner base malt. A 90-min. boil is recommended for pale lagers that use a majority of Pilsner malt, because this helps fully purge DMS from the wort.

## Acetaldehyde

Acetaldehyde production is often at odds with the other byproducts of yeast fermentation. It is produced early in the fermentation cycle as part of the ethanol production process and is reduced later by the yeast during maturation. Acetaldehyde typically arises due to rapid fermentation caused by warm temperatures (>60°F [>16°C]), or by overpitching and underaeration. It has been compared to the smell of green apples, raw pumpkin, wet grass, sliced avocado, or wet latex paint. Acetaldehyde should be cleaned up by the yeast during the diacetyl rest.

## Fusel Alcohols

Fusel alcohols are heavier alcohols (in terms of molecular weight) that can be produced by the yeast during a stressful fermentation. While a small amount of the total fusel alcohols can be esterified during maturation, it is a minor reaction pathway and not an effective means for reduction. Fusel alcohol levels are increased by warmer temperatures, excessive aeration, excessive levels of amino acids, and, conversely, also by underaeration and a lack of amino acids. To control fusel alcohol levels:

- increase the pitching rate to limit excessive yeast growth,
- pitch when the wort is cool,
- ferment at the lower temperatures in the suggested range for the yeast,
- don't add sucrose or other refined sugars to the wort.

## Esters

An ester is a compound formed by the yeast from an alcohol and a fatty acid. Most esters in beer are produced from ethanol, but some are formed from the esterification of fusel alcohols. Ester formation is promoted by rapid growth and yeast stress, which can be caused by a number of reasons, including underpitching, underaeration, pitching too warm, warm fermentation temperatures, high-gravity worts, and worts with a significant proportion of refined sugar. Ester production is discussed in more detail in chapter 7.

### Minimizing Off-Flavors in Lager

To summarize, a bad lager could have some, or all, the following flaws:

- a microwave butter popcorn aroma or flavor due to diacetyl;
- a sweet, buttery flavor from 2,3-pentanedione;
- an oxidized, cooked or creamed corn flavor from DMS
- a green apple or raw pumpkin or fresh paint aroma and flavor due to acetaldehyde;
- sharp solventlike aromas and flavors due to fusel alcohols;
- fruity aromas and flavors due to esters.

Here are some recommendations for producing a clean lager:

- Cool the wort to primary fermentation temperature before pitching the yeast.
- Pitch a relatively high quantity of yeast to limit yeast cell growth.
- Aerate the wort sufficiently, but not excessively.
- Do not chill or lager too soon—give the yeast time to finish the job.

## Brewing American Lager

Many people want to know how to brew their favorite American light lager, like Budweiser, Miller, or

Coors. First, I will tell you that it is difficult to do. Why? First, because these beers are brewed using all-grain methods that incorporate rice or corn (maize) as about 30% of the fermentables. The rice or corn adjunct must be pre-cooked to fully solubilize the starch and then added to the mash so that the enzymes can convert the starches to fermentable sugars. Chapters 15 and 16 go into more detail about mashing and adjuncts.

Second, there is no room in the light body of these beers for any off-flavors to hide. Your sanitation, yeast handling, and fermentation control must be rigorous for this type of beer to turn out right. The professional brewers at Bud, Miller, and Coors are very good at what they do—turning out a light beer, decade after decade, that tastes exactly the same.

You can brew either rice or corn-type lagers. See the "Typical American Lager Beer" recipe below. You need to find a corn (maize) or rice syrup that is high in maltose to brew these American lager styles most accurately. Rice extract is available in both syrup and powder form, and will produce a beer similar to Heineken or Budweiser. High maltose corn syrup will produce a beer similar to Miller or Coors. Corn syrup solids are different from the corn sugar (dextrose monohydrate) that is commonly sold for priming and bottling. High-maltose corn syrup and solids are made by mashing corn and still retain the essential corn flavor. Grocery store corn syrups, on the other hand, are often high in fructose to make them sweeter and contain additives, such as vanillin, and will not produce a good example of this style.

To brew a corn-type lager, substitute a high maltose corn syrup for the rice syrup in the recipe below. If you want to brew a classic American Pilsner, with a richer corn character, refer to the recipe, "Your Father's Mustache," in chapter 23 for appropriate OG and IBU levels. The cereal mash procedure using flaked corn or corn grits described in the "Your Father's Mustache" recipe will produce more corn character than extract methods using corn syrup. Other lager recipes can also be found in chapter 23.

## Typical American Light Lager

**Original gravity:** 1.042  
**Final gravity:** 1.010  
**IBU:** 21

**SRM (EBC):** 3 (6)  
**ABV:** 4%

| *Version: Extract and Steeping Grain* | | |
|---|---|---|
| **Wort A** | **Gravity points** | |
| 2.5 lb. (1.13 kg) Pilsner DME | 35 | |
| Boil gravity for 3 gal. | 1.035 | |
| **Hop schedule** | **Boil time (min.)** | **IBUs** |
| 1 oz. (30 g) Willamette 5% AA | 60 | 18 |
| 1 oz. (30 g) Willamette 5% AA | 10 | 3 |
| **Wort B (add after boil)** | **Gravity points** | |
| 1.5 lb. (680 g) Pilsner DME | 21 | |
| 1.5 lb. (680 g) rice syrup solids | 21 | |
| **Yeast strain** | **Pitch (billions of cells)** | **Fermentation temp.** |
| American lager* | 300 | 52°F (11°C) |

*Note: A good dry yeast option is Fermentis Saflager W-34/70.

# Brewing Strong Beers 12

It seems every new homebrewer goes through the same development arc: failure, success, over-confidence, innovation, and finally consistency. To put it another way: bad, good, strong, fruit or spiced, and classic style. The initial disappointment in the first batch usually generates more interest in figuring out how brewing actually works, the second batch is usually a rousing success and the homebrewer becomes world famous to his or her immediate friends and family. The next step on this development arc is usually, "I'm going to brew a really strong beer!"

What defines a strong beer? Opinions vary, but I consider strong beers to be greater than 1.075 OG. Some would argue that between 1.060 and 1.065 OG are strong beers as well, and in general I agree, but modern yeast and fermentation methods seem to handle these gravities without any trouble. It's when you get above 1.075 OG, and especially 1.090 OG and above, that you need to take a second look at the brewing process to ensure complete success.

Brewing high-gravity styles, or imperial versions of lower gravity styles, seems to be one of those rites of passage that everyone must go through and veteran homebrewers can only shrug and say, "Yes, try it. It should work." What they don't say, because it is hard to explain to a new homebrewer riding the wave of success, is that it is easy to brew a strong beer but it is hard to brew one that is actually good enough to want a second pint. Poorly brewed strong beers tend to be either very sweet and heavy, or incredibly bitter. Barleywines, Russian imperial stouts, and big IPAs tend to be the beers of choice for these endeavors. See the sidebar, "Kamoniwannaleiyah American Barleywine" for an example recipe.

The key to successfully brewing every beer, and strong beers in particular, is a properly managed fermentation resulting in thorough attenuation. As my good friend Jamil Zainasheff says, "Strong beers must be dry." This may seem like the wrong idea at first, but when you consider it, it makes sense. Big and strong beers have a lot of body. That body is made up of protein from the malt and unfermentable dextrins. Remember, most yeast only attenuate about 75% of the total soluble extract, and all the rest adds to the weight of the beer. If a strong beer is underattenuated, it feels twice as heavy. So, drier is better.

Strong beers need a strong fermentation. The best way to do that, as discussed in chapters 6 and 7, is to use a higher pitching rate. You could ferment a strong beer by switching to a different yeast with a higher tolerance for alcohol (i.e., wine or champagne yeast), and raising the temperature and aerating it more, but that would change the character of the beer. What you really want is the same beer flavor balance you are used to, but stronger. The answer therefore is usually just a higher pitching rate.

But before we address pitching rate, let's take a look at the initial problem, which is creating a stronger wort.

## Creating Higher Gravity

Basically, there are two ways to create high-gravity wort. The first is to simply add more malt extract to your recipe. If you don't have malt extract available, and many countries don't, you can add a couple of pounds or kilograms of sugar, or sugary adjuncts, to boost the gravity. However, be aware that fermented sugar doesn't taste the same as fermented extract.

The second way is to mash and lauter with more grain at a lower water-to-grist ratio to get high-gravity wort. As you will see in chapter 18, "Extraction and Yield," the more grain and less water you use in both mashing and sparging, the higher the concentration of sugar. The practical limit for maximum wort gravity from a mash is about 1.100 OG and the yield efficiency is low, about 50%. The hows and whys of these details will be discussed more in the mashing and lautering chapters, but the bottom line is that it takes double the grain (or more) to double the gravity, which requires either two subsequent mashes or twice as big a mash tun to get all of the wort for a full batch. This is why it is usually better in terms of time, effort, and beer quality to brew a middling-high-gravity wort (e.g., 1.060 OG) and add a few pounds of malt extract or sugar to the kettle to reach the higher gravity (e.g., 1.090 or more) you were looking for.

There is another option though, and that's doing what used to be called a "double mash." If you mash with the wort from a previous batch rather than water, the wort gravity is cumulative. For example, a water-to-grist ratio of 2 qt./lb. (4 L/kg) yields a wort specific gravity of 1.061. If you mash and drain that wort, you will get about 70% extract efficiency, and a wort volume of 1.5 qt./lb. (3 L/kg). This first wort can then be added to a new batch of grain for mashing, at the same water-to-grist ratio (but using the wort in place of water), same time and temperature, and the result will be a total specific gravity of about 1.120. Refer to chapter 18 for an idea of typical first wort gravities, in particular table 18.3.

However, this method is time consuming and leaves a lot of extract behind in the grain, reducing your total wort volume with each mash. Sparging after the second mash will help, but that will dilute your wort as well. Double mashing is an effective way of achieving high-gravity wort, even if it is not an efficient one, but you will use a lot of grain.

One last thing—extended boiling to concentrate ten gallons down to five gallons (for example) is not a good idea either. The extended boil generates a lot more Maillard products than a typical wort boil, and many of these products resemble those that are associated with extract twang. Either mash a big grain bill at a low water-to-grist ratio, or double mash, or find some good quality malt extract to add to your wort. Those methods will give better results.

## Pitching Rates for High-Gravity Brewing

See tables 12.1 and 12.2. The standard pitching rates for ales is 0.75 billion cells per liter per 4 gravity degrees (or 0.75 billion cells per liter per degree Plato [°P]), and 1.5 for lagers. (Standard pitching rates are discussed in chapter 7.) However, high-gravity fermentations tend to generate many byproducts, so a higher pitching rate of 1.0 or even 1.5 billion cells per liter per 4 gravity points (1.0–1.5 billion/L/°P) may be advisable for ales depending on how clean a character you want in the beer.

Lager pitching rates can be increased accordingly as well. Generally, it is very hard to overpitch[1] when using fresh, healthy yeast, because higher pitching rates result in less fermentation character and a cleaner tasting beer. I have tasted great beers that were 11% ABV that I would have sworn were only 6% or 7% ABV —they were that smooth, and it was all due to pitching rate and a properly managed fermentation.

## TABLE 12.1—PITCHING RATES FOR HIGH-GRAVITY WORTS IN BILLIONS OF CELLS PER GALLON AS A FUNCTION OF WORT GRAVITY

| | | Recommended pitching rate for style (billion cells/L/°P)[a] | | | | | |
|---|---|---|---|---|---|---|---|
| | | Ale | | | Lager | | |
| **SG** | **°P** | **0.75** | **1.0** | **1.25** | **1.5** | **1.75** | **2.0** |
| **1.075** | **18.2** | 52 | 69 | 86 | 103 | 121 | 138 |
| **1.080** | **19.3** | 55 | 73 | 91 | 110 | 128 | 146 |
| **1.085** | **20.5** | 58 | 77 | 97 | 116 | 135 | 155 |
| **1.090** | **21.6** | 61 | 82 | 102 | 122 | 143 | 163 |
| **1.095** | **22.7** | 64 | 86 | 107 | 129 | 150 | 172 |
| **1.100** | **23.8** | 67 | 90 | 112 | 135 | 157 | 180 |
| **1.105** | **24.9** | 71 | 94 | 118 | 141 | 165 | 188 |
| **1.110** | **25.9** | 74 | 98 | 123 | 147 | 172 | 196 |
| **1.115** | **27.0** | 77 | 102 | 128 | 153 | 179 | 204 |
| **1.120** | **28.1** | 80 | 106 | 133 | 159 | 186 | 212 |

*°P, degrees Plato; SG, specific gravity. (For details of relationship between °P and SG, refer to appendix A.)*

*[a] The recommended pitching rates have been increased by 0.25 billion cells/L/°P above the standard rates recommended in chapter 7.*

---

[1] However, it is easy to overpitch with old, tired, half-dead yeast, such as that left over from a previous strong beer fermentation you did last month. If you pitch with old, tired yeast, you will probably end up with many off-flavors, such as goaty, cheesy, or brothy. See the Simple Yeast Ranching section in chapter 7 for more information on revitalizing old yeast.

## TABLE 12.2—PITCHING RATES FOR HIGH-GRAVITY WORTS IN BILLIONS OF CELLS PER LITER AS A FUNCTION OF WORT GRAVITY

| | | Recommended pitching rate for style (billion cells/L/°P)[a] | | | | | |
|---|---|---|---|---|---|---|---|
| | | Ale | | | Lager | | |
| SG | °P | 0.75 | 1.0 | 1.25 | 1.5 | 1.75 | 2.0 |
| 1.075 | 18.2 | 14 | 18 | 23 | 27 | 32 | 36 |
| 1.080 | 19.3 | 14 | 19 | 24 | 29 | 34 | 39 |
| 1.085 | 20.5 | 15 | 20 | 26 | 31 | 36 | 41 |
| 1.090 | 21.6 | 16 | 22 | 27 | 32 | 38 | 43 |
| 1.095 | 22.7 | 17 | 23 | 28 | 34 | 40 | 45 |
| 1.100 | 23.8 | 18 | 24 | 30 | 36 | 42 | 48 |
| 1.105 | 24.9 | 19 | 25 | 31 | 37 | 43 | 50 |
| 1.110 | 25.9 | 19 | 26 | 32 | 39 | 45 | 52 |
| 1.115 | 27.0 | 20 | 27 | 34 | 41 | 47 | 54 |
| 1.120 | 28.1 | 21 | 28 | 35 | 42 | 49 | 56 |

°P, degrees Plato; SG, specific gravity. (For details of relationship between °P and SG, refer to appendix A.)

[a] The recommended pitching rates have been increased by 0.25 billion cells/L/°P above the standard rates recommended in chapter 7.

## Yeast Selection

Yeast selection for strong beers is important, but not so much for alcohol tolerance; most brewer's yeasts can ferment to 12% ABV without a problem. The primary consideration is selecting for fermentation character and maturation. Big beers do not lack flavor. They may lack restraint and complexity, but not flavor. Therefore, your choice of yeast should probably be a strain that is noted for having a clean character, high attenuation, and low to medium flocculation, because you want the yeast to maturate the beer thoroughly.

Of course, there are some styles, like Belgian tripel and Belgian strong ale, that use style-specific yeasts with phenolic characters, and those are totally appropriate. But if you are attempting to brew an "Imperial Whatzit" for the first time, you probably should choose a cleaner tasting yeast and see what that is like first.

## Scaling Your Recipes

A few years ago Jamil Zainasheff and I were talking about brewing strong beers on our podcast, *Brew Strong*,[2] and were discussing the grain bills for brewing doppelbock and eisbock beers. The color of these beers tends to be rather dark amber to dark brown, between 20 and 30 SRM

---

[2]   *Brew Strong*® is not actually about brewing strong beers. *Brew Strong* is a podcast about the nuts and bolts of brewing. We try to translate the technical aspects of brewing into more practical ones for our listeners.

(40–50 EBC). (See appendix B for a discussion of beer color.) Jamil noted that many brewers make the mistake of formulating these recipes with dark caramel or roast malts, which is the completely wrong character for these beers. Doppelbock's darker color comes from the higher gravity and the Maillard browning reactions that take place in the boil, not from dark malts. Likewise, when brewing an *eisbock*, the darker color comes from the beer being concentrated by freezing and removing the ice.

Most of the increase in gravity for your strong beers should come from the base malt. If you are scaling, for example, a brown ale recipe to an imperial-version brown ale recipe, you may want to decrease the percentages of the specialty malts a little bit, rather than doing a straight scale up. Sometimes you need to consider the total amount of a specialty malt in a beer, rather than just the percentage, because the balance of flavors is different at higher gravities. Base malts don't have a very strong flavor compared to most specialty malts, so simply scaling up with the percentage can result in comparatively more flavor from the specialty malt coming through. It is always important to balance the amount of flavor from your specialty malts to the beer as a whole. You can't have complexity without balance.

Sometimes, less is more, especially in terms of drinkability. Consider the difference between regular Irish stout and Russian imperial stout. Irish stout is a session beer; flavorful, yet not filling, a beer that you can drink several pints of throughout the evening. A good Russian imperial stout is a wonderfully complex explosion of flavor, but it is difficult to drink more than two pints of it at a time because it is so rich. A similar conundrum used to exist between the IPA and double IPA styles. Double IPA is the more bitter, higher alcohol beer, but many recipes are now formulated with 5%–10% simple sugar to cut the body and achieve better drinkability than a regular IPA. The point is don't just scale every recipe to make it an "imperial" style. Take a moment to consider how the higher gravity is going to affect the balance and drinkability of the beer, and don't be afraid to adjust the ingredients and proportions of the recipe to make it a better beer.

## Wort Aeration

Higher pitching rates need more oxygen than lower pitching rates. The oxygen saturation limit in the wort remains fairly constant, but the number of yeast cells using that oxygen goes up when you are using higher pitching rates for brewing strong beers. The solution is to aerate the wort a second time at the beginning of the high-growth phase to give that new generation of yeast the same oxygen and lipid reserves their mothers had. As described in chapter 6, the yeast cells will take up the initial oxygen quickly—within the first half hour after pitching—along with the wort FAN, and begin altering their cell membranes to adjust to the new nutrient environment. The yeast cells will then start dividing. You want to wait until the cells have undergone that first cell division before adding the additional oxygen; this step will typically be about 8–12 hours after pitching. This needs to be done before the yeast really gets going and truly enters the high-growth phase. Aerate the wort the same way and for the same amount of time as you did for the initial pitch.

If you're shaking or stirring to aerate, make sure there is lots of surface area for oxygen to be absorbed. However, although it is easy to saturate a starter wort with air by shaking, is just about impossible, perhaps even dangerous, to shake a full bucket or carboy. Rocking the vessel is easier than shaking, but you need a lot of head space to expose all of the wort to the air available and achieve saturation. It won't work with a full carboy.

Aeration with an airstone is much more effective, using either an aquarium air pump or an oxygen tank. Airstones typically come in two pore sizes: 0.5 and 2 microns (a micron is the same as a micrometer [μm]). The 0.5 micron size provides smaller bubbles and therefore more surface area, but the much smaller pore size creates much higher resistance to airflow (i.e., you will need a strong air pump), and the pores are prone to clogging if not scrupulously cleaned after every use. I find the 2 micron size to be more reliable, even though it does produce a fair amount of foam in the head space. Foam-reducing agents are very helpful for controlling this and will not affect the foam generation or retention in your beer. There are HEPA filters that can be placed between the air (or oxygen) source and the airstone. This will ensure you are not contaminating your beer during aeration.

*Bottom line:* aerate once, then aerate again 8–12 hours later.

## Wort Additions

Another technique applicable to very high-strength worts is staggered wort additions, where the first wort is pitched and allowed to attain high kräusen before a second wort (also aerated) is added to the fermentor. This technique is essentially using the first wort as a large starter for the second wort. The pitching rate should be based on the volume and gravity of the first wort, not the total wort. The second wort is not pitched. The second wort is added 24 hours after the first wort is pitched, at which point the yeast cells have gone through at least one round of cell division and roughly doubled in number, assuming there is some activity in the fermentor. A staggered wort addition is very helpful when the recipe contains a large portion of simple sugar, because the simple sugar can be added as the second wort.

The second wort needs to be aerated before it is added to the fermentor. Do not aerate the fermenting beer before adding the aerated second wort. In addition, don't aerate the combined wort a second time; that would be too much.

Adding zinc is useful with big beer fermentations, because this helps the yeast to reduce acetaldehyde and fully attenuate the beer. Ordinary fermentations do not usually need additional zinc; the yeast has enough left over from propagation and the trace amounts already present in the wort. For really big beers, 1.100 OG and above, increasing the pitching rate alone may not always be enough, which is why zinc additions will definitely help. However, be careful to not add too much (>1 ppm), because high zinc additions tend to produce goaty off-flavors. Also, a slightly lower fermentation temperature at the lower end of the range for the yeast at the beginning of fermentation will help limit yeast growth and retard acetaldehyde production. Mineral additions, including zinc, are discussed in chapter 6.

## Summary

Fermenting strong beers is not that different from fermenting regular beers. The key is managing the fermentation by managing the yeast. Managing the yeast means making sure you have sufficient pitching rates, aeration, and nutrients to get the job done in a controlled manner. Remember as well the principles covered in chapter 6 to ensure sufficient maturation of the beer.

## Kamoniwannaleiyah American Barleywine

*The following recipe was originally published in in my regular Q&A column in issue 34 of* Beer and Brewer *magazine (Spring issue, 2015).*

**Original gravity:** ~1.110 at 5 gal. (19 L)  **IBU:** 100+
**Final gravity:** ~1.028  **SRM (EBC):** ~18 (36)

| Grain bill | Gravity points | |
|---|---|---|
| 19.8 lb. (9 kg) pale ale malt | 71.5 | |
| 1.1 lb. (500 g) Munich malt | 3.5 | |
| 0.55 lb. (250 g) crystal 40°L | 1.5 | |
| 0.55 lb. (250 g) crystal 75°L | 1 | |
| 0.22 lb. (100 g) Special "B" | 0.5 | |
| 2.2 lb. (1 kg) pale DME | 15 | |
| Boil gravity for 6 gallons (23 L) | 1.093 | |
| **Hop schedule** | **Boil time** | **IBU** |
| 2.5 oz. (80 g) Warrior 16% AA | 60 min. | 94 |
| 1.4 oz. (40 g) Amarillo 9% AA | Steep 15 min. | 6 |
| 1.4 oz. (40 g) of Galaxy 14% AA | Steep 15 min. | 9 |
| **Yeast strain** | **Pitch (billions of cells)** | **Fermentation temp.** |
| California ale | 550 | 68°F (20°C) |

*Directions*

1.  *Mill and mash-in the grains at 1.5 qt./lb. (3 L/kg) water-to-grist ratio into a 10 gal. (38 L) or larger mash tun. The target mash temperature is 149°F (65°C) and the strike water temperature should be 160–162°F (71–72°C). Mash for 1 hour and stir occasionally to ensure that it is homogenous.*

2.  *After 1 hour, recirculate to clarify the wort and drain it completely to your boil kettle. You should end up with about 6 gal. (23 L) of wort with a specific gravity of about 1.078 in the kettle. Do not sparge.*

3.  *Add the 2.2 lb. (1 kg) of pale DME to the kettle and begin the boil. This wort will foam a lot, so be sure to wait for the hot break to occur (foam calms down) before adding the Warrior hop addition. Add Irish moss or other kettle finings 5 min. before the end of the boil.*

4.  *Boil the Warrior hop addition for 60 min., then turn off the flame and add the Amarillo and Galaxy hops. Let these hops sit in the hot wort for 15 min. before chilling the wort to pitching temperature. (The no-chill method of cooling the wort is fine; simply transfer the hot wort to the bin soon after adding the last hops.)*

5. When the wort temperature is 68°F (20°C), aerate the wort and pitch 550 billion cells of California ale yeast (or similar strain). Try to maintain the fermentation temperature between 68°F and 74°F (20–23°C). It is best if the temperature is kept at the cool end of this range (68°F [20°C]) for the first 36–48 hours after pitching. After that, the temperature may rise to 74°F (23°C) or even 77°F (25°C) without any flavor problems. In fact, a couple of degrees warmer after the first 2–3 days of fermentation will help the yeast clean up the beer flavors.

6. With proper pitching rates and fermentation, the beer should be done and ready to bottle or keg within 2–3 weeks. However, barleywine benefits from aging and the flavors will transition with time from hoppy to malty. Don't drink this beer too fast, give the bottles or keg a few months to mature. Enjoy the changes!

# Brewing with Fruits, Vegetables, and Spices 13

There comes a time in every homebrewer's development when they look at some odd thing and say to themselves, "Hey, I bet I could ferment that!" And they are usually correct. The sticking point is whether they can do it well. Fruits are fairly easy to incorporate into a beer; they are loaded with simple sugars and have nice aromas and flavors. Vegetables in beer tend to be alternative starch sources, and really don't contribute much in terms of flavor; in fact, they tend to be a novelty. The exception, of course, is the chili pepper, which is actually a fruit, and is very easy to overdo. The best chili beers are ones that have just a nuance or aftertaste of spice. Spice beers themselves tend to be overdone as well. The key to brewing with any fruit, vegetable, or spice is to brew a solid beer first, then use the fruit, vegetable, or spice as an accent. Restraint seems to be something that every homebrewer needs to learn for themselves.

## Brewing with Fruit

Think of the sensations when biting into a piece of fruit—generally there is a fruity aroma, followed by sweetness, followed by a cleansing acidity. We expect fruit flavor to be backed up by some sort of sweet or sour flavor structure. Do humans like sweet fruit? Yes. Do we like sour fruit? Yes, it can

## TABLE 13.1—TYPICAL pH AND BRIX VALUES FOR COMMON FRUITS AND VEGETABLES

| Fruit | Typical pH | Raw juice °Brix | Raw juice PPG (PKL) | Puree °Brix | Puree avg. PPG (PKL) |
|---|---|---|---|---|---|
| Apple | 3.3–3.9 | 13.3 | 6 (51) | | |
| Apricot | 3.3–4.0 | 14.3 | 7 (55 | 9–12 | 5 (40) |
| Banana | 4.5–5.2 | 10.0 | 5 (38) | | |
| Blackberry | 3.2–4.5 | 10.0 | 5 (38) | 9–16 | 6 (48) |
| Blueberry | 3.1–3.7 | 14.1 | 6 (54) | 10–16 | 6 (48) |
| Boysenberry | 3.0–3.5 | 10.0 | 5 (38) | 9–15 | 6 (48) |
| Cherry, tart | 3.2–3.8 | 14.3 | 7 (55) | 10–18 | 6 (54) |
| Coconut | 5.5–7.8 | 10.0 | 5 (38) | | |
| Crabapple | 2.9–3.0 | 15.4 | 7 (59) | | |
| Cranberry | 2.3–2.5 | 10.5 | 5 (40) | 6–9 | 3 (29) |
| Grapes (wine) | 2.8–3.8 | 21.5 | 10 (83) | | |
| Grapefruit | 3.0–3.8 | 10.2 | 5 (39) | 8–12 | 5 (38) |
| Guava | 5.5 | 7.7 | 4 (30) | | |
| Lemon | 2.0–2.6 | 8.9 | 4 (34) | | |
| Lime | 2.0–2.4 | 10.0 | 5 (38) | | |
| Mango | 3.4–4.8 | 17.0 | 8 (65) | | |
| Nectarine | 3.9–4.2 | 12.0 | 6 (46) | | |
| Orange | 3.1–4.1 | 11.8 | 5 (45) | | |
| Papaya | 5.2–5.7 | 10.2 | 5 (39) | | |
| Passionfruit | 2.7–3.3 | 15.3 | 7 (59) | | |
| Peach | 3.4–3.6 | 11.8 | 5 (45) | 9–12 | 5 (40) |
| Pear | 3.5–4.6 | 15.4 | 7 (59) | | |
| Pineapple | 3.3–5.2 | 14.3 | 7 (55) | 11–14 | 6 (48) |
| Plum | 2.8–4.6 | 14.3 | 7 (55) | 14–24 | 9 (73) |
| Pomegranate | 2.9–3.2 | 18.2 | 8 (70) | | |
| Raisin | 3.8–4.0 | 18.5 | 9 (71) | | |
| Raspberry | 3.2–3.7 | 10.5 | 5 (40) | 8–13 | 5 (40) |
| Strawberry | 3.0–3.5 | 8.0 | 4 (31) | 7–12 | 5 (36) |
| Tangerine | 3.3–4.5 | 11.5 | 5 (44) | | |
| Watermelon | 5.2–5.8 | 6.0 | 3 (23) | | |

## TABLE 13.1—TYPICAL PH AND BRIX VALUES FOR COMMON FRUITS AND VEGETABLES (CONTINUED)

| Vegetables | Typical pH | Raw juice °Brix | Raw juice PPG (PKL) | Puree °Brix | Puree avg. PPG (PKL) |
|---|---|---|---|---|---|
| Carrots | 4.9–5.2 | 12 | 6 (46) | | |
| Chili pepper | 4.9–5.2 | 5.0 | 2 (19) | | |
| Potato | 5.4–6.1 | 5.0 | 2 (19) | | |
| Pumpkin | 5.0–5.5 | 8.0 | 4 (31) | | |
| Sweet potato | 5.3–5.6 | 8.0 | 4 (31) | | |

Notes: All values are intended to be typical for average ripeness. Very ripe fruit may be higher. The sugar content of fruits and vegetables are usually given in °Brix, which is very similar to °P, and is essentially the weight percent of sugar in solution. You can use the conversion factor for sugar as 100% soluble extract, i.e., 46 PPG (384 PKL), to calculate the gravity contribution for any fruit or vegetable.

be refreshingly tangy. Do we like bitter fruit? Generally, no, because we instinctively associate bitter and fruity to mean that it is unripe and potentially hazardous to our health. Bitter and sour is an especially potent combination that instinctively warns us not to eat something. Therefore, construct your fruit beers the same way. Fruit beers can be sweet, or they can be sour, but they should generally not be bitter. Of course, there are exceptions, such as grapefruit in an IPA.

Think of fruit breads, pastries, and pies when you are looking for inspiration for fruit beers; dark breads and fruitcakes come to mind. Think of classic confections, such as candied orange slices, or chocolate-covered cherries. Try adding cherries to a chocolate porter or stout. Add some orange rind to a *dunkle weissbier* to complement the dark banana bread character. Pies have a toasty crust with a light honey or caramel flavor that complements fruit flavors, so pale beers with some residual sweetness work well, as do wheat beers. Remember, the yeast will consume all the simple sugars from the fruit, so the malt bill needs to supply the supporting sweetness that the drinker is expecting. Alternatively, the beer can support the fruit flavor with sourness, but be sure to reduce the hop bitterness of the original style so that it doesn't clash.

### Estimating Quantities and Gravity Contributions

The big question is always, "How much fruit should I use?" The answer usually varies from 0.5 to 2.0 lb./gal. (60–240 g/L), depending on the strength of the fruit's flavor. Raspberries and cherries carry through quite well, and 0.5 lb./gal. (60 g/L) will add a nice accent to the beer, whereas blueberries and strawberries are comparatively weak and may require 2–3 lb./gal. (240–360 g/L) to achieve sufficient flavor. Stone fruits usually have strong flavors that carry well into the beer, although peaches are an exception. Apricots are a better substitute for peach flavor, or they can be combined with peaches to support them. Fruit flavor extracts can help round out and support the flavor and aroma from more delicate fruits. Use real fruit to get the color, body, and intangibles, then punch up the aroma with an extract.

The juice from oranges and other citrus fruits can be used but it doesn't have the depth of flavor that you would expect. Use the peel or zest from citrus fruit instead. The peel of bitter orange (a.k.a.

Seville orange) has more flavor than navel (eating) oranges, but whichever you use, it is very important to avoid the white pith underneath the rind because it is astringently bitter.

The second question is usually, "How many gravity points is that?" This is easy to calculate if you are only using the juice. Simply place a drop of the juice on a refractometer and read the percentage of sugars as degrees Brix. Degrees Brix (°Brix) measure the density of sucrose in solution, and are discussed (along with how to use refractometers) in appendix A. For example, a fruit juice that is 10°Brix has 10% of its weight (mass) as soluble sugar. Pure sugar (sucrose) yields 100% of its weight as soluble extract and has a gravity contribution of 46 PPG (384 PKL). A juice that measures 10°Brix would be roughly 10% of that, or 4.6 PPG (38.4 PKL).

The gravity contribution question becomes more complicated for fruit purees, but the manufacturer often supplies this information. The assumption is that the fruit puree is homogenous and has been spread in a thin enough layer that the refractometer reading indicates the mass percentage of sugar in the puree. The contribution from whole (crushed) fruit is tricky, because, while the juice may be 15°Brix, the percentage of juice in the fruit by mass may only be 20%.

In the grand scheme of things though, calculating the gravity contribution is usually not the goal; the goal is sufficient fruit character, and that can be gauged simply by the mass of fruit used in the batch. If 1 lb./gal. was not enough, then 2 lb./gal. can be tried the next time. The gravity points from fruit are usually not included in OG declarations anyway.

### Tips for Brewing Fruit Beers

Here are a few tips and tricks that can help you make that perfect fruit beer.

*Break down the cell walls.* The fermentable sugars in fruit are generally locked inside a waterproof peel and often are further protected in vesicles (i.e., consider the segments of an orange). Freezing, thawing, and crushing the fruit—in other words, making a puree—helps break up those barriers and release the juice so the yeast can get to it.

*Secondary fermentation.* Fruit is best added after the primary fermentation, otherwise it will scrub out most of the aroma and character you were hoping to achieve. Most fruit beers are made as a secondary fermentation, by racking a young beer into a new fermentor with the fruit. The airlock should start bubbling again by the next day and the secondary fermentation may take several days, depending on the form and quantity of the fruit. Allow the beer to rest on the fruit for 2–4 weeks in the secondary fermentor before packaging to minimize the risk of residual fermentable sugars that can cause bottles to explode.

*Watch the pH.* As my good friend Randy Mosher says, "Don't brew flabby fruit beer." Fruit is naturally acidic and many of the common fruit flavors need adequate acidity to support them. Food that is at the wrong pH doesn't taste right, and this holds true for fruit and fruit flavor as well. The pH range for many fruit beers is 4.0–4.4, if not lower. The pH range for sour beers is generally 3.2–3.8. You may note that pH 3.8–4.0 is kind of a gray area, but a better description may be "tart." The natural pH range of a particular fruit or berry can indicate where the flavors or aromas of that fruit are best perceived and may be useful as a guide for adjusting the pH of the beer. Additionally, knowing the natural pH of the fruit or vegetable will help you predict the effect of adding that fruit or vegetable to the beer. For example, adding tart cherries (pH 3.2–3.8) will tend to lower the typical pH of a beer (4.2–4.6) by a few tenths to some new value. The ideal pH of any fruit beer is most likely somewhere between that of the fruit and the beer itself, and the fruit addition may accomplish this by itself. If necessary, citric or lactic acid can be added to a finished beer to lower the pH and brighten the fruit character. Do not add too much, just a couple of milliliters at a time, assuming a 5 gal. (19 L) batch size, and monitor the pH. Lowering the pH by one- or two-tenths should be sufficient. Adding acid is an option, not a necessity.

*Use pectinase.* Adding pectic enzymes (pectinase) will help you extract as much sugar from the fruit as possible and yield a clearer fruit beer. Most fruits have some level of pectin, although this varies, with typical jam-making fruits being the highest. Typical usage rate is one or two teaspoons per 5 gal. (19 L) of beer, depending on the pectin content in the fruit.

## Cherry Dubbel

This beer is one of my best, and the cherry flavor blends very well with the rich melanoidins of the malts, making it taste like cherry pie crust. One particular batch fermented too warm in the garage and I ended up souring it to get rid of a light phenolic flavor. It soured beautifully, using a blend of bugs on oak chips that I got at the National Homebrewers Conference from Vinnie Cilurzo of Russian River Brewery.

*Fruit Beer*

**Original gravity:** 1.070
**Final gravity:** 1.014
**IBU:** 21

**SRM (EBC):** 18 (36)
**ABV:** 7.5%

| Extract and Steeping Grain Version | | |
|---|---|---|
| **Wort A** | **Gravity points** | |
| 2.0 lb. (900 g) Pilsner DME | 30 | |
| 1.1 lb. (500 g) Munich DME | 16 | |
| 1.0 lb. (450 g) Caramunich® malt – Steep | 6 | |
| 1.0 lb. (450 g) aromatic malt 20°L – Steep | 6 | |
| Boil gravity for 3 gal. | 1.058 | |
| **Hop schedule** | **Boil time** | **IBUs** |
| 1 oz. (30 g) Aramis 8% AA | 60 min. | 21 |
| **Wort B (add at knockout)** | **Gravity points** | |
| 4.3 lb. (1.95 kg) Pilsner DME | 60 | |
| 1 lb. (450 g) Belgian dark candi syrup 90°L | 11 | |
| **Yeast strain** | **Pitch (billions of cells)** | **Fermentation temp.** |
| Trappist ale | 300 | 65°F (18°C) |
| **Fruit** | **PPG (PKL)** | **Fermentation temp.** |
| **Use either:** 6 lb. (2.7 kg) cherry puree or 1.5 lb. (0.5 L) tart cherry juice conc. | 7 (55) 31 (259) | 65°F (18°C) 65°F (18°C) |

## All-Grain Version

| Grain bill | Gravity points | |
|---|---|---|
| 11 lb. (5 kg) Pilsner malt | 44 | |
| 1 lb. (450 g) Munich malt | 4 | |
| 1 lb. (450 g) aromatic malt 20°L | 4 | |
| 1 lbs. (450g) Caramunich malt 60°L | 4 | |
| 1 lb. (450 g) Belgian dark candi syrup (90°L) | 5 | |
| Boil gravity for 7 gal. | 1.060 | |
| **Mash schedule** | **Rest temp.** | **Rest time** |
| Conversion rest – Infusion | 150°F (65°C) | 60 min. |
| **Hop schedule** | **Boil time** | **IBUs** |
| 1 oz. (30 g) Aramis 8% AA | 60 min. | 21 |
| **Yeast strain** | **Pitch (billions of cells)** | **Fermentation temp.** |
| Trappist ale | 300 | 65°F (18°C) |

| Fruit | PPG (PKL) | Fermentation temp. |
|---|---|---|
| **Use either:** 6 lb. (2.7 kg) cherry puree or | 7 (55) | 65°F (18°C) |
| 1.5 lb. (0.5 L) tart cherry juice conc. | 31 (259) | 65°F (18°C) |

| Recommended Water Profile (ppm) | | | Brew cube: Amber, Balanced, Medium | | |
|---|---|---|---|---|---|
| Ca | Mg | Total alk. | SO$_4$ | Cl | RA |
| 75–125 | 20 | 50–100 | 100–150 | 100–150 | 0–50 |

*Notes: If using cherry puree, place it in a secondary fermentor and rack the beer onto the fruit. Allow the beer to rest for 2–4 weeks after the secondary fermentation stops before packaging. The cherry juice concentrate can be added directly to the primary fermentor and the beer does not need to be racked.*

## Brewing with Vegetables

In my opinion, the main reason for brewing with vegetables is to provide alternative starch sources rather than flavors. Many of you will point to pumpkin beer, but pumpkin is actually a fruit rather than a vegetable and most of the flavor comes from spices. In fact, the best pumpkin beer I ever had was a 9% ABV strong ale that didn't contain any pumpkin at all; it was like eating pumpkin pie but all of that flavor came from the judicious use of spices.

Root vegetables are probably the most commonly used in brewing, because they are good sources of convertible starch but they don't have much flavor. Carrots and sweet potatoes can add some color too. Root vegetables should generally be cooked for a short period of time to ensure starch conversion in the mash, although mashed potato flakes can be added directly.

In 2015, my friends Drew Beechum and Denny Conn and I brewed a "Clam Chowder Saison" for the Club Night event at the AHA National Homebrewers Conference in San Diego, California. As strange as it sounds, we actually did use clam juice, potato flakes, and spices to brew this beer, and everyone agreed that is was surprisingly drinkable. I went back for seconds.

Note that the "Clam Chowder Saison" recipe that follows doesn't have an extract and steeping grain option.

## Clam Chowder Saison

*Vegetable and Spice Beer*

**Original gravity:** 1.055
**Final gravity:** 1.008
**IBU:** 30

**SRM (EBC):** 4 (8)
**ABV:** 6.3%

| *All-Grain Version* | |
| --- | --- |
| **Grain bill** | **Gravity points** |
| 5.0 lb. (2.27 kg) pale American two-row malt | 20 |
| 3.5 lb. (1.6 kg) pale ale malt | 14 |
| 1.0 lb. (450 g) wheat malt | 4 |
| 1.5 lb. (680 g) Thomas Fawcett Oat Malt | 5 |
| 1.0 lb. (450 g) potato flakes (plain, no butter or milk) | 4 |
| Boil gravity for 7 gal. | 1.047 |

| **Mash schedule** | **Rest temp.** | **Rest time** |
| --- | --- | --- |
| Conversion rest – Infusion | 153°F (67°C) | 60 min. |

| **Hop schedule** | **Boil time** | **IBUs** |
| --- | --- | --- |
| 0.75 oz. (23 g) US Magnum 13% AA | 60 min. | 28 |
| 0.5 oz. (15 g) US Fuggle 4.5% AA | 10 min. | 2 |

| **Spice schedule (add at knockout)** | **Steep time** | **IBUs** |
| --- | --- | --- |
| 1 bay leaf | 15 | … |
| 1 tsp. (3.5 g) black peppercorns | 15 | … |
| 4 sprigs fresh thyme (~2 g) | 15 | … |
| 1 tsp. table salt (6 g) | 15 | … |
| 8 oz. bottle (235 mL) clam juice | 15 | … |
| 0.5 lb. (225 g) lactose | 15 | … |

| **Yeast strain** | **Pitch (billions of cells)** | **Fermentation temp.** |
| --- | --- | --- |
| French saison | 250 | 67°F (19°C) |

| **Recommended Water Profile (ppm)** | | | **Brew cube:** Pale, Balanced, Medium | | |
| --- | --- | --- | --- | --- | --- |
| Ca | Mg | Total alk. | SO$_4$ | Cl | RA |
| 75–125 | 10 | 0–50 | 100–150 | 100–150 | –100–0 |

*Recipe used by permission. Previously published in* Homebrew All-Stars *by D. Beechum and D. Conn (Voyageur Press, Minneapolis, MN, 2016).*

Most commercial pumpkin beers tend to be either insipid or astringent. The following recipe from my friend Malcolm Frazer is neither. The spices are added at the end of the boil and are not removed. An "exBEERiment" on *Brülosophy* with this recipe demonstrated that a significant number of tasters couldn't distinguish between the recipe with pumpkin or without. I've included it in this version. You will need to either mash the pumpkin with 2 lb. (910 g) of base malt, or use a packet of Palmer's Instamash® brewing enzymes to convert the pumpkin's starch. Or you can simply leave it out.

## Pumpkin Beer

*Vegetable and Spice Beer*
*(Previously published on brulosophy.com by M. Frazer.[1] Used with permission.)*

**Original gravity:** 1.066      **SRM:** 8 (16)
**Final gravity:** 1.013      **ABV:** 6.5%
**IBU:** 18

| Extract and Steeping Grain Version | | |
| --- | --- | --- |
| **Wort A** | **Gravity points** | |
| 2 lb. (910 g) Munich DME | 30 | |
| 1 lb. (450 g) Vienna LME | 15 | |
| 1 lb. (450 g) Caramunich malt (45°L) – Steep | 6 | |
| 3.6 lb. (1.6 kg) pumpkin puree (2 cans) – Instamash® | 6 | |
| Boil gravity for 3 gal. | 1.051 | |
| **Hop schedule** | **Boil time** | **IBUs** |
| 0.5 oz. (15 g) US Magnum 13% AA | 60 | 18 |
| **Spice schedule (add at knockout)** | **Steep time** | **IBUs** |
| 4 g Vietnamese cinnamon, ground | 15 | ... |
| 2 g candied ginger, crystallized | 15 | ... |
| 1 g nutmeg, freshly ground | 15 | ... |
| 0.5 g allspice, whole, cracked | 15 | ... |
| **Wort B (add at knockout)** | **Gravity points** | |
| 0.6 lb. (270 g) pale ale DME | 8 | |
| 2.5 lb. (1.13 kg) Munich DME | 35 | |
| 2.2 lb. (1.0 kg) Vienna LME | 26 | |
| **Yeast strain** | **Pitch (billions of cells)** | **Fermentation temp.** |
| English ale | 275 | 67°F (19°C) |

[1] Malcolm Frazer, "It's the Great Pumpkin Xbmt – Pt.1: Does Pumpkin Make a Difference? | Exbeeriment Results!" Brülosophy, October 5, 2015, http://brulosophy.com/2015/10/05/its-the-great-pumpkin-xbmt-pt-1-does-pumpkin-make-a-difference-exbeeriment-results/.

| All-Grain Version | | |
|---|---|---|
| **Grain bill** | **Gravity points** | |
| 4.6 lb. (2.1 kg) pale ale malt | 19 | |
| 4.4 lb. (1.93 kg) Vienna malt | 17 | |
| 4.4 lb. (1.93 kg) Munich malt | 17 | |
| 1 lb. (450 g) Caramunich malt (45°L) | 4 | |
| 3.6 lb. (1.6 kg) pumpkin puree (2 cans) | 2 | |
| Boil gravity for 7 gal. | 1.059 | |
| **Mash schedule** | **Rest temp.** | **Rest time** |
| Conversion rest – Infusion | 153°F (67°C) | 60 min. |
| **Hop schedule** | **Boil time** | **IBUs** |
| 0.5 oz. (15 g) US Magnum 13% AA | 60 | 18 |
| **Spice schedule (add at knockout)** | **Steep time** | **IBUs** |
| 4 g Vietnamese cinnamon, ground | 15 | ... |
| 2 g candied ginger, crystallized | 15 | ... |
| 1 g nutmeg, freshly ground | 15 | ... |
| 0.5 g allspice, whole, cracked | 15 | ... |
| **Yeast strain** | **Pitch (billions of cells)** | **Fermentation temp.** |
| English ale | 275 | 67°F (19°C) |
| **Recommended Water Profile (ppm)** | **Brew cube:** Amber, Balanced, Medium | |

| Ca | Mg | Total alk. | SO₄ | Cl | RA |
|---|---|---|---|---|---|
| 75–125 | 10 | 50–100 | 100–150 | 100–150 | 0–50 |

## Brewing with Spices

The secret to brewing with spices is restraint and timing. Restraint to know not to use too much, and timing to know when to add, and remove, them from the brew. Do not boil your spices, even for just a short time; you will lose aroma and it will tend to add astringency. Hot steeping after the boil is always better. I have never found that the aroma or flavor of a spice was improved by boiling.

The steeping temperature has a large effect on the spice character, as was discussed in chapter 4. Cold steeping at room temperature gives a fresher aroma but less flavor from a spice, whereas hot steeping gives more flavor and a different aroma. The best example of this difference is between smelling a bag of freshly ground coffee and smelling a cup of freshly brewed coffee. The aroma is still coffee, but they are different. You may want to use both hot and cold methods to get the best expression of a spice, just like we do with hops (i.e., hop steeps and dry hopping).

Most spices, such as cinnamon, vanilla, ginger, and cloves, should be used in small amounts, for example, 1–10 g per batch. Delicate spices like vanilla should be used after fermentation. The exceptions to this guideline are flavorings, such as cocoa and coffee. These can be used in larger quantities, anywhere from a couple of ounces to a half pound (60–225 g) per 5 gal. (19 L) batch. Spices like leaves or bark should be placed in a mesh bag and removed before fermentation. Do your research on the Internet to see the quantity of spices other brewers use in their recipes. For more inspiration and advice on brewing beer with unusual ingredients, I recommend reading, among others, *Radical Brewing* by Randy Mosher (2004) and *Experimental Homebrewing* by Drew Beechum and Denny Conn (2014).

Coffee is a delightful addition to many beer recipes, and not just porters and stouts. It blends very well with the maltiness of many lagers, and even works in black IPA, although balancing the bitterness requires some skill. Flavored coffees, such as vanilla, hazelnut, and blueberry flavor, are also an option. However, coffee beers do not age well because of the oils, which oxidize pretty easily. I have judged coffee beers many times over the years and it can be rare to find a coffee beer that doesn't taste like it was sitting on the burner for a couple of days. The first solution to this problem is to not boil it. The second may be to only use cold-steeping methods and adding the coffee after fermentation, but I prefer the extra depth of flavor that you get from hot steeping. This is another case where a combination of hot and cold steeping may serve you best. Bottom line: coffee beers need to be consumed fresh.

For coffee beers, one other note of caution: when steeping the coffee for your beer be aware of the presence of 3-isobutyl-2-methoxypyrazine, a pyrazine compound that smells like green peppers or mashed green peas. It is not hazardous, but it can dominate the aroma of your beer, and that can be disappointing when unwanted. This pyrazine seems to be most prevalent in light- and medium-roast coffees that have been cold or hot steeped for long periods of time. Cold steeping is actually done at room temperature, and in this case a long period of time would seem to be more than 12 hours. Hot steeping should be only a matter of minutes, not hours, but this method seems to be more likely to pull astringent flavors from long steeps rather than 3-isobutyl-2-methoxypyrazine.[2]

Cocoa is a very popular brewing spice, but there are a couple things to know. First, the fats that are naturally part of the cocoa bean will oxidize and taste bad in the beer, so the less of those the better. Therefore, I don't recommend chopping up chocolate candy bars, or even blocks of baker's chocolate; these generally have too much fat. Cocoa nibs are very popular for adding chocolate flavor, because they are "minimally processed" and taste good. The problem is that cocoa nibs still have a fair amount of fat. "Minimally processed" is one of those buzzwords that make you think a foodstuff is automatically better, but there is a reason we process cocoa nibs into cocoa powder—to get rid of the fat and improve flavor stability. High quality cocoa powder is essentially concentrated cocoa nibs, which have been defatted by pressing. Interestingly, both cocoa nibs and cocoa powder can be added to the mash as well as the boil. I have never tried them in the mash, preferring to add cocoa at the end of the boil, but it seems like you would need to use more in the mash to get the same level of flavor. Cocoa powder doesn't dissolve into the wort, it becomes a suspension, and, if we are lucky, forms a colloid. Adding it to the boil for a short time right at the end of the boil, can help the powder disperse throughout the wort. Cocoa powder is the exception to the "don't boil" rule, but long boils will definitely degrade the flavor. Chocolate syrup is also a good option and won't settle out like cocoa powder can, but check the label for extraneous hydrogenated oils and emulsifiers.

Here is my recipe for "Viennese Mocha Stout," inspired by Jamil Zainasheff's Chocolate Hazelnut Porter. I don't use hazelnut coffee in this recipe, but that is certainly an option. This recipe uses almost twice the cocoa powder that most other chocolate beers do, but it demonstrably works here to create a rich and satisfying beer.

---

[2]    Drew Beechum, "Coffee and Jalapenos," *Experimental Brewing* (blog), March 27, 2014, https://www.experimentalbrew.com/blogs/drew/coffee-and-jalapenos.

## Viennese Mocha Stout

*Spice Beer*

**Original gravity:** 1.062
**Final gravity:** 1.016
**IBU:** 32

**SRM (EBC):** 30 (60)
**ABV:** 6.3%

| *Extract and Steeping Grain Version* | | |
|---|---|---|
| **Wort A** | **Gravity points** | |
| 2.2 lb. (1 kg) pale ale DME | 31 | |
| 1.0 lb. (450 g) caramel 40L Malt – Steep | 5.5 | |
| 1.0 lb. (450 g) caramel 80L Malt – Steep | 4.5 | |
| 0.5 lb. (225 g) roast barley (300L) – Steep | 4 | |
| 0.5 lb. (225 g) Briess Dark Chocolate malt (400°L) – Steep | 4 | |
| 0.5 lb. (225 g) Briess Carabrown® malt (55°L) – Steep | 2 | |
| Boil gravity for 3 gal. | 1.051 | |
| **Hop schedule** | **Boil time** | **IBUs** |
| 1 oz. (30 g) Centennial 10.5% AA | 60 min. | 32 |
| **Spice schedule (add at knockout)** | **Steep time** | **IBUs** |
| 10 g Ceylon cinnamon, ground | (add) | ... |
| 0.5 lb. (225 g) cocoa powder | (add) | ... |
| 0.5 lb. (225 g) coarse-ground espresso | 15 | ... |
| **Wort B (add at knockout)** | **Gravity points** | |
| 4.4 lb. (1.8 kg) pale ale DME | 56 | |
| **Yeast strain** | **Pitch (billions of cells)** | **Fermentation temp.** |
| English ale | 275 | 65°F (18°C) |

| *All-Grain Version* | |
|---|---|
| **Grain bill** | **Gravity points** |
| 11 lb. (5 kg) pale ale malt | 44 |
| 1 lb. (450 g) caramel 40°L malt | 2.5 |
| 1 lb. (450 g) caramel 80°L malt | 2 |
| 0.5 lb. (225 g) Briess Carabrown malt (55°L) | 1 |
| 0.5 lb. (225 g) Briess Dark Chocolate malt (400°L) | 2 |
| 0.5 lb. (225 g) roast barley (300°L) | 2 |
| Boil gravity for 7 gal. | 1.053 |

| Mash schedule | Rest temp. | Rest time |
|---|---|---|
| Conversion rest – Infusion | 153°F (67°C) | 60 min. |

| Hop schedule | Boil time | IBUs |
|---|---|---|
| 1 oz. (30 g) Centennial 10.5% AA | 60 min. | 32 |

| Spice schedule (add at knockout) | Steep time | IBUs |
|---|---|---|
| 10 g Ceylon cinnamon, ground | (add) | ... |
| 0.5 lb. (225 g) cocoa powder | (add) | ... |
| 0.5 lb. (225 g) coarse-ground espresso | 15 | ... |

| Yeast strain | Pitch (billions of cells) | Fermentation temp. |
|---|---|---|
| English ale | 275 | 65°F (18°C) |

| Recommended Water Profile (ppm) | | | Brew cube: Dark, Balanced, Medium | | |
|---|---|---|---|---|---|
| Ca | Mg | Total alk. | SO$_4$ | Cl | RA |
| 75–125 | 20 | 100–150 | 100–150 | 100–150 | 50–100 |

# Brewing Sour Beers 14

Sour beers have probably been around as long as beer itself. It is remarkable to realize that we only differentiated yeast and bacteria about 150 years ago, which is a drop in the bucket considering the roughly 9,000-year history of brewing. This is not to say that all historical beers were sour: the difference between a sour beer, a *Brettanomyces* beer, and a *Saccharomyces* beer are readily apparent and brewers have long known the difference; but without the understanding of what you are dealing with and a microscope to view them, they can be hard to separate.

A quick note before we begin: *Brettanomyces* is a yeast genus, like *Saccharomyces*, but it is most often used in conjunction with bacteria for producing sour beers, so that is why I am including it in this chapter. You can brew 100% non-sour beer using *Brettanomyces*, and many people do, but it is most often used with sours.

## A Note on Equipment
You should buy separate plastic fermentation equipment for brewing sour beers (fermentors, buckets, hoses, siphons, and airlocks) to prevent contamination of your regular equipment. This recommendation only applies to equipment that touches the bacteria and is not easily sanitized. Siphon hoses are notorious contamination sources. Stainless steel and glass are easier to aggressively sanitize than plastic, so those are usually not an issue. Be thorough if you don't want all of your beers becoming contaminated and eventually going sour.

## The Bugs *(Los Bichos)*

There are four main genera of microorganisms, "the bugs," commonly used to produce sour beers. These are the bacteria *Lactobacillus*, *Pediococcus*, and *Acetobacter*, and the yeast *Brettanomyces*. There are a few others, such as *Enterobacter*, which smells like vomit, but these are not encouraged and usually only encountered in wild fermentations. More microorganisms may be added to the list as time goes on, but for now, these are the main ones.

### *Lactobacillus*

*Lactobacillus* is the workhorse of the souring world and is used for making yogurt, sausages, probiotics, and sour beer. There are many different species and strains, but they all produce lactic acid to one degree or another. Lactobacilli bacteria tend to produce a clean sour character, generally soft, tangy, and tart as opposed to sharp, like vinegar (acetic acid), or puckeringly sour, like malic acid. Lactobacilli grow best between 90–115°F (32–46°C), but will still grow well at typical ale fermentation temperatures. Lactobacilli will typically only consume a small amount of sugars in the fermentation, about 2–4 gravity points.[1] *Lactobacillus* species only eat the simple sugars glucose, fructose and maltose; they do not consume maltotriose or raffinose. Most lactobacilli tend to be inhibited by hops and this is one reason that sour beers, such as Berliner *weiss*, usually only have 5–8 IBUs.

There are several common species of *Lactobacillus*, some of which are homolactic, meaning that they only produce lactic acid. Other species are heterolactic, meaning they are able to produce lactic acid or ethanol and carbon dioxide. Heterolactic strains do not produce significant amounts of diacetyl.[2] The homolactic strains produce more diacetyl, but not as much as yeast or *Pediococcus*.

*Lactobacillus* blends and single strains can be purchased from several of the brewing yeast companies, but it is possible to obtain lactobacilli elsewhere as well. Non-fat, "active culture" yogurt is a common source of lactobacilli, as are probiotic supplements sold at pharmacies and grocery stores. The only criticism of these off-the-shelf sources is that single-strain *Lactobacillus* fermentations are often described as being too clean, or lacking complexity, compared to blends that were developed especially for sour beers. However, probiotic supplement blends of multiple strains are now available and brewers have reported good results with these. The estimated cell count for probiotic supplements is often listed on the package.

The suggested pitching rate for *Lactobacillus* is between 100 and 200 billion cells per 5 gal. (19 L). Lower pitching rates are said to produce more off-flavors, such as cheese, sweaty feet, and apple juice. Pitching rates are not as critical for sour beers as yeast pitching rates are for regular beers, because attenuation is not the issue. The issue is flavor profile and the length of time it takes for the souring to occur, and this is where another consequence comes into play. Lactobacilli need amino acids to grow; some of these they can synthesize and others must be obtained by breaking down larger proteins present in the wort. In short, all of the popular brewing strains of *Lactobacillus* will excrete proteolytic enzymes (i.e., enzymes that break down proteins) during the souring process to facilitate their growth, and this can have demonstrable effects on the head retention of sour beers. There are two ways to combat this problem, both of which relate to the pH sensitivity of the proteolytic enzymes, which decrease in activity below a pH of 5 and cease

---

[1]   Elke Arendt, presentation, Belgian Brewing Conference, KU Leuven, Belgium, September 2015.

[2]   Matthew Humbard, "Physiology of Flavors in Beer – Lactobacillus Species," *A Ph.D in Beer* (blog), April 13, 2015, phdinbeer.com/2015/04/13/physiology-of-flavors-in-beer-lactobacillus-species/.

altogether below 4.5. The first method is to use a high pitching rate such that overall growth is reduced and the wort is soured below the pH threshold in a few hours rather than a few days. The second method is to pre-acidify the wort to a pH of 4.5–4.8 to inhibit the enzymes using off-the-shelf lactic acid, then pitching the *Lactobacillus* culture. The second method is more effective than the first, but the first may seem less like cheating to the brewing purists out there. Kettle souring and other fermentation methods will be discussed later in this chapter.

The following are endpoint pH values that are typically achieved by the various common strains of *Lactobacillus* during souring:

| *Lactobacillus* species | Fermentation type | Example endpoint pH |
| --- | --- | --- |
| L. brevis | Heterolactic | 3.3 |
| L. delbrueckii | Homolactic | 4.4 |
| L. buchnari | Heterolactic | 3.8 |
| L. plantarum | Heterolactic | 3.2 |

*Source: Endpoint pH data averaged from Matthew Humbard, "Beer Microbiology – Lactobacillus pH experiment," A Ph.D. in Beer (blog), August 5, 2015, http://phdinbeer .com/2015/08/05/beer-microbiology-lactobacillus-ph-expeirment/.*

## Pediococcus

*Pediococcus* is the other main souring bacteria. It is used for curing sausage and making sauerkraut, in addition to souring beer. *Pediococcus* differs from *Lactobacillus* in that it produces lactic acid more slowly, although it tends to sour the beer more (i.e., a comparatively lower pH) than *Lactobacillus*. Pediococci grow best at warm temperatures, 65–85°F (18–29°C), and will not grow above 95°F (35°C). The recommended pitching rate for pediococci is roughly 10 billion cells per 5 gal. (19 L). *Pediococcus* is generally resistant to hops and will grow in beers up to 30 IBU.

*Pediococcus* has two significant fermentation characteristics: it produces a lot of diacetyl and it can make the beer "sick," as the Belgians describe it. This sickness manifests as polysaccharide gel strands, which is why sick beer is also often described as "ropy." The polysaccharides are, in fact, beta-glucans that are normally produced by the bacteria as they grow, with more being produced toward the end of growth. These characteristics are the reason why *Brettanomyces* is often used in collaboration with *Pediococcus*, because *Brettanomyces* will clean up the diacetyl and help dissolve the ropy strands.

Even though *P. damnosus* is stated to be homolactic, *Pediococcus* in general is said to produce small amounts of acetic acid as well, and this may account for the general opinion that *Pediococcus* produces a sharper, more complex sour than does *Lactobacillus*. Perhaps this character actually comes from *P. claussenii*, which is heterolactic, or perhaps some strains of *P. damnosus* are actually heterolactic as well. In his book, *American Sour Beers*, Michael Tonsmeire states,

> *Many brewers describe the acid character from* Pediococcus *as being more aggressive and sharp than that from* Lactobacillus. *This could be due to either the lower pH generated by* Pediococcus, *or the sub-threshold presence of acetic acid from the* Brettanomyces *that would not always be found in beers soured with* Lactobacillus. *(p. 57)*

*Pediococcus* is the primary souring bacteria in Belgian *lambic* beers, and also in American sour beers, such as those from Russian River Brewing Company.

The following are endpoint pH values that are typically achieved by the various common strains of *Pediococcus* during souring:

| *Pediococcus* species | Fermentation type | Example endpoint pH |
| --- | --- | --- |
| P. damnosus | Homolactic | <3.0 |
| P. claussenii | Heterolactic | <3.0 |

### Brettanomyces

*Brettanomyces* is a genus of yeast that loves to live in wood barrels. *Brettanomyces* is typically not an acid producer, unless there is a lot of oxygen, in which case it will produce a small amount of acetic acid. It mainly produces "funk," which are unusual phenolics and esters that contribute aromas and flavors ranging from barnyard and leather, to spice and tropical fruit. *Brettanomyces* produces amylase enzymes, such as alpha-glucosidase, that allow it to break the bonds in dextrins and other sugars normally considered unfermentable for brewer's yeast. It also is a strong diacetyl reducer, and some brewers have reported that it breaks down DMS as well. *Brettanomyces* can add nice complexity to saisons and American sour beers.

There are five species within the *Brettanomyces* genus: *B. anomalus*, *B. bruxellensis*, *B. custersianus*, *B. nanus*, and *B. naardenensis*. The most commonly available types are the *B. bruxellensis* and *B. anomalus*. *Brettanomyces* is a very contrary creature compared to *Saccharomyces*. It will produce very funky flavors when added in small amounts to a normal *Saccharomyces* fermentation, but can produce very clean, light bodied, thirst-quenching beers in "100% Brett" fermentations. The pitching rate for 100% Brett fermentations is 1.0–1.25 billion cells per liter, between that of ale and lager pitching rates. *Brettanomyces* tends to prefer warmer fermentations, between 70°F and 80°F (21–26°C), and ferments slowly. A 100% brett fermentation may take 2–6 weeks to finish. (See cautionary sidebar.)

*Brettanomyces* is often pitched in conjunction with *Pediococcus* to clean up the copious diacetyl that *Pediococcus* produces. For more information on *Brettanomyces* and the common souring bacteria, you should visit the wiki at *Milk The Funk* (http://www.milkthefunk.com/wiki).

### CAUTION FOR BREWING WITH BRETT

*Brettanomyces* can ferment larger sugars than *Saccharomyces* yeast. Never add Brett to a beer immediately before bottling; it must added to the fermentor and given sufficient time (weeks) to ferment to terminal gravity before bottling. That gravity will vary with the recipe and must be determined firsthand by measurement with a hydrometer. Failure to do so may result in exploding bottles.

### Microorganisms from Other Sources

There are two other readily available sources for bacteria that you can use to make sour beers: malt and the great outdoors. Adding a couple of handfuls of high quality malt to a 1.040 gravity starter works well, if you pre-acidify the starter with a couple of milliliters of 88% lactic acid and purge the head space of oxygen with some carbonated water before affixing the airlock. Detailed instructions are presented below.

The other source is wild yeast and bacteria from the great outdoors. The romanticism of this method may appeal to many brewers, but it is important to understand that you are dipping your net into a dirty pond and hoping to pull out a tasty fish. Many brewers will build a coolship (basically a long shallow open fermentor) and lay their wort out overnight in hopes of getting lucky.

Amazing as it seems, many brewers will trust to luck and prepare themselves to like whatever it is they eventually create, but it doesn't have to work that way. Instead, you can create a small starter wort and place that outside, let it start, and then smell or taste the result to see if it is something you actually want to brew with. I first heard of this procedure at the 2016 National Homebrewers Conference in Baltimore, Maryland, and it was a head-slapping moment. John Wilson and Brian Wolf presented "Brewing Wild," in which they described their wild inoculation process and then served us a beer made from it. It was fantastic, and it was simply because the wild yeast starter had been screened before pitching to ensure it was worth brewing with. Simple as that!

John and Brian recommended placing the inoculation wort in a garden or by fruit trees—somewhere out of direct sun where it would have a better chance of picking up a favorable wild yeast (or bacteria). It stands to reason that a garden is a better location than a parking garage. The starter is exposed for a day and night and then brought inside and capped with aluminum foil or an airlock to see what develops.

## Making a Wild Inoculation Wort

*Step 1:* Create a 1.5 L starter of 1.040 gravity wort using 175 g of DME dissolved in 1.5 L of water. Boil it, cover with aluminum foil, and allow it to cool to room temperature.

*Step 2:* Acidify the wort with 2–3 mL of 88% lactic acid, or about 50 mL of pineapple juice. (Pineapple juice contains citric acid and has a natural acidity of 3.5 pH.) Lowering the pH of the wort to 4.5 will inhibit undesirable and potentially hazardous microorganisms from growing. (See sidebar, "Caution for Wild Fermentations.") Verify the pH with a calibrated pH meter before continuing. Do not trust pH test strips.

*Step 3:* Place the wort in an open jar or container where it is likely to pick up favorable microorganisms. You can cover the jar with window screen mesh or a piece of cheesecloth to keep the flies out.

*Step 4:* After 24 hours bring the wort inside and attach an airlock, or pour it into a flask that can take an airlock. Hopefully the wort will start fermenting within a couple of days. Smell the starter to see if you want to taste it, and taste it to see if you want to brew with it. Good luck!

## Culturing Lactobacilli from Malt

*Step 1:* Make an inoculation wort, including acidification, as described above for "Making a Wild Inoculation Wort."

*Step 2:* Add two handfuls of fresh Pilsner malt to the jar or flask. Select your malt carefully, it should smell dry and fresh, without any hints of mildew or mold. Old malt may carry a greater variety of microbes than you want to deal with.

*Step 3:* Top up the flask with carbonated water[3] (250–500 mL) to leave only an inch or so (2–3 cm) of head space. The carbonated water will fizz and should purge the headspace of oxygen, reducing the growth of unwanted microbes from the malt.

*Step 4:* Attach an airlock and let it grow. *Lactobacillus* grows best at 100–110°F (38–43°C), but it will also grow at room temperature, around 68-77°F (20–25°C). At 110°F (43°C), the growth should

---

[3]   This advice taken from Derek Springer, "Lactobacillus Starter Guide," *Five Blades Brewing,* April 19, 2015, http://www.fivebladesbrewing.com/lactobacillus-starter-guide.

be done in about 2–3 days. The gravity of the wort will not significantly change, but the pH will have dropped to between 3.2 and 3.8.

## CAUTION FOR WILD FERMENTATIONS

A wild fermentation can be very wild indeed. We often say that fermenting beer will not harbor pathogenic (harmful) bacteria, but that statement does not include wild fermentations. A spontaneous or wild fermentation can include members of the pathogenic enteric bacteria group such as *Escherichia coli* (*E. coli*). These bacteria can produce undesirable metabolites such as biogenic amines that can cause severe allergic reactions. Pre-acidification to a pH of 4.5 will prevent these types of bacteria from growing in your inoculation wort or starter.

## Brewing Sour Beers

There are three options for brewing a sour beer:
1. Yeast first, bacteria second.
2. Bacteria first, yeast second.
3. Yeast and bacteria together at the same time.

The first option is the more traditional method used by homebrewers for many years and it generally produces a soft complex sour beer over a few months. The souring usually proceeds slowly because there aren't many residual sugars and carbohydrates left for the bacteria to eat. This is a good method because it's like the tortoise and the hare—slow and steady will win the race. This method also has the advantage of low maintenance; you set it and forget it. It can take several months to produce a pleasing level of sourness, but it is difficult to over-sour the beer.

The second option involves a method that has traditionally been done as a sour mash, by allowing the mash to sit over a night (or two) to allow the lactic acid bacteria on the malt to proliferate and produce the sour character. The problem with this method is that there is a lot more microbial variety on the malt than just lactobacilli, and you can end up with a variety of stinky fatty acids in the mix, such as butyric acid that smells and tastes like vomit. The popular fix to this problem is kettle souring, where you mash, lauter, and drain to your boil kettle as usual (see chapter 20 for instructions on brewing with grain), and then pitch a prepared bacterial culture to sour it before the yeast fermentation. This method is fast, like 18–36 hours fast, but you need to shoot the rabbit when the race is done. I will lay out the specific steps shortly, but to summarize: you boil the wort briefly to kill any microbes and spores left in the wort from the mash, pitch the prepared bacterial culture, monitor the pH as it drops to about 3.8, then boil the wort (with hops) to stop the souring, and then transfer to your fermentor and pitch the yeast. This method has the advantage of consistency, if it is properly controlled. Many commercial brewers have adopted this method because it allows them to produce a consistent sour product in just a few more days than it takes to do a non-sour fermentation.

The third option, pitching both yeast and bugs together, has been a popular method with homebrewers for several decades, but it can be inconsistent depending on which culture has the higher pitching rate. The brewer's yeast generally ferments the simple sugars in the normal timeframe, while the slower-growing bacteria "chew on" the remaining carbohydrates after the beer fermentation finishes. The two types of microorganism aren't really competing for the same resources, because we know the yeast will work much faster on wort simple sugars, but can't process the complex carbohydrates.

If you do your homework and pay attention to pitching rates and all the fermentation details, you will produce a great sour beer. Many styles are routinely produced this way, but it takes skill. Yeast companies sell several bacteria and yeast blends for styles like Berliner weiss, and Flanders red, just to name a few. You can also (and likely will need to) select and combine your own choices of yeast and bacteria strains to pitch to make your own sour style. Most homebrewers (and commercial brewers) used this method to produce their sour beers until kettle souring came along.

## Kettle Souring

Kettle souring is the compromise between patience and skill, the difference between a pony ride at the state fair and breaking in your own mustang. You still need to hold tight to the reins but there is a clearly marked trail guide to help you arrive at your destination. The following method uses two steps to prevent the growth of unwanted bacteria. The first is a short pre-boil to kill any bacteria and spores that have survived the pasteurizing effects of the mash. The second is pre-acidifying the wort to a pH of 4.5, which will inhibit the growth of *Enterobacter* and *Acetobacter*, if you are going to be pitching a mixed culture. If you are going to use a single culture of *Lactobacillus*, then inhibiting the other bacteria is not an issue.

*Step 1. Prepare the culture.* You may need to make a starter for your bacterial culture. You can use the inoculation wort starter method described for "Culturing Lactobacilli from Malt" above (but without the grain). Generally, 30–60 mL of yogurt in a 1–2 L starter will be sufficient for a 6 gal. (23 L) batch. If you are using probiotics, then 1–2 capsules into the starter will work as well.

*Step 2. Produce your wort.* Mash and lauter using your normal methods to produce your wort, and transfer it to your boiling kettle.

*Step 3. Short boil.* Boil the wort for 10–15 min. to thoroughly sanitize it. Chill the wort to 100–110°F (38–43°C). You can chill to room temperature, but it works more rapidly and more consistently when it's warm.

*Step 4. Pre-acidify.* There are two reasons to pre-acidify your wort, (a) because it will help inhibit any unwelcome microorganisms, and (b) because it will inhibit the proteolytic enzymes secreted by the *Lactobacillus* and thereby preserve your head retention. Acidifying with 88% lactic acid to a pH of 4.5–4.8 should be sufficient. The amount needed will vary but it should be in the neighborhood of 5 mL. Use a pH meter instead of color strips to be sure of an accurate reading. You don't want to overshoot, or there will be little point in proceeding with your culture!

*Step 5. Pitch the bacteria.* Pour the bacterial culture starter into the wort and say the magic words.

*Step 6. Souring the wort.* The wort should take 18–36 hours to sour to a pH of 3.5-ish, depending on the temperature (cooler wort will take longer) and other factors. Measure the pH at least every eight hours to gauge the progress. Taste it as well to judge the sour character and help you decide when it is sour enough. You are looking for a pH of 3.2–3.8, depending on the type of bugs you are pitching. A pure *Lactobacillus* culture has a soft acidity, and a lower pH toward 3.2 seems to taste best. In mixed cultures, such as those grown from the grain, the mix of acids is more biting and a higher pH toward 3.8 tastes best.

*Step 7: Boil and ferment the wort.* After the wort has soured to the desired level, boil it with your hops (according to the recipe), then chill, aerate, and pitch your yeast as you normally would. You should experience a normal fermentation and decrease in gravity. In addition, the pitching rate for the yeast fermentation should be higher than normal. Therefore, for ales that would normally pitch 0.75 billion cells per liter per 4 gravity points (or 0.75 billion/L/°P; see chapter 7) you would instead pitch 1.0 billion cells. Brewers report that European ale strains seem to be more acid tolerant than American ale strains and are better able to fully attenuate under these conditions.

## PELLICLES AND THE ZEN OF SOUR BEERS

Some bacterial fermentations (other than kettle souring) will produce a pellicle, which is usually a whitish gelatinous film, or mat, on top of the beer. The pellicle is typically a fraction of an inch (a few millimeters) thick and may have some texture to it, such as bubbles or ripples. The surface may appear powdery but will have a very organic look to it. It does not look like the foamy yeast kräusen.

The formation of a pellicle indicates that oxygen is getting to the beer, and it is thought that the pellicle is the microorganisms' response to the presence of oxygen, that is, aerobic versus anaerobic fermentation. We really don't know what pellicles do or how they affect the beer. Therefore, when in doubt, don't mess with it. This is the advice of professional sour brewers as well, "Leave it alone." If you want to sample the beer, gently move it to the side, or poke a single small hole through it with a wine thief. A small hole will quickly repair itself.

The pellicle may dissolve back into the beer, or it may settle to the bottom, or it may be fully intact when it is time to bottle the beer. In every case, ignore it. Rack out from underneath it if you have determined that the beer has reached terminal gravity and best flavor. There is no need to include part of the pellicle in the bottles or keg. It is recommended when bottle conditioning to lay the bottles on their side to decrease the likelihood of pellicle formation.

One final note: pellicles are not black or green or hairy; those are signs of mold and the beer should be discarded.

The following three recipes cover a few European styles and an American sour ale. The fermentation temperatures given for these recipes are general recommendations. Check the yeast or bacteria supplier's packaging for specific temperature recommendations. Kettle souring for extract versions should be done before boiling with hops. Procedure: Boil wort A for 15 minutes without hops, conduct the kettle sour procedure, then boil with hops according to recipe, add wort B, then chill and ferment as usual.

## Die Weiße Königin

*Berliner Weisse*

**Original gravity:** 1.031
**Final gravity:** 1.008
**IBU:** 5

**SRM (EBC):** 2 (4)
**ABV:** 3%

| Extract Version | |
| --- | --- |
| **Wort A** | **Gravity points** |
| 2 lb. (910 g) wheat DME | 28 |
| Boil gravity for 3 gal. | 1.028 |

| Hop schedule | Boil time | IBUs |
|---|---|---|
| 0.35 oz. (10 g) Hersbrucker 4% AA | 60 min. | 5 |
| **Microbe strain** | **Pitch (billions of cells)** | **Fermentation temp.** |
| Kettle sour with *Lactobacillus* | 150 | 104°F (40°C) |
| German ale or Kölsch yeast | 175 | 65°F (18°C) |
| **Wort B** | **Gravity points** | |
| 1 lb. (450 g) wheat DME | 14 | |
| 1 lb. (450 g) Pilsner DME | 14 | |

### *All-Grain Version*

| Grain bill | Gravity points | |
|---|---|---|
| 3.3 lb. (1.5 kg) wheat malt | 14 | |
| 3.3 lb. (1.5 kg) Pilsner malt | 13 | |
| Boil gravity for 7 gal. | 1.027 | |
| **Mash schedule** | **Rest temp.** | **Rest time** |
| Conversion rest – Infusion | 150°F (65°C) | 60 min. |
| **Hop schedule** | **Boil time** | **IBUs** |
| 0.35 oz. (10 g) Hersbrucker 4% AA | 60 min. | 5 |
| **Microbe strain** | **Pitch (billions of cells)** | **Fermentation temp.** |
| Kettle sour with *Lactobacillus* | 150 | 104°F (40°C) |
| German ale or Kölsch yeast | 175 | 65°F (18°C) |

| Recommended Water Profile (ppm) | | | Brew cube: Pale, Balanced, Soft | | |
|---|---|---|---|---|---|
| **Ca** | **Mg** | **Total alk.** | **$SO_4$** | **Cl** | **RA** |
| 50–100 | 10 | 0–50 | 0–50 | 50–100 | –50–0 |

## Oud Geestigbier

*Belgian Witbier*

**Original gravity:** 1.050  
**Final gravity:** 1.011  
**IBU:** 20

**SRM (EBC):** 5 (10)  
**ABV:** 5%

### *Extract and Steeping Grain Version*

| Wort A | Gravity points |
|---|---|
| 2.5 lb. (1.3 kg) wheat DME | 35 |
| 1.1 lb. (500 g) flaked oats – Instamash® | 9 |
| Boil gravity for 3 gal. | 1.044 |

| Hop schedule | Boil time | IBUs |
|---|---|---|
| 0.6 oz. (17g) Mandarina Bavaria 8% AA | 60 min. | 15 |
| **Microbe strain** | **Pitch (billions of cells)** | **Fermentation temp.** |
| Kettle sour with *Lactobacillus* | 150 | 104°F (40°C) |
| Belgian witbier or Kölsch yeast | 275 | 68°F (20°C) |
| **Wort B** | **Gravity points** | |
| 1.5 lb. (680 g) Pilsner DME | 21 | |
| 2.0 lb. (910 g) wheat DME | 28 | |
| ~1.75 oz. (50 g) orange peel – Steep 15 min. | | |
| ~0.5 oz. (15 g) crushed coriander seed – Steep 15 min. | | |
| ~0.1 oz. (3 g) dried chamomile flower – Steep 15 min | | |

## All-Grain Version

| Mashing option | Gravity points | |
|---|---|---|
| 5 lb. (3.2 kg) Pilsner malt | 20 | |
| 5 lb. (3.2 kg) wheat malt | 21 | |
| 1.1 lb. (500 g) flaked oats | 4 | |
| ~1.75 oz. (50 g) orange peel – Steep 15 min. | | |
| ~0.5 oz. (15 g) crushed coriander seed – Steep 15 min. | | |
| ~0.1 oz. (3 g) dried chamomile flower – Steep 15 min. | | |
| Boil gravity for 7 gal. | 1.045 | |
| **Mash schedule** | **Rest temp.** | **Rest time** |
| Conversion rest – Infusion | 153°F (67°C) | 60 min. |
| **Hop schedule** | **Boil time** | **IBUs** |
| 0.6 oz. (17g) Mandarina Bavaria 8% AA | 60 min. | 15 |
| **Microbe strain** | **Pitch (billions of cells)** | **Fermentation temp.** |
| Kettle sour with *Lactobacillus* | 150 | 104°F (40°C) |
| Belgian witbier or Kölsch yeast | 275 | 68°F (20°C) |

| Recommended Water Profile (ppm) | | | | Brew cube: Pale, Malty, Soft | |
|---|---|---|---|---|---|
| Ca | Mg | Total alk. | SO$_4$ | Cl | RA |
| 50–100 | 10 | 0–50 | 0–50 | 50–100 | −100–0 |

*Notes*

*1. Any bittering hop will work, the goal is 15 IBU of clean bitterness. Mandarina Bavaria has a nice orange or tangerine character that blends well with this style.*

*2. Wash two medium-sized oranges and use a vegetable or apple peeler to carefully remove the outer rind. This should be about 50 g total. Do not peel too deeply—the peel should be thin, bright orange, and without white pith.*

*3. After the boil, turn the heat off and place the orange peel, crushed coriander, and chamomile in a mesh bag and allow them to hot steep for 15 min. before chilling. They should be removed before fermentation. Additional spices (same proportions) can be cold steeped after fermentation for extra aroma, if desired.*

## Sour Bastard

*American Sour Ale*

**Original gravity:** 1.070
**Final gravity:** 1.016
**IBU:** 25

**SRM (EBC):** 18 (36)
**ABV:** 7.5%

| *Extract and Steeping Grain Version* | | |
|---|---|---|
| **Wort A** | **Gravity points** | |
| 3.5 lb. (1.6 kg) pale ale DME | 53 | |
| 1.0 lb. (450 g) Weyermann Caraaroma® malt – Steeped | 3 | |
| 0.5 lb. (225 g) Briess Victory malt – Steeped | 3 | |
| Boil gravity for 3 gal. | 1.059 | |
| **Hop schedule** | **Boil time** | **IBUs** |
| 0.9 oz. (25 g) Centennial 10.5% AA | 60 min. | 25 |
| **Microbe strain** | **Pitch (billions of cells)** | **Fermentation temp.** |
| Kettle sour with mixed* culture | 250 | 104°F (40°C) |
| German ale or Kölsch yeast | 350 | 65°F (18°C) |
| *Brettanomyces bruxellensis* yeast | 200 | 65°F (18°C) |
| **Wort B** | **Gravity points** | |
| 4.5 lb. (2.04 kg) pale ale DME | 67.5 | |

| *All-Grain Version* | | |
|---|---|---|
| **Mashing option** | **Gravity points** | |
| 13.78 lb. (6.25 kg) pale ale malt | 55 | |
| 1.0 lb. (450 g) Weyermann Caraaroma malt 150°L | 3.5 | |
| 0.5 lbs. (225g) Victory malt 20°L malt | 1.5 | |
| Boil gravity for 7 gal. | 1.060 | |
| **Mash schedule** | **Rest temp.** | **Rest time** |
| Conversion rest – Infusion | 153°F (67°C) | 60 min. |
| **Hop schedule** | **Boil time** | **IBUs** |
| 0.9 oz. (25 g) Centennial 10.5% AA | 60 min. | 25 |

| Microbe strain | Pitch (billions of cells) | Fermentation temp. |
|---|---|---|
| Kettle sour with mixed* culture | 250 | 104°F (40°C) |
| German ale or Kölsch yeast | 350 | 65°F (18°C) |
| *Brettanomyces bruxellensis* yeast | 200 | 65°F (18°C) |

| Recommended Water Profile (ppm) | | | Brew cube: Amber, Balanced, Medium | | |
|---|---|---|---|---|---|
| Ca | Mg | Total alk. | SO$_4$ | Cl | RA |
| 75–125 | 20 | 75–125 | 100–200 | 100–150 | 0–50 |

*Notes*

*\* You can use a mixed souring culture, such as Wyeast's Roeselare Blend (3763) or Whitelab's Belgian Sour Mix 1 (WLP655), and pitch all at the same time, or you can kettle sour. If kettle souring, you can grow your own pediococci/lactobacilli blend from an inoculation starter, or you can use two specific strains of Pediococcus and Lactobacillus.*

*Cherry juice option: You can make this a nice sour cherry beer by adding 1.5 lb. (680 g) of cherries, or 0.5 L of sour cherry juice concentrate (68°Brix, 31 PPG [259 PKL]) after the second day of yeast fermentation. This equates to about 0.5 lb./gal. (60 g/L) of cherry puree.*

# Section II

## All-Grain Brewing

# Understanding Malted Barley and Adjuncts

## What is Barley and Why Do We Malt It?

Barley (*Hordeum vulgare*) is a member of the grass family Poaceae, and the fourth-largest cultivated cereal crop in the world. It was domesticated around the same time as wheat, meaning it has been cultivated for about 12,000 years. There are three types of barley, two-row, four-row, and six-row, the terms referring to the arrangement of the kernels around the shaft of the ear. Only two- and six-row barley are used for brewing. The kernels of six-row barley are usually physically smaller than those of two-row barley, but higher in protein. Two-row is considered to be superior to six-row for malting and brewing, but modern malting varieties of six-row barley still make excellent beer.

Barley is harvested, sorted, dried, cleaned, and stored before it is malted. The purpose of malting is to make the barley starch more accessible to the brewer.

The malting process begins when the highest-grade barley—brewing grade—is steeped in water (fig. 15.1) until it has absorbed almost 50% of its initial weight in water and rootlets, or "chits,"

**Figure 15.1.** The barley is steeped for a total of 38–46 hours. (Photos on page 220 and 222 courtesy of Briess Malt & Ingredients Co.)

begin to appear at the base of each barley grain (fig. 15.2). The barley is then drained and moved to a germination room where the actual malting process occurs. The barley is held at a controlled humidity level and periodically turned and moved to keep the temperature in the grain bed uniform (fig. 15.3). At this stage, it is referred to as green malt. After germination, the green malt is moved to a kiln where it is carefully dried at temperatures between 122°F and 158°F (50–70°C) to about 4% moisture. This malt is typically referred to as base malt or lager malt.

The malting process allows each barley grain to partially germinate, making the seed's resources available to the growing shoot (acrospire). During germination, enzymes in the aleurone layer (fig. 15.4) are released that break down the endosperm's protein-carbohydrate matrix into smaller carbohydrates, amino acids, and lipids, and free the seed's starch reserves.

The endosperm contains large and small starch granules that are packed in layers, like bags of mixed jellybeans and hard candy separated by cardboard in a box. The cell walls (bags and cardboard) within the matrix that contain the large and small starch granules (jellybeans and hard candy, respectively) are primarily composed of beta-glucans (a type of cellulose), some pentosans (gummy polysaccharides), and some protein. The box in this metaphor is the outer husk. The point is that there are several barriers that the maltster and brewer have to overcome to get to the candy.

Malt "modification" is the degree to which the enzymes tear open the bags and start unpacking the starch granules (i.e., break down the endosperm) for use by the growing plant, or, in our case, the brewer. One visual indicator that a maltster uses to judge the degree of modification is the length of the acrospire, which grows underneath the husk. The length of the acrospire in a fully modified malt will typically be 75%–100% of the grain length. More often, maltsters judge

**Figure 15.2.** The end of steeping is signaled by the emergence of the rootlets, or "chits." Once the barley has chitted, it must be moved to the germination tank where it will have more oxygen. (Photo courtesy of Briess Malting & Ingredients Co.)

**Figure 15.3.** After steeping, the malt is moved to a germination tank where it is aerated and turned over for several days to obtain uniform growth. The malt spends about four days in the germination tank before it is dried in a kiln room. (Photo courtesy of Briess Malting & Ingredients Co.)

modification by squeezing the moist kernel between their fingers; they can judge whether some areas of the kernel are not softened, and are able to feel if the kernel is too soft as well.

If the seed were allowed to continue to germinate, a plant would grow, and all of the starches that the brewer hoped to use would be used by the plant. So, the maltster gauges the germination carefully and judges when the proper balance between resources made available for the acrospire and resources consumed by the acrospire has been reached. At this point, the maltster stops the process by drying.

After modification, the grain is dried and the rootlets are knocked off by tumbling. The kiln drying of the new malt denatures (destroys) a lot of the enzymes activated during germination, but several types of enzyme remain, including the ones necessary for starch conversion. The amount of enzymatic starch conversion potential that a malt has is referred to as its *diastatic power*. A malt that contains sufficient diastatic power to convert itself in the mash is called a base malt. Base malts can be made from barley, wheat, rye, and oats, although barley and wheat are by far the most common. Mashing is the hot water soaking process that provides the right conditions for the enzymes to convert the grain starches into fermentable sugars.

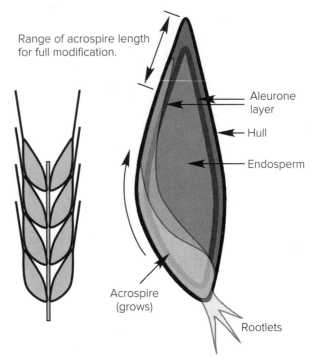

**Figure 15.4.** A simplified diagram of a barley kernel during malting, showing a progressive picture of how the acrospire (the plant shoot) grows along one side of the kernel. The aleurone layer releases pre-existing and newly created enzymes that "modify" the endosperm (the protein-carbohydrate matrix starch reserve) for the acrospire to use as it grows.

Malted barley is the principal source of the sugars (principally maltose) that will be fermented into beer. From a brewer's point of view, there are basically two kinds of malts: base malts and specialty malts. Base malts make up the bulk of the wort's fermentable sugars. Some base malts, such as Vienna and Munich malt, are kilned at higher temperatures to create warmer, breadier flavors. The higher temperatures denature some of the enzymes that contribute to the diastatic power of these malts.

The diastatic power of a particular base malt will also vary with the type of barley it is made from. Two-row barley varieties are the preferred type for all-malt beers (i.e., no adjuncts), having a slightly higher yield per pound, lower protein levels, and, it is claimed, a more refined flavor than six row. However, six-row barley has a slightly higher diastatic power than two-row. Historically, in the US, the higher protein level of six-row barley (which can produce a very heavy bodied and hazy beer) led brewers to thin the wort with low-protein grains like corn and rice. Brewers were able to take advantage of six-row barley's higher diastatic power to achieve full conversion of the mash in spite of these non-enzymatic starch sources. As explained below, these non-enzymatic starches are called *adjuncts*.

Specialty malts are non-enzymatic malts that have been kilned and/or roasted to higher temperatures to generate a variety of flavors to enhance the beer. There are three types of specialty malts. The first are the highly-kilned malts, which are essentially toasted base malts, such as biscuit and amber malt. These specialty malts have flavors ranging from toasted bread, to cookie or biscuit, to graham cracker and pie crust.

The second type of specialty malts are the caramel or crystal malts, which are not dried after malting but kilned while still wet, converting the starches to sugars right inside the hull. Afterward, these sugary malts are roasted at higher temperatures to generate a variety of unfermentable sugars and flavors, from honey to caramel to toasted marshmallow. These crystal malts are available in different colors (the color denoted by the unit Lovibond [°L]), each having a different degree of fermentability and characteristic sweetness. These malts are typically called crystal malts in the UK and caramel malts in the US, and referred to casually as, for example, C40, C60, or C120. (i.e., crystal 40°L, caramel 60°L, or crystal 120°L).

The third type of specialty malts are the roasted malts. These malts are produced by roasting dry base malt at high temperatures, giving them a deep red-brown or black color, flavors ranging from heavily toasted bread to cocoa to coffee. The color of roasted malts is also denoted in °L, for example, chocolate malt 350°L.

Specialty malts generally don't need to be mashed, they can simply be steeped in hot water to release most of their character. These grains are very useful to the extract brewer, making it easy to increase the color and complexity of the wort without much effort.

Lastly, there are fermentables not derived from malted barley—the adjuncts. Adjuncts are defined as being unmalted, and include raw or refined sugars, corn, rice, and flaked rye, wheat, and barley. Adjuncts should not to be scorned. Some, like unmalted wheat and roasted barley, are essential to certain beer styles. Some classic beer styles, such as Belgian wit, American lager, and Irish stout, depend on the use of adjuncts. Adjuncts made from unmalted grains must be mashed with enzymatic malts (i.e., malts that possess diastatic power) to convert their starches to fermentable sugars. Roasted barley is an exception to this rule, because its starches have been converted by high roasting, and it can simply be steeped.

## Malt Flavor Development

Maltsters usually divide the malt world into four types: base malts, kilned malts (including highly kilned), stewed malts, and roasted malts. Varying the moisture level, time, and temperature develops the characteristic flavors and colors of each specialty malt (fig. 15.5). Caramelization and Maillard reactions both play an important role in the wide variety of flavors that develop in these malts and in the beers made from them.

Caramelization is the thermal decomposition and chemical reaction of sugars that occurs via pyrolysis (see table 15.1). Caramelization is a non-enzymatic browning

### TABLE 15.1—SUGAR CARAMELIZATION TEMPERATURES

| Sugar | Minimum temperature |
|---|---|
| Fructose | 230°F (110°C) |
| Galactose | 320°F (160°C) |
| Glucose | 320°F (160°C) |
| Maltose | 356°F (180°C) |
| Sucrose | 320°F (160°C) |

**Figure 15.5.** A mosaic of malts, produced by different kilning and roasting conditions.

**Figure 15.6.** After kilning, specialty malts like caramel and chocolate are roasted at high temperatures to produce caramelization and Maillard reactions for distinctive flavors. (Photo courtesy of Briess Malt & Ingredients Co.)

reaction that requires high temperatures and low moisture to occur. Maillard reactions, on the other hand, are reactions between amino acids and sugars, and can occur across a wide range of temperatures and moisture conditions, starting as low as 120°F and occurring up through 450°F (49–232°C). Maillard reactions produce a wide range of volatile, low molecular weight heterocyclic flavor compounds, and higher molecular weight compounds, like reductones and melanoidins. Reductones can oxidize and bind oxygen to improve flavor stability. Melanoidins are the color aspect of the Maillard reaction.

Both Maillard reactions and caramelization can generate some of the same flavors, like toffee, molasses and raisin, but generally caramelization reactions are responsible for the toffee-sweet caramel flavors in malt, while Maillard reactions are responsible for the malty, toasty, biscuit flavors associated with baking. The low-temperature, high-moisture Maillard reactions produce malty and fresh bread flavors, and the high-temperature, low-moisture Maillard reactions produce the toasty and biscuit flavors.

After drying, lightly-kilned base malts like pale ale malt are kilned at low temperatures (120–160°F [49–71°C]) to retain their diastatic power. The flavors expressed are lightly grainy with hints of warm bread and toast. Vienna malt is dried to about 5%–10% moisture and then lightly kilned at 194°F (90°C) to develop its color and flavor while retaining most of its diastatic power. More highly-kilned base malts like Munich, and aromatic malts are dried to relatively higher moisture levels (15%–25%), and treated to a schedule of higher temperatures (195–220°F, [90–105°C]), to produce rich, malty and bready flavors. Only Maillard reactions are involved at these temperatures.

Stewed malts, such as caramel 60°L and caramel 120°L, are produced by roasting green malt, that is, malt that was not dried by kilning after germination. These malts are put into a roaster and heated to the starch conversion range of 150–158°F (66–70°C). The converted sugars are in a semi-liquid state inside the kernel. After conversion, these malts are roasted at higher temperatures of 220–320°F (105–160°C), depending on the degree of color wanted.

Roasting at these temperatures causes the sugars inside the grains to caramelize, breaking them down and recombining them into less-fermentable forms. Maillard reactions also occur, causing further browning. The lighter caramel malts have a light honey and caramel flavor, while the darker caramel malts have a richer caramel and toffee flavor, with hints of burnt sugar and raisin at the darkest roasts.

The roasted malts include amber, brown, chocolate, and black malt. These malts are dried to a low moisture content (4%–6%) before roasting (fig. 15.6). Amber malts are produced by roasting fully kilned pale ale malt at temperatures up to 335°F (168°C). These high temperatures give amber malt its characteristic toasty, nutty, and biscuit flavors. Brown malts are roasted longer than amber malts and achieve a very dry dark toasted flavor, with color equal to that of the caramel malts.

The roasting process for chocolate malt begins at about 165°F (74°C) and is steadily increased to over 325°F (163°C), at which point the malt begins fuming. As the temperature is raised further to 420°F (216°C), the fumes turn blue and the malt develops chocolatey flavors. Some degree of caramelization occurs, but the majority of the flavors are from Maillard reactions. Black malts are roasted to slightly higher temperatures of 428–437°F (220–225°C), producing coffeelike flavors. The malt will actually burn at temperatures exceeding 480°F (249°C), so the trick is to spray the roasted malt with water at the critical point in time, and this was the basis of the invention of black "patent" malt. Roast barley is produced in a similar manner, but the difference is that it is never malted to begin with.

To summarize, kilning produces breadlike flavors from the low-temperature Maillard reactions. Roasting dry malts increases the Maillard reactions and accentuates the malt flavors of biscuit and toast. Roasting green malt causes both Maillard and caramelization reactions that produce toffee-sweet flavors. Kilning and roasting of dry malt at high temperatures produces the chocolate and coffeelike flavors.

## Common Malt Types and Usages

Note: there are a few trademarked products in the following malts listed. I have listed these products because they best represent a particular style of malt that is commonly used for a particular flavor or purpose. Please note that this is an incomplete list—every malting house has its own specialties and I don't come close to listing every malt here. Typical Lovibond color values are listed as °L.

In addition, it is important to understand that every batch of malt is unique, in the same way that every batch of beer is unique, and that the same type of malt (like the same style of beer), will be different between different maltsters (and brewers). Read the maltster's websites for the best description about a particular malt type and product.

### Base Malts

(The following base malts should be mashed.)

*Lager or Pilsner malt (1–2°L).* This type of base malt is the palest of the pale. The name comes from the fact that pale lagers are the most common style of beer and this is the type of malt used to produce them. Lager malt can be used as the base malt for brewing nearly every other style as well. After germination, lager malt is carefully heated in a kiln to 90°F (32°C) for the first day, withered at 120–140°F (49–60°C) for 12–20 hours, and then cured at 175–185°F (79–85°C) for 4–48 hours, depending on the maltster. This produces a malt with fine, mild flavor and excellent diastatic power.

The names lager malt and Pilsner malt should be synonymous, and either one can be produced from most any barley variety and on any continent, but the moniker "Pilsner malt" is generally associated with low-protein European varieties or indicates a premium lager malt. The "Pilsner" label may also be used to indicate a malt with lower modification compared to other base malt products from the same maltster.

*Pale malt (a.k.a. two-row base malt) (2–3°L).* There is a subtle difference between European and North American base malts; pale malt is essentially North America's version of lager malt, generally made with North American barley varieties. It has slightly higher protein levels and diastatic power than European Pilsner malt, but can be used for any beer style.

*Pale ale malt (3–4°L).* Pale ale malt is kilned at higher temperatures than pale malt, giving a warmer, toastier malt flavor well suited to English style pale ale and will produce a golden to pale amber beer. Pale ale malt can be mimicked by using a combination of Pilsner or pale malt and Munich 10°L, but pale ale malt is often preferred for its own unique flavors, such as the pale ale malt produced using Maris Otter two-row barley.

*Wheat malt (3°L).* Wheat has been used for brewing beer nearly as long as barley has, and it has equal diastatic power. Malted wheat is used for 5%–70% of the grain bill, depending on the style. Wheat has no outer husk and therefore has fewer tannins than barley. Wheat kernels are generally smaller than barley kernels and contribute more protein by weight to the beer, aiding in head retention. But this higher protein content and lack of a husk causes wheat mashes to be stickier than barley mashes, which may cause lautering problems. A protein rest during the mash or adding rice hulls to the mash (or both) can help with lautering when using a high proportion of wheat.

*Rye malt (3°L).* Malted rye is not common, but is gaining in popularity. It can be used as 5%–10% of the grain bill for a "spicy" rye note. Rye is even stickier in the mash than wheat and should be handled accordingly.

*Smoked malt (2-6°L).* Smoked malts are a family of two-row base malts that have been cured over wood during the kilning process or have been smoked afterwards. Smoked malt can be used as 100% of the grain bill for a traditional *rauchbier*, or in smaller amounts (e.g., 20%) to add accent. Various woods, such as apple, beech, cherry, or mesquite, can be used for the smoking to give distinct characters.

*Acidulated malt (2°L).* Acidulated malt is a specialty product from Weyermann Malting and is prepared by spraying base malt with lactic acid produced from a soured wort. It is an all-natural product that complies with the *Reinheitsgebot* and allows German brewers to lower their mash pH without using brewing salts or commercial acid. It is reported to lower the mash pH by 0.1 for every 1% by weight used in the grain bill.

### Kilned Base Malts

(Should be mashed.)

Kilned base malts are commonly produced by increasing the moisture content and curing temperatures used for base malt production.

*Vienna malt (4°L).* Vienna malt has a lighter flavor than Munich malt and is a principal ingredient of light amber beers. Retains more than enough diastatic power to convert itself, but is often used with a base malt in the mash. Typically used as 10%–40% of the grain bill, depending on beer style, although it can be used at 100% for Vienna style lagers. Contributes a warm malt flavor without excessive sweetness or bread crust flavor.

*Munich malt (10°L).* Munich malt has an amber color and gives a very malty flavor. This malt has just enough diastatic power to convert itself, but is usually used in conjunction with a base malt in the mash. Munich malt is used as 10%–60% of the grain bill for Oktoberfests, bocks, and many others, including pale ales. It can be used at 100% for Munich dunkles. Munich malt is the primary tool for imparting rich maltiness to many styles. Its flavor is like toasted bread crust.

*Aromatic malt (15–25°L).* Aromatic malt (a.k.a. melanoidin malt) is similar to a dark Munich 20°L, and in some cases probably is literally that. It has a very low diastatic power, but gives wonderful rich malt flavor and aromas, like dark bread crust. Aromatic malt contributes a deep amber or walnut-brown color to beer. Use as 5%–10% of the grain bill for accent.

*Amber malt (20–40°L).* Amber malt, a.k.a. biscuit or Victory malt, is a fully toasted, lightly roasted malt used to give the beer a warm flavor like fresh baked cookies (biscuits in the UK). It is typically used as 10% of the grain bill. Gives a deep amber color to the beer. No diastatic power.

*Brown malt (60°L).* Brown malt is getting hard to find, because it is only used in a couple of styles, such as old ale, porter, and stout. It has a very dry, bitter roasted character that falls somewhere between amber and chocolate malt, and it is not sweet. Kind of like concentrated bread crust. Use as 5%–10% of the grain bill, depending on style. No diastatic power.

### Stewed Malts

(May be steeped or mashed.)

*Caramel malts.* Caramel malts (a.k.a. crystal malts) have undergone a special heat process that "stews" the malt, converting the starch and liquefying the sugar inside the kernels. These malts are roasted at various temperatures to caramelize the sugars to different degrees, which yields a range of flavors, from honey sweet to toffee to dark caramel. The same color rating from different maltsters can have different flavors due to individual techniques; malting is as much an art as brewing.

Caramel malts are used to some degree in most beer styles. They are ideal for adding aroma and body to extract beers by steeping, but it is possible to overdo it and make the beer cloyingly sweet. Caramel malts are typically added as 5%–15% of the total grain bill.

*Dextrin malt (3°L).* Dextrin malt is typically used as 1%–5% of the grain bill, and enhances the body, mouthfeel, and foam stability of the beer without affecting the color or flavor. This glassy malt is very hard and difficult to crush. Consequently, dextrin malt does not give a good yield from steeping even though it is fully converted the same as other caramel malts. Two examples are Carapils® from Briess Malt & Ingredients Co. and Carafoam® from Weyermann.

*Caramel 10 malt (10°L).* Caramel 10 malt adds a light honeylike sweetness and some body to the finished beer.

*Honey malt (25°L).* Also known as brumalt, honey malt has a rich honey flavor that is very versatile.

*Caramel 40 malt (40°L).* The additional color and light caramel sweetness of caramel 40 malt is perfect for pale ales and amber lagers.

*Caramel 60 malt (60°L).* Caramel 60 malt is the most commonly used caramel malt, and is also known as medium crystal malt. It is well suited for pale ales, English-style bitters, porters, and stouts. Caramel 60 malt adds a full caramel taste and body to the beer. However, this malt is reported to oxidize (go stale) more quickly than other caramel malts, and some brewers opt to use a combination of 40°L and 80°L caramel malts instead.

*Caramel 80 malt (80°L).* Caramel 80 malt is used for making reddish-colored beers and gives a lightly bittersweet caramel flavor.

*Caramel 120 malt (120°L).* Caramel 120 malt adds a lot of color. It has a toasted, bittersweet caramel flavor, with hints of burnt sugar and raisin. Caramel 120 malt is useful in small amounts to add complexity, or in greater amounts for old ales, barleywines, and doppelbocks.

*Special "B" malt (150°L) (Castle Malting).* Special "B" is a unique Belgian malt that has a definite roasty or toasty flavor consisting of dark caramel, toasted marshmallow, and raisin. Used in moderation in the grain bill (1%–5%), it is very good in brown ales, porter, and doppelbocks. Larger amounts, >5% of the grain bill, will lend a prune-like flavor to abbey ale styles like dubbel. Several other maltsters make a similar 150°L–180°L product under various names, but Special "B" was the original.

## Roasted Malts
(May be steeped or mashed.)

This class of highly roasted malts contribute bitter chocolate, coffee, or burnt toast flavors, to brown ales, porters, and stouts. Obviously, these malts should be used in moderation, typically 1%–5% of the grain bill (e.g., this equates to 0.25–0.5 lb./5 gal., or 115–225 g/19 L). Some brewers recommend that they be added toward the end of the mash to reduce the acrid bite that these malts can contribute. This practice will produce a smoother beer for people brewing with naturally "soft" or low-bicarbonate water. Roasted malts are typically used in small amounts and may be finely ground to achieve a better color contribution with a smaller addition.

*Pale chocolate (200–250°L).* Pale chocolate malt is used in small amounts (0.5 lb./5 gal., or 225 g/19 L) for brown ale, porter, and stout. This malt has a medium roast coffee flavor and contributes a rich ruby-brown color. Using too much will dominate the character of the beer.

*Chocolate malt (300–400°L).* Chocolate malt is used in small amounts (0.5 lb./5 gal., or 225 g/19 L) for brown ale, but used extensively (1 lb./5 gal., or 450 g/19 L) in porter and stout. This malt has a bittersweet chocolate-coffee flavor, pleasant roast character, and contributes a deep ruby-black color.

Using a lot of chocolate malt will not make the beer taste like chocolate! I recommend not exceeding 1 lb. in 5 gal. (450 g/19 L). Too much will give the beer an unpleasant inky aftertaste.

*De-bittered black malt (500°L).* This special roast malt has had the husk removed prior to malting, creating a much smoother roasted character in the beer. This malt is often used for color adjustment, but only in small amounts (roughly 1.5–3 oz. in 5 gal., or 50–80 g in 19 L). It can be used in larger amounts, such as 5% of the grain bill, to contribute the roast character for porters and stouts without as much bite as a traditional black malt.

*Black wheat malt (500–550°L).* These are, obviously, black malts made from wheat instead of barley. Wheat doesn't have a husk, so these malts can be highly roasted to give a smoother coffee-like flavor with less bitterness, like the de-bittered black malts. They can be used interchangeably with black malt for all styles.

*Roast barley (500°L).* Roast barley is not actually a malt, but highly roasted plain barley. It has a dry, distinct coffee taste and is the signature flavor of Irish and dry stouts. Use about 0.5–1 lb. per 5 gal. (230–450 g/19 L) for stout.

*Black ("patent") malt (500–600°L).* Black "patent" malt is the blackest of the black and is mainly used for color. It should be used sparingly, generally less than 0.5 lb. per 5 gal. (225 g/19 L). As well as contributing color, black malt is useful for setting a "limit" on the sweetness of other beer styles that use a lot of caramel malt; 1–2 oz./5 gal. (30–60 g/19 L) is useful for this purpose.

**Figure 15.7.** Some of the common flaked adjuncts. *Clockwise from top,* flaked corn, flaked wheat, flaked rye, and flaked oats.

## Other Grains and Adjuncts

Note: If you intend to use more than 10% of any of the following adjuncts in your recipe, you may want to conduct a cereal mash to better utilize them and achieve better flavors. Mashing of adjuncts is described in chapter 17. Adjuncts are frequently rolled and flaked (fig. 15.7).

*Oatmeal.* Oats are wonderful in a porter or stout. Oatmeal lends a smooth, silky mouthfeel and a creaminess to a stout that must be tasted to be understood. Oats are available whole, steel-cut (i.e., grits, or pinhead), rolled, and flaked. Rolled and flaked oats have had their starches gelatinized (made soluble) by heat and pressure, and are usually readily available as "instant oatmeal" in the grocery store. Whole oats, steel-cut oats, and "old-fashioned rolled oats" have not had the degree of gelatinization that instant oatmeal has had, so these must be cooked before adding to the mash. "Quick cooking" oatmeal has had more gelatinization than "old fashioned" oatmeal, but still benefits from being cooked before adding to the mash.

Cook the oatmeal according to the directions on the box (but add more water) to ensure that the starches will be fully utilized. Use 0.5–1.5 lb. per 5 gal. batch (225–680 g/19 L). Oats need to be mashed with barley malt (and its enzymes) for conversion. Oat malt is available as well, including Golden Naked Oats®, which is a caramel-type malt from Simpsons Malt.

*Flaked corn (maize).* Flaked corn (maize) is a common adjunct in British bitters and milds, and was also used extensively in American light lager in the past (although today corn grits are more common). Properly used, corn will lighten the color and body of the beer without overpowering the flavor. Use 0.5–2 lb. per 5 gal. batch (225–910 g/19 L). Corn must be mashed with base malt.

*Flaked barley.* Unmalted flaked barley is often used in stout to provide protein for head retention and body. It can also be used in other strong ale styles. Use 0.5–1 lb. per 5 gal. batch (225–450 g/19 L). Flaked barley must be mashed with base malt.

*Flaked wheat.* Unmalted flaked wheat is a common ingredient in wheat beers, and is essential to styles like Belgian lambic and wit. It can add starch haze and higher levels of protein than malted wheat. Flaked wheat also imparts more wheat flavor "sharpness" and a thicker mouthfeel than malted wheat. Use 0.5–2 lb. per 5 gal. batch (225–910 g/19 L), or up to 50% of the grain bill in classic witbier or lambic recipes. Flaked wheat must be mashed with base malt.

*Flaked rice.* Rice is the other principal adjunct used in American and Japanese light lagers. Rice has very little flavor and makes for a drier tasting beer than corn. Use 0.5–2 lb. per 5 gal. batch (225–910 g/19 L). Flaked rice must be mashed with base malt. Whole rice needs to be cooked in a cereal mash to effectively utilize it in the mash.

*Oat and rice hulls.* Not an adjunct per se, the hulls of oats and rice are not fermentable, but they can be useful in the mash. The hulls provide bulk and help prevent the mash from settling and becoming stuck during the sparge. This can be very helpful when making wheat or rye beers with a low percentage of barley malt and barley husks. Use 2–4 qt. (~2–4 L) of oat or rice hulls for 6–10 lbs. (~3–5 kg) of wheat, if doing an all-wheat beer. The barley hull is 5% of the kernel weight, so 5% of the adjunct weight is a good place to start. Do not exceed 3% by weight of the total grain bill or you will start tasting them as astringency in the beer.

## How to Read a Malt Analysis Sheet

Every batch of malt is unique, so every lot is tested, sometimes multiple times to check the consistency of large batches. The requirements differ across the various types of malt, depending on primary usage and individual customer needs. At a minimum, each lot is tested for color, soluble extract yield, and percentage moisture (these are usually listed as Color, Yield, and % Moisture). There are two ways of measuring the soluble extract yield, either percent extract (%Extract) or hot water extract (HWE).

The other parameters typically tested for malt lots are: size characterization, protein levels, modification, and diastatic power. Example values for various malt types are given at the end of this chapter in table 15.2 for comparison.

## Percent Extract–Fine Grind, Dry Basis

A typical malt analysis sheet does not give the malt's yield in gravity points per pound per gallon (PPG) or liter degrees per kilogram equivalent to PKL—see explanation below. What you will most likely see instead for North American and European malts is a value called "% Extract–Fine Grind, Dry Basis" (FGDB). This percentage is the maximum soluble extract by weight that the malt will yield when mashed, and is typically 80% for base malt. This soluble extract percentage equates to 37 PPG, or 309 PKL.

When a malting house analyzes a malt sample to determine its extract yield, it conducts a "Congress mash," named for the European Brewery Convention (EBC) of 1975 that standardized the procedure. A Congress mash (see sidebar) consists of a multistep infusion mash using a standard weight of finely ground malt (i.e., flour). The mash is continually stirred over a two-hour period and then drained for another hour. These times may not seem remarkable until you consider that the malt test sample is only 50 grams! This procedure yields the maximum soluble extract as a percentage of the original sample weight.

This yield is known as the percent extract–fine grind, as-is (FGAI). It is called "as-is" because properly kilned malt contains about 4% moisture by weight. To compare different lots of malt with different moisture levels, this weight needs to be accounted for in the extract calculation. Therefore, the basis of comparison, and the number you will most consistently see on an analysis, is the "%FGDB," the fine grind, dry basis value corresponding to a malt sample that has been oven-dried to zero percent moisture. Extract yield will be discussed further in chapter 18.

The moisture content for the lot will be listed on the certificate of analysis, and should be 2%–4% for base and kilned malts. Caramel and roast malts typically have more moisture at 5%–6%, but the moisture level should always be less than 6%.

## DETERMINING PERCENT EXTRACT USING A CONGRESS MASH

In a Congress mash, 50 g of finely ground malt is infused in a beaker with 200 mL of warm distilled water to a temperature of 45°C (113°F). This mash beaker is placed in a warm water bath to maintain that temperature for 30 min. The mash is then heated at a rate of 1°C/min. to 70°C (158°F) and infused with 100 mL of 70°C water. The mash is held at 70°C for 60 min., then gradually cooled to room temperature by the addition of cold water. The total weight of the mash is adjusted to 450 g with more distilled water, and drained through filter paper. The wort is measured for specific gravity to an accuracy of 0.00001 and then converted to percent extract–fine grind, as-is (%FGAI) by means of the American Society of Brewing Chemists (ASBC) "Tables for Extract Determination of Malt and Cereals." The percent moisture is determined from another sample of malt from the same lot, and that measurement is used to calculate the percent extract–fine grind, dry basis (%FGDB).

### Percent Extract–Coarse Grind, As-Is and Dry Basis

The same Congress mash method is also used to determine the percent extract–coarse grind, as-is (%CGAI), and the moisture is measured to calculate the percent extract–coarse grind, dry basis value (%CGDB). Coarse grind represents a mill setting that is closer to what many professional breweries would use. The %CGAI is a slightly more realistic number for gauging the extract potential of a malt, but it's still a maximum that few professional breweries attain.

Percent extract–coarse grind is not measured for most specialty malts due to the extra time and effort it takes. Professional brewers are not as concerned about the yield of specialty malts because they usually only represent a small percentage of the grain bill. Thus, the standard parameter of %FGDB is usually the only value determined for specialty malts, like caramel, chocolate, and roast malts.

### Fine/Coarse Difference

The fine/coarse (F/C) difference value is simply the difference between the fine and coarse grind percentages (both as-is and dry basis—same difference, respectively). The F/C difference value allows the brewer to quickly convert between the two parameters. For example, looking at the numbers for Munich malt in table 15.2, the %CGDB is 2% less than the %FGDB, and this is indicated by the F/C value given in the next row. The F/C difference also serves as an indicator of the degree of modification, although the soluble-to-total protein ratio is most often used (see section further below). An F/C difference of 1% is typical for highly modified base malts. An F/C difference of 2% indicates a less modified malt, and a value of less than 1% would be over-modified.

### Hot Water Extract

The hot water extract (HWE) parameter may be seen on malt analyses from the UK and Australia, where maltsters utilize a single temperature infusion mash method that differs from the ASBC and EBC Congress mash methods. The main difference is that the malt sample is mashed at 149°F (65°C) for 1 hour. The HWE is "as-is" and is measured as liter degrees per kilogram (L·°/kg).

Note the units for HWE, where a *liter degree* is a unit of extract yield, abbreviated to L·° or simply L°, and should not be confused with degrees Lovibond (°L). The degree in L° refers to the number of degrees, or points, of specific gravity. You may realize at this stage that a value in L°/kg can also be expressed as gravity points/kg/L—in other words, PKL (e.g., 300 L°/kg = 300 PKL). Therefore, HWE expressed as PKL is equivalent to PPG when the metric conversion factors for volume and weight are applied. (Also note: gravity points/lb./gal. = gal.°/lb.). The conversion factor is PKL = 8.345 × PPG.

### Color

Degrees Lovibond (°L) was a unit created in 1883 by J.W. Lovibond to denote the color of beer and brewing malts. His system consisted of glass slides of various shades that could be combined to produce a range of colors and compared against samples of wort or beer. In 1950, the American Society of Brewing Chemists (ASBC) set a standard using optical spectrophotometers to measure the absorbance of a specific blue wavelength of light (430 nanometers) through a standard-sized sample, and this was called the Standard Reference Method (SRM). The SRM aligned nicely with the Lovibond scale and the two can be considered nearly identical for most of their range. However, the resolution of a spectrophotometer diminishes greatly with darker worts, when very little light can penetrate the sample to reach the detector.

For this reason, the Lovibond scale is still in use today, in the form of precision visual comparators. The use of comparators is most prevalent in the malting industry, particularly for roast specialty malts, and thus the color of malts is given as °L, while beer color is given as SRM, though the reference standard (absorbance at 430 nm) is the same. See the SRM color samples on the inside front cover of this book.

Prior to 1990, the EBC used a different wavelength for measuring absorbance, and conversion between the SRM and EBC scales was an approximation. Today, the EBC scale uses the same 430 nm wavelength for measurement, but uses a smaller sample glass. Color measurements using the current EBC scale work out to be about twice the SRM rating (actually, EBC = SRM × 1.97). See appendix B for more information on beer color.

## Size

The average size of the kernels and the size distribution is important to the brewer, because it affects how well the malt is crushed by the roller mills. If a significant proportion of the kernels are small, then those kernels will not be crushed well and the extract from the mash and lauter will decrease. Kernel size and distribution are measured by sieving. The ASBC method uses standard sieves with mesh sizes of 7/64", 6/64", and 5/64". Kernels that pass through the 5/64" sieve are caught in a pan and classified as "thru" on the analysis sheet. The sum of the percentages captured by the 7/64" and 6/64" sieves is often described on the malt analysis sheet as "% plump." Typically, malt is required to have 80% or 90% of the batch be plump. The percentage that passes through the 5/64" sieve is often labeled "% thin." The requirement for malt is typically a maximum 2% of the batch be thin.

In Europe and the UK, the sieve sizes are very slightly larger, being 2.8 mm, 2.5 mm, and 2.2 mm.

## Protein

The protein measurement in malt is actually an approximation, based on chemical analysis of the total amount of nitrogen in a malt sample. Every 1% of nitrogen is assumed to represent 6.25% of protein. You may see "total nitrogen" on an analysis instead of "total protein."

American barley varieties are usually higher in protein than European varieties. The range of protein for two-row varieties is 11%–13% for North American barley, whereas European and Australian two-row is usually 9.5%–12%. Six-row varieties average a little higher at 12%–13.5%. Barley with total protein measuring over 13.5% is not used for malting.

## Soluble-to-Total Protein Ratio

The soluble-to-total protein (S/T) ratio, also known as the Kolbach Index, is the most commonly used indicator of malt modification. During the malting process, the proteolytic enzymes in barley cleave the large insoluble proteins into smaller soluble proteins. About 38%–45% of the malt protein (as measured by nitrogen as total protein above) is converted to soluble protein, including enzymes, foam-positive proteins (i.e., proteins that form foam in the wort and beer), haze-forming proteins, and amino acids. The S/T ratio for the malt describes the extent of modification of the endosperm. To generalize, an S/T ratio of 36%–40% is a less-modified malt, 40%–44% is a well-modified malt, and 44%–48% is a highly modified malt. Soluble protein levels below 35% can result in low extraction due to the starches in the protein-carbohydrate matrix still being relatively inaccessible, and difficulty in lautering due to higher beta-glucan levels. Soluble protein levels exceeding 55% will lead to excessive darkening during wort boiling, beer haze, and loss of body in the beer.

## Diastatic Power

The diastatic power of a malt is a measure of the starch conversion capability in degrees Lintner (°Lintner).[1] Diastatic power is measured by evaluating the effects of all the diastatic enzymes in the malt, that is, those enzymes that can convert starch. The diastatic enzymes in malt are degraded by kilning, and thus the diastatic power of highly kilned malts, like Munich and Vienna, is less than that of lager malt. Malts with diastatic power of 40°Lintner or greater are able to convert themselves. Munich malt is typically 40–50°Lintner, pale ale malt is about 80°Lintner. Lager malt is generally 100–140°Lintner, and wheat malt and six-row brewer's malt can be as high as 165°Lintner. High diastatic power is most useful when brewing with starch adjuncts. You can determine the conversion potential of an adjunct mash by calculating the dilution of the enzymatic malts and their diastatic power. In other words, a six-row brewer's malt could be part of a grain bill that is two-thirds adjuncts and still have an equivalent diastatic power of 55°Lintner for the mash. The only caveat is that low-diastatic power mashes will take longer to convert, and there is the risk that all of the beta-amylase will be denatured by the mashing temperature before conversion is finished.

## Summary

The malting process allows the barley grain to partially germinate, making the seed's resources available to the brewer. Malted barley is the principal source of the sugars that are fermented into beer. From a homebrewer's point of view, there are basically two kinds of malts, those that need to be mashed and those that don't. Mashing is the hot water soaking process that provides the right conditions for the enzymes within the grain to convert the grain starches into fermentable sugars. The base malts, such as Pilsner, pale ale, Vienna, and Munich malts, have sufficient diastatic power to convert their starches into fermentable sugars.

Specialty malts are non-enzymatic malts, that is, they have no diastatic power whatsoever following heat treatment. Specialty malts are used for flavor and coloring, and can be divided into three groups, kilned, stewed, and roasted. The kilned malts, such amber and brown malts, do not have significant diastatic power and need to be mashed with base malts. Caramel malts have had their starches converted to sugars by heat and moisture right inside the hull, and can be steeped or mashed to release their character. The sugars in caramel malts have a pleasant caramel-like sweetness. The starches in roasted malts have been converted by high heat into soluble melanoidin compounds that have bitter chocolate and coffeelike flavors. These malts can also be steeped or mashed.

Lastly, there are non-enzymatic fermentables that are not derived from malted barley, which are called adjuncts. Adjuncts include refined sugars, corn (maize), rice, unmalted rye and wheat, and unmalted barley. Adjuncts made from grains must be mashed with enzymatic malts to convert their starches to fermentable sugars.

---

[1]    Probably named after Carl Lintner (1828–1900), a director of the brewing school at Weihenstephan. Although °Lintner can also be represented as °L, I have not used this abbreviation here to avoid confusion with degrees Lovibond.

## TABLE 15.2—REPRESENTATIVE MALT ANALYSIS NUMBERS

| Malt type | Two-row lager | Six-row brewer's | Pale ale | Munich | Amber | Caramel 60 | Chocolate | Black (patent) | Roast barley |
|---|---|---|---|---|---|---|---|---|---|
| %Mealy | 98 | 95 | 98 | 95 | 95 | 0 | ... | ... | ... |
| %Half | 2 | 5 | 2 | 5 | 5 | 5 | ... | ... | ... |
| %Glassy | 0 | 0 | 0 | 0 | 0 | 95 | ... | ... | ... |
| **Size** | | | | | | | | | |
| 7/64" | 60 | 45 | 60 | 55 | 55 | 40 | ... | ... | ... |
| 6/64" | 20 | 30 | 20 | 25 | 25 | 40 | ... | ... | ... |
| 5/64" | ... | ... | ... | ... | ... | ... | ... | ... | ... |
| thru | 2 | 3 | 2 | 2 | 5 | 2 | ... | ... | ... |
| % moisture | 4 | 4.5 | 4 | 3.3 | 2.5 | 5.5 | 6 | 6 | 5.5 |
| %FGDB | 80.5 | 78 | 80 | 78 | 73 | 73 | 73 | 70 | 72 |
| %CGDB | 79.5 | 76.5 | 78.5 | 76 | ... | ... | ... | ... | ... |
| F/C | 1 | 1.5 | 1.5 | 2 | ... | ... | ... | ... | ... |
| Protein | 12 | 13 | 11.7 | 11.7 | ... | ... | ... | ... | ... |
| S/T | 42 | 40 | 42 | 38 | ... | ... | ... | ... | ... |
| DP (°Lintner) | 140 | 160 | 85 | 40 | 30 | ... | ... | ... | ... |
| Color (°Lovibond) | 1.8 | 1.8 | 3.5 | 10 | 28 | 60 | 350 | 500 | 300 |

CGDB, coarse grind, dry basis; DP, distatic power; F/C, fine/coarse (difference); FGDB, fine grind, dry basis. %Mealy/%Half/%Glassy describes the friability of the malt, i.e., how it crushes.

# How the Mash Works 16

The technology behind malting and brewing is one of the oldest in the history of humankind. Through time, brewing scientists helped develop the microscope, pH, pasteurization, and a whole host of other technologies. And yet, it could be argued, we know more about flying than we do about the biochemistry of beer. Making beer is so easy that people have been just doing it for thousands of years; they didn't need to know how it worked, it just worked. And if it works, what more is there to know?

The short answer to that question is given below and, if you want, you can just read the "Mashing in a Nutshell" section and skip to chapter 20 to brew your first all-grain batch. This chapter and the three that follow will teach you how to control and manipulate the mashing process to fine-tune your beer's character, adapting the process to your recipe (or vice versa), and to optimize your yield.

## Mashing in a Nutshell
Mashing is actually a continuation of the malting process, that is, making the barley starches accessible to enzymes and converting them to fermentable sugars. In fact, most breweries historically had malting houses attached, they weren't separate facilities. Farmers would deliver the barley and the brewery would malt it and mash it to make the beer.

Mashing is the term for the hot water steeping process that hydrates the malt, gelatinizes its starches, releases its enzymes, and leads to the conversion of the starches into fermentable sugars. It is the core of the brewing process.

During mashing, malt is crushed to facilitate hydration, and infused with 160–165°F (71–74°C) water at a ratio of 2 qt. per pound of grain (4 L/kg) to achieve a mash temperature of 149–155°F (65–68°C). The mash is typically held at that temperature for one hour, although a half hour is usually sufficient for conversion. The mash is slowly drained of wort and rinsed with hot sparge water. This step is called *lautering,* and *sparge* means to sprinkle (from the German). When the spent grains are sufficiently rinsed, the lautering is stopped and the wort is ready to be boiled.

Almost any temperature between 149°F and 162°F (65–72°C) will make wort, although it is important to understand that the temperature does have a significant effect on fermentability. To generalize, warmer temperatures will produce a more dextrinous wort, lower temperatures will make the wort more fermentable. The temperature of your mash can drift downward during the hour, but as long as it doesn't get below 140°F (60°C) it's fine. The starches will be converted to fermentable sugars, and you will have made wort.

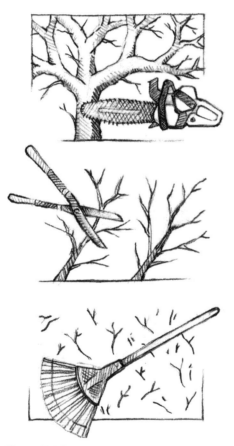

**Figure 16.1**. The tree is broken down into small pieces during the mash by the various tools (enzymes).

## An Allegory of a Mash

### Cast of Characters
**You:** Amylase, the starch converter.
**Brother:** Beta-Glucanase, the gum breaker.
**Sisters:** Proteinase & Peptidase, they share the work.
**Dad:** Limit Dextrinase, cuts the branch points.
**Mom:** Gelatinization Temperature, tells you when you can start.

### Our Story
There has been a big windstorm that has blown down a big tree and a lot of other branches in the backyard. Your parents decide that some yard work will build character—yours. Your task is to cut as much of it as you can into two-inch lengths and haul it out to the road. You have two tools to do this with, a hedge trimmer and a pair of hand-held clippers. The hedge trimmer is in the garage, but the last time anyone saw the clippers they had been left outside in the grass, which has since grown knee high. Plus, there are a lot of brambles growing around the tree that will make access difficult. Fortunately, your dad has decided that your sisters and older brother should take part in this too, and will send them out there with two lawn mowers and a weed whacker right now to mow the grass, clear away the brambles, and find the clippers. Likewise, Dad will fire up the chainsaw and be ready

to cut through the big limbs at the joints as you work. This will ensure that you won't leave any long pieces behind. As soon as the grass and brambles are cut, you can find your clippers and get to work.

Your tools are rather limited for the amount of work you have to do. The hedge trimmer will be really useful for cutting all of the twig ends off, but will quit working once you get back toward the thicker branches. The clippers will be useful then—they will be able to cut the middles of all the branches, but aren't strong enough to cut through the joints, and that's where your dad will help. When you are done, there will be a lot of odd branched pieces left over in addition to your little pieces. Your success will be measured by how many little pieces make it out to the road. A large part of your success is going to depend on how well your siblings do their job in making the tree accessible to you. If you leave a lot of the tree behind, and if the stuff you get out to the road consists of a lot of big pieces instead of the small pieces your dad wanted, then you won't have done the job correctly. You had better plan your activity carefully.

Okay, your brother and sisters have done their work, and your dad is ready and waiting, but just as you are ready to get started your mom says that you have to wait until it warms up to 65 degrees outside, because she doesn't want you to catch a cold. It's more likely to get too hot to finish the job later on, but you still have to wait until it warms up to 65 degrees before you can get started.

## Defining the Mash

The allegory above attempted to illustrate all the activity that goes into converting malt starches into fermentable sugars (fig. 16.1). There are several key enzyme groups that take part in the conversion (more details follow in the "Starch Conversion, or Saccharification, Rest" section below, and in table 16.2). During malting, beta-glucanase (weed whacker) and proteolytic enzymes (lawn mowers) do the majority of their modification work, opening the protein matrix and giving access to the starches for hydration and gelatinization. A small amount of further modification can occur during the mash, but the main event is supposed to be the conversion of starch molecules into fermentable sugars and non-fermentable dextrins by the diastatic enzymes (your clippers and hedge trimmer, and Dad's chainsaw).

Each of these enzyme groups can be influenced by different temperature and pH conditions; temperature is the primary influence, and pH secondary. You can adjust the mash temperature to favor each successive enzyme's function and thereby customize the wort to your taste and purpose.

Both malted and unmalted grains have their starch reserves locked in a tightly packed protein-carbohydrate matrix that prevents the enzymes from being able to physically access the starches for conversion. The starches must be gelatinized and liquefied before the starches can be efficiently converted to sugars. Crushing or rolling the grain helps expose the starch granules to hydration during the mash. Once hydrated, the starches will begin to gelatinize (i.e., swell as they become hydrated) from a combination of heat and enzyme action. Alpha-amylase is able to work on the surface of ungelatinized starch, but that is not very effective.

Gelatinized starch is much more accessible to diastatic enzyme action. The average temperature range for barley starch gelatinization is between 140°F (60°C) and 149°F (65°C), but it can occur between 131°F and 153°F (55–67°C) depending on barley variety and growing conditions. Gelatinization is a gradual process, starting at the bottom of the temperature range and finishing at the top. One reason for the range of temperatures for gelatinization is that there are two types of starch granules embedded in the matrix, small granules and large granules. The large granules are more easily gelatinized (i.e., at lower temperatures), while the small granules are harder and require higher temperatures. Going back to the jelly beans in a box metaphor (see chapter 15), the large granules are the jelly beans, and the small granules are the hard candies. Jelly beans dissolve more easily in hot water than hard candies.

In addition, some cereal starch adjuncts, like oats and corn, contain a small amount of lipids (fats) that are associated with the starch, making the starch more resistant to degradation, that is, it requires a longer time at temperature to fully gelatinize. The best way to ensure accessibility of high-gelatinization temperature adjuncts is by cooking the grain before adding it to the mash to pre-gelatinize the starch. This can be done by steam cooking and rolling (as is done for flaked oats), or by simply boiling. Table 16.1 lists starch gelatinization temperatures for barley and various unmalted grains.

After gelatinization, alpha-amylase is better able to break up the long starch chains into smaller starch chains (dextrins), which greatly reduces the mash viscosity. This stage is called "liquefaction." These dextrins are now fully accessible to the other diastatic enzymes in the malt (beta-amylase, limit dextrinase, and alpha-glucosidase), so conversion of these shorter gelatinized starches begins.

### TABLE 16.1—STARCH GELATINIZATION TEMPERATURE RANGES

| | | |
|---|---|---|
| Barley | 136–149°F | 58–65°C |
| Wheat | 136–147°F | 58–64°C |
| Rye | 135–158°F | 57–70°C |
| Oats | 135–162°F | 57–72°C |
| Sorghum | 156–167°F | 69–75°C |
| Corn (maize) | 162–172°F | 72–78°C |
| Rice | 158–185°F | 70–85°C |

Notes: Barley, wheat, oats, and rye can be gelatinized in the mash, because their temperature range is below or mostly below the saccharification temperature range. Corn and rice need to be pre-gelatinized by cooking or hot rolling into flakes before they can be utilized in the mash. The degree of gelatinization depends on how hot the starch gets during the rolling or flaking process. The gelatinization temperature of a starch does not change significantly with malting, but is a natural property of the grain that can vary due to botanical variety and growing conditions.

Source: Hertrich (2013).

### Acid Rest

Before the turn of the nineteenth century (and perhaps even the twentieth), when the interaction of malt and water chemistry was not well understood, brewers used an acid rest in the temperature range of 86–126°F (30–52°C) to help acidify the mash when using pale malts. There are two reasons why this temperature rest helped. First, these temperatures are ideal for growing lactobacilli bacteria, which are commonly found coating the outside of all base malts. A few hours at these temperatures was sufficient to grow the lactobacilli such that they produce a sufficient, but tasteless, quantity of lactic acid to lower the pH by one or two tenths, e.g., from 5.8 to 5.6, (although bear in mind the pH scale as we know it today was not invented until 1924). The problem with growth of "wild" *Lactobacillus* species is that it is difficult to predict, and there are often other, less beneficial, bacteria present as well, such as *Pediococcus*.

The second reason for using these temperatures is the further production of phosphates from malt phytin (a.k.a. phytate, or phytic acid). Malt is rich in phytin, which is the primary storage form of phosphorus for the barley plant. During malting, the enzyme phytase breaks down phytin into phosphates and *myo*-inositol (a vitamin) that could be used by the developing plant. In the mash, the phosphates react with calcium and magnesium in the brewing water to form insoluble phosphate compounds, releasing two hydrogen ions in the process that lower the mash pH. Actually, this reaction will occur with or without the phytase enzyme, but the enzyme helps as long as there is sufficient calcium in the water.

The acid rest is not used nowadays, because the desired mash pH range can be achieved from the outset through knowledge of water chemistry and appropriate mineral and acid additions.

### Doughing-In

To the best of my knowledge, the acid rest is no longer used for lowering the mash pH by any commercial brewery. However, that same temperature regime is sometimes used by brewers for "doughing-in." Doughing-in involves mixing the grist with 95–113°F (35–45°C) water to allow time for the enzymes to be distributed and help liquefy the mash. The use of a short rest at temperatures near 104°F (40°C) can improve the total yield by a couple of points, but is considered to be optional.

However, there is a problem when doughing-in at those temperatures, because long-chain fatty acids can be oxidized by the enzyme lipoxygenase. These oxidized fatty acids can cause the formation of oxidized flavors in the beer later on (e.g., *trans*-2-nonenal, which is an oxidation product that has the taste and aroma of old paper). This reaction is part of a phenomenon known as "hot side aeration," which is thought to cause faster staling in beer. Doughing-in at a higher temperature, 140°F (60°C) or greater, will denature lipoxygenase and its effects will be greatly reduced. The whole issue of hot side aeration and its effects on flavor stability is still being investigated at the time of this writing. Several researchers believe that its effects can be mitigated by a strong fermentation.

Current opinion across many sources is that mashing-in at lower temperatures with modern, well-modified malts may have more negative consequences than positive.

### Beta-Glucanase Rest

The other enzymes that are active at lower temperatures are the beta-glucanases and cytases, part of the cellulase enzyme family, which can be used to break up the beta-glucans (non-starch polysaccharides) in unmalted barley, rye, oatmeal, and wheat. These beta-glucan polysaccharides (i.e., the allegorical brambles) are natural gums that are partly responsible for the stiffness of the mash; therefore, if not broken down, beta-glucans will cause lautering difficulties. Most of the beta-glucan in barley is degraded during malting (from 4%–6% by weight to less than 0.5%), so beta-glucan viscosity is usually not a problem for well-modified malts. The same applies to malted wheat, oats, and rye. Oats and rye typically have 2%–3% beta-glucan by weight, and wheat typically has 1%–2% by weight. Corn and rice do not contain significant levels of beta-glucan compared to the other cereal grains.

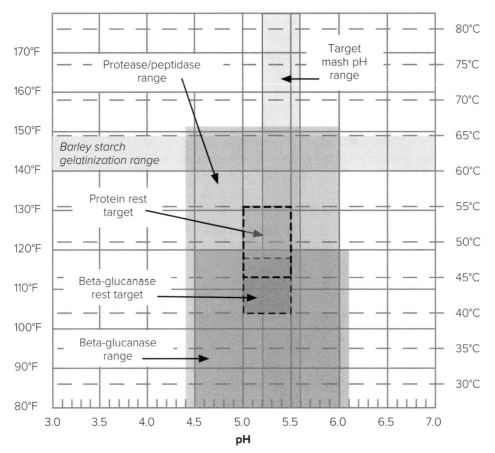

**Figure 16.2. Temperature and pH ranges for beta-glucanases and proteases.** A graphical representation of the activity ranges for the proteolytic enzymes and beta-glucanases. The boxed areas within each range indicate a preferred or higher activity region.

Resting the mash at 104–118°F (40–48°C) for 20 min. will break down these gums (fig. 16.2). This rest is recommended to improve lautering when using more than 20% of unmalted or flaked barley, oatmeal, rye, or wheat in the grain bill. A beta-glucan rest is optional when using 10%–20% unmalted adjunct, and amounts less than 10% can usually be handled by just increasing the temperature at lautering time (mash-out). In addition, beta-glucan in the wort acts as a non-Newtonian fluid, meaning the viscosity decreases as you stir it, just like ketchup. So, if your mash is not lautering well, give it a few stirs to thin it. However—and this is a big however—if you overstir at the starch conversion temperatures, that is, if you stir too vigorously for too long, the beta-glucan polysaccharide molecules will straighten out, link up, and become very viscous.

## Protein Rest and Modification

Barley contains many amino acid chains (peptides) that are used to synthesize the proteins needed by the germinating plant. During malting and mashing, amino acids are cleaved from these peptide chains by proteolytic enzymes, and these liberated amino acids are subsequently used by the yeast for their own growth and development during fermentation.

The two main proteolytic enzyme groups are the proteases and peptidases. There are at least 40 different enzymes that belong to these two groups. Some work by cleaving the large, generally insoluble protein chains into smaller soluble proteins, which can enhance the head retention of beer but are also involved in haze formation. Other proteolytic enzymes "snip" amino acids from the ends of the protein chains to produce small peptides and individual amino acids, which are the wort nutrients that can be used by the yeast. These enzymes do the majority of their work during the malting process.

Most base malt in use in the world today is well-modified or even highly modified. See the sidebar, "Malt Modification in a Nutshell," for a discussion of the distinctions. Modification is the term that describes the degree of breakdown of the cell walls and protein-carbohydrate matrix of the endosperm during malting. Moderately modified malts benefit from a protein rest to allow the proteolytic enzymes to break down any remaining large proteins into smaller proteins and amino acids, as well as to release even more starch from the endosperm. Well-modified malts have already made use of these enzymes and do not benefit from the protein rest regime. In fact, using a long (>30 min.) protein rest at 122°F (50°C) on well-modified malts tends to remove some of the body from the final beer and reduce its foam stability, but this warning tends to be overstated.

Moderately modified malts allow the brewer to take more control of the mashing process to tailor the fermentability and body of the wort to their own specifications. Craft brewers claim these malts allow them to produce fuller, maltier flavors than single-temperature mashing with well-modified malt. Moderately modified malts work better with multiple-temperature rest and decoction mashing than well-modified or highly modified malts.

The active temperature and pH ranges for protease and peptidase enzymes overlap. Both types of enzyme are active enough between 113°F and 152°F (35–67°C) that talking about an "optimum range" for each is irrelevant. At one time it was thought that protein rests at higher temperatures favored proteases and lower temperatures favored peptidases, but more recent studies have shown that is not true. All the protein degradation reactions happen concurrently.

There are a several different enzymes within each group, and these specific enzymes are active across a wide range of temperature and pH. The total activity of the proteolytic enzymes is highest at lower pH levels (3.8–4.5), but the difference is only about 15% more than the activity at a typical mash pH of 5.2–5.6.[1]

The recommended temperature and time for a protein rest is 122°F (50°C) for 15–30 min., but the enzymes will still be active for some time in the conversion temperature range of 140–155°F (60–67°C).[2]

Historically, the main purpose of a protein rest was to help free up more of the starch from the endosperm. The secondary purpose was to provide free amino nitrogen (FAN) to the wort. Less-modified and moderately modified malts have less soluble protein than well-modified malts. Unmalted grains have the least of all, with only small amounts. Thus, worts made from a large proportion of unmalted grains, particularly corn (maize) and rice but also flaked barley, can be deficient in FAN due to lack of soluble protein for peptidases to act on.

Raw barley has less than 13.5% total protein by weight. During malting, about 50% of this total is solubilized, and 20% or less is solubilized later during the mash. More importantly, only about 3% of the total solubilized protein will be converted to FAN by the end of malting and mashing. A

---

[1]   Jones and Budde (2005).
[2]   Jones (2005).

protein rest need only be used for moderately modified malts, or when using well-modified malts with more than 20% of unmalted wheat, rye, or oats (unmalted wheat contains twice as much high molecular weight protein as barley malt). A beta-glucanase rest can be combined with a protein rest by resting at 113–122°F (45–50°C) for 15–30 min. to help break up the highly viscous beta-glucans in unmalted grains (fig. 16.2).

The necessity of a protein rest to improve head retention has been overstated for many years. A short protein rest will aid head retention, but it will also promote the formation of haze-active proteins. The proteins that really enhance foam (i.e., head retention) are released at temperatures greater than 140°F (60°C). Unless you are using less-modified malt or moderately modified malt, the clarity and head retention of your beer will be better without a protein rest.

---

## MALT MODIFICATION IN A NUTSHELL

One topic that new all-grain brewers will often hear about, and one that even experienced brewers may not have a clear understanding of, is malt modification. The more the malt has been modified, the easier it is for the amylase enzymes to access and convert the starches to fermentable sugars.

The most commonly used indicator of malt modification is the soluble-to-total protein (S/T) ratio, also known as the Kolbach Index (KI). During malting, proteolytic enzymes break up the large proteins in the endosperm protein-carbohydrate matrix. In addition to exposing the starch, this breakdown of protein creates soluble amino acids, measured as free amino nitrogen, otherwise known as FAN. The S/T ratio for the malt describes the extent of breakdown of the endosperm.

- To generalize, an S/T ratio of 36%–40% is a moderately modified malt, 40%–44% is a well-modified malt, and 44%–48% is a highly modified malt. Historically, base malts commonly used a couple hundred years ago had a ratio of 30%–35%, and are nowadays considered less-modified.

- The yield from less-modified malts can be improved by decoction mashing—where boiling portions of the mash and multiple temperature rests, including a protein rest, help to fully release, solubilize, and convert the starches.

- The yield from moderately modified malts can be improved by utilizing a protein rest during mashing, but almost all of the extract can be obtained without one simply by using a conversion rest at 149–155°F (65–68°C).

- The yields from well-modified and highly modified malts do not benefit significantly from protein rests during the mash and they can be easily converted using a single temperature rest at 149–155°F (65–68°C).

However, note that we are talking about conversion and yield, not fermentability. Conversion and yield are two sides of the same coin—the better the conversion, the better the yield. Yield is the sum total of sugar, while fermentability is a description or measure of the types of sugars that are created during conversion. Conversion, to a very large extent, is driven by the maltster, whereas fermentability is entirely driven by the brewer manipulating the beta-amylase and alpha-amylase conversion rest temperatures and times.

---

### Starch Conversion, or Saccharification, Rest

Finally, we come to the main event, which is converting the starch reserves into sugars, a process known as *saccharification*. To understand the starch conversion, or saccharification, process, it

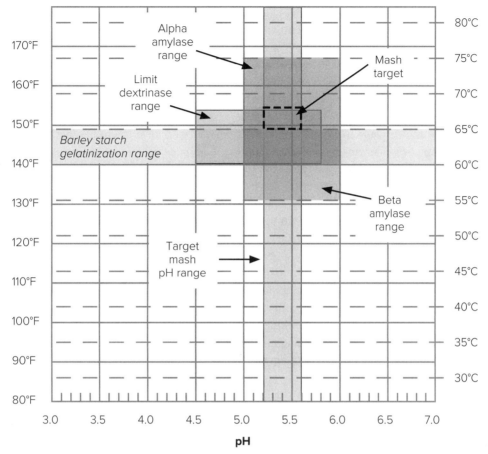

**Figure 16.3. Temperature and pH ranges for Amylase and Dextrinase.** A graphical representation of the activity ranges of the diastatic enzymes. The boxed areas within each range indicate a preferred or higher activity region.

helps to remember that the basic building blocks of starch are glucose molecules that are chemically bonded together and arranged in long chains. A single straight-chain starch molecule is called an amylose, and these amylose chains are typically hundreds or thousands of glucose units long. An amylopectin is a very large molecule, which can be considered to be composed of a multitude of amylose chains branching off from one another. Actually, amylose molecules can be branched as well, but the difference is essentially a couple pieces of string tied together versus the head of a mop.

Breaking the chemical bond between any two of the glucose building blocks requires the addition of two atoms of hydrogen and one of oxygen, in other words, $H_2O$. Therefore, the breaking of these bonds to separate the molecules is known as hydrolysis. The diastatic enzymes break up starch molecules into sugars by hydrolyzing different parts of a starch molecule, as will be explained.

## Diastatic Enzymes

The various enzymes that break down starch are collectively known as diastatic enzymes. There are a grand total of four types of diastatic enzyme that hydrolyze starches into sugars: alpha-amylase, beta-amylase, limit dextrinase, and alpha-glucosidase.[3] Each of these enzymes has several forms. Each form of an enzyme may have a slightly different pH and temperature range.

Let's go back to our yardwork allegory for a second. You have three tools to make sugars with: a pair of clippers (alpha-amylase), a hedge trimmer (beta-amylase), and a chainsaw (limit dextrinase). While beta-amylase and some limit dextrinase are pre-existing, alpha-amylase and more limit dextrinase are synthesized within the aleurone layer during malting. In other words, we start with the hedge trimmer and chainsaw in the garage ready to grab, but the clippers are out in the grass and brambles somewhere. In a well-modified malt, the combined work of your brother (beta-glucanase) and sisters (proteolytic enzymes) is completed during malting, which allows you to get all of your tools together in order to cut the branches into little pieces. All of the diastatic enzymes are available in the mash, and act concurrently to break down the starches.

The amylase enzymes work by hydrolyzing the bonds between the individual glucose molecules that make up the amylose and amylopectin straight chains, but they work differently. Beta-amylase can only work on "twig" ends of the chain, not the "root" end. Beta-amylase removes one maltose sugar unit (a disaccaharide, i.e., two glucose molecules joined together) at a time, so it works sequentially down the starch chains. On an amylopectin, with its many branches, there are many twig ends available, and beta-amylase can remove a lot of maltose very efficiently, like our allegorical hedge trimmer. However, due to the size and structure of the enzyme itself, beta-amylase cannot get close to the branch joints. It will stop working about three glucose units away from a branch joint, leaving behind a small, branched sugar chain. Because this small branched piece represents the limit of beta-amylase's activity it is called "beta-amylase limit dextrin."

Unlike beta-amylase, alpha-amylase can attack the bonds between glucose units anywhere along the chains that make up amylose and amylopectin, much as you can with a pair of clippers. Alpha-amylase is instrumental in breaking up large amylopectins into smaller amylopectins and amyloses, creating more ends for beta-amylase to work on. Alpha-amylase is able to get within one glucose unit of an amylopectin branch, which leaves behind an "alpha-amylase limit dextrin."

The branched joints of the limit dextrins that alpha- and beta-amylase cannot break can be hydrolyzed by the enzyme limit dextrinase. The action of limit dextrinase serves to cut up the branched sections into smaller chains, much like the allegorical chainsaw can be used to make more manageable pieces of tree branch. Once the limit dextrinases have done their work, we are left with small unbranched chains that are now more accessible to the alpha- and beta-amylases. This further round of action by the amylases leads to the formation of glucose, maltose, and maltotriose units. In other words, much of the original starch has now been converted to fermentable sugars.

The other diastatic enzyme, alpha-glucosidase, makes glucose out of both starches and dextrins. Alpha-glucosidase does not seem to play a significant role in the overall conversion, although it is more heat stable than beta-amylase and will act along with alpha-amylase to produce fermentable sugars after beta-amylase has been denatured. In all likelihood, alpha-glucosidase is very useful to a growing barley plant, but it doesn't seem to help us make beer.

### Enzyme Thermostability in the Mash

The temperatures most often quoted for starch conversion are in the range 149–153°F (65–67°C). This is a compromise between the completion of starch gelatinization and thermal denaturing of the beta-amylase and limit dextrinase enzymes. Collectively, the diastatic enzymes work best from

---

[3]    I didn't include alpha-glucosidase in the yard work allegory, but if I had, it would be a pair of elementary school scissors.

131–149°F (55–65°C), but remember, the generally accepted range for gelatinization is between 140–149°F (60–65°C) and may go as high as 153°F (67°C) depending on barley variety and growing conditions. Alpha-amylase works best at 140–158°F (60–70°C), while beta-amylase works best between 131–149°F (55–65°C).

## TABLE 16.2—MAJOR ENZYME GROUPS INVOLVED DURING MALT MODIFICATION AND STARCH CONVERSION

| Enzyme | Active temp. range | Preferred temp. range | Active pH range | Preferred pH range | Function |
|---|---|---|---|---|---|
| Phytase[a] | 86–126°F 30–52°C | 95–113°F 35–45°C | 5.0–5.5 | 4.5–5.2 | Helps lower the mash pH, but not required. |
| Beta-glucanase[b,c] | 68–122°F 20–50°C | 104–118°F 40–48°C | 4.5–6.0 | 4.5–5.5 | Best gum breaking rest for unmalted adjuncts. |
| Proteases[d] | 68–149°F 20–65°C | 113–131°F 45–55°C | 4.5–6.0 | 5.0–5.5 | Solubilize insoluble barley storage proteins. |
| Peptidases[d] | 68–153°F 20–67°C | 113–131°F 45–55°C | 4.5–6.0 | 5.0–5.5 | Produce free amino nitrogen (FAN) from soluble proteins. |
| Alpha-glucosidase[e] | 140–158°F 60–70°C | Unknown | 4.5–6 | 5.0–5.5 | Cleaves maltose and larger sugars into glucose. Negligible effect on total yield. |
| Limit dextrinase[f] | 140–153°F 60–67°C | 140–149°F 60–65°C | 4.5–5.8 | 4.8–5.4 | Cleaves limit dextrins. |
| Beta-amylase[c] | 131–149°F 55–65°C | 131–149°F 55–65°C | 5.0–6.0 | 5.4–5.5 | Produces maltose. |
| Alpha-amylase[c] | 140–167°F 60–75°C | 140–158°F 60–70°C | 5.0–6.0 | 5.6–5.8 | Produces a variety of sugars and dextrins, including maltose. |

*Note: The pH ranges are quoted at 25°C. The active temperature range for each enzyme indicates substantial measured enzyme activity under laboratory conditions. In the case of the diastatic enzymes, the preferred range begins where the starch becomes soluble during gelatinization so that it is most accessible for efficient enzyme action, without being inactivated. The enzymes will be active outside the indicated ranges, but will become denatured as the temperature increases above each range.*

*Sources: a. Lee (1990); b. Muller (1995); c. Kunze (2014); d. Jones and Budde (2005); e. MacGregor and Lenoir (1987); f. Stenholm and Home (1999).*

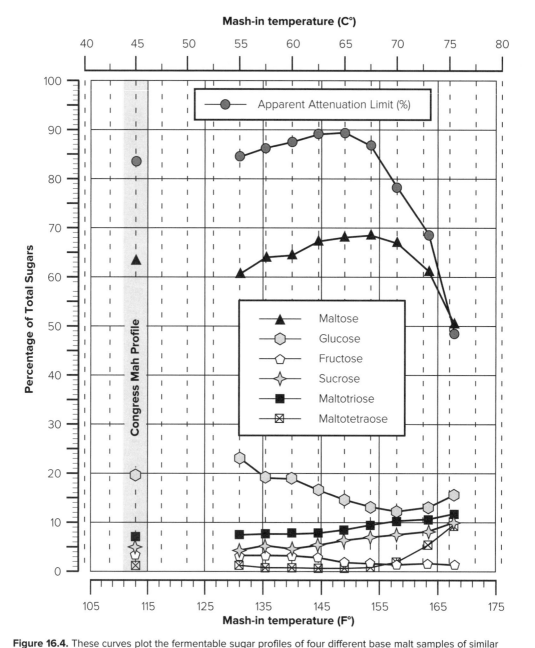

**Figure 16.4.** These curves plot the fermentable sugar profiles of four different base malt samples of similar diastatic power tested under the same procedure. Test mashes were conducted at nine different mash-in temperatures and compared to the fermentable sugar profile and apparent attenuation limit (AAL) of a Congress mash for the same malts. Each data point is the mean of four different base malts.

Notes: The Congress mash consists of mashing-in at 45°C (113°F) for 30 min., then increasing at 1°C/min. to 70°C (158°F), and holding for 60 min. before cooling. The samples with mash-in temperatures ≤70°C (≤158°F) were held at the noted temperature for 50 min. and then raised to 70°C for 10 min. before cooling. The samples with mash-in temperatures >70°C (>158°F) were held isothermally for 60 min.

(Data extracted from Evans *et al.* [2005].)

The higher the temperature, the faster an enzyme will work, and the faster it will denature as it exceeds its preferred temperature range. Denaturing means that the shape of the enzyme changes and it will no longer function as the "lock" to the enzyme target's "key." (Author's note: I know this sounds backwards, but enzyme molecules are usually larger than the target molecules.) This is true of all enzymes, so even though beta-amylase is denaturing at 149°F (65°C), it is also working faster. The specific rate of both activity and denaturing for beta amylase will depend on the particular malt and mashing conditions, but as an example, it has been shown that in typical brewery mashing conditions the initial level of beta-amylase activity was reduced by 75% after 30 min. at 149°F (65°C) and by 90% of its original activity after 60 min.[4]

While there appears to be much less limit dextrinase in the mash than beta-amylase, studies examining limit dextrinase in actual mashes suggest that it is more heat and pH stable than beta-amylase, working better at lower mash pH (4.8–5.4) and maintaining 60% of initial activity after an hour at 149°F (65°C). Between beta-amylase and limit dextrinase activities, maltose generally makes up between 60% and 70% of the wort sugars.

Generally, many of the thermostability and thermal optima numbers cited in textbooks and scientific papers were made using purified enzymes acting on a suitably buffered substrate, not in an actual mash. The information presented above comes from work published in the last 20 years using actual barley starch in real-world mashing conditions. This helps explain the discrepancy between what past laboratory data told us shouldn't work versus what we have been able to do for the past 5,000 years.

Understanding the thermostability and temperature optima of diastatic enzymes (see fig. 16.3 and table 16.2) allows us to customize the fermentability of the wort by changing the final proportions of fermentable sugars. A lower mash temperature of 144–149°F (62–65°C) favors beta-amylase and yields a lighter-bodied, more attenuated beer. A higher mash temperature of 154–162°F (68–72°C) favors alpha-amylase and yields a more dextrinous, less attenuated beer. Temperatures in between produce a range of fermentability.

As a practical example, see figure 16.4, which illustrates data from a study[5] that looked at the fermentable sugars profiles produced by conducting two-step mashes, where the primary temperature rest was conducted at temperatures ranging from 131°F to 169°F (55–76°C) for 50 min., and the second rest was at 158°F (70°C), for 15 min., followed by forced cooling. If the mash-in temperature was greater than or equal to 158°F (70°C), the mash was simply held at that same temperature for the total time of 1 hour. The sugar profiles were compared to that of a Congress mash, which is a standard malt test for determining the maximum percentage of extract by weight (see chapter 15). While the mash profile of the Congress mash has a basis in actual lager brewing practice, it was originally designed for moderately-modified malts, and utilizes a 30 min. rest at 113°F (45°C) that combines a beta-glucanase and protein rest. However, the Congress mash schedule is the reference standard for total soluble extract, and thus its fermentable sugar profile is included for comparison.

The chart in figure 16.4 shows that the percentage of maltose and degree of attenuation is highest at 149°F (65°C) and that this percentage falls as the primary mash temperature increases. The presence of maltose for primary mash temperatures >158°F (>70°C) is best explained by the action of alpha-amylase and limit dextrinase, because beta-amylase is rapidly denatured at 158°F (70°C). The degree of fermentability was expressed as the apparent attenuation limit (AAL). The AAL was determined by measuring the OG and FG of the worts, which were subjected to an agitated fermen-

---

[4]    Stenholm and Home (1999).
[5]    Evans *et al.* (2005).

tation with a high pitching rate according to industry standards (apparent attenuation is discussed in chapter 7). A maximum AAL of about 90% is typical. It is interesting to note how fast the AAL falls when mash temperatures go above 149°F (65°C) and beta-amylase is denatured. It doesn't take a large shift in mash temperature to produce a big change in wort fermentability. Based on this data (which is consistent with my experience), mashing in at 158°F (70°C) instead of 149°F (65°C) would raise a 1.050 OG wort's FG from 1.005 to 1.011.

## Mash-Out

Even though the brewing process does not require it, many brewers perform a mash-out before the wort is drained from the mash and the grain is rinsed (sparged) of residual sugars. Mash-out is the term for raising the temperature of the mash to 170°F (77°C) prior to lautering. This stops all of the enzyme action, preserving your fermentable sugar profile, and makes the grain bed and wort more fluid. For most mashes with a ratio of 1.5–2.0 qt. of water per pound of grain (3–4 L/kg), a mash-out is not needed, because the grain bed is loose enough to flow quite well. For a thicker mash, or a mash composed of more than 25% of wheat, rye, or oats, a mash-out may be helpful to prevent a set mash or stuck sparge, that is, where there is no flow. A mash-out helps prevent this by making the sugars more fluid, like the difference between cold and warm honey. If your mash has cooled during the hour and dropped below 140°F (60°C), beta-glucans, pentosans, and any unconverted starches will turn gummy and make lautering very difficult. The mash-out step can be done using external heat or by adding hot water according to the multi-rest infusion calculations (multi-rest infusion mashes are covered in chapter 17). Most homebrewers tend to skip the mash-out step without consequence, but if you do have lautering problems a mash-out is the first thing to try.

## Other Factors Affecting Starch Conversion

There are four other factors besides temperature that affect amylase enzyme activity to a lesser extent. These are the mash pH, degree of crush, water-to-grist ratio, and mash time.

### Mash pH

Mash pH can have a significant effect on both yield and fermentability. Problems can occur when mash pH is <5.0 or >6.0. When mash pH is <5.0, and especially when <4.5, beta-amylase activity is severely diminished, and wort clarity can be a real problem as well. When mash pH is >6.0 silicates and tannin extraction from the malt husks increases substantially, affecting beer flavor.

Mash pH changes within the 5.0–6.0 range can also result in significant changes to yield. For example, one experiment, conducted at Ballast Point Brewery and Tap Room (San Diego, CA) by myself and Aaron Justus, using the same recipe for two pilot batches, demonstrated an 8% increase in yield where the only difference was a change in mash pH from 5.5 to 5.3, caused by an increase in the calcium ion level of the water from 54 to 120 ppm. Generally speaking, across a mash pH range of 5.2–5.8:
- the fermentability will not significantly change;
- the FAN content will improve by a small but significant amount at the low end of the range, all else being equal;
- a lower mash pH (5.2) reduces the effect of water-to-grist ratio on yield and fermentability, all else being equal;

- and a lower mash pH (5.2) reduces the activity of lipoxygenase and so reduces the rate of oxidation of fatty acids, which should improve long-term flavor stability, all else being equal.

Brewing salts can be used to raise or lower the mash pH, but these salts should only be used to a limited extent because they can also affect the flavor. Water treatment is an involved topic and will be discussed in more detail in chapters 21 and 22. If you are a beginner at mashing, it is often best to let the pH do what it will and work the other variables around it, as long as your water is not extremely soft or hard. In most situations, malt selection can do as much (or more) to influence the mash pH as brewing salt additions. The pH of the mash or wort runnings should be checked with a pH meter for best accuracy. The pH test papers that are sold at brewshops can give you a rough estimate, but even the best of them are not very accurate in wort. Testing pH will be discussed more in chapters 21 and 22.

## Degree of Crush

Basically, the finer the crush, the better the enzymes will be able to act on the starches, and the faster the conversion will be. Do you get better yield from a finer crush? No, not really. Modern well-modified and highly modified malts differ in yield by a percentage point or two between fine grind and coarse grind, as in, the difference between 79% and 80%. The main difference is the length of time it will take for complete conversion. See below for more information on mash times.

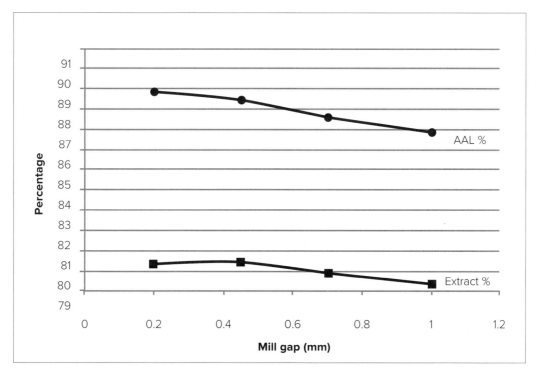

**Figure 16.5. Changes in AAL and Yield vs. Mill Gap.** Data from mashes studied in the laboratory indicate that the difference in extract yield between a very fine grind (0.2 mm gap) and a typical homebrewing roller mill (1 mm gap) is about 1.5% at most. The difference in apparent attenuation limit (AAL) is about 2% at most. (Data taken from Evans *et al.* [2011].)

Figure 16.5 shows the effect of a finer crush on two different lager malt varieties, and indicates that the degree of crush will only improve AAL by about 2% and yield by about 1%–1.5%. Changes in mashing temperature have a much larger effect on fermentability and AAL than degree of crush. Generally speaking, a finer crush:

- will convert faster,
- will yield more extract by a small amount,
- won't significantly improve fermentability,
- and won't increase FAN.

## Water-to-Grist Ratio

The water-to-grist ratio (usually in quarts per pound, or liters per kilogram) is the least significant factor influencing the performance of the mash (e.g., see fig. 16.6). A thinner mash of >2.0 qt./lb. (>4L/kg) dilutes the relative concentration of the enzymes, which slows the conversion and leads to quicker denaturing, but could ultimately lead to a more fermentable mash because the enzymes are not inhibited by a high concentration of sugars. A stiff mash of <1.25 qt./lb. (<2.5 L/kg) is better for malt protein degradation, but the resultant wort is less fermentable and will result in a sweeter, maltier beer.

According to the results of a study published in 2013,[6] varying the water-to-grist ratio across a range of 1-2 qt./lb. (2–4 L/kg) changed the AAL and yield by less than 5%. Very thick mashes of less than 1 qt./lb. (2 L/kg) had the worst performance, requiring twice as much time, 40 versus 20 min-

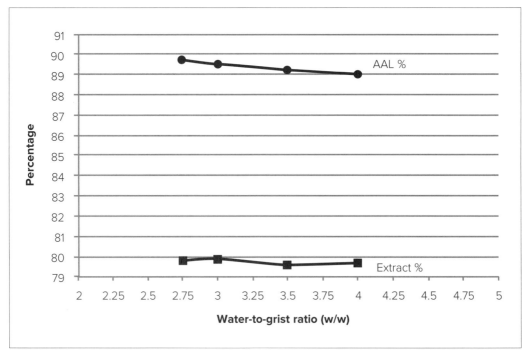

**Figure 16.6. Changes in AAL and Yield vs. Water to Grist Ratio.** Laboratory mash studies indicate that changes in water-to-grist ratios across the range of 1–2 qt./lb. (2–4 L/kg) have only a small effect on fermentability and yield. (Data taken from Evans *et al.* [2011].)

---

6    De Rouck *et al.* (2013).

utes, to fully convert to the same AAL as the thinner mashes. Anecdotal evidence from homebrewers suggests that higher water-to-grist ratios (2–4 qt./lb. [4–8 L/kg]) used by single vessel brewing methods, such as brew-in-a-bag, do not seem to significantly effect AAL or yield, although mashes may take longer due to enzyme dilution.

There is so much interaction of the various factors in the mash that it is difficult to generalize, especially with such a weak variable as water-to-grist ratio. When it comes to the mechanics of mashing, however, a thicker mash is better for multi-rest infusion mashes (see chapter 17), because it is gentler to the enzymes and easier to step up from rest to rest due to the lower heat capacity of grain compared to water.

## Mash Time

Depending on the mash pH, water-to-grist ratio, and temperature, the time required to completely convert the mash can vary from under 30 min. to over 60 min. Enzyme activity is highest during the first 20 min. of the mash and tapers off thereafter, with a steep drop after 60 min. (typically). There is a lot of talk on homebrewing forums about mashes that are finished in 20 min., as verified by an iodine test for starch. But the iodine test only tells you that you have degraded starch, it does not tell you to what extent or the resulting fermentability, which will depend upon which sugars have been released. Generally, more time is needed to achieve a higher degree of fermentability, and I recommend mashing for 60 min. to be sure.

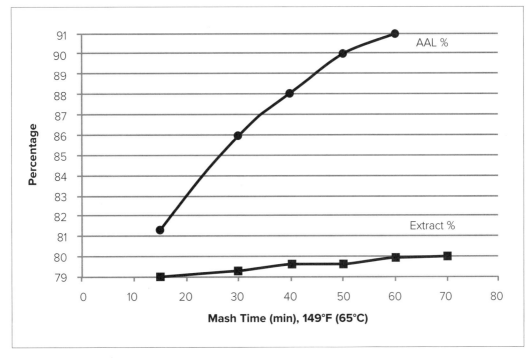

**Figure 16.7. Changes in AAL and Yield vs. Time.** Laboratory mash studies illustrating changes in apparent attenuation limit over time. Data shows the responses from two different malts, Gairdner (squares) and Flagship (triangles). (Data taken from Evans et al. [2011].)

A laboratory mashing study indicated that most of the starch conversion is achieved in the first 15 min., reaching 80% AAL (compared to a typical maximum AAL of 90%) in that time. The study was designed to mimic current brewing procedures in terms of degree of crush, single temperature infusion, mash pH, and water-to-grist ratio. A graph of the data is shown in figure 16.7, which includes the responses from two different malts. The change in fermentability (i.e., AAL) between 30 and 60 min. is about 5%. The change in total soluble extract (yield) is about 1%.

## Summary

For modern well-modified malts (S/T or KI = 40% or higher), maximum extract yield and fermentability can be achieved by a single temperature infusion mash with a water-to-grist ratio between 1.25 and 2 qt./lb. (2.6–4.2 L/kg), mashed at or slightly above the gelatinization temperature of the malt, which is typically about 149°F (65°C). I would recommend going a degree or three higher, i.e., 152°F (67°C) to assure full starch solubilization, at least initially. More than half of the starch conversion and dextrinization will occur within 15 min., and the majority (>75%) will be complete after 30 min. Small gains in extract (1%) and fermentability (5%) can be realized with longer mashing times.

In mixed lots of base malt, there will be more variation in the gelatization temperatures, and thus a slightly higher infusion temperature or a second mash rest at the alpha-amylase rest temperature (or both) may be beneficial for both total extract and fermentability.

A compromise of all factors yields the standard mash conditions for most homebrewers. These conditions provide the best combination of high yield, normal fermentability, sufficient FAN, better foam, less haze, and reduced fatty acid oxidation with modern well-modified malts:

- A water-to-grist ratio of about 1.5–2.0 qt./lb. (3–4 L/kg).
- A mash pH of 5.2–5.6, measured at 77°F [25°C].
- A single mash temperature rest between 149–155°F (65–68°C) for 30–60 min. Personally, I would recommend 153°F (67°C) for 60 minutes as a baseline; you can adjust from there as you gain experience with your equipment and recipes.

Alternatively, you can use a two step mash to achieve higher fermentability than a single temperature rest:

- The first temperature rest at 147°F (64°C) for 30–40 minutes followed by 162°F (72°C) for 15–20 minutes. This schedule should achieve slightly higher fermentability, depending on your malts.

# The Methods of Mashing

In chapters 15 and 16 you learned about the biochemistry of malting and mashing. In this chapter we will discuss how to physically manipulate the mash to create desired characters in the wort and the beer. There are two basic schemes for mashing: a single temperature mash, which is a compromise temperature for all the mash enzymes; and a multi-rest mash, where two or more temperatures are used to favor different enzyme groups. There are two basic ways of heating the mash as well, by the addition of hot water (infusion) or by heating the mash tun directly. These two heating methods can also be combined into a third method, called decoction mashing. In decoction mashing, part of the mash itself is heated on the stove and added back (infused) to the main mash to raise the temperature.

All of these mashing schemes are designed to achieve saccharification (conversion of starches to fermentable sugars). However, the method used to achieve that goal can have a considerable influence on the overall wort character. Some malts, adjuncts, and beer styles need a particular mash procedure to arrive at the right wort for the style. First, let's take a look at the entire grain brewing process.

## Overview of the Grain Brewing Process

The grain brewing process is quite simple:

*Step 1.* Heat water.
*Step 2.* Crush the grain.
*Step 3.* Soak the crushed grain in the hot water for an hour. (Mashing.)
*Step 4.* Drain the wort from the grain. (Lautering.)
*Step 5.* Rinse the grain to extract more wort. (Sparging.)
*Step 6.* Boil the wort and ferment as usual.

The hot water in step 1 is usually heated in the hot liquor tank (HLT). Traditionally, each step has its own vessel, or tun. The mash would be conducted in a mash tun, then transferred to a lauter tun for lautering and sparging, and then the wort is boiled in the boil kettle. If you combined the mash and lauter tun into a single vessel, as homebrewers most often do, and use gravity instead of a pump to get the hot water into the mash tun and the wort into the boil kettle, then you have the common three-vessel, gravity-fed system used by most homebrewers all over the world (fig. 17.1). The vessels can be large stockpots (i.e., kettles), or two pots and an insulated cooler, or one pot and two coolers. There are many ways to accomplish these basic steps.

**Figure 17.1.** A conceptualized diagram from ancient Egypt of a three-tier, gravity-fed system.

A relatively recent method that only uses a single vessel for mashing and lautering is "brew-in-a-bag" (BIAB), sometimes pronounced "bob." In BIAB, the boil kettle is large, typically 15 gal. for a 5 gal. batch (57 L for a 19 L batch), and is used to heat all of the water for the brew, as in, the combined volumes of the mash and sparge water. The grist is put into a large mesh bag and submerged in the pot and left there for an hour. The result is a water-to-grist ratio that is double that of a traditional mash, but it still works. Additionally, the mash is not sparged—the bag is simply hoisted and drained at the end of the mash, leaving the entire boil volume of wort in the kettle. This method will be discussed further in chapters 19 and 20.

The grain brewing process as a whole is not very complicated; after all, we have been doing it for thousands of years. There are a few details, however, that warrant due consideration when you start looking closely at the individual steps. Fortunately, we invented writing thousands of years ago to help us remember them.

> *And God said, "Let there be beer." And there was beer, and God looked down upon the beer and saw that it was good. And God said, "You probably should write this down..." (According to John's Book, chap. 17, para. 7)*

The mashing method you use will depend in large part on the malts that you are mashing. Nowadays, base malts are manufactured to convert quickly and easily in the mash, because time is money to large industrial breweries. The more easily these brewers can convert and extract the sugars, the more wort they can make each day. The result is that for the last 50 years malt modification and diastatic power has been increasing, and today's base malts can convert quickly and easily with a single temperature infusion mash in about 30 min. This is good for industrial brewers of pale lager using a high proportion (~30%) of unmalted adjuncts, like corn and rice. However, many of today's craft brewers, brewing all-malt beers, decry the high modification and high diastatic power, because it makes the malts convert too fast and restricts the brewers' options to manage wort fermentability and body.

The most commonly used indicator of malt modification is the soluble-to-total protein (S/T) ratio, also known as the Kolbach Index. The S/T ratio for a malt describes how well the endosperm has been opened up to expose the starches to the amylase enzymes. To recap from chapter 15, a ratio of 36%–40% is a moderately modified malt, 40%–44% is a well-modified malt, and 44%–48% is a highly modified malt. Historically, malts of a couple hundred years ago had S/T ratios between 30% and 35%, and would be rated as less-modified malt by today's standards. These historical malts usually needed a protein rest to finish breaking down the endosperm, and they definitely needed a multi-rest or decoction mash to improve the yield. Today's well-modified and highly modified malts usually don't need a protein rest or multi-rest and decoction mashes for efficient conversion and high yield, but we may use these techniques for other reasons, as you will see.

You can't really judge a chocolate cake only by its sugar-to-fat ratio, and the same goes for S/T ratio with regards to modification. However, the malt modification and diastatic power numbers tell you a lot about how a specific batch of malt will respond to a specific mash scheme. For example, using a multi-rest decoction mash on a highly modified malt is most likely a waste of time and could result in the degradation and loss of beer characters, such as foam and body, but a single temperature decoction mash done correctly, solely for melanoidin development, would produce desired flavors that other methods cannot.

Therefore, plan your mashes accordingly; look at the base malt's S/T ratio, the F/C difference, and other information to determine which mashing process is best for your ingredients and intended beer style. Chapters 15 and 16 discuss malt modification and temperature rests in more detail.

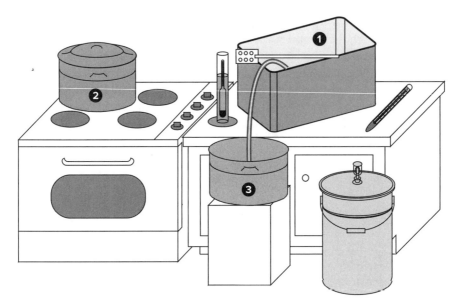

**Figure 17.2.** Mashing in the kitchen. The grist is added to the cooler (1) and infused with the strike water from the hot liquor tank (2) to bring the mash temperature to the desired rest temperature. Additional hot water can be added to raise the temperature to a second rest, if desired. During the mash, sparge water is heated in the hot liquor tank (2). After mashing, the first runnings are drained to the boiling pot (3) and the mash is sparged for the second runnings. The full wort (3) is then placed on the stove and boiled with the hops.

## Single Temperature Infusion

A single temperature infusion mash is the simplest method, and does the job for most beer styles (table 17.1). All of the crushed malt is mixed (infused) with hot water (the "strike water") to achieve a mash temperature in the range 150–155°F (65–68°C). The strike water temperature can be adjusted to hit either the low end or the high end of the mash temperature range, depending on the style of beer being made. The strike water temperature also varies with the water-to-grist ratio being used for the mash, but in general the strike water temperature should be 10–15°F (5–8°C) above the target mash temperature. The equation for calculating the temperature is listed below in the section, "Infusion Calculations." The mash should be held at the saccharification temperature for 30–60 min., hopefully cooling no more than a couple of degrees. The goal is to maintain a steady temperature in the range 150–155°F (65–68°C).

The best way to maintain a steady mash temperature is to use an ice chest or picnic cooler as the mash tun. This is the method I recommend throughout this book, but you can also use a kettle.

Generally, I recommend a water-to-grist ratio of 1.25–2 qt./lb. (2.5–4 L/kg), and a strike water temperature of 160–165°F (71–74°C). It may help to start out at the low end of the water-to-grist ratio in case you undershoot or overshoot the target temperature. If at first you don't succeed, you can add more hot (or cold) water according to the infusion calculations to adjust the temperature. It is always a good idea to heat more water than you think you'll need in case your mash temperature comes out lower than expected. Pre-heating the mash tun with hot water will also help you achieve your predicted temperatures more consistently.

## TABLE 17.1—SUGGESTED SINGLE TEMPERATURE INFUSION MASH SCHEDULES

| Schedule description | Temperature | Time (min.) | Comments |
|---|---|---|---|
| High fermentability / best yield | 149°F (65°C) | 30–60 | Highest yield and fermentability, but least body of the three options. |
| Medium fermentability | 153°F (67°C) | 30–45 | Good yield, good fermentability, and good body. The most common mash temperature for most styles. |
| More dextrinous | 158°F (70°C) | 30 | Still very good yield, but more body and lower fermentability. Good for brewing low-alcohol light ales, or rich, heavy-bodied beers. |

## Multi-Rest Mashing

A multi-rest mash requires you to add heat to raise the temperature of the mash to the various temperature rests. There are three basic ways of doing it: direct heat, infusion, or decoction. If you are using your boiling kettle as a mash tun, you can heat it directly using the stove or a stand-alone burner. Infusion is the addition of hot water (typically near boiling) to raise the mash temperature by a few degrees. Decoction is similar, except that instead of adding additional water to the mash, a portion of the mash (typically 20%–40%) is removed from the tun and brought to a boil, and the decocted mash is added back to the main mash to raise the temperature. These methods will be discussed more later on in this chapter.

### TO REST OR NOT TO REST, THAT IS THE QUESTION. . .

When should you do a protein rest?

A protein rest is done for two reasons: to improve the breakdown of the endosperm in less-modified malts to improve yield, and to increase the FAN and foam in high-adjunct worts. Malts with an S/T ratio (or Kolbach Index) of 36%–40% benefit from a short protein rest to further degrade the protein matrix around the starches of the endosperm. If the S/T ratio is less than 36%, then a longer protein rest and the boiling action of a decoction mash will probably be necessary to fully release the malt starches into the mash.

You can also do a protein rest with well-modified malts (S/T ratio >40%) in high-adjunct mashes that use unmalted wheat, oats, or rye. These adjuncts contain little soluble protein, but have a lot of insoluble protein that can be accessed with the application of a protein rest. Corn (maize) and rice contain very little protein at all and will not benefit from a protein rest; the purpose of the rest in this case is to obtain more soluble protein from the barley.

If you do use a protein rest on a well-modified all-malt beer without adjuncts, you will not ruin it. Protein rests of 15–30 min. are common practice in well-modified, high-adjunct commercial brewing, and those are not ruined beers. Be aware that too long a rest—more than 30 min.—can diminish the head retention, but your beer will not be "ruined." Relax. Don't worry.

## Heating the Mash

Direct heating is the simplest method. The first temperature rest is usually achieved by infusion using the single temperature mash method described below. The subsequent rest(s) are achieved by carefully adding heat using the stove and with constant stirring to heat the mash uniformly. After the conversion, the mash is carefully poured or ladled from the mash tun into the lauter tun and lautered. If the mash tun has a false bottom, it can be lautered directly without transfer.

*Note:* A good way to prevent hot spots and scorching with enamelware and thin stainless steel pots is to use a "flame tamer" under the pot. A flame tamer is an ⅛" (3 mm) thick aluminum or copper plate that spreads the heat more uniformly across the bottom of the pot due to its high heat conductivity.

If you are using a picnic cooler or ice chest for your mash tun, multi-rest infusion mashes are a bit trickier. You need to start out with a stiff mash (e.g., 0.75–1.0 qt./lb., or 1.5–2.0 L/kg) to leave yourself enough room in the mash tun for the additional infusions of hot water. Usually, only two additional temperature rests (after the initial rest) are possible with this method, because each addition adds more water and increases the total mass that needs to be heated to the next rest. Reaching a third rest is possible if the change in temperature is only a couple of degrees, but by that time the mash tun may be pretty full. For example, raising the mash temperature for 8.0 lb. (3.6 kg) of grain from 150°F to 158°F (66°C to 70°C) at a water-to-grist ratio of 2 qt./lb. (4 L/kg) would require approximately 2.7 qt. (2.5 L) of boiling water.

## Choosing a Multi-Rest Mash Schedule

From chapter 16, "How the Mash Works," we know that several types of enzymes are at work, liquefying the mash and gelatinizing the starches in the endosperm. Varying the times spent at the beta-amylase rest, 140°F (60°C), and alpha-amylase rest, 158°F (70°C), allows you to adjust the fermentable sugar profile (see figs. 16.2, 16.3, and 16.4). For example, a 20 min. rest at 140°F (60°C) combined with a 40 min. rest at 158 °F (70°C) produces a sweeter, more dextrinous beer, while switching the times at those temperatures would produce a drier, more attenuated beer from the same grain bill. You can also change the rest temperatures to change the profiles. For example, you could rest at 145°F (63°C) and 155°F (68°C) to improve gelatinization and beta-amylase activity to make a more attenuable wort more quickly than the previous schedule.

If you use a moderately modified malt (e.g., S/T ratio 37%) a multi-rest mash with a protein rest will produce a better yield than a single temperature mash. One recommended schedule is a 122°F, 145°F, 158°F mash (50°C, 63°C, and 70°C mash) with half hour rests at each temperature. The length of the protein rest can be adjusted depending on the degree of modification. This schedule is often used for brewing continental lager beers from moderately modified malts.

A useful multi-rest schedule for brewing Bavarian wheat beers is the 104°F, 145°F, 158°F (40°C, 63°C, 70°C) mash, using a half hour rest at each temperature. This mash schedule produces high yields and good fermentability. The time at 104°F (40°C) serves as a ferulic acid rest to enhance the phenolic character in this style. You can include a protein rest in the schedule at 122°F (50°C) if you want to thin the body of the beer or are including unmalted wheat in the grain bill.

The schedules mentioned above and in table 17.2 are intended as guidelines; almost any combination of time and temperature will produce wort, so don't freak out if you miss a rest by a few degrees. It's OK if the mash cools by a couple of degrees during the rest. If it cools more than 10°F (5°C) that may be more of a problem, but wort will still happen. Don't obsess over minutiae. Play with the times and temperatures and have fun.

## TABLE 17.2—SUGGESTED MULTI-REST MASH TEMPERATURES AND TIMES

| Schedule description | Temperatures | Time (min.) | Comments |
|---|---|---|---|
| Traditional with max. fermentability | 140°F (60°C) 158°F (70°C) | 15–30 15–30 | The beta-amylase rest temperature is at the lowest end of the range, giving the longest enzyme life, but a large proportion of the malt starches are not fully soluble until 149°F (65°C). |
| Max. fermentability with highest yield | 145°F (63°C) 158°F (70°C) | 15–30 15–30 | These temperatures will convert highly modified malts more quickly than the 140°F (60°C) traditional rest. |
| Protein rest plus beta- and alpha-amylase rests | 122°F (50°C) 145°F (63°C) 158°F (70°C) | 15–20 15–30 15–30 | Generally, today's malts only a require a short protein rest for maximum extract. Protein rests are more common when brewing with unmalted wheat or a high proportion of adjuncts. |
| Protein rest plus beta-glucanase rest or ferulic acid rest | 104°F (40°C) 122°F (50°C) 145°F (63°C) 158°F (70°C) | 10–20 10–20 15–30 15–30 | This schedule is similar to the previous, but includes the 104°F (40°C) rest, which can be used to break down beta-glucans in flaked oats, wheat or rye, or it can be used to create more ferulic acid in a wheat mash to enhance the clove character of Bavarian wheat beers. |

## RECIRCULATION MASH METHODS

There are two additional methods for multi-rest mashing that won't be covered in this book, because they require additional equipment and are more appropriate for the rabid hobbyist than a novice. These methods are recirculation infusion mash system (RIMS) and heat exchanger recirculation mash system (HERMS). Both setups require a wort pump to recirculate the wort from the mash past a heat source, either gas-fired or electric, or through a heat exchanger (copper coil) that sits in a tank of hot water. Both of these systems can be fitted with electronic controllers that allow the brewer to more precisely control the mash temperature(s). There are many resources on the Internet for building or buying these systems, but it is more important at this point in time for you to learn the fundamentals of mashing using a more manual approach.

## Infusion Calculations

Infusions are calculated based on thermal mass, in other words, how much the temperature of water mass "A" will affect the temperature of water mass "B." You cannot use volume because the density of water (and therefore the volume) varies with temperature. However, once we have solved the infusion equation for the weight of the infusion, we can convert that weight to volume by dividing the result by the density of water at that temperature range, which is 2.055 lbs./qt. (0.985 kg/L). (See the sidebar on *R* for explanation.)

## *R*, THE WEIGHT TO WEIGHT WATER-TO-GRIST RATIO

Infusion calculations are based on weight, not volume. Therefore, to accurately calculate infusion temperatures and amounts, we need to determine the water-to-grist ratio, *R*, in terms of weight-to-weight.

Normally, homebrewers discuss the water-to-grist ratio of the mash by volume, as either quarts per pound or liters per kilogram, and assume that the conversion factor between these units is 2, that is, 1 qt./lb. equals 2 L/kg. The actual conversion factor is 2.0864, which gives a more precise conversion, but this is still only a volume-to-weight ratio. We will call this ratio *Rv*.

To help illustrate this conversion difference, consider the following typical values for *Rv* and how they convert between US standard units and metric units:

1.5 qt./lb. × 2.0864 = 3.13 L/kg

3 L/kg / 2.0864 = 1.44 qt./lb.

Homebrewers use the water volume-to-grist ratio to calculate how much water to mash with by multiplying *Rv* by the grist (grain bill) weight, *G*. The weight units cancel each other, leaving the volume, *V*.

$V = G \times Rv$

To convert water volume to water weight, you need to multiply the water volume by the density of water, ρ.

Weight of water for the mash = $G \times Rv \times \rho$

Most people learn in school that one liter of water weighs one kilogram, and that the density of water is a constant at 1 kg/L (about 2 lb./qt.). The problem with this assumption is that water density is not constant—assuming that it is will either overestimate the pounds of water needed or underestimate the kilograms of water needed, by about 4% in each case. The error is small at low water-to-grist ratios, but becomes larger with multiple infusions and higher water-to-grist ratios.

Therefore, I don't recommend assuming that the density of water is 1 kg/L (or 2 lb./qt.) when calculating mash infusions. The density of water varies with temperature, and at mash temperatures the density varies from 0.995 to 0.975 kg/L. I have chosen an average value of 0.985 kg/L (equal to 2.055 lb./qt.) to simplify the calculations.

Therefore, the volume of an infusion is equal to the mass of the infusion divided by the density of water at the infusion temperature::

$Rv = R / \rho$

where *R* is the water weight to grist weight ratio, *Rv* is the water volume to grist weight ratio, and ρ is the average density of water at mash temperatures in either lb./qt. (2.055) or kg/L (0.985).

Use table 17.3 for the conversion of *Rv* (volume to weight) to actual *R* (weight to weight) values, and use the *R* values in the infusion equations section that follows this sidebar. Professional brewers always consider the mash ratio, *R*, to be the weight to weight water-to-grist ratio.

## TABLE 17.3—WEIGHT TO WEIGHT WATER-TO-GRIST RATIO *(R)* VALUES DERIVED FROM TYPICAL VOLUME TO WEIGHT WATER-TO-GRIST RATIO *(Rv)* VALUES[a]

| For *Rv* in qt./lb. | *R* (lb./lb.) | For *Rv* in L/kg | *R* in (kg/kg) |
|---|---|---|---|
| 1.00 | 2.06 | 2.00 | 1.97 |
| 1.50 | 3.08 | 3.00 | 2.96 |
| 2.00 | 4.11 | 4.00 | 3.94 |
| 2.50 | 5.14 | 5.00 | 4.93 |
| 3.00 | 6.17 | 6.00 | 5.91 |
| 3.50 | 7.19 | 7.00 | 6.90 |
| 4.00 | 8.22 | 8.00 | 7.88 |
| 4.50 | 9.25 | 9.00 | 8.87 |
| 5.00 | 10.28 | 10.00 | 9.85 |

[a] $R = Rv \times \rho$, where $\rho$ is 0.985 kg/L (2.055 lb./qt.)

The calculations assume that no heat will be lost to the surroundings (which is not true), but we can minimize this error by pre-heating the tun (such as a picnic cooler or ice chest) with some boiling hot water before adding the grist and mash water. If you are mashing using the boil kettle as the mash tun, and heating the water in the kettle directly, then these considerations don't apply.

To pre-heat the mash tun, pour a gallon or two (4–8 L) of boiling water into your tun and swirl it around. Let the tun sit for a few minutes with the lid on, and then dump it and replace the lid. Preheating the tun just before you add your grist and strike water will help the infusion heat to go into the grist rather than the tun, and your infusion estimates will be more accurate.

The subsequent infusions also assume no heat is lost to the tun, and my experiments have shown that if I calculate the infusion for about 1.5°F (1°C) higher than my actual target, then I will hit my target with the calculated amount. I was using a 10 gal. cylindrical beverage cooler. Your equipment will have its own offset.

The calculation for first infusion of the dry grain only depends on your initial grain temperature, the target mash temperature, and the weight to weight ratio (*R*) of water to grist (see sidebar and table 17.3). The amount of grain in the calculation is taken into account by the ratio, *R*. The typical water-to-grist ratio for single infusion mashing is 1.5 qt./lb. (i.e., roughly 3 L/kg). On brew day, you most likely won't be very precise and will be measuring your infusions by eyeballing the pitcher to the nearest half quart or liter anyway. Always err on the side of caution and heat a little more water than you think you will actually need, just in case. Creating a mash is not rocket science or neurosurgery and being within one or two degrees of your target temperature is really close enough, but I wanted to explain it correctly first, and then let you make up your own mind on how precise you want to be.

## QUICK AND EASY MASHING

If you don't like math, this combination of temperature and volume generally works, assuming that your grain is at room temperature (70°F [21°C]), and you have pre-heated your mash tun:

- Strike water temperature 11°F (6°C) above your target mash temperature.
- Water volume to grist weight ratio of 1.5 qt./lb., or 3 L/kg.
- Always round up to the nearest whole quart or liter.

For example, if you were going to mash 10 lb. (4.5 kg) of malt at a target temperature of 153°F (67°C), you would heat 15 qt., or 14 L, of water to 164°F (73°C) and mix it with your crushed grain. You should hit your target temperature. Always heat a bit more water than you think you need, just in case, but don't add it unless you need it.

### Dry Grain Infusion Calculations

The equations for calculating the temperature and volume of the strike water are:

Strike water temperature, $Tw = [(S/R) \times (T2 - T1)] + T2$

Strike water volume, $Vw = (G \times R)/\rho$, or $Vw = G \times Rv$

where:

$T1$ = temperature of dry grain.

$T2$ = target temperature of mash.

$G$ = grist weight (mass).

$\rho$ = density of water, 2.055 lb./qt. or 0.985 kg/L.

$R$ = the weight per weight ratio of water to grist. See sidebar, "$R$, the Weight to Weight Water-to-Grist Ratio," for explanation.

$Rv$ = the volume per weight ratio of water to grist. Hence, $Rv \times \rho = R$.

$S$ = the heat capacity of grain relative to water, which is 0.4 (i.e., 40%), regardless of unit system.

### Single Infusion Example

Before we begin, it is important to understand that the two common $Rv$ values used in the example, 1.5 qt./lb. and 3 L/kg, are similar but not equal. A mash ratio of 1.5 qt./lb. more precisely equals 3.13 L/kg; similarly, 3 L/kg equals 1.44 qt./lb. Therefore, the water volumes that will be calculated from the following example will not be equal—similar, but not equal.

We will assume that the dry grain is at room temperature, i.e., 70°F (21°C). We are going to mash 10 lb. (4.54 kg) of grain at a target temperature of 153°F (67°C). *A reminder: you may want to add one or two degrees to the target temperature to account for heat lost to the tun, but I will not include it for these examples.*

**1.** First, calculate the infusion volume (Vw) from the grist weight and the water-to-grist ratio, $Rv$, and then calculate the actual $R$ value (see sidebar that discusses $R$.)

$$Vw = 10 \text{ lb.} \times 1.5 \text{ quarts per pound} = 15 \text{ qt.}$$

$$Vw = 4.54 \text{ kg} \times 3 \text{ L/kg} = 13.6 \text{ L}$$

therefore,

$$R \text{ (for pounds)} = 1.5 \times 2.055$$
$$= 3.08$$

and

$$R \text{ (for kilograms)} = 3 \times 0.985$$
$$= 2.96$$

**2.** Using the single infusion equation, the strike water temperature (Tw) is:

$$Tw = [(0.4/R) \times (T2 - T1)] + T2$$

$$= [(0.4/3.08) \times (153 - 70)] + 153 = 163.7°F, \text{ or } 164°F$$

$$= [(0.4/2.96) \times (67 - 21)] + 67 = 73.2°C, \text{ or } 73°C$$

## Wet Grain Infusion Calculations

Weight of water, $Wm = G(S + R) \times [(T2 - T1)/(Tw - T2)]$

and

Volume of water, $Wv = Wm/\rho$

where:

$R$ = weight ratio of water to grain currently in the mash. When you perform multiple infusions, the value of $R$ changes after each infusion and needs to be recalculated before calculating the next infusion volume.

$Wm$ = weight of hot water added (in pounds or kilograms).

$Wv$ = volume of hot water added (in quarts or liters).

$\rho$ = density of water (2.055 lbs/qt. or 0.985 kg/L).

$G$ = weight of grain in the mash (in pounds or kilograms).

$T1$ = initial temperature (°F or °C) of the mash.

$T2$ = target temperature (°F or °C) of the mash.

$Tw$ = actual temperature (°F or °C) of the infusion water.

*A reminder: You may want to add a degree or two to the target temperature (T2) to account for heat that will be lost to the tun. Also, the infusion water does not have to be boiling, brewers often use the sparge water at 170°F (77°C). In that case, Tw becomes 170°F and more water (Wv) will be needed to make up the additional heat.*

### Multiple Rest Infusion Example

This example will use three rests. We are going to mash 10 lb. (4.54 kg) of grain at an initial volume ratio (*Rv*) of 1 qt./lb., through a 122°F, 149°F, 158°F (50°C, 65°C, 70°C) multi-rest mash schedule. For the purposes of this example, we will assume that the temperature of the dry grain is 70°F (21 °C). The first infusion will need to take the temperature of the mash from 70°F to 122°F (21°C to 50°C). *Remember, you may want to add one or two degrees to these targets to account for heat lost to the tun, but I will not do that for these examples.*

**1.** We will start with an initial water-to-grist ratio (*Rv*) of 1 qt./lb., which equates to *R* of 2.06 (see *R* sidebar and table 17.3). This means that we will use about 10 qt., or about 9.5 L, for the volume of the initial infusion. Note that *R*, which is weight-to-weight, is unitless and does not need to be converted when switching to liters and kilograms.

$$Vw = (G \times R)/\rho = (10 \times 2.06)/2.055 = 10 \text{ qt.}$$

$$Vw = (G \times R)/\rho = (4.54 \times 2.06)/0.985 = 9.5 \text{ L}$$

Using the single infusion equation, the strike water temperature is:

$$Tw = [(0.4/R)(T2 - T1)] + T2$$

$$= [(0.4/2.06)(122 - 70)] + 122 = 132°F$$

$$= [(0.4/2.06)(50 - 21)] + 50 = 55.6°C$$

**2.** For this example, we will assume our mash has not lost any heat during the rest and is still at 122°F (50°C). In reality, you should measure the temperature and use that value as T1 in the next equation. Our mash currently contains 10 qt. (9.5 L) at a water-to-grist ratio (*R*) of 2.06. We need to use the mash infusion equation for the second infusion to raise the temperature to 149°F. We will assume that our boiling hot water for the infusions has cooled somewhat to 205°F (96°C).

$$Wm = G(S + R) \times [(T2 - T1)/(Tw - T2)]$$

$$Wv = Wm/\rho$$

$$Wm = 10(0.4 + 2.06) \times [(149 - 122)/(205 - 149)]$$

$$= 11.86 \text{ lb. of hot water}$$

$$Wv = Wm/\rho$$

$$= 11.86/2.055 = 5.8 \text{ qt.}$$

For metric:

$$Wm = 4.54(0.4 + 2.46) \times [(65 - 50)/(96 - 65)]$$

$$= 5.4 \text{ kg of hot water}$$

$$Wv = Wm/\rho$$

$$Wv = 5.4/0.985 = 5.48 \text{ L}$$

**3.** For the third infusion, the total water volume is now 10 + 5.8 = 15.8 qt. We need to recalculate R. The current mash temperature is 149°F (65°C), and the target is 158°F (70°C). The temperature of the infusion water is still 205°F (96°C).

$$R = (Wv \times \rho)/G = (15.8 \times 2.055)/10 = 3.25$$

$$Wm = G(0.4 + R) \times [(T2 - T1)/(Tw - T2)]$$

$$Wv = Wm/\rho$$

$$Wm = 10(0.4 + 3.25) \times [(158 - 149)/(205 - 158)]$$

$$= 6.99 \text{ lb. of hot water}$$

$$Wv = 6.99/2.055 = 3.4 \text{ qt.}$$

For metric:

$$Wm = 4.54(0.4 + 3.25) \times [(70 - 65)/(96 - 70)]$$

$$= 3.19 \text{ kg of hot water}$$

$$Wv = 3.19/0.985 = 3.24 \text{ L}$$

The total volume of water required to perform this schedule is:

$$10 + 5.8 + 3.4 = 19.2 \text{ qt., or 4.8 gal.}$$

Remembering that the weight of grain is 10 lb., the final water volume-to-grain ratio has increased to 1.92 qt./lb. (i.e., 19.2/10).

The final weight-to-weight mash ratio is:

$$\text{Final } R = [(10 \times 2.055) + 11.86 + 6.99]/10$$

$$= 39.4/10$$

$$= 3.94$$

Note to liters, kilograms, and degrees celsius natives: If you had been conducting this mash from the beginning using the same amount of grain (4.54 kg) but using liters, your nominal $Rv$ would have been 2 L/kg and your R (weight-to-weight) would have been 1.97 (table 17.3). Your R for the second infusion would have been 3.11, and the final R would have been 3.79. These differences demonstrate that while 1 qt./lb. is very similar to 2 L/kg, they are not exactly the same.

## Decoction Mashing

**Equipment needed:**

- picnic cooler mash tun
- 4-gallon (15–16 L) heavy stockpot
- 1-quart (1 L) Pyrex glass measuring cup
- thermometer

## BREWING TIP: USE DECOCTIONS TO FIX INFUSIONS

Let's suppose your infusions didn't quite work—the mash temperature is too low. If you have added all the hot water the mash tun can take, or your mash ratio is over 3 qt./lb., you can use a small decoction to add heat without adding any more water!

Calculate the necessary decoction volume from the equations in the next section. Use a saucepan to pull out a thick portion (mostly grain, but soupy) and heat it to conversion at 155–158°F (68–70°C), hold it there for about 15 min., and then boil it for about five minutes. Return this hot decoction to the mash tun and stir it in evenly to raise the mash temperature without creating hotspots.

Check your mash temperature. If it is still low, the decoction process can be repeated.

Decoction mashing was developed to get the best extraction from the old-time northern European barley strains that depended on overwintering to germinate, and were more difficult to malt and modify. Decoction mashing provided for better breakdown and solubilization of the starches and better extraction from those less-modified malts. Beer connoisseurs claim better malt flavor and aroma from decoction mashing of those malts.

These days less-modified malts are hard to find, but decoction mashing is still useful for extracting that extra bit of malt character for German weissbier, bocks, and Oktoberfest style lagers. In addition, decoction mashing tends to increase the hot break and clarity of the wort. The pH of decoction mashes has been shown to be 0.10–0.15 pH units lower than the same recipe wort from an infusion mash, most likely due to Maillard reaction acidification.

Decoction mashing is a good way to conduct multi-step mashes without adding additional water or applying heat to the mash tun. It involves removing a portion of the mash to another pot, heating it to the conversion rest on the stove, then boiling it, and returning it to the mash to raise the rest of the mash to the next temperature rest. The portion removed should have a water-to-grist ratio between 1 and 1.25 qt./lb., or 2 to 2.5 L/kg. One way to tell if it is too dry is to watch how it backfills as you stir—it should backfill immediately, there shouldn't be any delay and stirring should never uncover the bottom of the pot.

Decoction mashing is not some mysterious, arcane lore passed from master to apprentice for hundreds of years. Well, all right it was, but the reason that decoction works is that you are simply boiling the grist to solubilize stubborn starches and create malty flavors while leaving most of the wort with the enzymes behind in the mash tun. The number one question new decocters ask is, "How much grist do I decoct?" It helps to realize that this question was not relevant to the old time brewers using under-modified malt. Those brewers intended to boil all of it to get the best yield, and that is why a traditional triple decoction regimen used one third of the grist for each step. They added the hot decoction back to the mash tun bit by bit until they reached their next temperature rest, and then pulled the next one. They knew that with a nominal water-to-grist ratio of about 2 qt./lb., or 4 L/kg, decocting one-third of the volume would raise the mash temperature about 10°C (18°F). It's easy to see how a 50°C, 60°C, 70°C mash progression became standard practice for old-time lagers.

Today, using well-modified malts, we don't need to boil all of the grist and can get by with one or two smaller decoctions. The easiest way to do it is use the old rule of thumb of scooping one-third of the mash from the bottom of the tun (this will have a thicker water-to-grist ratio.) Alternatively, you can calculate a more precise volume using the equations below. Either way, simply dip out the grist

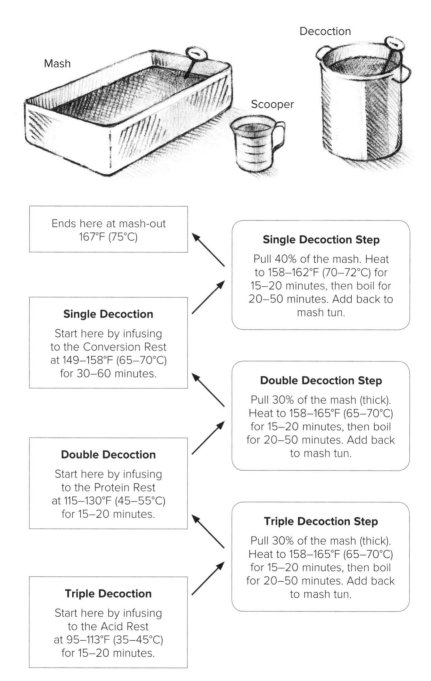

**Figure 17.3.** Use this diagram to help you plan your decoction mash. For example, if you are going to do a Zweimaischverfahren, start at the Double Decoction box and infuse your mash with hot water to achieve a protein rest or the beta-amylase rest, depending on the beer. Next, pull your first decoction according to the decoction step description, and add it back to your mash to achieve the conversion rest. Then pull your second decoction according to its description to achieve mash-out. Use the measuring cup or a saucepan to transfer the decoctions. You can use the decoction calculations presented on the next pages or just wing it and watch your thermometer.

into another pot and heat it to conversion temperature, about 158°F (70°C), where starch conversion happens quickly. Heat it quickly but gently and stir to avoid hot spots. Hold the decoction mash at conversion temperature for 15 min. and then heat it to boiling. Allow the mash to boil for 10–30 min. to develop the appropriate color and flavor; shorter boils for pale beers, longer for dark beers. Add most, but not all, of the decoction back to the mash while stirring. Mix thoroughly and then check the temperature. Add the remainder if you are low. If you have achieved your target temperature and there is some decoction left over, simply wait for it to cool to your target temperature before adding it back in.

You can use a decoction to move to any temperature rest you want, but it has traditionally been used to move the main mash from mash-in to the phytase/beta-glucanase/ferulic acid rest at 95–113°F (35–45°C), the protein rest at 113–131°F (45–55°C), the conversion rest at 149–158°F (65–70°C), or to mash-out at 167°F (75°C). The three-step decoction process is called *Dreimaischverfahren*. As lager malts became more modified, the dough-in stage was dropped, and the main mash was initially infused to the protein rest temperature. The two-step decoction for taking the mash from protein rest to conversion rest, and then to mash-out, is called *Zweimaischverfahren* (see fig. 17.3). A double decoction can also be used for transitioning between the beta- and alpha-amylase rests. You would infuse to the beta-amylase rest (20–30 min.) and then decoct to achieve the alpha-amylase rest and mash-out.

The single decoction mash is usually done from conversion temperature to mash-out. According to Greg Noonan, author of *Seven Barrel Brewery Brewers' Handbook*, the important thing when brewing for extra malt character is not the number of decoctions, but the time spent boiling one of the decoctions to develop the Maillard reactions and flavors. He recommends boiling for 20 to 45 minutes. What this says to me is that you can use triple decoctions with less-modified malts, double decoction with moderately modified malts, and single decoctions with well-modified malts, to achieve the same degree of extract and (hopefully) the same sorts of malt flavors simply by adjusting the boiling time of the main decoction. See figure 17.3 for a diagram of the decoction process and keep your thermometer handy.

For recipes and more insight on when to use decoction mashing, I encourage you to read books such as *New Brewing Lager Beer* by Greg Noonan and some of the *Classic Beer Styles Series* books by Darryl Richman, Eric Warner, and George Fix.

### Decoction Calculations

The key difference with decoction mashing is in the way heat is added to the system. With infusion mashing, hot water is always being added to the system. Decoction differs in that a volume of mash is removed from the tun, heated to boiling on the stove, and returned to the tun to raise the mash to the next temperature rest.

In essence, the amount of heat needed raise the temperature of the main mash, minus the decoction, has to equal the amount of decoction multiplied by its heat. The equation for estimating the volume of the decoction to pull and add back to the mash (Vd) is given by:

$$Vd = Vm \times [\ \frac{(T2 - T1)}{(Td - T1)}\ ]$$

where,

Vm = volume of the main mash

Vd = volume of the decoction

Td = temperature of the decoction

T2 = target temperature of the next rest

T1 = current temperature of the mash

Note that this equation calculates the volume of the decoction and not the weight. Thermodynamically, weight is the proper factor, but this equation calculates a conservative estimate that provides slightly more decoction to work with than if we had calculated a precise amount based on the relative water-to-grist ratios of the mash and weight of the decoction. The big unknown in the equation is the total volume of the mash, which can be measured or calculated using the grain weight and the water volume-to-grist ratio ($Rv$):

$$Vm = G \times (Rv + 0.38) \text{ for pounds and quarts}$$

$$Vm = G \times (Rv + 0.8) \text{ for kilograms and liters}$$

where,

Vm = total volume (quarts or liters)

G = dry weight of grain (pounds or kilograms)

$Rv$ = water-to-grain ratio of the mash (qt./lb. or L/kg)

## Adjunct Mashing Procedure

I wanted to open this section with the Monty Python line: "And now for something completely different…" But that's wrong. Adjunct mashing is actually something completely the same, being just a combination of some of the methods we have already discussed. To brew with starch adjuncts, you need to hydrolyze and gelatinize the starches so the amylase enzymes can break them into fermentable sugars. Accessibility is the key. You can gelatinize the starches just by boiling them, but you can do it more effectively by using a combination of enzymes and heat.

What about flaked adjuncts, like corn (maize) and oats? Aren't they already pre-gelatinized? Yes, to a degree. Gelatinization is not an all-or-nothing state, it's more like cooking. Actually, it is cooking; the starches can be partially, mostly, or fully gelatinized depending on how long they were cooked. Instant oats are more gelatinized than old-fashioned rolled oats. Also, just because an adjunct is flaked and pre-gelatinized does not mean that it is fully accessible to the mash enzymes. It helps to grind or break up the rolled flakes too, especially the big flakes like barley, oats, rye, and wheat.

Rice and corn contain very little beta-glucan and protein. There is no need to do a beta-glucanase rest when mashing these grains. Unmalted barley has a lot of beta-glucan, as does unmalted rye, oats, and wheat, and a beta-glucanase rest is necessary for good lauterability. If you are using the malted form of these grains you don't need a beta-glucanase rest, but you may want to include a protein rest. A protein rest at 120–130°F (50–55°C) is a good idea to break up the relatively high levels of high-molecular-weight proteins these malts contribute, if you are using more than 20% in your grain bill.

Lastly, you may want to include rice hulls in the mash to help with lautering. The husk of malted barley constitutes about 5% of the weight, so if you are using a lot of wheat (which has no husk) or rye (which is really sticky) in your brew, you would want to add at least 5% of its weight in rice hulls to make up for it. Rice hulls are very helpful with wheat beers, rye beers, and high-adjunct beers, such as American lager. Corn and rice don't have the beta-glucan that makes lautering difficult, but the high proportion of no-hull adjuncts will affect the lauterability all the same. I had a Classic American Pilsner recipe turn into porridge; it couldn't be lautered and was finally dumped. Using rice hulls the next time took care of the problem. I have not needed rice hulls for oatmeal stouts.

## Conducting a Cereal Mash

To conduct a cereal adjunct mash, I recommend that you use a heavy stockpot that can hold at least 4 gal. (15 L).

### Step 1.

If the adjunct is not flaked, then grind it a few times in your roller mill or use small coffee grinder, or a hammer, but you need to break it down for the best results.

### Step 2.

Combine your cereal grist with about 25% by weight of base malt, and infuse it at a mash ratio of 2 qt./lb. (4 L/kg) to the first temperature rest. Barley, oats, rye, and wheat should be started at 113°F (45°C) for a combined beta-glucanase and protein rest (you can cover all your bases that way.) You can start corn and rice at the beta-amylase rest of 145°F (63°C). Try not to exceed a 3:1 ratio of adjunct to malt to avoid diluting the enzymes too much.

### Step 3.

Hold the mash at the beta-glucanase rest for about 15 min., and then heat it slowly, stirring constantly to get to the conversion rest. Barley, wheat, oats, and rye can be fully converted at 155–158°F (68–70°C). Corn and rice will need higher temperatures to assure gelatinization, 165–172°F (74–78°C), but the barley enzymes will convert any starch pre-gelatinized from rolling or flaking.

### Step 4.

Next, bring the mash to a gentle boil for about 10–15 min. to fully gelatinize all the starch. There should be few, if any, crunchy bits left.

### Step 5.

The main mash can be infused and waiting at whatever temperature rest is appropriate for your recipe. You can use this hot starch soup as a decoction for your main mash to reach the next rest temperature, or wait for it to cool to the saccharification temperature if you are doing a single rest mash. Keep in mind that you may need a short protein rest for your main mash to generate more FAN if you are using a high proportion of low-protein adjunct.

### Step 6.

You can conduct further decoctions as necessary to finish the mash and then add rice hulls as necessary to help lautering. Good Brewing!

## Summary

There you have it: the two or three methods of mashing, and the calculations to take out the guess-work. Most brewers keep it simple and use a single rest infusion, which is the easiest method for producing all-grain wort. Decoction mashing used to be regarded as the hallowed domain of expert all-grain brewers, but it really is just another tool that any homebrewer can use to gain an extra malty edge on a pale wort. The most common homebrewing mash schedule consists of a water-to-grain ratio of 1.5 qt./lb. (3 L/kg), and holding the mash between 150°F and 155°F (65–68°C) for 1 hour. Probably 90% of the beer styles in the world today can be produced with this method.

# Extraction and Yield

## Or What to Expect from Your Mash

So far, we have discussed the malts, the enzymes, and the various ways of conducting the mash, but we haven't discussed the quality and quantity of wort that you actually get from the mash. This chapter will discuss the extract and yield that you get from the mash, and how to calculate your mashing efficiency to enable you to plan malt quantities for any recipe original gravity (OG).

One note before we begin. The discussions in this chapter are based on traditional mashing, lautering, and continuous sparging methods in which the mashed grain is uniformly rinsed of the sugars. This method takes the longest, but gives the maximum yield from the grain. In the next chapter we will look at several different lautering and sparging methods that are a little less efficient (i.e., use more grain), but don't take quite as much time.

## Malt Analysis Sheet—a Review

### Percent Extract–Fine Grind, As-Is and Dry Basis

A malt analysis sheet does not give the malt's yield in points per pound per gallon (PPG) or points per kilogram per liter (PKL). Instead, what you will most likely see is a weight percentage called percent extract–fine grind, dry basis (%FGDB). This percentage is the maximum soluble extract that the malt will yield when mashed under laboratory test conditions. The %FGDB is typically 80%–82% by weight, although wheat malt can be up to 85% because it doesn't have a husk like barley malt; the husk makes up roughly 5% of the weight of a barley kernel.

Malting houses analyze yield using the EBC Congress mash (described in chapter 15). A Congress mash consists of a multiple step infusion mash and rinsing using a standard weight of finely ground malt, which yields the maximum soluble extract as a percentage of the original sample weight. This yield is known as the percent extract–fine grind, as-is (%FGAI). As we saw in chapter 15, %FGAI includes the moisture content of the malt by percentage weight. Properly kilned malt contains about 4% moisture by weight, although it can range from 2%–6%. To compare different lots of malt with different moisture levels, this weight needs to be accounted for in the extract calculation. Therefore, the basis of comparison, and the number you will most consistently see on an analysis, is the %FGDB, which corresponds to a malt that has been oven-dried to zero percent moisture.

But what does 80% soluble extract mean to you as a brewer; how can you use it to calculate how much malt to use to hit your target wort gravity?

### Converting Percent Extract to PPG or PKL

The wort that is collected from the Congress mash is precisely measured for both gravity and volume. The wort gravity (points) multiplied by the volume is the total extract for that malt sample and it is reported as its percent extract (i.e., %FGDB) and is the maximum yield for that malt. The maximum yield is typically 60%–80% of the dry weight. For example, 80% FGDB means that 80% of the malt's dry weight is soluble and extracted by the laboratory mash and lauter (the other 20% represents the husk and insoluble proteins and starches). In the real world, most brewers will never hit this target, but it's the basis for comparison.

The reference for comparison is pure sugar (sucrose), because sucrose yields 100% of its weight as soluble extract when dissolved in water, and has no moisture of its own. One pound of sucrose will yield a specific gravity of 1.046 when dissolved in one gallon of water. To calculate the maximum yield for different malts and adjuncts the %FGDB for each malt is multiplied by the reference number for sucrose, which is 46 PPG (equivalent to 384 PKL).

For example, let's consider a typical two-row base malt with a %FGDB of 80.7% by weight. So, if we know that sucrose will yield 100% of its weight as soluble sugar and that raises the gravity of the wort by 46 PPG (384 PKL), then the maximum increase in gravity we can expect from this base malt, at 80.7% FGDB, is 80.7% of 46 = 37 PPG (80.7% of 384 = 310 PKL).

### Hot Water Extract

The hot water extract (HWE) parameter may be seen on malt analyses from the UK, where the Institute of Brewing (IOB) utilizes a single temperature infusion mash that differs from the ASBC and EBC Congress mash methods. Specifically, "Method 2.3—Hot Water Extract of Ale, Lager, and

Distilling Malts" in the *Recommended Methods of the Institute of Brewing* uses an hour-long mash at 149°F (65°C) to measure the maximum extract. As we saw in chapter 15, HWE is "as-is" and is measured as liter degrees per kilogram (L°/kg), a value that is equal to the PKL. Therefore, HWE in L°/kg can be converted to PKL and thence to PPG when the metric conversion factors for volume and weight are applied. The conversion factor is HWE (or PKL) = 8.345 × PPG.

## TABLE 18.1— TYPICAL EXTRACT ANALYSIS SHEET DATA FOR SEVERAL MALT TYPES

| Parameter | Malt type | | | | | | |
|---|---|---|---|---|---|---|---|
| | Two-row base | Pale ale (UK) | Munich 10L | Caramel 15L | Caramel 75L | Choco-late | Roast barley |
| % moisture | 4.4 | 3.5 | 4.0 | 7.9 | 4.8 | 3.5 | 3.3 |
| %FGAI | 78.1 | ... | 78.7 | 73.3 | 75.7 | 74.3 | 64.5 |
| %FGDB | 81.7 | 316 (HWE) | 82.0 | 79.6 | 79.5 | 77.0 | 66.7 |
| %CGAI | 77.1 | ... | 77.6 | ... | ... | ... | ... |
| %CGDB | 80.7 | 312 (HWE) | 80.9 | ... | ... | ... | ... |
| F/C difference | 1.0 | 1.0 | 1.1 | ... | ... | ... | ... |

CGAI, coarse grind, as-is; CGDB, coarse grind, dry basis; F/C, fine/coarse; FGAI, fine grind, as-is; FGDB, fine grind, dry basis; L, Lovibond (i.e., 15L = 15°L).

## Crush and Extract Efficiency

A good crush of the grist produces a range of particle sizes that provides a compromise between the speed of extract conversion and your ability to extract it. Coarse particles allow for better wort flow and lautering, but are not converted as quickly by the enzymes. Fine particles are converted faster by the enzymes, but if all the grain were finely ground you would end up with a porridge that could not be lautered. The ASBC method mashes a small sample of malt for two hours in a multi-rest mash. It is then lautered for another one to two hours! A homebrewer would be viewed as a real diehard if they spent half that much time on their mash and lautering.

If you look back at the discussion of malt analysis sheets in chapter 15, we discussed how the extract yield of a malt is measured based on whether the crush is fine grind and coarse grind. The degree of crush can have a large effect on the speed of starch conversion by the enzymes, but only a small effect on the total amount of extract. Usually there is only a 1% difference between ASBC Congress mash values for fine grind and coarse grind. A difference greater than 1.5% indicates a problem with low modification in the malt. However, if you compare the pictures between home-crushed malt and ASBC fine and coarse grinds (fig. 18.1 and fig. 18.2) you can see there is a big difference in the typical particles sizes between them.

### Yields between Different Crushes

How much does the degree of crush affect the total extraction for homebrewers? When a new all-grain homebrewer complains on an Internet forum that his yield from the mash was low, the

**Figure 18.1.** A well crushed grist with a good mix of large and small particles and unshredded husk.

**Figure 18.2.** Here is a picture of fine grind (top) and coarse grind (bottom) of the same malt sample prepared according to the ASBC "Methods of Analysis, Malt, #4–Extract." Note that the fine grind has the consistency and particle size of cornmeal or flour. The barley husks have been ground up as well. The coarse grind sample below has much larger particles than the fine grind, but overall particle size is still smaller than the home-grind pictured in figure 18.1. Also note that the husks in the coarse grind sample have been broken up and even shredded to some extent.

experienced homebrewers often question the crush. Did he buy the malt pre-crushed or did he crush it himself? Was it coarse? Was it crushed at all? (It can happen.)

Increasing the degree of crush seems to improve the yield for home-mashing systems, but it doesn't have much effect on the F/C difference from the Congress mash. Why the difference in results?

I conducted an experiment on crushing and yield with the help of Briess Malt and Ingredients Co. We compared the maximum extract values for four different grinds of the same lot of malt. I was hoping to show the difference in yield that homebrewers usually experience with their own systems compared to commercial brewers' systems and the coarse grind numbers reported on malt analysis sheets. However, the results were surprising. The extraction results and the particle size assay (sieve analysis) for samples prepared by ASBC fine grind, and ASBC coarse grind, versus one pass and two passes through a homebrewing two roller mill are shown in table 18.2. A sieve analysis is done by passing the grist through successively smaller mesh sizes, until the finest sized particles collect in the pan on the bottom.

## TABLE 18.2—COMPARISON OF PARTICLE SIZE DISTRIBUTION AND YIELD AS A FUNCTION OF CRUSH

| Particle sizes | Size distribution (%) | | | |
|---|---|---|---|---|
| | 1 pass | 2 passes | Coarse grind | Fine grind |
| **14 Sieve** | 70.4 | 58.4 | 10.6 | 1.0 |
| **20 Sieve** | 16.2 | 23.2 | 61.0 | 7.8 |
| **60 Sieve** | 6.2 | 8.6 | 13.6 | 52.8 |
| **Sieve pan** | 7.2 | 9.8 | 14.8 | 38.4 |
| | Yield, dry basis (%) | | | |
| | 79.4 | 80.1 | 79.7 | 80.9 |

It's interesting that the yield values are all within 1% of each other; in fact, this level of variation is not significant. In other words, the difference in yield is not necessarily due to the difference in the particle size distribution, but may simply be due to normal measurement error. Statistically, all the values could be the same, but we don't have enough data here to prove it.

So how can we account for the difference in homebrewer yields between different degrees of crush? One factor is time. Smaller particles will definitely convert faster than coarser particles, as discussed in chapter 16. The experiment above was conducted under laboratory conditions where the samples were mashed and lautered over a longer time period than many homebrewers are perhaps willing to spend. There is a trade-off between the crush size, the time required to achieve starch conversion, and the time required for lautering to get that extract out. This is an important consideration.

The insoluble grain husks are very important for a good lauter. The grain bed forms its own filter from the husk and other insoluble grain material. The husks allow wort and water to flow through the bed to extract the sugar, and also prevent the grain bed from compacting. The wort is drawn out through the bottom of the bed by means of a false bottom, or manifold, which has openings that allow the wort to be drawn off while preventing the grain from being sucked in as well. Usually these openings are narrow slots, or holes up to ⅛" (3 mm) in diameter. If you use too fine a screen or too narrow a slot, or you try to drain the wort too fast during lautering, you can clog part of the manifold or false bottom. It can also cause channeling (areas of preferential flow) through the grain bed, which leads to a loss of extract because you aren't collecting wort uniformly across the whole grain bed (see appendix E for more information on lauter flow). Bottom line: I think lautering and sparging problems are a bigger cause of poor yield than crush size.

## RICE HULLS AND SUCCESSFUL LAUTERING

You may want to add rice hulls to the mash to help with lautering. The husk of malted barley con- stitutes about 5% by weight of the malt, so if you are adding a fair percentage of malted wheat (which has no husk) or rye (which is very viscous) to your mash, you should add at least 5% of the adjunct's weight in rice hulls to make up for it. Rice hulls are often helpful with wheat beers, rye beers, and high-adjunct beers like American lager. Corn (maize) and rice don't contain the beta-glucans that make lautering difficult, but a high proportion of no-hull grain will affect lauter- ability all the same. I once had a Classic American Pilsner recipe seize up like a bowl of porridge; I made sure to use rice hulls the next time. I have not needed rice hulls for oatmeal stouts.

**Figure 18.3.** Examples of homebrewing two-roller and three-roller mills. These are designed to sit on top of a typical 6 gal. bucket as you crush the grain.

A good crush is described as thoroughly crushing (not just cracking) the endosperm while leav- ing the husks as intact as possible, which helps keep the wort flowing easily during the lauter. Roller mills are the industry standard, crushing the malt between two rollers like a clothes wringer. Roller mills are available in two- or three-roller configurations (fig. 18.3), with either fixed or adjustable gap settings, and give a consistent crush to the grist. A three-roller mill produces more fines than a two-roller mill, but the value of that for a homebrewing setup is debatable, as you saw in the experiment above (but I have one anyway.) Most homebrewing supply shops have a mill and will crush the grains for you, though it is worth noting that many shops tend to crush the grain on the coarse side of the range to minimize lautering problems for their customers.

## Water-to-Grist Ratio and Initial Wort Gravity

The initial gravity of your wort is primarily dependent on the water-to-grist ratio of the mash. (It also depends to a small extent on the degree of crush, but that is a secondary effect.) The model for predicting first wort gravity from $R$ is based on the maximum weight percentage of soluble extract in base malt and I have confirmed the numbers by experiment. The net wort volume in table 18.3 below is the volume of wort that can be drained from the mash, assuming that a proportion of the wort will be retained by the wet grain—typically 0.5 qt. of wort per pound of grain, or 1 L of wort per kilogram of grain. We can refer to this retained wort ratio as $r$. Therefore, the net volume is the original water-to-grist ratio ($R$) minus the retained wort ratio ($r$). In other words, if you mash in with an $Rv$ of 1.5 qt./lb., then 0.5 qt./lb. will be retained by the wet grain, and you will get 1 qt. of wort per pound of grist out of the mash into your boil kettle.

Figure 18.4 shows the curve for wort gravity as a function of the weight per weight water-to-grist ratio, $R$ (see chapter 17 for a detailed discussion of this ratio). The model for these curves is based on the conversion equation for specific gravity (SG) as a function of percent soluble extract in solution (i.e., degrees Plato)[1] at 59°F (15°C). The model assumes that 75% of the total unit weight of the mash (i.e., $R + 1$) is released into the wort as soluble extract, and that percentage is used in place of the value for °P.

$$SG = 259 / (259 - °P)$$

$$= 259 / \{259 - [75 / (R + 1)]\}$$

Four curves are shown in figure 18.4: the first is for sucrose (white table sugar), the 100% soluble extract standard; the second, third, and fourth curves are for 80%, 75%, and 70% total soluble extract by weight, respectively. The 80% curve corresponds with FGDB data. The 75% curve matches typical roller mill crush, single infusion mash data. The 70% curve corresponds to less-modified malts, or grain bills with high proportions of specialty malts (i.e., less soluble extract overall). Note that this model is for total soluble extract, and predicts the wort gravity as it exists in the mash tun before lautering. The numbers given below are typical; your results may vary, but they will not vary by much.

Common volume per weight water-to-grist ratios ($Rv$, see again chapter 17) and their corresponding initial wort gravity readings (as predicted by the appropriate $R$ value) are listed in table 18.3. These numbers form the basis for predicting the yields from the different lautering and sparging methods that will be discussed in chapter 19.

---

[1] From Martin P. Manning, "Understanding Specific Gravity and Extract," *Brewing Techniques,* September/October 1993, 30.

## TABLE 18.3—TYPICAL INITIAL WORT GRAVITY AND NET YIELD AS A FUNCTION OF WATER-TO-GRIST RATIO

| Water-to-grist ratio (Rv) | | Typical initial wort gravity (OG) | Typical net volume[a] qt./lb. (L/kg) |
|---|---|---|---|
| qt./lb. | L/kg | | |
| 0.75 | 1.5 | 1.131 | 0.25 (0.5) |
| 1.0 | 2.0 | 1.107 | 0.5 (1.0) |
| 1.25 | 2.5 | 1.090 | 0.75 (1.5) |
| 1.5 | 3.0 | 1.078 | 1.0 (2.0) |
| 1.75 | 3.5 | 1.069 | 1.25 (2.5) |
| 2.0 | 4.0 | 1.061 | 1.5 (3.0) |

[a] Assuming 0.5 qt./lb. or 1 L/kg of wort retained by wet grain.

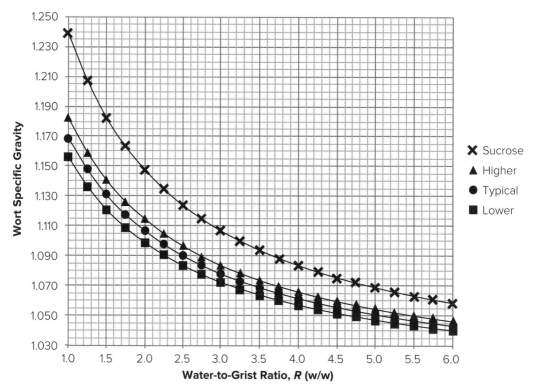

**Figure 18.4. First Runnings Gravity vs. Ratio.** Graph of initial wort gravity (OG) as a function of *R* and typical yield. Sucrose represents 100% yield as soluble extract. Higher is 80% (i.e., FGDB), Typical is 75% soluble extract, and Lower is 70% soluble extract by weight. Coarse grind, dry basis would correspond to about 79% extract.

## Extract Efficiency and Typical Yield

The maximum yield is just that, a value you might get if all the mash variables (e.g., pH, temperature, time, viscosity, crush, phase of the moon, and whatever else) lined up in your favor and 100% of the starches were converted to sugars and were then completely separated from the mash tun into the boil kettle. But most brewers, even commercial brewers, don't get that value from their mashes and lauters. Most homebrewers get 70%–80% of the maximum possible yield (e.g., 70% of a malt's FGDB of 81.7%). You may be wondering how useful the maximum yield number of a malt can be if you can never expect to hit it. The answer is to apply an efficiency factor to the maximum yield and derive a number that we will usually achieve, in other words, a typical yield. This efficiency factor is called the *brewing efficiency* and it is the percentage of the brewer's actual yield to the malt's maximum yield (FGDB). Every brewery is unique and your efficiency will depend on your methods and equipment.

When all-grain homebrewers get together to brag about their brewing prowess or equipment and say something like, "I got 28 (PPG) from my mash schedule," they are referring to their yield in terms of the volume and gravity of wort they collected from the mash and lauter before the boil. You need to measure these parameters accurately to calculate your efficiency accurately.

It is important to realize that the total amount of sugar is constant, but the concentration (i.e., gravity) changes depending on the volume. To understand this, let's look at the units of PPG and PKL. These are units of concentration, so the unit is always expressed in reference to one gallon or one liter, respectively. Another way of writing these units is gallon degrees per pound (gal.°/lb.) and liter degrees per kilogram (L°/kg). You are collecting a specific volume of wort of a particular gravity that was produced from a specific weight of malt. To calculate your total extraction in terms of PPG or PKL, you need to multiply the volume of wort you collected by the gravity points (i.e., the last two digits in a hydrometer reading "1.0__") and divide that by the weight of malt that was used. Depending on the units system you use, this will give you the total extract per weight of malt in either PPG or PKL. Let's look at an example using a grain bill for "Palmer's Short Stout."

To calculate our brewing efficiency, we will assume that 8.5 lb. (3.86 kg) of malt was mashed and lautered to yield precisely 6.0 gal. (22.7 L) of wort to the boil kettle with a gravity of 1.038. We can now work out the total sugar extraction (yield) for this batch and calculate your brewing efficiency.

$$\text{total extract per weight} = (\text{volume wort} \times \text{gravity points}) / \text{weight malt}$$

Using US standard units:

$$\text{PPG} = (6 \text{ gal.} \times 38) / 8.5 \text{ lb.}$$

$$= 228 \text{ gal.}° / 8.5 \text{ lb.}$$

$$= 27 \text{ PPG}$$

Using metric units:

$$\text{PKL} = (22.7 \text{ L} \times 38) / 3.86 \text{ kg}$$

$$= 862.6 \text{ L}° / 3.86 \text{ kg}$$

$$= 223 \text{ PKL}$$

> ### GRAIN BILL FOR PALMER'S SHORT STOUT
>
> **Original gravity = 1.048 for 5 gal. (19 L)**
>
> 6.5 lb. (2.95 kg) lager malt
>
> 0.5 lb. (225 g) Caramel 15
>
> 0.5 lb. (225 g) Caramel 75
>
> 0.5 lb. (225 g) chocolate malt
>
> 0.5 lb. (225 g) roast barley
>
> **Total: 8.5 lb. (3.86 kg)**

Comparing these numbers to lager malt's maximum yield of 37 PPG (311 PKL) gives us a good approximation of our brewing efficiency:

$(27/37) \cdot 100 = 73\%$, or $(223/311) \cdot 100 = 72\%$.

If we use the maximum yields (in PPG or PKL) given in table 18.4 for each of the recipe's individual malts, we can calculate our actual brewing efficiency from the wort gravity achieved divided by the maximum gravity.

There are naturally some rounding errors in the conversions that may lead to marginal differences between results, but the point is that these are typical efficiencies you can refer to the next time you brew.

## TABLE 18.4—CALCULATING BREWING EFFICIENCY USING MAXIMUM YIELD GRAVITY

| Quantity of malt | Max. yield (PPG) | Max. yield (PKL) |
|---|---|---|
| 6.5 lb. (2.95 kg) lager | 37 × (6.5/6) = 40.08 | 309 × (2.95/22.7) = 40.16 |
| 0.5 lb. (225 g) Caramel 15 | 35 × (0.5/6) = 2.92 | 292 × (0.225/22.7) = 2.89 |
| 0.5 lb. (225 g) Caramel 75 | 34 × (0.5/6) = 2.83 | 284 × (0.225/22.7) = 2.82 |
| 0.5 lb. (225 g) chocolate | 34 × (0.5/6) = 2.83 | 284 × (0.225/22.7) = 2.82 |
| 0.5 lb. (225 g) roast barley | 36 × (0.5/6) = 3.00 | 300 × (0.225/22.7) = 2.97 |
| *Max. total yield (as-is)* | *51.66 points, or 1.052* | *51.66 points, or 1.052* |
| *Actual gravity/Max. gravity* | *(38/52)•100 = 73%* | *(38/52)•100 = 73%* |

Typical malt yields in terms of gravity are given in table 18.5. Note that the values in table 18.5 are based on an extract efficiency of 75%, which is considered typical for homebrewers. A few points less yield (e.g., 70% extract efficiency), is still considered to be a good extract efficiency. A large commercial brewery would typically get 85% extract efficiency, and would see the 5% reduction as significant because they are using thousands of pounds of grain a day. But, for a homebrewer, adding 5% more grain per batch to make up for the difference in efficiency is a pittance.

## TABLE 18.5—TYPICAL MALT YIELDS IN PPG AND PKL

| Malt or adjunct type | Max. yield FGDB | Max. PPG | Typical PPG | Max. PKL | Typical PKL |
|---|---|---|---|---|---|
| Two-row lager malt | 81% | 37 | 28 | 311 | 233 |
| Two-row pale ale malt | 80% | 37 | 28 | 307 | 230 |
| Biscuit malt | 75% | 35 | 26 | 288 | 216 |
| Victory malt | 75% | 35 | 26 | 288 | 216 |
| Vienna malt | 75% | 35 | 26 | 288 | 216 |
| Munich malt | 75% | 35 | 26 | 288 | 216 |
| Brown malt | 70% | 32 | 24 | 269 | 202 |
| Dextrin malt | 70% | 32 | 24 | 269 | 202 |
| Light caramel malt (10–15L) | 75% | 35 | 26 | 288 | 216 |
| Pale caramel malt (25–40L) | 74% | 34 | 26 | 284 | 213 |
| Medium caramel malt (60–75L) | 74% | 34 | 26 | 284 | 213 |
| Dark caramel malt (120L) | 72% | 33 | 25 | 276 | 207 |
| Special "B" malt | 68% | 31 | 23 | 261 | 196 |
| Chocolate malt | 74% | 34 | 26 | 284 | 213 |
| Roast barley | 78% | 36 | 27 | 300 | 225 |
| Black patent malt | 68% | 31 | 23 | 261 | 196 |
| Wheat malt | 85% | 39 | 29 | 326 | 245 |
| Rye malt | 82% | 38 | 28 | 315 | 236 |
| Oat malt | 75% | 35 | 26 | 288 | 216 |
| Oatmeal (flaked) | 70% | 32 | 24 | 269 | 202 |
| Corn (flaked) | 84% | 39 | 29 | 323 | 242 |
| Barley (flaked) | 70% | 32 | 24 | 269 | 202 |
| Wheat (flaked) | 77% | 35 | 27 | 296 | 222 |
| Rice (flaked) | 82% | 38 | 28 | 315 | 236 |
| *Maltodextrin powder* | *100%* | 46 | 35 | 384 | 288 |
| *Corn sugar* | *92%* | 42 | 32 | 353 | 265 |
| *Cane sugar (sucrose)* | *100%* | 46 | 35 | 384 | 288 |

Notes: Typical PPG and PKL values are based on an extract efficiency of 75%, which is common for homebrewers. Malt yield data is percent extract–fine grind, dry basis (%FGDB) obtained and averaged from several sources.

## Planning Malt Quantities for a Recipe

### Using PPG to Calculate Malt Quantities

We use the brewing efficiency concept in reverse when designing a recipe to achieve a targeted OG.
*How much base malt do we need to produce 5 gal. of 1.050 wort using the Palmer's Short Stout recipe?*
You can build a grain bill "top-down" or "bottom-up," meaning that you can plan the bulk of your fermentables from the base malt first and adjust the specialty grains to make up the rest, or you can plan your specialty grain additions first and use the base malt to complete the OG. I generally use the bottom-up approach and, for this example, I am going to use 0.5 lb. of each specialty malt, and then calculate how much base malt I need to hit my target gravity.

**1.** First, we multiply the target gravity by the recipe volume to get the total gravity points of sugar.

$$5 \text{ gal.} \times 50 \text{ gravity points (degrees)} = 250 \text{ gal.}°$$

**2.** Then we need to assume an efficiency (e.g., 75%, or use what you have calculated from previous batches, if you are consistent) to calculate an anticipated yield in PPG based on the base lager malt, which has 81% FGDB. Remember that percent extract yield is based on 46 PPG from sucrose, which is considered the equivalent of 100% FGDB:

$$\text{anticipated yield (PPG)} = 81\% \text{ FGDB} \times 75\% \text{ eff.} \times 46 \text{ PPG}$$

$$= 0.81 \times 0.75 \times 46$$

$$= 27.9 \text{ (round up to 28)}$$

**3.** Dividing the total gravity points (250 gal.°) by our anticipated yield (28 PPG) gives the weight of malt required.

$$\text{weight malt required} = 250 \text{ gal.}° / 28 \text{ PPG}$$

$$= 8.9 \text{ lb. (I generally round up to the nearest half pound, i.e., 9 lb.)}$$

**4.** So, about 9 lb. of malt will give us our target 1.050 OG in 5 gal. Using the specific malt PPG values for 75% extract efficiency in table 18.5, we can figure out how much of each malt to use to make up our recipe. The OG contributions for each specialty malt at 75% efficiency are:

| | |
|---|---|
| Caramel 15 | $26 \times (0.5/5) = 2.6$ |
| Caramel 75 | $26 \times (0.5/5) = 2.6$ |
| Chocolate | $26 \times (0.5/5) = 2.6$ |
| Roast barley | $27 \times (0.5/5) = 2.7$ |
| Total gravity: | 10.5 points (out of 50) |

To calculate how much base malt is required, subtract the total specialty malt contribution from the target OG (50 points in this case), multiply that amount by the recipe volume and divide that by the base malt's 75% PPG number (27).

$$\text{base malt required} = \frac{[(\text{target gravity} - \text{specialty contr.}) \times \text{volume}]}{\text{base malt PPG @75\%}}$$

$$= [(50 - 10.5) \times 5]/28$$

$$= 7.0 \text{ lb.}$$

I generally round to the nearest quarter or half pound for convenience sake (i.e., 7.0 lb. in this case).

Thus, the grain bill for Palmer's Short Stout, based on these malts and a 75% extract efficiency is:

| | |
|---|---|
| Two-row lager malt | 7.0 lb. |
| Caramel 15 | 0.5 lb. |
| Caramel 75 | 0.5 lb. |
| Chocolate | 0.5 lb. |
| Roast barley | 0.5 lb. |
| **Total:** | 9.0 lb. |

Remember though that this is the OG—the post-boil gravity. Calculate your target pre-boil gravity by multiplying the post-boil gravity by the post-boil volume to get the total gravity points (gallon degrees or liter degrees). Then, divide the total points by the pre-boil volume target to get the necessary pre-boil gravity. You can check your progress during the lauter at any time by measuring the collected wort volume and checking the wort gravity. For example, to have 5 gal. of 1.050 wort after boiling, you would need to collect a starting volume of (at least):

250 gal.°/6 gal. = 41.67, or 6 gal. of 1.042 wort

or

250 gal°/7 gal. = 35.71, or 7 gal. of 1.036 wort

## Using HWE or PKL to Calculate Malt Quantities

The concept described above for designing a grain bill using PPG works the same with HWE and PKL. Recall that because HWE is given in L°/kg it is really the same thing as PKL in points/kg/L. Let's use the Palmer's Short Stout recipe, but substitute pale ale malt for the lager malt and make the volume 20 L (instead of 19 L). The maximum PKL value (FGDB) for pale ale malt in table 18.5 is equivalent to HWE for our purposes, so the number is 307 L°/kg .

1. First, we multiply the target gravity by the recipe volume to get the total gravity points of sugar.

20 L × 50 gravity pts. (degrees) = 1000 L°.

**2.** Then we need to assume an extract efficiency (we will take 75%, as before) and calculate an anticipated yield in PKL based on the pale ale malt HWE FGDB value (307 L°).

$$\text{anticipated yield (PKL)} = 307 \text{ L}° \times 75\% = 230 \text{ PKL}$$

**3.** Dividing the total gravity points (1000 L°) by our anticipated yield (230 PKL) gives the weight of malt required in kilograms.

$$\text{weight malt required} = 1000 \text{ L}° / 230 \text{ PKL} = 4.35 \text{ kg}$$

**4.** As noted in the PPG example, I generally build recipes from the bottom up, so here we will plan the specialty malt amounts and then make up the rest with base malt to get to the target OG. Assuming 1 pound is 450 grams, then one half pound is 225 grams. The OG contributions for each specialty malt at 75% efficiency (table 18.5) are:

| | |
|---|---|
| Caramel 15 | $216 \times (0.225 / 20) = 2.4$ |
| Caramel 75 | $213 \times (0.225 / 20) = 2.4$ |
| Chocolate malt | $213 \times (0.225 / 20) = 2.4$ |
| Roast barley | $225 \times (0.225 / 20) = 2.5$ |
| Total contribution: | 9.7 points (out of 50) |

To calculate how much base malt is required, subtract the total specialty malt contribution from the target OG (50 points in this case), multiply that amount by the recipe volume and divide that by the base malt's 75% PKL number (230).

$$\text{base malt required} = \frac{[(\text{target gravity} - \text{specialty contr.}) \times \text{volume}]}{\text{base malt PKL @75\%}}$$

$$= [(50 - 9.7) \times 20] / 230$$

$$= 3.5 \text{ kg}$$

Thus, the grain bill for Palmer's Short Stout, scaled up (from 19 L at 1.048 OG) to 20 L at 1.050 OG, and 75% extract efficiency is:

| | |
|---|---|
| Pale ale malt | 3.5 kg |
| Caramel 15 | 225 g |
| Caramel 75 | 225 g |
| Chocolate | 225 g |
| Roast barley | 225 g |
| **Total:** | 4.4 kg. |

Remember though that this is the OG—the post-boil gravity. Calculate your target pre-boil gravity by multiplying the post-boil gravity by the post-boil volume to get the total gravity points (gallon degrees or liter degrees). Then, divide the total points by the pre-boil volume target to get the necessary pre-boil gravity. You can check your progress during the lauter at any time by measuring the collected wort volume and checking the wort gravity. For example, to have 20 L of 1.050 wort after boiling, you would need to collect a starting volume of (at least):

$$1000 \text{ L}° / 23 \text{ L} = 43.48, \text{ or } 23 \text{ L of } 1.043 \text{ wort}$$

or

$$1000 \text{ L}° / 26.5 \text{ L} = 37.74, \text{ or } 26.5 \text{ L of } 1.038 \text{ wort}$$

### Using Degrees Plato to Calculate Malt Quantities

You may have a hydrometer or refractometer that measures in degrees Plato (°P) instead of specific gravity (SG). Refractometers are based on the Brix scale, which is functionally equivalent to °P. You have two options for calculating your brewing efficiency and malt quantities when measuring in °P:

- Convert the °P to SG (see table A.2 in appendix A) and estimate your malt quantities and efficiency using the PPG and PKL methods described above.
- Use the °Plato and the extract weight-percent method to calculate these quantities.

The first option is pretty self-explanatory. The following will describe how to use the second option, the extract weight-percent method.

Degrees Plato measures the percentage extract (as sucrose) in wort by weight. In other words, a wort that measures 10°P has 10% soluble extract, that is, 100 g of wort contains 10 g of soluble extract. Commercial breweries use °P more often than PPG or PKL to figure out malt quantities, because it gives them better visibility to their malt usage, being malt weight-oriented rather than wort volume-oriented.

To calculate your brewing efficiency using °P, the equation is:

$$\text{brewing efficiency} = \frac{\text{weight of actual extract}}{\text{wt. of maximum extract}}$$

$$= \frac{(\text{wort volume} \times \text{wort density} \times \% \text{ extract in the wort})}{(\text{weight malt} \times \text{maximum yield})}$$

So what we are doing is calculating the actual weight of the extract in the wort and comparing it to the maximum extract we could have gotten from the grain bill. The weight of the actual extract is calculated by multiplying the volume of wort by its density to get the total weight of the wort, and then using the measured °P to say how much of that total weight is extract. The density of the wort is equal to the density of the water (0.985 kg/L, or 8.22 lb./gal. at mash temperatures) multiplied by the SG of the wort. (Specific gravity is actually a ratio of the wort's density to the density of pure water.) To get the SG value for the wort, you either have to measure it separately with another hydrometer or convert the °P reading to specific gravity using the reference table in appendix A (table A.2). The conversion from °P to SG for worts less than 13°P is simply the gravity points multiplied by 4 (i.e., 10°P = 1.040 SG), although the error increases at higher gravities.

The weight of the maximum extract is simply the weight of malt multiplied by the maximum yield (e.g., 80% FGDB). Then the previous equation becomes:

$$\text{brewing efficiency} = \frac{[(\text{wort volume} \times \text{water density} \times \text{SG}) \times °P]}{(\text{weight malt} \times \%\text{FGDB})}$$

For example, let's say we used 4 kg of base malt to brew 20 L of wort that measures 12°P (i.e., the wort is 12% soluble extract by weight). Using the rule of thumb for worts less than 13°P, the SG is 1.048. The brewing efficiency equation is:

$$\text{brewing efficiency} = \frac{[(\text{wort volume in L} \times \text{water density in L/kg} \times \text{SG}) \times °P]}{(\text{weight malt in kg} \times \%\text{FGDB})}$$

$$= \frac{[(20 \times 1 \times 1.048) \times 12\%]}{(4 \times 80\%)}$$

$$= 78.6\%$$

The equation can also be re-arranged to calculate the grain bill necessary to brew 20 L of a 12°P wort. For this example, let's use our standard brewing efficiency of 75%. The equation becomes:

$$\text{wt. malt in kg} = \frac{[(\text{wort volume in L} \times \text{water density in L/kg} \times \text{SG}) \times °P]}{(\%\text{FGDB} \times \% \text{ brewing efficiency})}$$

$$= \frac{[(20 \times 1 \times 1.048) \times 12\%]}{(80\% \times 75\%)}$$

$$= \frac{(20.96 \times 0.12)}{(0.8 \times 0.75)}$$

$$= 2.52 / 0.6$$

$$= 4.19 \text{ kg, or 4.2 kg of malt}$$

**Notes:**
1. The calculations can also be conducted in gallons and pounds in place of liters and kilograms by using the water density constant of 8.32 lb./gal.
2. I should mention that commercial brewers are most likely to use the coarse grind, as-is (CGAI) number from the malt lot analysis sheet, because that number takes the moisture into account, and helps them more accurately plan their malt usage and maintain better process consistency. On a homebrewing scale, we can be less rigorous and base our efficiency on the more readily available FGDB number. To take moisture into account, multiply the dry basis number by the percentage of dry weight. For example, 80% FGDB at 4% moisture becomes:

   80% × 96% = 76.8% FGAI

   If the F/C difference of the malt is 1.2%, then we can calculate the CGAI from the FGAI:

   76.8% − 1.2% = 75.6% CGAI

3. Substituting this CGAI value for FGDB into the malt quantity calculation above gives 4.4 kg malt required versus 4.2 kg.

4.  Bear in mind that if you recalculate your brewing efficiency using 4.2 kg and 75.6% CGAI, your brewing efficiency jumps up to 79.2%. It depends on whether you want to take the moisture of each lot of malt into account when assessing your brewing efficiency, or just assume that the moisture content is going to be fairly consistent across different lots of malt, and accept that this will be reflected as a lower overall brewing efficiency.

## Summary

So there you have it, the keys to understanding malt yield, extract efficiency, and determining your grain bill for all-grain brewing. A malt analysis sheet will list the maximum yield as %FGDB, and you can convert that percent by weight measure to PPG, PKL, or °P. You can determine your extract efficiency by comparing the collected wort gravity with maximum yield, and by knowing your extract efficiency, you can go on to calculate a grain bill for any wort you want to brew.

# Getting the Wort Out (Lautering)

19

Okay, let's see where we are: we have mashed the grist to create the wort; we have discussed how we will measure the gravity of the wort and calculate our extract efficiency; and now it is time to extract the wort by lautering and sparging the wort to the boil kettle.

## The Lautering Process

The word "lauter" comes to us from the German word, *läuter*, which means to clean, clarify, or purify. It also has the secondary meaning to make something more virtuous. Therefore, lautering is the improvement of our wort as we separate it from the mash. A lauter tun consists of a large vessel with a false bottom, or manifold, which allows the wort to drain out while retaining the grain. Lautering usually consists of three steps: recirculation, first runnings, and sparging for the second runnings.

### Recirculation

After the mash, the grain bed needs to be readied for draining and sparging. If you were to simply throw the valve of the lauter tun wide open, the grain bed would get sucked down onto the false bottom, no wort would flow, and you would have a stuck sparge. Therefore, at the beginning of the lauter, the wort flow is started slowly so the grain bed settles gently and uniformly and doesn't stick or clog. The first few quarts or liters of wort are collected in a pitcher or small bucket and poured back into the lauter tun on top of the grain bed. The first few quarts are always cloudy with excess proteins and grain debris, and recirculation helps filters out these materials from your boil kettle. Recirculation is also known as the *vorlauf*, or more properly, *vorläufig*, which means preliminary. In other words, it is the preliminary lautering step.

The wort should clear fairly quickly. After the wort starts running clear (it will still be dark and a little bit cloudy, but chunk free), you are ready to collect the wort in the kettle and sparge the grain bed. Recirculation can be done anytime the grain bed is disturbed and bits of grain and husk appear in the runoff, though if your grain bed has good depth disturbing it is unlikely. Recirculation is not entirely necessary, but most brewers do it. It is not used in the brew-in-a-bag process. See the section below on sparging methods for more information.

### First Runnings

The first runnings is the highest gravity and richest tasting wort that will come from the mash. The gravity and volume of the first runnings are entirely dependent on the water-to-grist ratio, as was discussed in chapter 18. A high water-to grist ratio will produce a lower gravity first runnings but larger volume, and a lower water-to-grist ratio will produce the opposite. Typically, in commercial brewing practice, the entire first runnings, or first wort, are drained to the kettle after the recirculation step, and then the grain bed is sparged with a rotating shower arm to rinse the grainbed for the second runnings. This method would generally result in grainbed compaction and poor extraction in a homebrewing setup. However, modern commercial lauter tuns are often a separate vessel from the mash tun, and are fitted with revolving rakes and sparge arms that can stir the grain bed just above the false bottom to prevent any channeling and keep the extraction more uniform. Homebrewing lauter tuns don't have rakes and sparge arms; draining the first wort completely requires the grain bed to be stirred and recirculated before it can be drained of the second runnings.

### Sparging for the Second Runnings

Sparging is the rinsing of the grain bed to extract as much of the sugars from the grain as possible without extracting astringent tannins from the grain husks. *Sparging* means "to sprinkle" and you may have seen sparge arms, or sprinklers, in the lauter tuns of commercial breweries. In commercial breweries, the lauter tun is quite large, and it is fitted with revolving rakes and sparge arms to ensure that every cubic inch of the grain bed is rinsed of all the extract. We don't need to be that economical on the homebrewing scale, in fact, the smaller scale makes it easier to rinse the grain bed of the wort, but without the rakes we need to take more care to maintain fluidity and uniform flow through the grain bed.

The temperature and pH of the sparge water is important. Hotter water is more effective at extracting residual starch and converting it to sugar via residual alpha-amylase, but it increases the risk of tannin extraction. This risk can be offset by maintaining the runnings at a pH of less than 5.6. (Chapter 21 has a more detailed discussion of water chemistry and pH.) Generally, the sparge water should be no more than 175°F (79°C), as husk tannins become more soluble above this temperature, especially when the wort pH gets to 6.0 or above. This will lead to astringency in the beer. Higher temperatures will also denature any remaining alpha-amylase, increasing the risk of residual starch in the wort.

## Methods of Sparging

There are several sparging methods practiced by homebrewers, and each has its pros and cons.

### Continuous Sparging

Continuous sparging, also known as fly sparging, usually results in the best yield. In continuous sparging, the wort is recirculated and drained until just under an inch (2.5 cm) of wort remains above the grain bed. The sparge water is then slowly added at the same rate as the wort is drained, such that you achieve a steady flow into and out of the grain bed. The idea is that the grain bed is never actually drained; instead, the wort is gradually exchanged with sparge water. The soluble extract yield is highly dependent on the uniformity of fluid flow through the grain bed to ensure that every grist particle is fully rinsed. Sparge time varies (30–150 min.) depending on the amount of grain and grist particle size.

The sparge is stopped when the gravity from the runnings is ≤1.008, or when enough volume of wort has been collected, whichever comes first. Continuous sparging demands more attention by the homebrewer, but can produce a higher yield per unit weight of malt. Sprinkling is not required; the only requirement is to maintain at least an inch of water above the grain bed so that it stays fluid.

The sparge water can be poured in with a pitcher, or fed in through a hose. If you are going to use a pitcher, then lay a small plate or plastic lid on top of the grain bed and pour onto that to avoid disturbing the bed too much. If you are going to use a hose, make sure it is long enough to lay on top of the grain bed along the perimeter. What you don't want to do is have the hose pointing down at the grain bed so that it drills a hole in it. You would end up collecting sparge water instead of wort. You always want to maintain about an inch of sparge water above the grain to ensure that the grain bed doesn't compact when using continuous sparging.

The continuous sparging method is actually the easiest to use if you have a three-tier mashing system, where a hot liquor (water) tank is set up above the mash-lauter tun with a valve and hose to slowly feed sparge water into the tun as the wort is drained to the boil kettle. Using valves makes it relatively easy to adjust flow rates so that you are adding sparge water at the same rate as you are draining wort to the kettle.

### Batch Sparging

Batch sparging, another common homebrewing practice, is where a large volume of sparge water is added to the mash all at once instead of gradually. It is most often used with large ice chests or picnic coolers. In practice, the mash is completely drained of the first runnings, and then a second batch of water is mixed in and mashed for a short period of time (e.g., 15 min.). The grain bed is allowed to settle, recirculated for clarity, and the second runnings wort is then drained off.

Batch sparging is most efficient when both runnings are the same volume, in other words, each runnings is half of the intended boil volume. The volume of the first runnings consists of the mash water minus the amount of water retained by the wet grain. If the first runnings volume is less than half of the target boil volume, then more water should be added to the mash to make up the difference before draining. The wet grain retention factor ($r$) is about 0.5 qt./lb. or 1 L/kg.

For example, if you were mashing 12 lb. of grain, you would expect 6 qt. (1.5 gal.) of the first runnings to be retained by the wet grain. Assuming that you are mashing with a water-to-grist ratio ($Rv$) of 1.5 qt./lb., that would equal 18 qt., or 4.5 gal., of potential first runnings, minus the 1.5 gal.

retained, for 3 gal. net first runnings into the boil kettle. If you were planning on collecting and boiling 7 gal. of wort total, you would add an additional half gallon of water to the mash before draining the first runnings, and then sparge with 3.5 gal. of hot water to get the second half of the boil volume as the second runnings.

Batch sparging differs from the English parti-gyle method (see next) in that the separate runnings are combined to produce a single beer. Batch sparging was originally developed from parti-gyle to make large quantities of porter, and was known as "entire." It is slightly less efficient than continuous sparging (typically 75% versus 80% efficiency), but it is convenient if you don't have a large hot liquor tank or gravity flow setup for continuous sparging. Much of the difference in efficiency between batch and continuous sparging comes down to not fully draining the mash of the first runnings before adding the next batch of sparge water for the second runnings.

### Parti-Gyle

The parti-gyle method was common in brewing in England before the nineteenth century, and it allows the brewer to make two or more beers from the same mash by blending the different strength worts. A large mash is produced and the first wort is drained completely before more water is added to the grist for a second mash and drained again. The first runnings typically had a gravity of about 1.080 and were used for making an "aging" beer. The second runnings were lighter in gravity and could be used for making a "running" or table beer, and the mash was often "capped" with some additional grain strewn onto the grain bed to produce a low- gravity "small beer." The worts could be hopped and boiled separately and blended into different products before fermentation, or they could be blended before the boil. Parti-gyle gives the brewer several options for producing a range of products from a single large mash, but you need multiple boil kettles and fermentors to fully utilize it.

### No-Sparge

The no-sparge method is the least efficient in terms of the amount of malt used, but it's easy and has the benefit of being immune to tannin extraction that may occur during the sparge. Like batch sparging, no-sparge is a draining rather than a rinsing method, and the beer is produced entirely from first runnings, resulting in a smoother, richer tasting wort at the expense of efficiency. There are a couple of ways of working this method to produce all the wort needed for the boil. First, you can create a mash large enough, with a specific water to grist ratio, to yield the entire volume of wort at the target gravity needed for your boil. These calculations will be discussed later in this chapter. Alternatively, you can produce high-gravity wort at a smaller volume than your target boil volume and gravity, and dilute that wort in the kettle with water to reach your target. Homebrewers will often turn a no-sparge brew into a parti-gyle brew, because they can't stand to waste the extract potential left over in the mash tun. Decisions, decisions . . .

### Brew-in-a-Bag

The brew-in-a-bag (BIAB) method turns the traditional process on its head by removing the grain from the mash instead of removing the wort from the grain. It is a combination of batch sparge and no-sparge where the entire boil volume plus the retained water volume is used for the mash (i.e., there is a very high water-to-grist ratio). The grist is contained in a mesh bag and at the conclusion of the mash the bag is lifted out of the tun (typically the boil kettle), allowed to drain, and the grain is discarded. The wort is brought to a boil and brewing proceeds as usual. The grain bag is usually not sparged, only allowed to drain. The amount of wort retained by the grain is typically half that of normal lautering due to the weight of the grain in the bag squeezing more wort out; a typical retention rate is 0.25 qt./lb. (0.5 L/kg). The BIAB method can use a much finer crush for better yield, because it doesn't depend on wort flow through the grain bed for good extraction. However, the very high water-to-grist ratio, near 4.3 qt./lb. (9 L/kg) in low-gravity worts (≤1.040), means that these mashes are more sensitive to water chemistry and pH than other methods. But generally BIAB works well, and it uses less equipment, so there is less to clean up at the end of the day.

One thing to consider for BIAB is that you will probably need a small block and tackle to hoist the wet grain bag out of the kettle and hold it there for several minutes while it drains. (My kid's arms got tired.)

## Rinsing versus Draining

Commercial breweries use continuous sparging with large rotating sparge arms and grain rakes, because that is the most efficient way to rinse the grain in a lauter tun that is twenty feet across. As homebrewers, the difference in scale is in our favor. Large commercial systems need to engineer solutions to problems we just don't have. We can spend an extra dollar on malt to make up for a lack of efficiency, and pour the sparge water directly onto the grain bed, whereas a large commercial brewery would have to spend an extra five hundred dollars and monitor uniformity of the flow.

Continuous sparging is a rinsing process that depends on uniform flow through the grain bed to achieve the best yield. The first runnings are rich in sugar, making a dense first wort. As the sparge water moves through the grain bed, this first wort is displaced by the less dense sparge water, so the grain does not float as well. This causes the grain bed to compact, which can lead to a stuck sparge. You will also get a stuck sparge if the runoff rate is too fast, because this creates a partial vacuum under the false bottom, or around the manifold, which compacts the grain against it. The maximum recommended runoff rate for continuous sparging is about 1 qt./min. (roughly 1 L/min.). After the heavy first wort has been displaced, the remaining sugars in the grist particles will diffuse into the sparge water. This diffusion process takes time, which is another reason to go slowly, otherwise the sparge water will simply fill the boil kettle without having extracted much of anything.

The grain bed can be a few inches to a couple of feet deep (between 5 and 60 cm), but the optimum depth depends on the overall tun geometry and the total amount of grain being mashed. If the grain bed is very shallow, for example, from lautering too little grain in too large a tun, then the filter bed will be inadequate, the wort won't clear, and you will get hazy beer. A minimum useful depth is probably about 4 in. (10 cm) but a depth of about 8 in. (20 cm) is preferable. In general, deeper is better, but if the grain bed is too deep then it is more easily compacted, making uniform lautering nearly impossible. Since fluids always follow the path of least resistance, compaction can lead to preferential flow, in which some regions of grain are completely rinsed while others are not rinsed at all (i.e., channeling). Non-uniform flow is a major cause of poor efficiency or low yield.

Batch sparging, no-sparge, and BIAB get around these problems by draining instead of rinsing. By draining, you are simply extracting what is already there, rather than trying to extract more. It's like picking all the low-hanging fruit—you get a rich wort with little effort. The problem is that quite a bit of wort can be left behind. Batch sparging solves this problem by mixing another batch of water into the mash tun to obtain a second runnings. The second wort in batch sparging typically has a gravity ≥1.016, so tannin extraction due to rising mash pH, as can happen in continuous sparging, is usually prevented (see chapter 21 for more discussion of this issue). The BIAB method handles it a little differently by essentially pre-sparging, initially using more water so that the wort retained by the grain bed is at a much lower gravity than that retained from the first runnings of batch sparging.

## Efficiencies of Sparging Methods

The extract yields given for the various malts in chapter 18, "Extraction and Yield," are based on the traditional mash, lauter, and continuous sparging techniques that are used in breweries all over the world. Continuous sparging takes a little longer, but is the most efficient of the methods. Batch sparging, no-sparge, and BIAB can take less time and use less equipment, but lose a little bit in efficiency, requiring a bit more grain to produce the same volume and gravity of wort. For homebrewers, where grain is relatively inexpensive and time and equipment are more so, these less efficient but quicker options are attractive. Let's compare these methods by using the following example.

Table 19.1 is a comparison of grain weights and efficiencies for a simple 1.050 OG brown ale recipe brewed using different sparge methods. We will assume that we are collecting 7 gal. (27 L) of wort at 1.043 from the mash, to boil down to 6 gal. (23 L) at 1.050 to go into the fermentor, to finally yield at least 5 gal. (19 L) into our keg or bottles. We will assume 80% brewing efficiency for the standard continuous sparge method, which is about 30 PPG (250 PKL). The same comparisons are made in table 19.2, this time using metric units. The two sets of examples in tables 19.1 and 19.2 are not exact conversions of each other due to the small difference between brewing based on quarts per pound versus liters per kilogram.

### TABLE 19.1—GRAIN WEIGHTS AND BREWING EFFICIENCIES COMPARED BETWEEN SPARGE METHODS

| Grain bill | Continuous sparging | Batch sparging | No-sparge | BIAB |
|---|---|---|---|---|
| Pale ale malt (lb.) | 8.65 | 9.45 | 9.8 | 8.88 |
| Biscuit malt (lb.) | 0.5 | 0.55 | 0.56 | 0.51 |
| Crystal 60 malt (lb.) | 0.5 | 0.55 | 0.56 | 0.51 |
| Chocolate malt (lb.) | 0.5 | 0.55 | 0.56 | 0.51 |
| Total weight (lb.) | 10.15 | 11.1 | 11.5 | 10.4 |
| Water-to-grist ratio (Rv) | 1.8 | 1.8 | 2.9 | 2.9 |
| Brewing efficiency | 80.0% | 73.2% | 70.7% | 78.0% |
| Total mash volume (gal.) | 5.5 | 5.9 | 9.5 | 8.6 |

BIAB, brew-in-a-bag.

Note: Grain bill in each case is for the same 1.050 OG brown ale recipe, assuming 7 gallons of wort collected and boiled to 6 gallons, based on a 5 gal. batch size.

Each sparge method produces the same 7 gal. of 1.043 wort, but using differing quantities of grain. Similarly, the size of the mash differs: 5.9 gal. for batch sparging and 9.5 gal. for no-sparge, versus 5.5 gal. for continuous sparging. The values for BIAB don't follow the same trend as batch sparging and no-sparge, because the retention factor for BIAB is about half that for the other processes due to the suspended weight of the grain squeezing more wort out. This means BIAB can deliver 7 gal. of 1.043 wort with less grain and higher efficiency.

## TABLE 19.2—GRAIN WEIGHTS AND BREWING EFFICIENCIES COMPARED BETWEEN SPARGE METHODS, METRIC UNITS

| Grain bill | Continuous sparging | Batch sparging | No-sparge | BIAB |
|---|---|---|---|---|
| Pale ale malt (kg) | 4.0 | 4.33 | 4.46 | 4.1 |
| Biscuit malt (g) | 225 | 244 | 251 | 229 |
| Crystal 60 malt (g) | 225 | 244 | 251 | 229 |
| Chocolate malt (g) | 225 | 244 | 251 | 229 |
| Total weight (kg) | 4.7 | 5.1 | 5.2 | 4.8 |
| Water-to-grist ratio (Rv) | 3.7 | 3.7 | 6.2 | 6.2 |
| Brewing efficiency | 80.0% | 73.9% | 71.7% | 78.6% |
| Total mash volume (L) | 21.0 | 22.6 | 36.4 | 33.2 |

BIAB, brew-in-a-bag.

Note: Grain bill in each case is for the same 1.050 OG brown ale recipe, assuming 27 L collected and boiled to 23 L, based on a 19 L batch size.

## Continuous Sparging Efficiency

For continuous sparging, brewing efficiency typically varies little, usually being between 75% and 80%. The brewing efficiency is most dependent on the lautering equipment and uniformity, topics discussed in detail in appendices E and F.

When using continuous sparging, calculating the grist weight and volumes is as discussed in chapter 18, being simply the OG and recipe volume (or boil gravity and boil volume) divided by either 30 PPG or 250 PKL (depending on units), assuming 80% brewing efficiency in this case. If the brewing efficiency is taken to be 75%, then the values are about 28 PPG and 234 PKL.

The water-to-grist ratio ($R$) can vary at the discretion of the brewer, although ratios in the 2.5–4.0 range are most common.

## Batch Sparging Efficiency

The key to understanding the various efficiencies of all these methods is to understand figure 18.4, which shows the gravity of the first wort as a function of $R$ and typical yield. We will refer to the gravity of the first runnings as SG1. If you refer back to the sidebar on the weight per weight water-to-grist ratio in chapter 17, you will remember that $R$ (the weight to weight water-to-grist ratio) is related to $Rv$ (the volume to weight water-to-grist ratio) by the density of the water, represented by $\rho$ (rho). As explained in chapter 17, the value of $\rho$ at mash temperatures is taken to be 2.055 lb./qt. (0.985 kg/L). Knowing $Rv$ and $R$ allows us to calculate the gravity of the wort in the mash according to the equation for initial wort gravity as a function of $R$, given in chapter 18:

$$SG1 = 259 / \{259 - [75 / (R+1)]\}$$

where,

$$R = Rv \times \rho$$

*Note: SG1 is calculated as specific gravity, for example, 1.077, but only the degrees of gravity portion (i.e., gravity points) will be used in subsequent calculations, which in this example is 77.*

Batch sparging is most efficient when the volume of the first (SG1) and second runnings (SG2) are identical, meaning they are each equal to half of the target boil volume. The ratio, $Rv$, is calculated from half of the boil volume and the grist weight according to the equation:

$$Rv = [Vb + (G \times r)] / 2G$$

where,

$Vb$ is the boil volume,

$G$ is the grist weight,

$r$ is the wet grain wort retention factor, taken to be about 0.5 qt./lb., or 1 L/kg.

The above equation reduces to:

$$Rv = (Vb / 2G) + r$$

The grist weight ($G$) is equal to the total degrees of gravity (i.e., OG × post-boil volume, or BG × boil volume) divided by your expected yield in PPG or PKL based on the extract efficiency.

$$G = \frac{(\text{original gravity} \times \text{recipe volume})}{(\text{max. yield} \times \% \text{ extract efficiency})}$$

or,

$$G = \frac{(\text{boil gravity} \times \text{boil volume})}{(\text{max. yield} \times \% \text{ extract efficiency})}$$

where,

Max. yield = maximum malt yield in PPG or PKL based on %FGDB.

In the solutions worked out for the sparge efficiencies in tables 19.3-19.8, located at the end of this chapter, the maximum yield is taken as 80% FGDB soluble extract by weight, which is about 37 PPG, or 307 PKL (see table 18.5). The maximum yield is multiplied by the brewing efficiency, which can range anywhere from 50% to 85% depending on $Rv$. The batch sparging efficiency must be solved by iteration, as I will explain below.

The second runnings is the wort that is retained in the wet grain and then diluted by the volume of sparge water (remember this is equal to $Vb/2$). The gravity of the second runnings, which we will refer to as SG2, is related to the gravity of the first runnings by the ratio of the two ratios: $r$ and $Rv$.

$$SG2 = SG1 \bullet (r/Rv)$$

*Remember: SG1 must be input as degrees of gravity (gravity points) instead of specific gravity (i.e., 56 instead of 1.056).*

The combined gravity of the collected wort, as gravity points, in the boil kettle is:

$$\text{boil gravity} = (SG1 + SG2)/2$$

The total mash volume ($Vm$) is given by:

In quarts: $Vm = G(R + 0.38)$

In liters: $Vm = G(R + 0.8)$

The problem with these equations is that solving for the grist weight and $Rv$ is indeterminate; there are many combinations of gravity, volume, and grist weight that work together, but there is only one solution that gives us our target runnings volume and combined boil gravity. This solution must be solved iteratively with the aid of a spreadsheet or by using higher math. I chose a spreadsheet. The result is table 19.3 and table 19.4, and also the tables further on that deal with alternative sparge methods (see relevant sections below). The nice thing about these tables is that the numbers are robust; they depend on malt modification, not degree of crush, and if your volume and weight measurements are accurate, you will hit your numbers every time.

*Note: The numbers for the same OG between tables 19.3 and 19.4 are slightly different, because 6 and 7 gal. are not exact equivalents of 23 and 27 L.*

## No-Sparge Efficiency

The model for no-sparge is basically the first half of the batch sparging model described above, we just have to scale up the amount of grist and the water-to-grist ratio ($Rv$) to get all of our boil volume and gravity in the first runnings, SG1. The numbers for no-sparge are given in tables 19.5 and 19.6. The equation for the boil gravity as a function of $R$ is the same as for the first runnings in batch sparging, but note that the equation for $Rv$ in no-sparge is not divided by 2.

$$\text{boil gravity} = 259/\{259 - [75/(R+1)]\}$$

where,

$$Rv = (Vb/G) + r$$

$r = 0.5$ qt./lb., or 1 L/kg.

$$G = \frac{(\text{original gravity} \times \text{recipe volume})}{(\text{max. yield} \times \text{\% extract efficiency})}$$

or,

$$G = \frac{(\text{boil gravity} \times \text{boil volume})}{(\text{max. yield} \times \text{\% extract efficiency})}$$

*Note: The numbers for the same OG between tables 19.5 and 19.6 are slightly different, because 6 and 7 gal. are not exact equivalents of 23 and 27 L.*

## Brew-in-a-Bag Efficiency

The model for BIAB, tables 19.7 and 19.8, is very similar to no-sparge described in the section above, except that the wort retention factor, $r$, is roughly half that of the other methods because the suspended weight of the grain bag squeezes out more wort from the grist. The value of $r$ is taken to be roughly 0.25 qt./lb. (0.5 L/kg) when using BIAB. This changes the runnings volume and therefore the efficiency and amount of grain needed to hit the boil gravity. The water-to-grist ratio ($Rv$) is the same as for no-sparge, but remember that the retention factor is different.

*Note: The numbers for the same OG between tables 19.7 and 19.8 are slightly different, because 6 and 7 gal. are not exact equivalents of 23 and 27 L.*

## TABLE 19.3—BATCH SPARGING TARGETS USING POUNDS, QUARTS, AND GALLONS

| OG | 1.030 | 1.035 | 1.040 | 1.045 | 1.050 | 1.055 | 1.060 | 1.065 | 1.070 | 1.075 | 1.080 | 1.085 | 1.090 |
|---|---|---|---|---|---|---|---|---|---|---|---|---|---|
| Grist weight (lb.) | 5.9 | 7.1 | 8.3 | 9.7 | 11.1 | 12.6 | 14.2 | 15.9 | 17.8 | 19.9 | 22.1 | 24.5 | 27.1 |
| PPG | 30.5 | 29.7 | 28.8 | 28.0 | 27.1 | 26.2 | 25.3 | 24.5 | 23.6 | 22.7 | 21.8 | 20.8 | 19.9 |
| Brewing efficiency | 82.4% | 80.2% | 77.8% | 75.6% | 73.2% | 70.9% | 68.5% | 66.1% | 63.7% | 61.2% | 58.8% | 56.3% | 53.9% |
| Mash volume (qt.) | 17.0 | 17.5 | 18.2 | 18.8 | 19.5 | 20.3 | 21.1 | 22.0 | 22.9 | 23.9 | 25.0 | 26.2 | 27.5 |
| Runnings volume (qt.) | 14.0 | 14.0 | 14.0 | 14.0 | 14.0 | 14.0 | 14.0 | 14.0 | 14.0 | 14.0 | 14.0 | 14.0 | 14.0 |
| First runnings gravity, SG1 | 1.044 | 1.050 | 1.056 | 1.061 | 1.067 | 1.072 | 1.077 | 1.082 | 1.086 | 1.091 | 1.095 | 1.099 | 1.103 |
| Second runnings gravity, SG2 | 1.008 | 1.010 | 1.013 | 1.016 | 1.019 | 1.022 | 1.026 | 1.030 | 1.034 | 1.038 | 1.042 | 1.046 | 1.051 |
| Boil gravity | 1.026 | 1.030 | 1.034 | 1.039 | 1.043 | 1.047 | 1.051 | 1.056 | 1.060 | 1.064 | 1.069 | 1.073 | 1.077 |
| Rv | 2.9 | 2.5 | 2.2 | 1.9 | 1.8 | 1.6 | 1.5 | 1.4 | 1.3 | 1.2 | 1.1 | 1.1 | 1.0 |
| R | 5.9 | 5.1 | 4.5 | 4.0 | 3.6 | 3.3 | 3.1 | 2.8 | 2.6 | 2.5 | 2.3 | 2.2 | 2.1 |
| Mash volume (gal.) | 4.8 | 5.1 | 5.3 | 5.6 | 5.9 | 6.3 | 6.6 | 7.0 | 7.4 | 7.9 | 8.4 | 8.9 | 9.5 |

OG, original gravity; PPG, (gravity) points per pound per gallon; R, wt./wt. water-to-grist ratio; Rv, vol./wt. water-to-grist ratio; SG, specific gravity.

Notes: Boil volume is 7 gal. to yield 6 gal. to the fermentor at the stated OG. For discussion of calculations used, refer to the section "Batch Sparging Efficiency."

## TABLE 19.4—BATCH SPARGING TARGETS USING KILOGRAMS AND LITERS

| OG | 1.030 | 1.035 | 1.040 | 1.045 | 1.050 | 1.055 | 1.060 | 1.065 | 1.070 | 1.075 | 1.080 | 1.085 | 1.090 |
|---|---|---|---|---|---|---|---|---|---|---|---|---|---|
| Grist weight (kg) | 2.7 | 3.2 | 3.8 | 4.4 | 5.1 | 5.7 | 6.5 | 7.3 | 8.1 | 9.0 | 10.0 | 11.0 | 12.2 |
| PKL | 255 | 248 | 241 | 234 | 227 | 220 | 213 | 206 | 199 | 192 | 184 | 177 | 170 |
| Brewing efficiency | 82.8% | 80.6% | 78.4% | 76.1% | 73.9% | 71.6% | 69.3% | 66.9% | 64.6% | 62.2% | 59.9% | 57.5% | 55.1% |
| Mash volume (L) | 16.2 | 16.7 | 17.3 | 17.9 | 18.6 | 19.2 | 20.0 | 20.8 | 21.6 | 22.5 | 23.5 | 24.5 | 25.7 |
| Runnings volume (L) | 13.5 | 13.5 | 13.5 | 13.5 | 13.5 | 13.5 | 13.5 | 13.5 | 13.5 | 13.5 | 13.5 | 13.5 | 13.5 |
| First runnings gravity, SG1 | 1.044 | 1.050 | 1.056 | 1.062 | 1.067 | 1.072 | 1.077 | 1.082 | 1.087 | 1.091 | 1.096 | 1.100 | 1.104 |
| Second runnings gravity, SG2 | 1.007 | 1.010 | 1.012 | 1.015 | 1.018 | 1.022 | 1.025 | 1.029 | 1.033 | 1.037 | 1.041 | 1.045 | 1.049 |
| Boil gravity | 1.026 | 1.030 | 1.034 | 1.038 | 1.043 | 1.047 | 1.051 | 1.055 | 1.060 | 1.064 | 1.068 | 1.072 | 1.077 |
| Rv | 5.99 | 5.16 | 4.54 | 4.06 | 3.67 | 3.35 | 3.09 | 2.85 | 2.67 | 2.50 | 2.35 | 2.22 | 2.11 |
| R | 5.90 | 5.09 | 4.48 | 4.00 | 3.62 | 3.30 | 3.04 | 2.82 | 2.63 | 2.46 | 2.32 | 2.19 | 2.07 |
| Mash volume (L) | 18.4 | 19.3 | 20.4 | 21.4 | 22.6 | 23.8 | 25.1 | 26.6 | 28.1 | 29.7 | 31.5 | 33.4 | 35.5 |

OG, original gravity; PKL, (gravity) points per kilogram per liter; R, wt./wt. water-to-grist ratio; Rv, vol./wt. water-to-grist ratio; SG, specific gravity.

Notes: Boil volume is 27 L to yield 23 L to the fermentor at the stated OG. For discussion of calculations used, refer to the section "Batch Sparging Efficiency."

## TABLE 19.5—NO-SPARGE TARGETS USING POUNDS, QUARTS, AND GALLONS

| OG | 1.030 | 1.035 | 1.040 | 1.045 | 1.050 | 1.055 | 1.060 | 1.065 | 1.070 | 1.075 | 1.080 | 1.085 | 1.090 |
|---|---|---|---|---|---|---|---|---|---|---|---|---|---|
| Grist weight (lb.) | 6.0 | 7.3 | 8.6 | 10.0 | 11.5 | 13.1 | 14.8 | 16.6 | 18.6 | 20.8 | 23.2 | 25.7 | 28.5 |
| PPG | 29.8 | 28.9 | 28.0 | 27.1 | 26.2 | 25.3 | 24.4 | 23.5 | 22.5 | 21.6 | 20.7 | 19.8 | 18.9 |
| Extract efficiency | 80.6% | 78.1% | 75.6% | 73.2% | 70.7% | 68.3% | 65.8% | 63.4% | 60.9% | 58.5% | 56.0% | 53.6% | 51.1% |
| Mash volume (qt.) | 31.0 | 31.6 | 32.3 | 33.0 | 33.7 | 34.5 | 35.4 | 36.3 | 37.3 | 38.4 | 39.6 | 40.9 | 42.3 |
| Wort volume (qt.) | 28.0 | 28.0 | 28.0 | 28.0 | 28.0 | 28.0 | 28.0 | 28.0 | 28.0 | 28.0 | 28.0 | 28.0 | 28.0 |
| Wort SG | 1.026 | 1.030 | 1.034 | 1.039 | 1.043 | 1.047 | 1.051 | 1.056 | 1.060 | 1.064 | 1.069 | 1.073 | 1.077 |
| Rv | 5.1 | 4.4 | 3.8 | 3.3 | 2.9 | 2.6 | 2.4 | 2.2 | 2.0 | 1.8 | 1.7 | 1.6 | 1.5 |
| R | 10.6 | 8.9 | 7.7 | 6.8 | 6.0 | 5.4 | 4.9 | 4.5 | 4.1 | 3.8 | 3.5 | 3.3 | 3.0 |
| Mash volume (gal.) | 8.3 | 8.6 | 8.9 | 9.2 | 9.5 | 9.9 | 10.3 | 10.7 | 11.1 | 11.6 | 12.1 | 12.7 | 13.3 |

OG, original gravity; PPG, (gravity) points per pound per gallon; R, wt./wt. water-to-grist ratio; Rv, vol./wt. water-to-grist ratio; SG, specific gravity.

Notes: Boil volume is 7 gal. to yield 6 gal. to the fermentor at the stated OG. For discussion of calculations used, refer to the section "No-Sparge Efficiency."

## TABLE 19.6—NO-SPARGE TARGETS USING KILOGRAMS AND LITERS

| OG | 1.030 | 1.035 | 1.040 | 1.045 | 1.050 | 1.055 | 1.060 | 1.065 | 1.070 | 1.075 | 1.080 | 1.085 | 1.090 |
|---|---|---|---|---|---|---|---|---|---|---|---|---|---|
| Grist weight (kg) | 2.8 | 3.3 | 3.9 | 4.5 | 5.2 | 5.9 | 6.7 | 7.5 | 8.4 | 9.4 | 10.4 | 11.5 | 12.8 |
| PKL | 250 | 243 | 235 | 228 | 221 | 213 | 206 | 199 | 191 | 184 | 177 | 169 | 162 |
| Extract efficiency | 81.2% | 78.8% | 76.4% | 74.0% | 71.7% | 69.3% | 66.9% | 64.5% | 62.1% | 59.8% | 57.4% | 55.0% | 52.6% |
| Mash volume (L) | 29.8 | 30.3 | 30.9 | 31.5 | 32.2 | 32.9 | 33.7 | 34.5 | 35.4 | 36.4 | 37.4 | 38.5 | 39.8 |
| Wort volume (L) | 27.0 | 27.0 | 27.0 | 27.0 | 27.0 | 27.0 | 27.0 | 27.0 | 27.0 | 27.0 | 27.0 | 27.0 | 27.0 |
| Wort SG | 1.026 | 1.030 | 1.034 | 1.038 | 1.043 | 1.047 | 1.051 | 1.055 | 1.060 | 1.064 | 1.068 | 1.072 | 1.077 |
| Rv | 10.8 | 9.1 | 7.9 | 6.9 | 6.2 | 5.6 | 5.0 | 4.6 | 4.2 | 3.9 | 3.6 | 3.3 | 3.1 |
| R | 10.6 | 9.0 | 7.8 | 6.8 | 6.1 | 5.5 | 5.0 | 4.5 | 4.1 | 3.8 | 3.5 | 3.3 | 3.1 |
| Mash volume (L) | 32.0 | 33.0 | 34.0 | 35.2 | 36.4 | 37.7 | 39.1 | 40.5 | 42.1 | 43.9 | 45.7 | 47.8 | 50.0 |

OG, original gravity; PKL, (gravity) points per kilogram per liter; R, wt./wt. water-to-grist ratio; Rv, vol./wt. water-to-grist ratio; SG, specific gravity.

Notes: Boil volume is 27 L to yield 23 L to the fermentor at the stated OG. For discussion of calculations used, refer to the section "No-Sparge Efficiency."

## TABLE 19.7—BIAB TARGETS USING POUNDS, QUARTS, AND GALLONS

| OG | 1.030 | 1.035 | 1.040 | 1.045 | 1.050 | 1.055 | 1.060 | 1.065 | 1.070 | 1.075 | 1.080 | 1.085 | 1.090 |
|---|---|---|---|---|---|---|---|---|---|---|---|---|---|
| Grist weight (lb.) | 5.7 | 6.8 | 8.0 | 9.2 | 10.4 | 11.7 | 13.1 | 14.5 | 16.0 | 17.5 | 19.2 | 20.9 | 22.7 |
| PPG | 31.4 | 30.8 | 30.1 | 29.5 | 28.8 | 28.2 | 27.6 | 26.9 | 26.3 | 25.7 | 25.0 | 24.4 | 23.7 |
| Extract efficiency | 84.9% | 83.2% | 81.4% | 79.7% | 78.0% | 76.3% | 74.5% | 72.8% | 71.1% | 69.3% | 67.6% | 65.9% | 64.2% |
| Mash volume (qt.) | 29.4 | 29.7 | 30.0 | 30.3 | 30.6 | 30.9 | 31.3 | 31.6 | 32.0 | 32.4 | 32.8 | 33.2 | 33.7 |
| Wort volume (qt.) | 28.0 | 28.0 | 28.0 | 28.0 | 28.0 | 28.0 | 28.0 | 28.0 | 28.0 | 28.0 | 28.0 | 28.0 | 28.0 |
| Wort SG | 1.026 | 1.030 | 1.034 | 1.039 | 1.043 | 1.047 | 1.051 | 1.056 | 1.060 | 1.064 | 1.069 | 1.073 | 1.077 |
| Rv | 5.1 | 4.4 | 3.8 | 3.3 | 2.9 | 2.6 | 2.4 | 2.2 | 2.0 | 1.8 | 1.7 | 1.6 | 1.5 |
| R | 10.6 | 8.9 | 7.7 | 6.8 | 6.0 | 5.4 | 4.9 | 4.5 | 4.1 | 3.8 | 3.5 | 3.3 | 3.0 |
| Total volume (gal.) | 7.9 | 8.1 | 8.3 | 8.4 | 8.6 | 8.8 | 9.1 | 9.3 | 9.5 | 9.8 | 10.0 | 10.3 | 10.6 |

BIAB, brew-in-a-bag; OG, original gravity; PPG, (gravity) points per pound per gallon; R, wt./wt. water-to-grist ratio; Rv, vol./wt. water-to-grist ratio; SG, specific gravity.

Notes: Boil volume is 7 gal. to yield 6 gal. to the fermentor at the stated OG. For discussion of calculations used, refer to the section "Brew-in-a-Bag Efficiency."

# TABLE 19.8—BIAB TARGETS USING KILOGRAMS AND LITERS

| OG | 1.030 | 1.035 | 1.040 | 1.045 | 1.050 | 1.055 | 1.060 | 1.065 | 1.070 | 1.075 | 1.080 | 1.085 | 1.090 |
|---|---|---|---|---|---|---|---|---|---|---|---|---|---|
| Grist weight (kg) | 2.6 | 3.1 | 3.6 | 4.2 | 4.8 | 5.3 | 6.0 | 6.6 | 7.3 | 8.0 | 8.7 | 9.5 | 10.3 |
| PKL | 263 | 258 | 252 | 247 | 242 | 237 | 232 | 226 | 221 | 216 | 211 | 206 | 200 |
| Extract efficiency | 85.3% | 83.7% | 82.0% | 80.3% | 78.6% | 76.9% | 75.2% | 73.5% | 71.8% | 70.1% | 68.4% | 66.7% | 65.1% |
| Mash volume (L) | 28.3 | 28.6 | 28.8 | 29.1 | 29.4 | 29.7 | 30.0 | 30.3 | 30.6 | 31.0 | 31.4 | 31.8 | 32.2 |
| Wort volume (L) | 27.0 | 27.0 | 27.0 | 27.0 | 27.0 | 27.0 | 27.0 | 27.0 | 27.0 | 27.0 | 27.0 | 27.0 | 27.0 |
| Wort SG | 1.026 | 1.030 | 1.034 | 1.038 | 1.043 | 1.047 | 1.051 | 1.055 | 1.060 | 1.064 | 1.068 | 1.072 | 1.077 |
| Rv | 10.8 | 9.1 | 7.9 | 6.9 | 6.2 | 5.6 | 5.0 | 4.6 | 4.2 | 3.9 | 3.6 | 3.3 | 3.1 |
| R | 10.6 | 9.0 | 7.8 | 6.8 | 6.1 | 5.5 | 5.0 | 4.5 | 4.1 | 3.8 | 3.5 | 3.3 | 3.1 |
| Total volume (L) | 30.4 | 31.1 | 31.7 | 32.4 | 33.2 | 33.9 | 34.7 | 35.6 | 36.5 | 37.4 | 38.3 | 39.4 | 40.4 |

BIAB, brew-in-a-bag; OG, original gravity; PKL, (gravity) points per kilogram per liter; R, wt./wt. water-to-grist ratio; Rv, vol./wt. water-to-grist ratio; SG, specific gravity.

Notes: Boil volume is 27 L to yield 23 L to the fermentor at the stated OG. For discussion of calculations used, refer to the section "Brew-in-a-Bag Efficiency."

# Brewing Your First All-Grain Batch

20

The comment you will most often hear from homebrewers after their first all-grain brew is, "I didn't realize it would be so easy!" Making wort from malted grains is very easy. You mix the crushed grain with hot water, check the temperature and wait a while . . . and you have wort. I have talked to hundreds of people over the years that won't try mashing because they worry and hesitate and try to read everything about it beforehand. They will plan every excruciating detail and failure scenario and still want someone to hold their hand before they try it themselves. My response? "Just do it!" Don't overthink this.

After all, what's the worst that can happen when you mix some crushed grain with hot water? It's not going to explode. At worst, it will just sit there. The most likely failure scenario is that you will have mediocre conversion from not being in the right temperature range. So what?! You still made wort! You will do better next time; every batch is a learning experience.

I hope that you have done several extract-and-specialty grain batches by now. You should know to have your ingredients and brewing water ready, with everything clean and sanitized. Unless you have purchased a roller mill, have the grain crushed for you at the brew shop. Crushed grain will stay fresh for about two weeks if kept cool and dry.

There are several options that you can use for your first all-grain batch, but to keep it simple I will only describe two in this chapter. These are brew-in-a-bag (BIAB) and a combined mash and lauter tun (MLT) made from a picnic cooler or ice chest. In addition, you have the option of doing a partial mash with either of these methods, conducting a smaller mash to make half the wort and making up the difference with malt extract. The partial mash option is just like the partial-boil Palmer Brewing Method used for extract and specialty grains, but now Wort A is the mash. This is a good option if you are limited on kettle size.

## Mash and Lauter Tun or Brew-in-a-Bag?

How will you conduct the mash and lauter? It's a good question. The MLT method is more traditional and very similar to commercial brewing practice. In MLT, separate vessels are required, the most common configuration being a hot liquor tank to heat the brewing water, a single tun used to conduct both the mash and lauter, and a boil kettle. An insulated mash-lauter tun made from an ice chest with a manifold or false bottom works really well, because the insulation keeps it from losing more than a couple of degrees per hour. A mash-lauter tun needs to be sparged to get the best efficiency, but that is simply a matter of pouring in a pitcher of hot water every few minutes as the wort drains to your kettle. Or you can batch sparge and add it all at once.

Brew-in-a-bag appeals to many first time mashers because of its simplicity—it's just like steeping, except you are mashing. You do everything in a single kettle and you don't have to sparge. Mash, drain the bag, and boil. The downside is that you need a larger kettle and you have to heat more water at a time than you would with MLT. The other thing to consider is whether you are strong enough to lift and hold 20–30 lb. (9–14 kg) of hot wet grain, for 15 min. while it drains (many people use a rope and pulley).

There isn't much difference in the cost of equipment between the two methods. The BIAB method needs a larger kettle than the MLT method plus a large grain bag, and the cost difference between the two kettle sizes plus the grain bag is about the same as a suitable ice chest and false bottom. I tend to prefer the MLT method myself. Maybe it's just because that's the way I learned to brew, but I also believe it's important to learn the corners before you starting cutting them off. If you have faithfully read all the preceding chapters to this point, then you know what the corners are, so choose whichever method you like.

### Additonal Equipment Needed

For the MLT method you will need:
- an 8–10 gal. (30–38 L) boil kettle;
- a 10 gal. (38 L), or larger, ice chest (or another kettle) for the mash-lauter tun;
- a 5 gal. (20 L) kettle for the hot liquor tank.

For the BIAB method you will need:
- a 10–15 gal. (38–56 L) boil kettle;
- a grain bag (same size as kettle).

*Larger boil kettle.* You will need a larger boil kettle for either all-grain method, because you are going to be boiling the whole batch. The BIAB method needs a larger kettle than the MLT method, but a 15 gal. (56 L) vessel would work for both. All of the mash and sparge water is heated at the same time for BIAB, because there is no sparging step and it must be able to contain the entire mash. See tables 19.7 and 19.8 for estimated total volumes as a function of OG for nominal 5 gal. (19L) batch sizes.

Using the MLT method means that each of the vessels can be smaller than the single large kettle in BIAB. The boil kettle and mash-lauter tun only need to be about 10 gal. (37L). The sparge water can be heated in your old 5 gal. (20 L) kettle, which now serves as the hot liquor tank. You will probably use 4–8 gal. (15–30 L) of water for the mash, depending on OG, and another 3–4 gal. (11–15 L) of water for the sparge, so be prepared with the right sized vessels.

*Grain bag.* The grain bag for BIAB can be made from muslin, cheesecloth, or mosquito netting. Commercially made bags are also available and these are sturdy enough to be reused many, many times. Your grain bag should be about the same diameter as your brewing kettle and tall enough to either close with a drawstring or drape over the rim of the kettle. For best results, you want the bag to be big enough so that the grain can move freely during the mash. Do not try to use a pillow case; they don't drain well and bits of grain husk might get stuck in other clothes during washing. (For the record, I have not tried this and I don't know if it can cause irritation of family members. It could have been sawdust from the boy's clothes.)

*Mash-lauter tun.* Instructions for building a mash-lauter tun are given in appendices E and F. A 36–48 qt. (34–45 L) rectangular chest cooler or 10 gal. (38 L) circular beverage cooler are probably the best choice for most 5 gal. (19 L) batches. The tun needs to be outfitted with some sort of straining device and a valve or siphon to separate the wort from the grain. Several options are discussed in appendices E and F. These devices are fun to make by yourself, but can be purchased as well.

*Propane or natural gas burner.* See chapter 9 for a discussion of appropriate burners.

*Thermometer.* Last, and most importantly, you will need a good thermometer to check the mash temperature quickly and accurately. A quick-read digital thermometer is probably best. Some brewing kettles come with thermometers mounted in the side, but a handheld thermometer is always, well . . . handy.

## Suggested Recipe

For this all-grain batch, we will make a brown ale, using five malts and an adjunct in a single-temperature infusion mash. I will take you through the entire all-grain brewing procedure, then go back and discuss some options for various steps. Of course, if there is another beer style that you prefer, you are welcome to use one of the other recipes from chapter 23.

## Oatis A. Brown Ale

**Original gravity:** 1.052
**Final gravity:** 1.013
**IBU:** 36

**Color:** 17 SRM (34 EBC)
**ABV:** 5.2%

### All-Grain Version

| Grain bill | Gravity points |
|---|---|
| 8.5 lb. (3.9 kg) pale ale malt | 35 |
| 1.0 lb. (450 g) Carabrown malt | 3 |
| 0.5 lb. (225 g) flaked oats | 2 |
| 0.5 lb. (225 g) crystal 80°L malt | 1 |
| 0.5 lb. (225 g) aromatic 20°L malt | 2 |
| 0.33 lb. (150 g) chocolate malt | 1 |
| Boil gravity for 7 gal. (27 L) | 1.044 |
| Original gravity for 6 gal. (23 L) | 1.052 (see notes) |

| Mash Schedule | Rest temp. | Rest time |
|---|---|---|
| Conversion Rest – Infusion | 153°F (67°C) | 60 min. |

| Hop schedule | Boil time | IBUs |
|---|---|---|
| 0.75 oz. (21 g) Horizon 12% AA | 60 min. | 28 |
| 0.5 oz. (14 g) East Kent Goldings 5% AA | 15 min. | 4 |
| Total IBUs | | 32 |

| Yeast strain | Pitch (billions of cells) | Fermentation temp. and time |
|---|---|---|
| English ale | 225 | 68°F (20°C) for 2 weeks. |

**Directions for Mash Schedule**

MLT Method: Single temperature infusion mash with strike water at 161°F (72°C) and water-to-grist ratio of 2 qt./lb. (4 L/kg).

BIAB Method: Single temperature infusion mash with strike water at 158°F (70°C) and water-to-grist ratio of 2.85 qt./lb. (5.96 L/kg).

Target mash temperature of 153°F (67°C). Mash time of 1 hour. No mash-out.

*Notes: Target boil gravity of 1.044 for 7 gal. (27 L)—this will equal 1.052 for 6 gal. (23 L) after boiling. The extra gallon (3.8 L) provides for wort soaked up by the hops and break material in the kettle and fermentor, giving you at least five gallons (19 L) of clean beer.*

*Adjust the amount of chocolate malt between 0.25 and 0.5 lb. (110–225 g) depending on how dark you want it.*

*You can use any hops you want to; the key is the boiling times and total IBUs so that the hops don't distract from the malt character. I've used English hops, but you could use American, European, orPacific instead (see table 5.3 for comparison.)*

| Partial Mash Option |
| --- |
| **Wort A** |
| 2.0 lb. (0.9 kg) pale ale malt |
| 1.0 lb. (450 g) Carabrown malt |
| 0.5 lb. (225 g) crystal 80°L malt |
| 0.5 lb. (225 g) flaked oats |
| **Wort B** |
| 4 lb. (1.8 kg) pale ale LME |
| 0.33 lb. (150 g) chocolate malt |

## PARTIAL MASH OPTION

Not everyone can jump right into full-scale mashing. You can still use the partial-boil Palmer Brewing Method from chapter 1 with a small mash to provide wort complexity and freshness, plus a couple pounds of malt extract to provide the bulk of the fermentables. This option is particularly attractive for brewers in small apartments with little room in the kitchen for a lot of equipment. I used a partial mash when I first started mashing and I was extremely pleased with the results.

A partial mash is carried out just like a full-scale mash, only the volume is smaller. You will collect 3 gal. (11.4 L) as Wort A, using either the BIAB or MLT method. The advantage over an extract & steeped specialty grain procedure is that you're actually mashing, which means you're not limited to just specialty grains, like caramel and roasted malts. Wort A will consist of base malt and most of the specialty malts, and Wort B will be malt extract, although a portion of the specialty malts could be used (steeped) there as well, such as the roast malts. It would be steeped and removed before cooling.

You can mash in either your 5 gal. (19 L) kettle on the stove, or buy a smaller cooler (12–16 qt., or 12–15 L) and build a small manifold for it. You probably have a small beverage cooler already that would work well with a drop-in manifold like those discussed in appendix E.

## The MLT Method

Basically, you will heat the mashing water in your larger boil kettle, pour that into your mash-lauter tun with the grain to mash, and then place the larger boil kettle underneath the tun to receive the wort (fig. 20.1). Your smaller boil kettle will be used to heat the sparge water. At the end of the mash, some of the wort is recirculated through the mash-lauter tun to clarify it, and drained to the large kettle. The sparge water is then added to the tun, mixed, recirculated, and the wort is drained again. This is the batch sparging option. Continuous sparging is similar, but the sparge water is added in small batches, or continuously with a hose, while the wort is drained slowly to the kettle. Chapter 19 explains sparging methods in more detail.

### Starting the Mash
#### 1. Heat the Mash Water
Heat up enough water to conduct the mash. Always make a little more hot water than you calculate is necessary, you will often need it. At a water-to-grist ratio of 2 qt./lb. (4 L/kg), the volume needed

**Figure 20.1.** The larger of the two brew kettles is used to heat the mash water, and the smaller kettle is used for heating the sparge water.

would be 21 qt. (20 L), or just over 5 gal. Therefore, heat up about 6 gal. (23 L) in the larger of your two boil kettles. At this water-to-grist ratio (or mash ratio), the initial infusion temperature should be 161°F (72°C) to create a mash temperature of 153°F (67°C). Depending on the amount of heat lost to the tun, the strike water could be as hot as 165°F (74°C), but that would (theoretically) create a mash temperature of 156°F (69°C). At 156°F, the wort would become more dextrinous than we intended, but it would still be a fermentable wort, so don't worry if your infusion temperature is a couple of degrees off. Chapter 17 explains the infusion calculations.

*Note on brewing salts:* If you are going to add brewing salts to the water, this would be the time to do it. Remember that calcium sulfate actually dissolves better in cold water. You could also add brewing salts to the mash instead. See chapter 22 for more information on adjusting your brewing water for a recipe.

**2. Preheat the Tun**
Preheat the cooler with some boiling water, about a gallon or roughly four liters. Swirl the boiling water around to heat up the cooler and then pour it back to your sparge water kettle. Preheating will prevent initial heat loss from the mash to the tun, which can throw off your infusion calculations.

### 3. Mash-In

When you mash-in, you can add the grain to the water, or the water to the grain. Thermodynamically, it's easier on the enzymes to add the water to the grain, but that tends to create dry spots on the bottom. I find it is easier to measure the water into the mash-lauter tun first and then add the grain. Stir as you pour in the grain. Mix thoroughly to make sure all the grain is fully wetted and break up any doughballs (fig. 20.2).

### 4. Check the Temperature

Check the temperature of the mash (fig. 20.3) to see if it has stabilized at the target temperature of 152°F (67°C), or at least in the range of 150–155°F (65–68°C). If the temperature is too low (e.g., 145°F [63°C]), add some more hot water. If it is too high (e.g., 160°F [71°C]), then add cold water to bring it down. For this recipe, 156°F (69°C) is the highest we would want the mash temperature to go. The target temperature range will yield a sweet, medium-bodied wort with good attenuation.

### 5. Adjust the Temperature

OK, the mash temperature came out a little low at 148°F (64°C), so I added 1.5 qt. (1.4 L) of boiling water to bring it up to 152°F (67°C). Stir as you add the hot water to evenly distribute the heat (fig. 20.4).

**Figure 20.2.** Mixing the grist and mash water. I find that adding the grist to the water is more convenient, although it can be done the other way. Stir thoroughly to make sure the grist is fully wetted and the temperature is uniform throughout the mash.

**Figure 20.3.** Checking the temperature of the mash after the infusion.

**Figure 20.4.** Adjusting the temperature of the mash. In this case, adding 1.5 qt. (1.4 L) of boiling hot water to raise the mash temperature another 4°F (3°C), to 152°F (67°C).

**Figure 20.5.** A picture of the mash at *t* = 0. The wort is cloudy with starch.

**Figure 20.6.** A picture of the mash at *t* = 60 minutes. Some of the larger grist particles are floating around, as are some husks. Notice the wort has cleared; it is no longer cloudy with starch, and it smells great.

## Monitoring the Mash

### 6. Monitor the Mash

Stir the mash every 15–20 min. during the hour (fig. 20.5) to prevent cold spots and help ensure a uniform conversion. Monitor the temperature each time you stir. If the temperature drops by less than 5°F (3°C) over the hour-long mash, nothing further needs to be done. Cover the mash tun with the cooler lid between stirrings and let it sit for a total of 60 min. If you notice that the temperature drops below 145°F (62°C) within the first half hour, you can add more water to bring the temperature back up (fig. 20.4).

### 7. Heat the Sparge Water

Meanwhile, heat up your sparge water in the smaller of your two boil kettles (fig. 20.7). You will need about 3.5 gal. (13.25 liters) for the batch sparge volume. The sparge water temperature should be less than boiling, preferably 165–175°F (73–80°C). If the sparge water is too hot, the probability of tannin extraction from the grain husks increases substantially.

## Conducting the Lauter

Okay, the hour has gone by and the mash should look a little bit different, the wort appearing clearer and smelling great (fig. 20.6).

### 8. Recirculate (Vorlauf)

Open the valve of your mash tun slowly and drain about 2 qt. (roughly 2 L) of the first runnings into a pitcher. This wort will be cloudy with bits of grain. Gently pour the wort back into the grain bed, recirculating the wort. Repeat this procedure until the wort exiting the tun is mostly clear (like unfiltered apple cider). It will be dark-amber colored, hazy, but not chunky. You should only have to recirculate a couple of pitchers to clarify the wort (fig. 20.8).

### 9. Lauter

Once the wort has cleared through recirculation, you are ready to lauter (drain) the first runnings into your boil kettle (fig. 20.9). Fill the kettle slowly at first; if you drain the mash-lauter tun too quickly you will compact the grain bed and get a stuck sparge. Drain the tun to collect about 3.5 gal. (13.25 L) of wort in the boil kettle.

### 10. Add the Sparge Water

Close the valve and add your sparge water to the mash-lauter tun using the pitcher until you can just dump in the rest straight from the kettle (be careful, it's hot). Stir the grist thoroughly to get as much residual extract into the wort as possible (fig. 20.10). There is a chance of dissolving unconverted starch into the wort at this stage, so it doesn't hurt to let the mash sit for 15 min. to allow residual alpha-amylase to convert it to sugars. Recirculate this new wort and then drain to your boil kettle. You should now have 7 gal. (27 L) total in your boil kettle.

### 11. Stuck Sparge?

If the wort stops flowing then you have a stuck sparge. There are two common ways to fix it: blow back into the outlet hose to clear any obstruction of the manifold; or close the valve and add more hot water, stirring to resuspend the mash. You will need to recirculate again. Stuck sparges can result from

**Figure 20.7.** Heating the sparge water in the smaller brew kettle while the mash is going on.

**Figure 20.8.** Recirculating the wort by draining about 2 qt. (roughly 2 L) of the first runnings into a pitcher. The wort will be cloudy with bits of grain. Then, slowly pour this wort back into the grain bed, recirculating the wort. Repeat this procedure until the wort exiting the tun is mostly clear (like unfiltered apple cider)—you just want to eliminate the bits of grain coming through. The wort will be dark amber, hazy, but not chunky.

**Figure 20.9.** Starting draining the lauter tun of the first runnings.

**Figure 20.10.** Adding the sparge water a pitcher at a time until the rest can just be dumped in. Stir the mash thoroughly to refloat the grain bed and get all of the remaining extract.

grinding your grist too fine, or using too much adjunct, but it shouldn't happen very often. Stirring in a couple of handfuls of rice hulls will usually fix the problem if the first two methods don't.

### 12. Calculate Your Brewing Efficiency

Measure the gravity of the wort in the boil kettle (stir it first) and multiply the gravity points by the number of gallons (or liters) you collected. Then divide by the number of pounds (or kilograms) of grain you used. The result should be somewhere around 28 PPG (or 235 PKL), which is about 75% efficiency. Your yield and efficiency will decrease as the OG of your recipe increases because higher concentration means less volume, but that is not an issue in this recipe. If your yield efficiency is low, for example, 65%, you may not be getting good conversion in the mash, which could be caused by too coarse a grist, the wrong temperature, not enough time, or your mash pH is a factor. However, low efficiency is more often due to poor lautering, such as poor mixing of the batch sparging water, or draining the tun too fast. These issues are discussed more thoroughly in appendices E and F.

### 13. Done!

Okay, throw the spent grain on the compost pile and you are done! You have produced your first all-grain wort! The rest of the brewing process is just like extract brewing. Boil the wort, add hops, chill, and ferment. All-grain brewing produces more break material than extract brewing, so you will probably want to add Irish moss at the end of the boil to help with coagulation during the cold break and aid clarity. See appendix C for more information on beer clarity.

## The BIAB Method

Basically, with BIAB, you are mashing with a giant tea bag in your large boil kettle. All of the water that would be used for the mash and sparge in the MLT method is combined here for the mash. At the end of the mash, the grain bag is lifted out of the kettle and allowed to drain. There should be a full 7 gal. (27 L) of wort at the target gravity in the kettle. There are no sparging or recirculation steps. The grain is thrown away and the boil can commence. This method creates a turbid (i.e., cloudy) wort that nevertheless usually produces a clear beer.

**Notes when using BIAB for the Oatis A. Brown recipe:**
- BIAB has higher yield than batch sparging, so the pale ale malt quantity will be reduced from 8.5 lb. (3.86 kg) to 8 lb. (3.63 kg).
- Do not grind the grist extra fine. Use a standard crush, it will drain easier.
- The water-to-grist ratio is higher with BIAB, so the strike temperature of the infusion water is lower compared to that for typical water-to-grist ratios used in the MLT method.

## Starting the Mash

### 1. Put the Grist in the Bag

Crush the grain and put it in your grain bag. The bag should be large enough that the grain is not compacted, but can move loosely within the bag during the mash. The grain should be loose enough to stir.

**Figure 20.11.** The kettle is filled with 7.75 gallons (29.5 L) of water, and the grain bag is ready to be immersed.

**Figure 20.12.** Immersing the grain bag when the water reaches strike temperature.

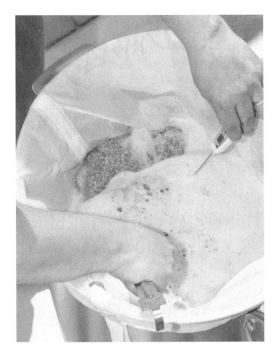

**Figure 20.13.** Checking the temperature of the BIAB mash.

**Figure 20.14.** The bag is lifted while the burner is on to avoid scorching or melting the bag while I raise the mash temperature a few degrees.

## 2. Heat the Brewing Water

Heat up enough water to conduct the mash. Use the BIAB efficiency tables in chapter 19 (tables 19.7 and 19.8) to help you determine how much water to use to yield your target OG. For this recipe, with an OG of 1.052, we will be using 31 qt. (7.75 gal.) for a grain weight of 10.85 lb., or 29.5 L for 4.9 kg of grain (fig. 20.11). The water-to-grist weight ratio is about six (i.e., $R \approx 6$). From the dry grain infusion calculations we discussed in chapter 17, we can work out what the strike temperature should be:

$$Tw = [(S/R) \times (T2 - T1)] + T2$$

This calculation gives us:

$$Tw = [(0.4/6) \times (152 - 70)] + 152$$

$$= 157.6°F$$

Obviously, this temperature of 157.6°F (69.7°C) can be rounded up to 158°F (70°C).

*Note on brewing salts:* If you are going to add brewing salts to the water, this would be the time to do it. Remember that calcium sulfate actually dissolves better in cold water. You could also add brewing salts to the mash instead. See chapter 22 for more information on adjusting your brewing water for a recipe.

## 3. Immerse the Grain Bag

When the water reaches the strike temperature, which is 158°F (70°C) in this instance, turn off the burner and immerse the grain bag (fig. 20.12). It helps to stir both inside and outside the bag to make sure the grain is fully wetted.

**Figure 20.15.** I built a simple wooden frame with a hook to hold the bag for draining, and obtained my full boil volume of 7 gallons (27 L) without squeezing it.

## 4. Check the Temperature

Check the temperature once the grain is wetted (fig. 20.13). The mash temperature should be about 152°F (67°C). Anywhere between 150–155°F (66–68°C) is fine. Stir to make sure the temperature is uniform, and then cover the mash with a lid to help retain the heat. If it is cold outside, you may want to wrap the kettle with a heavy blanket or sleeping bag.

## Monitoring the Mash

### 5. Heat as Needed

Check the temperature of the mash about every 15 min. If the temperature falls below 140°F (60°C), then you will need to turn on the burner for a few minutes to warm it up. Raise the bag a few inches off the bottom of the kettle while heating (fig. 20.14). Stir, or bob the bag up and down, to distribute the heat while heating. Turn off the burner after a couple of minutes and check the temperature again. Stir the mash thoroughly to make sure the temperature is uniform.

## Lautering—BIAB-style

### 6. Drain the Grain Bag

The mash will take 60 minutes total. At the end of the hour, lift the grain bag out of the kettle and let it drain into the kettle (fig. 20.15). The weight of the grain in the bag will squeeze out more wort than you get from batch sparging in the MLT method. You should have 7 gal. (27 L) of 1.043 wort in the kettle.

*Note on squeezing:* I do not recommend manually squeezing the bag to get out every drop; the weights and volumes for the recipe have been calculated without it, and my personal experiments with roasted grains have demonstrated harsh flavors when the bag is squeezed.

### 7. Done!

Discard the grain (not the bag!) and start the boil as usual. The BIAB method is essentially a no-sparge process. Today's highly modified malts with high diastatic power seem to have no problem with the BIAB method. It's simple, and there is less to clean up at the end of the day.

# Residual Alkalinity, Malt Acidity, and Mash pH

## Or, Everything You Ever Wanted to Know about Mash pH but Were Afraid to Ask

Ignorance is bliss for the beginning masher, because most any water will produce wort when mixed with crushed malt at the right temperature. If you remember the top five brewing priorities from chapter 1, water is not on that list. Let's be realistic, if you are going to heat up a can of condensed soup, make a box of macaroni and cheese, or make a pot of coffee, you probably won't give much thought to the water that you are going to make it with. As long as the water tastes good, the food will taste good.

However, let's say you wanted to brew a truly exceptional cup of coffee; now the taste and composition of the water becomes more important. There is a reason why the foods of master chefs and the beers of master brewers are more highly praised than the rest, and those reasons are attention to detail and control. Their attention to detail and understanding of those details allows them to control the results of the process. Beer is liquid food. We need to understand how the water can affect the brewing process and the taste of the beer in order to be able to exert control. This is where brewing water can get complicated, but I will do my best to lay it all out in a clear, logical stream.

There are six key concepts for understanding how water affects your beer:

1. Beer and brewing are food and cooking.
2. Know your water source.
3. Residual alkalinity is the cornerstone of mash pH.
4. Mash pH is the result of the interaction between water chemistry and malt chemistry.
5. The mash pH is the cornerstone of beer pH.
6. Beer pH is the cornerstone of beer flavor.

## Beer and Brewing is Food and Cooking

Beer is food, and the brewing of beer is both a cooking and fermentation process. You, as brewmaster, have the same ability to brew a great beer as a chef does to create a great meal. The key is attention to detail in every aspect of the brewing, or cooking, process: the quality of ingredients, the proportions, and the techniques. There are no shortcuts anymore—everything you do is going to affect the flavor of the beer you make.

The first thing you need to do is consider the flavor of your beer. What aspects or properties control the flavor of food? The answer is pH and seasoning. I like to use spaghetti sauce as an example. A standard commercial spaghetti sauce from the grocery store is not very exciting; it's probably sweet and bland, and lacks complexity. Children will probably love it, but adults with more refined tastes probably won't. The other extreme is a spaghetti sauce that was made fresh that morning at a paleo-food Italian restaurant. It will have such bright tomato acidity that you can't really taste anything else! To make the best spaghetti sauce with a full expression of flavors the pH of the sauce must be balanced, which creates a balance between tomato acid brightness and rich tomato flavor. You cannot have complexity without balance. Once you have balanced the flavors of the sauce with pH, you can fine-tune them further with seasonings to accentuate the specific flavors you prefer.

This is how you need to think about your beer. The pH of the beer controls the way flavors are expressed to your palate, while the water profile—the minerals in the water—are the seasonings for your beer. There are two ways that the minerals season your beer; one is the relative proportions of sulfate and chloride, and the other is the total amount of minerals in the water. We will discuss these aspects more toward the end of the chapter.

## Know Your Water Source—a Review

The source of the water dictates the mineral profile of the water. Surface water sources, such as lakes, rivers, and streams, will typically be low in minerals but high in organic matter, such as algae, fish, and lots of microorganisms. Groundwater from aquifers underground tends to be low in organic matter but high in dissolved minerals, such as calcium, sodium, and bicarbonate. Depending on where you live, your source water may be surface water, groundwater, or a mixture of both. The source could change seasonally, such as surface water during the summer and groundwater during the winter. This change in mineral profile can have a big impact on the pH of your mash, wort, and beer.

Surface water sources usually need more microbiological control than groundwater sources, and during the summer a water utility will often increase the chlorine treatments, which can be a problem. However, the low mineral content of surface water makes it easy to add salts to adjust the pH and flavor effects. Groundwater sources often cause calcium carbonate scaling problems for the utility, so they may use lime or other softening treatments to reduce the hardness in the water, which is the opposite of what brewers want. To review, water hardness is determined by the levels of calcium and magnesium ions dissolved in the water, both ions being beneficial for the brewing and fermentation processes. Permanent hardness is calcium and magnesium ions derived from highly soluble salts, such as sulfates and chlorides, because these ions stay in solution. Temporary hardness comes from calcium and magnesium ions derived from carbonate salts, because these are likely to precipitate out as carbonate scale (limescale) when the water is heated or boiled.

Home water softeners remove the calcium and magnesium ions from water but don't remove the bicarbonate ions, which determine the alkalinity. This is good for the water utility's pipes, because an alkaline pH (7.0–9.0) is less corrosive than an acidic pH, but it is not as good for our mash, which has a pH target of 5.2–5.6. Alkalinity drives the mash pH up, which can cause the extraction of unpleasant tannin flavors, a harsher bitterness from the hops, and a high beer pH that dulls the overall flavor. This is why you will often see references in brewing text books saying to remove the temporary hardness from the water; it is not because we want to remove the hardness, but because we want to remove the alkalinity that is associated with it.

Alkalinity is the sum of the dissolved carbonate species in the water. Alkalinity is measured by acid titration, that is, by measuring the amount of acid of known concentration required to lower the water's pH to 4.3, where all of the carbonate species have converted to dissolved carbon dioxide and carbonic acid. This measure of alkalinity by acid titration is called either "total alkalinity," "general alkalinity," or "M alkalinity." By convention, total alkalinity is usually reported in the form "total alkalinity, ppm as $CaCO_3$" (as calcium carbonate), which refers to the potential for calcium carbonate scale (which is important from a water company's point of view). Both total hardness and total alkalinity are reported as ppm as $CaCO_3$ for this reason. See the sidebar for a discussion of units for mineral concentration.

## UNITS OF MINERAL CONCENTRATION

When minerals are dissolved in water, their associated ions are measured by concentration, usually as milligrams per liter (mg/L) or parts per million (ppm). These units are basically the same, because there are 1 million milligrams in a kilogram and 1 liter of water weighs (about) 1 kilogram. This means 1 milligram of anything in 1 liter of water is one part per million. Therefore, a concentration of 35 mg/L = 35 ppm.

When discussing alkalinity, the unit "ppm as $CaCO_3$" is a little bit different. Let me explain. Ion concentration can also be discussed in terms of reactivity. For example, 1 mole of sodium will react with 1 mole of chlorine to create 1 mole of sodium chloride.

$$Na + Cl \rightarrow NaCl$$

To understand this, you need to know that one "mole" of any element or compound contains exactly the same number of particles (i.e., atoms or molecules). One mole of any substance is $6.02 \times 10^{23}$ of that chemical's atoms or molecules. The mole unit can also be applied to the charged particles (i.e., ions) of a substance when dissolved in water. Sodium has an ionic charge of +1 and chloride has an ionic charge of −1. A more accurate way to describe the reaction above is that 1 mole of sodium ions ($Na^+$) reacts with 1 mole of chloride ions ($Cl^-$); each therefore has a reaction equivalent of 1.

$$Na^+ + Cl^- \rightarrow NaCl$$

In this way, we say 1 mole of sodium ions = 1 equivalent; likewise, 1 mole of chloride ions = 1 equivalent. An equivalent is 1 mole of charge, i.e., 1 mole of hydrogen ions ($H^+$) or 1 mole of electrons ($e^-$), in a chemical reaction.

Now let's look at what happens with equivalents when a substance has more than a single charge. Calcium has an ionic charge of +2. One mole of calcium ions reacts with 2 moles of chloride ions to form 1 mole of calcium chloride.

$$Ca^{2+} + 2Cl^- \rightarrow CaCl_2$$

The charge of +2 on a calcium ion ($Ca^{2+}$) means it can react with two chloride ions that only have a charge of −1 each. Thus, we say 1 mole of calcium ions has 2 equivalents. The compound calcium chloride is also considered to have 2 equivalents per mole. Two equivalents of calcium and 2 equivalents of chloride; not 4 altogether.

The equivalent weight of an ion or compound is the atomic or molecular weight divided by the amount of charge, which is often, but not always, equal to the valence. The atomic weight of chlorine is 35.5, thus, the equivalent weight of chloride is simply 35.5/1 = 35.5. The equivalent weight of calcium works out as half of its atomic weight, because the atomic weight of calcium is 40 but the calcium ion provides 2 equivalents (i.e., a +2 charge); thus, the equivalent weight of calcium is 40/2 = 20.

Calcium carbonate is treated the same way as calcium chloride. Calcium ions have a +2 ionic charge ($Ca^{2+}$), carbonate has a −2 ionic charge ($CO_3^{2-}$). The molecular weight of calcium carbonate is 100, therefore, the equivalent weight of calcium carbonate is 100/2 = 50.

The total alkalinity of a water sample is measured by how many milliequivalents of acid per liter of water are required to acidify it to a pH of 4.3 (where the pH indicator methyl orange turns from yellow to red). For example, let's say that it took 3 milliequivalents of acid per liter of water to reduce the pH to 4.3. Therefore, the total alkalinity of that water sample is 3 mEq/L. The number of acid milliequivalents is converted to ppm as $CaCO_3$ by multiplying by calcium carbonate's equivalent weight of 50, so that 3 mEq of acid would equal 150 ppm as $CaCO_3$. (This is where we get "ppm as $CaCO_3$.")

Calcium and magnesium concentrations are usually listed as ionic concentration in mg/L or ppm, but these can also be given as calcium carbonate milliequivalents. The equivalent weight of calcium is 20 and the equivalent weight of magnesium is 12.1. If we had a water sample with a calcium ion concentration of 60 ppm and a magnesium ion concentration of 12 ppm, we can calculate the calcium hardness and total hardness by multiplying the number of equivalents of each ion by the equivalent weight of calcium carbonate as follows:

$$\text{ion equivalents (mEq)} = \text{ion concentration (ppm)} / \text{equivalent weight of ion}$$

$$= 60 \text{ ppm} / 20$$

$$= 3 \text{ mEq Ca}^{2+}$$

$$\text{hardness as ppm } CaCO_3 = \text{ion equivalents} \times \text{equivalent weight } CaCO_3$$

$$= 3 \text{ mEq} \times 50$$

$$= 150 \text{ ppm as } CaCO_3$$

The same calculations for magnesium hardness would be 12 ppm/12.1 equals 1 mEq $Mg^{2+}$, and therefore would equal 50 ppm as $CaCO_3$. The total hardness is the sum of the calcium and magnesium hardness: 3 mEq + 1 mEq = 4 mEq. Therefore, the total hardness would be 200 ppm as $CaCO_3$.

In order to control your water and its effects on your beer, you need to know what is in it. There are six ions that determine how your brewing water affects your beer and they are: calcium, magnesium, total alkalinity (as $CaCO_3$), sulfate, chloride, and sodium. Let's look at each in more detail.

### Calcium

- Atomic weight = 40
- Equivalent weight = 80
- Recommended range = 50–150 ppm

The calcium ion ($Ca^{2+}$) is the most important ion in brewing. It is a cofactor in many biochemical reactions in the mash and fermentation. It stabilizes alpha-amylase in the mash at high temperatures and pH. It improves beer clarity via trub coagulation and yeast flocculation. It does not have a flavor but beers taste watery without it. Calcium ions react with malt phosphates to reduce mash pH.

The minimum recommended level for calcium ions in brewing water is 50 ppm for light lagers and ales. Calcium ion levels of 100–150 ppm are generally preferred for good mash and lauter pH stability, although these levels may be too robust for some light beer styles. Calcium ion levels in excess of 200 ppm tend to taste minerally. These concentration values are for the calcium ion itself; for calcium hardness as $CaCO_3$, divide these ppm levels by 20 and multiply by 50.

## Magnesium

- Atomic weight = 24.3
- Equivalent weight = 12.15
- Recommended range = 0–40 ppm

The magnesium ion ($Mg^{2+}$) is the sidekick to the calcium ion, participating in many of the same reactions. It a vital yeast nutrient with a minimum required level of 5 ppm, but usually the malt supplies all the magnesium that the yeast would need. Magnesium can be displaced by calcium in the yeast cell if the calcium ion level in the water is too high, although references do not say what those conditions are. It may be beneficial to add magnesium salts to the water if no magnesium is present. Magnesium ion concentrations greater than 80 ppm are said to contribute a sour bitter flavor to beer, although lower levels of 20–40 ppm are said to enhance the flavor of dark beer styles, such as porter and stout. Magnesium hardness as $CaCO_3$ is equal to the magnesium ion concentration divided by 12.15 and multiplied by 50.

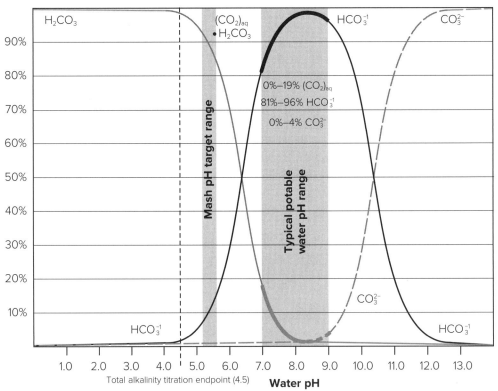

**Figure 21.1.** The carbonate system consists of carbonate ions ($CO_3^{2-}$) at high pH levels, bicarbonate ions ($HCO_3$) at medium pH levels, and very low levels of carbonic acid ($H_2CO_3$) in chemical equilibrium with dissolved carbon dioxide ($CO_2$)$_{aq}$ at low pH levels.

### Total Alkalinity as CaCO₃

- Molecular weight = 100
- Equivalent weight = 50
- Recommended range = 0–100 ppm

Alkalinity in brewing water is undesirable when brewing most pale beers, but advantageous in some darker brews. Carbonates in water exist in three species (chemical forms), which are carbonate ions ($CO_3^{2-}$), bicarbonate ions ($HCO_3^-$), and carbonic acid ($H_2CO_3$). (Only a very small amount of carbonic acid actually forms; most of it exists as dissolved carbon dioxide.) The relative proportions of these species depend on the water pH. Potable water has a pH between 7.0 and 9.0, often falling between 8.0 and 8.5. The predominant carbonate species in potable water is the bicarbonate ion ($HCO_3^-$), which makes up 98% of total carbonate in the pH 8.0–8.5 range. Figure 21.1 shows a graphical representation of the system.

Alkalinity in the form of carbonate activity is the primary buffering system in water, and a large factor in determining the buffering of the mash and wort. Generally, it raises mash pH, which can cause problems for reasons already described. However, this alkalinity can also prevent the mash pH from going too low due to mashing with dark specialty malts, which are naturally acidic. Low levels of carbonate alkalinity (~50 ppm) will also prevent the beer from tasting watery. Higher alkalinity levels (100–150 ppm) can buffer the pH of dark beer styles, such as porter and stout, and prevent them from tasting acrid.

## WHAT IS A BUFFER?

A buffer is a chemical compound that dissociates or associates in response to a change in pH, and thereby resists or counteracts that change in pH. For example, bicarbonate is the most common buffer in drinking water. Bicarbonate ($HCO_3^-$) will react by absorbing a hydrogen ion in acidic conditions to form carbonic acid ($H_2CO_3$), or will give off a hydrogen ion in alkaline conditions to form carbonate ($CO_3^{2-}$), according to the reaction:

$$H_2CO_3 \longleftrightarrow HCO_3^- \longleftrightarrow CO_3^{2-}$$

### Sulfate

- Atomic weight = 96
- Equivalent weight = 48
- Recommended range = 50–150 ppm for most beer styles
- = 150–400 ppm for pale ale and IPA styles

Sulfate ions ($SO_4^{2-}$) accentuate the hop character in beer, making it taste drier and crisper. Sulfate ions work better with some hop families and beer styles than others. Pale ales and IPA styles will often use 150 ppm or higher to make the bitterness very assertive and dry. Other styles, such as German *helles* and Kölsch, must use lower amounts (50–75 ppm) to prevent the bitterness from dominating the soft malt palate of the beer. Noble hops and sulfate ions do not get along well together—sulfate ions seem to bring out very sulfuric flavors in these hops.

Sulfate ions do not affect mash or wort pH.

## Chloride

- Atomic weight = 35.4
- Equivalent weight = 35.4
- Recommended range = 50–150 ppm

The chloride ion (Cl$^-$) is not the same as elemental chlorine, and chloride in brewing water will not cause the same sort of flavor problems residual chlorine or chloramine can. Chloride accentuates the maltiness of the beer, making the beer taste fuller and sweeter. Chloride levels greater than 150 ppm can make the beer taste underattenuated or cloying. High concentrations (>300 ppm) of chloride can hurt clarity and flavor stability and cause corrosion of your equipment.

Chloride ions do not affect mash or wort pH.

## Sodium

- Atomic weight = 22.9
- Equivalent weight = 22.9
- Recommended range = 0–100 ppm

Sodium ions (Na$^+$) accentuate the malt character of beer, rather like chloride, but you can easily oversalt your beer. Sodium ions can taste metallic in combination with other ions such as calcium, magnesium, bicarbonate, and sulfate. The sodium ion level should be kept to less than 100 ppm in the brewing water. Reverse osmosis is just about the only way to remove sodium from water.

Sodium ions do not affect mash or wort pH.

## Water pH

The pH of the water is not important. The pH of the water describes the chemical activity of the water; but that is not what we are interested in—we are interested in the chemical activity and pH of the mash. The only information the water pH can give us is an indication of the balance between the hardness and alkalinity in the water. You can have the same water pH for two completely different mineral profiles that would have very different effects on the mash pH, and subsequently the beer. A high water pH does indicate that the water has more alkalinity than acidity, but it doesn't tell you what that alkalinity level actually is. It is like the difference between having two children or two gorillas on a seesaw (teeter-totter); children are a lot easier to move (less buffering power) than gorillas. Water pH is useful to know, but you cannot make any real judgments about the suitability of your brewing water, or the effect on mash pH, because of it. It is the actual mineral concentrations in the water that are important; as we will find out in the next section about residual alkalinity.

## Residual Alkalinity is the Cornerstone of Mash pH

Water hardness helps lower mash pH, and water alkalinity raises mash pH. The combined effect of this is called *residual alkalinity* (RA). Residual alkalinity is the most important parameter for understanding how your water is going to affect your beer, because it has a direct effect on the mash pH, and therefore the wort pH and beer pH. Perhaps Dr. David Taylor said it best,

*The key point for control of pH throughout the brewing process is during mashing. This is due to the major influence that can be exerted at this stage on the content and format of the buffer systems that will operate subsequently in the wort and beer. (Taylor, 1990, p.135)*

## TABLE 21.1—CONVERSION FACTORS FOR ION CONCENTRATIONS

| To get | From | Do this |
|--------|------|---------|
| $Ca^{2+}$ mEq/L | $Ca^{2+}$ ppm | Divide by 20. |
| $Ca^{2+}$ ppm | $Ca^{2+}$ mEq/L | Multiply by 20. |
| $Ca^{2+}$ ppm | $Ca^{2+}$ hardness as $CaCO_3$ | Divide by 50 and multiply by 20. |
| $Ca^{2+}$ hardness as $CaCO_3$ | $Ca^{2+}$ ppm | Divide by 20 and multiply by 50. |
| $Ca^{2+}$ hardness as $CaCO_3$ | Total hardness as $CaCO_3$ | Estimate by assuming that the calcium is often about ⅘ of the total hardness. |
| $CaCO_3$ mEq/L | $CaCO_3$ ppm | Divide by 50. |
| Alkalinity as $CaCO_3$ | $HCO_3^-$ ppm | Divide by 61 and multiply by 50. |
| $HCO_3^-$ mEq/L | $HCO_3^-$ ppm | Divide by 61. |
| $HCO_3^-$ ppm | Alkalinity as $CaCO_3$ | Divide by 50 and multiply by 61. |
| $Mg^{2+}$ mEq/L | $Mg^{2+}$ ppm | Divide by 12.1. |
| $Mg^{2+}$ ppm | $Mg^{2+}$ mEq/L | Multiply by 12.1. |
| $Mg^{2+}$ ppm | $Mg^{2+}$ hardness as $CaCO_3$ | Divide by 50 and multiply by 12.1. |
| $Mg^{2+}$ hardness as $CaCO_3$ | $Mg^{2+}$ ppm | Divide by 12.1 and multiply by 50. |
| $Mg^{2+}$ hardness as $CaCO_3$ | Total hardness as $CaCO_3$ | Estimate by assuming the magnesium is often about ⅕ of the total hardness. |
| Total hardness as $CaCO_3$ | $Ca^{2+}$ as $CaCO_3$ and $Mg^{2+}$ as $CaCO_3$ | Add them. |

*$Ca^{2+}$, calcium ion; $CaCO_3$, calcium carbonate; $HCO_3^-$, bicarbonate ion; mEq/L, milliequivalents per liter; $Mg^{2+}$, magnesium ion; ppm, parts per million.*

## What Does the Mash pH Do?

As was discussed in chapter 16, the mash pH is the second most important factor, after temperature, for enzyme activity in the mash. The activity rate of enzymes in the mash typically follows a bell-shaped curve about their pH and temperature optima. Enzymes can be denatured by either high temperatures, too high or too low pH, or a combination of both. Denaturing means that the three-dimensional shape of the enzyme irreversibly changes, with the result that the enzyme can no longer interact with the substrate it is supposed to act on, such as starches or proteins. Therefore, mash pH can have a significant impact on starch conversion, soluble and total nitrogen levels, lauterability, fermentability, and yield.

And, as Dr. Taylor notes, it sets up the composition of the wort, including sugars, proteins, and pH buffers, that will affect the performance of all the brewing steps that come after.

## Optimum Mash pH

What is the optimum mash pH? Well, it depends. It depends on what parameters you are trying to optimize. There are many different enzyme groups in the mash, each category has several types, and each of which have their own optima. The temperature and pH optima for best protein degradation is not necessarily going to give you the best lauterability or fermentability. The best fermentability may not give you the best yield. Each of these properties will have its own optima. For example, Briggs *et al.* (1981, 279) state that the best yield is at pH 5.45–5.65, while Bamforth and Simpson (1995) found that it was at pH 5.55–6.05 (note both ranges are measured on room temperature samples). In *Technology Brewing and Malting* (Kunze 2014, 227), it is stated that the optimum for both alpha- and beta-amylase activity is pH 5.5–5.6, but then later the author says that for the best beer the pH should be 5.2, although I am speculating that the author may have been speaking in the context of pale lagers.

The optimum can vary with temperature in two different ways. First, the pH of every aqueous solution will change with temperature; higher temperatures cause molecules to dissociate more, meaning more ionization occurs and therefore the pH changes. In the case of wort, the generally accepted change is a decrease of 0.3 pH between room temperature and mash temperature (temperature goes up = mash pH goes down). But consider why the mash pH is changing: the constituents of that wort—the ions, peptides, and acids—are changing their activity in response to temperature. It stands to reason that different worts from different styles of beer will experience different amounts of change due to temperature. Experiments have indicated that the pH change between room temperature (68–77°F [20–25°C]) and mash temperature (149–158°F [65–70°C]) can range from 0.27–0.38.[1] Therefore, brewing scientists make a point of either declaring what temperature the pH readings were taken at, or measuring the pH at room temperature. All of the professional brewing organizations have standard procedures for measuring mash, wort, and beer pH that specify to measure it at "room temperature," which is frequently taken to mean 68–77°F (20–25°C).

Second, the optimum pH for a particular process can vary with the temperature the process takes place at. For example, Kolbach and Haase[2] determined that the pH optima for starch extraction (from a particular malt) changed non-linearly with temperature; the pH optima changed from 4.9–5.3 at 122°F (50°C), to 5.1–5.5 at 140°F (60°C), and to 5.5–5.9 at 149°F (65°C).

Finally, aside from temperature, the optimum can vary with the agent you are using to adjust the mash pH. Bamforth and Simpson (1995) found that optimum pH for best lauterability changed

---

[1]   A.J. deLange (unpublished data); Hansen and Geurts (2015).
[2]   P. Kolbach and G.W. Haase, *Wochenschrift für Brauerei* 56 (1939): 143.

from 4.4–4.6 when the mash pH was adjusted using sulfuric acid, and to 5.1–5.5 when adjusted using calcium chloride. It is reasonable to assume that a combination of the two agents would have an optimum somewhere in between.

So, what is the answer? There is no single correct answer, it is a range that depends on several factors and the brewer's priorities. Generally, when we ask what is the optimum mash pH, we are thinking in terms of optimum yield. Notable references agree that the optimum mash pH for yield (conversion and extraction) would seem to fall in the 5.5–5.8 range. However, notable sources also agree that better beer flavor, clarity, and flavor stability are obtained at lower mash pHs. Therefore, the commonly agreed target range for mash pH is 5.2–5.6, as measured at room temperature (68–77°F [20–25°C]). Your specific target within that range will depend on your brewing processes and the recipe or style. Experience[3] has demonstrated that pale beers seem to taste better with lower mash pH of 5.2–5.4, while dark beers tend to taste better with a slightly higher mash pH of 5.4–5.6.

## MEASURING MASH PH, PH METERS, AND ATC

Mash pH should be measured by dipping out a small sample of wort, somewhere between 5 and 10 min. into the saccharification rest, and pouring it onto a shallow saucer to cool to near room temperature. The sample should be measured with a calibrated pH meter that has a resolution of two decimal places and an accuracy of at least ±0.05. As brewers, we need to be able to reliably differentiate between values of 0.1 pH in our mashes. pH test papers are really only accurate enough to tell you that you are between 5 and 6. Cheap pH meters with a resolution of one decimal place are not much better.

It is important to understand that the mash pH will decrease slowly throughout the hour-long mash, that conversion happens relatively quickly, and that enzymes can be rapidly denatured by a combination of non-optimum pH and high temperature. In other words, attempting to fix a mash pH that is too low or too high may take too long to be effective. You will probably do better by planning ahead with water or recipe changes for the next batch. Use what you have made, it is still wort, just understand that it is probably not the best it can be.

Temperature affects a pH measurement in two ways: 1) the electrochemical response of the probe in the pH meter changes with temperature, and 2) the chemical activity of the wort changes with temperature. Both of these factors change the reading you obtain with the pH meter.

When you calibrate a pH meter to measure acidic solutions, you use buffered calibration solutions with pH values of 4 and 7. Those solutions are buffered to be most accurate to their declared pH at room temperature, 60–77°F (20–25°C).

Modern pH meters have a feature called automatic temperature compensation (ATC). This feature compensates for the change in electrochemical response of the probe caused by higher temperatures. In other words, ATC maintains calibration of the probe away from the calibration temperature. However, ATC does nothing to account for any actual change in pH of the solution due to temperature. Thus, brewers have standardized mash, wort, and beer pH measurements and citations to be at room temperature unless explicitly stated otherwise.

You can measure the pH of your mash hot, but understand that doing so will shorten the useful life of the probe and that the amount of change in pH with temperature (roughly 0.3 on the pH scale) will vary between individual worts. The standard procedure for brewers is to measure and report the pH of mash, wort, or beer with respect to room temperature unless explicitly stated otherwise.

---

[3] John Palmer, Colin Kaminski, Martin Brungard, and AJ DeLange.

## Controlling Mash pH

How can we control mash pH? One way is by adjusting the residual alkalinity (RA) of our brewing water using salt and acid additions.

In 1953, German brewing scientist Paul Kolbach determined that 3.5 equivalents (Eq) of calcium ions react with phosphate ions from the malt to release 1 Eq of hydrogen ions that can "neutralize" 1 Eq of alkalinity in water. Magnesium ions, the other ion species contributing to water hardness, also work in the same way as calcium ions, but to a lesser extent, needing 7 Eq to neutralize 1 Eq of alkalinity in water. Neutralization of alkalinity by the chemical reactivity of calcium and magnesium ions does not require enzyme activity or an acid rest. The remaining alkalinity that is not neutralized is the residual alkalinity. As mentioned earlier, residual alkalinity is the result of the competing effects of water hardness and alkalinity on mash pH. When compared to a mash conducted with distilled water (RA = 0), a positive RA value increases mash pH, and a negative RA value (i.e., more hardness than alkalinity) decreases mash pH.

On a per volume basis, this relationship between hardness and alkalinity can be expressed as:

$$RA \text{ (in mEq/L)} = \text{alkalinity (in mEq/L)} - [(\frac{Ca^{2+} \text{ mEq/L}}{3.5}) + (\frac{Mg^{2+} \text{ mEq/L}}{7})]$$

where mEq/L is defined as milliequivalents per liter.

The same relationship can also be expressed in terms of ppm as $CaCO_3$:

$$RA \text{ (in ppm as } CaCO_3) = \text{total alkalinity ppm as } CaCO_3 -$$
$$[(\frac{Ca^{2+} \text{ hardness ppm as } CaCO_3}{3.5}) + (\frac{Mg^{2+} \text{ hardness ppm as } CaCO_3}{7})]$$

And it can be expressed in the units most often seen on municipal water reports, "Total alkalinity as $CaCO_3$," "Ca ppm," and "Mg ppm."

$$RA \text{ (in ppm as } CaCO_3) = \text{total alkalinity ppm as } CaCO_3 - [(\frac{Ca^{2+} \text{ ppm}}{1.4}) + (\frac{Mg^{2+} \text{ ppm}}{1.7})]$$

However, Kolbach's experiments measured the effects of brewing water on wort pH (i.e., after the mash and sparge), using a 1.048 wort with a water-to-grist ratio of about 5 L/kg. More recent experiments[4] indicate that the hardness equivalency factor (i.e., the 3.5 Eq value for calcium ions) in the mash not only varies with water-to-grist ratio but also with different types of malt, typically ranging from 2.2 to 3.5. Using a factor of 3.5 represents a more conservative value, that is, the rate at which pH decreases with increasing hardness is less.

The amount of residual alkalinity it takes to change the pH by 0.1 also varies with water-to-grist ratio and malt type. My own experiments with sample mashes using two different base malts and a porter recipe (giving 55 data points including replicates), indicated that the amount of change in RA to effect a change in pH of 0.1 ranged from 25 ppm as $CaCO_3$ at $Rv$ = 8 L/kg (4 qt./lb.) to 100 ppm at $Rv$ = 2 L/kg (1 qt./lb.). The relationship was inversely proportional to $Rv$, and very nearly equal to 200/$Rv$. Therefore, a typical number for 0.1 pH change at $Rv$ = 4 L/kg (2 qt./lb.) is about 50 ppm as $CaCO_3$, according to my data.[5] Your results may vary.

The Kolbach equation as given above is commonly used to predict the effects of residual alka-

---

[4]    Troester (2009), Barth and Zaman (2015), and Palmer (2016).
[5]    Palmer (2016). Recall that $Rv$ is the volume per weight water-to-grist ratio, as detailed in chapter 17.

linity changes to the mash pH, and has been incorporated in many brewing software applications. Future studies may build on Troester's and Barth's work to make a more refined model, but until then the basic Kolbach equation serves to help us understand the basic mechanism we can use to adjust the mash pH. In essence, if you paddle this way, you go forwards; if you paddle that way, you go backwards. The important things to realize are that a) you are in a boat, b) you have a paddle, and c) you can move the boat. Don't get hung up worrying too much about how fast you are going.

## Adjusting Residual Alkalinity

Residual alkalinity is the combined effects of water alkalinity and water hardness on the mash pH. We can adjust either of these quantities with the addition of salts, such as calcium sulfate or sodium bicarbonate, or by adding strong acids or bases.

The following table lists the various brewing salts that can be used to adjust RA. These salts can be added to the brewing water beforehand, or added to the mash. Generally, I prefer to add them to the brewing water beforehand, but some, such as gypsum, can be difficult to dissolve (they need lots of stirring) unless they are added later on to the mash where the pH is lower.

### Adjusting Residual Alkalinity with Salt Additions

Salt additions are easy to calculate. Look at your water report to determine the starting concentration of each of the six ions in your source water. You may want to test the water yourself to get more current numbers (see sidebar, "Testing Your Water," for advice). Table 21.2 gives a list of brewing salts and their ion contributions in both grams per liter (g/L) and grams per gallon (g/gal.). An addition of 1 g/gal. means that you would be adding 1 g of a brewing salt for each gallon of water you are adjusting, so if you have 10 gal. of water then you would be adding 10 g of the brewing salt. The concentrations simply add together. For example, and referring to table 21.2, if your water report lists the calcium ion concentration as 36 ppm, adding 2 g/gal. of calcium chloride ($CaCl_2$) would give you a final concentration of:

> For $x$ g/gal. added, and taking values from table 21.2:
>
> final concn. = ($x$ × concn. at 1 g/gal.) + existing concn.
>
> $$= (2 \times 72 \text{ ppm Ca}^{2+}) + 36 \text{ ppm}$$
>
> $$= 180 \text{ ppm Ca}^{2+} \text{ (which is probably a bit too much).}$$

Also, keep in mind that by adding calcium chloride you will be increasing the chloride concentration as well. Keeping with the above example of adding 2 g/gal. of calcium chloride and assuming, for example, you had 50 ppm of chloride in the water to start with, your total chloride concentration would be:

> final concn. $Cl^-$ = (2 × 127.4 ppm $Cl^-$) + 50
>
> $$= 304.8 \text{ ppm Cl}^- \text{ (which is too much).}$$

Note in the above example that we are adding 2 g of calcium chloride for every gallon of water. Alternatively, you may wish to only add a total of 2 g of calcium chloride to the entire 7 gal. batch of water. The calculation is much the same, except you need to divide the ion concentration by the appropriate volume when adding. For example, when we add 2 g of calcium chloride to source water with existing concentrations of 36 ppm $Ca^{2+}$ and 50 ppm of $Cl^-$:

For calcium:

$$[(2 \times 72 \text{ ppm})/7] + 36 \text{ ppm} = 20.57 + 36$$

$$= 56.57 \text{ ppm } Ca^{2+}$$

For chloride:

$$[(2 \times 127.4 \text{ ppm})/7] + 50 \text{ ppm} = 36.4 + 50$$

$$= 86.4 \text{ ppm } Cl^-$$

This process is easier when using de-mineralized water (either by distillation or reverse osmosis) as your water source because the mineral ion concentrations are essentially zero. If you dilute your tap water with distilled water, the mineral ion concentrations are effectively divided by whatever proportion of distilled water you use. For example, if you are diluting using one part distilled to one part tap water (i.e., a 50/50 dilution), then your tap water mineral concentrations are cut in half, and any salt additions you make to that diluted brewing source water would be added onto those new values.

## TESTING YOUR WATER

There are several products on the market for testing the mineral content of your water supply, but only a few are made for testing the specific ranges found in brewing water. My recommendation is to use the BrewLab® kit from Lamotte Company, which I helped develop specifically for this purpose. Each kit can test a minimum of 50 water samples and is fast and easy to use. BrewLab kits are available from many homebrew supply shops, or from the company's website at www.lamotte.com.

Another option is to send a sample to your local water utility or to Ward Laboratories, Inc. in Kearney, Nebraska. Their web page has instructions at www.wardlab.com.

## TABLE 21.2—ION CONTRIBUTIONS FOR BREWING SALT ADDITIONS

| Brewing salt | Concn. at 1 g salt per liter (1 g/L) | Concn. at 1 g salt per gallon (1 g/gal.) | Comments |
|---|---|---|---|
| **Calcium carbonate** $CaCO_3$ mw = 100 eqw = 50 | 400 ppm $Ca^{2+}$ 600 ppm $CO_3^{2-}$ 20 mEq/L alkalinity | 106 ppm $Ca^{2+}$ 158 ppm $CO_3^{2-}$ 5.3 mEq/gal. alkalinity | Will not dissolve in water, but will in mash. However, does not raise mash pH effectively. |
| **Sodium bicarbonate** $NaHCO_3$ mw = 84 eqw $Na^+$ = 23 eqw $HCO_3^-$ = 61 | 273.7 ppm $Na^+$ 710.5 ppm $HCO_3^-$ 11.8 mEq/L alkalinity | 72.3 ppm $Na^+$ 188 ppm $HCO_3^-$ 3.04 mEq/gal. alkalinity 150 ppm as $CaCO_3$ | Dissolves readily, and effective at raising alkalinity and mash pH. Best added to water before mashing. |

*Table 21.2 (continued)*

| Brewing salt | Concn. at 1 g salt per liter (1 g/L) | Concn. at 1 g salt per gallon (1 g/gal.) | Comments |
|---|---|---|---|
| **Calcium hydroxide** $Ca(OH)_2$ mw = 74.1 eqw $Ca^{2+}$ = 20 eqw $OH^-$ = 17 | 541 ppm $Ca^{2+}$ 459 ppm $OH^-$ 27 mEq/L alkalinity $\Delta RA$ = 19.3 mEq/L | 143 ppm $Ca^{2+}$ 121 ppm $OH^-$ 7.1 mEq/gal. alkalinity $\Delta RA$ = 5.1 mEq/gal. 255 ppm as $CaCO_3$ | Dissolves readily in water. Food-grade pickling lime seems to be good purity. |
| **Sodium hydroxide** NaOH mw = 40 eqw $Na^+$ = 23 eqw $OH^-$ = 17 | 575 ppm $Na^+$ 425 ppm $OH^-$ 25 mEq/L alkalinity | 152 ppm $Na^+$ 112.3 ppm $OH^-$ 6.6 mEq/gal. alkalinity 330 ppm as $CaCO_3$ | Dissolves readily. Raises alkalinity. Caution! Hazardous material! Consult material safety data sheet (MSDS) before use. |
| **Calcium sulfate** $CaSO_4 \cdot 2H_2O$ mw = 172.2 eqw $Ca^{2+}$ = 20 eqw $SO_4^{2-}$ = 48 | 232.8 ppm $Ca^{2+}$ 557.7 ppm $SO_4^{2-}$ | 61.5 ppm $Ca^{2+}$ 147.4 ppm $SO_4^{2-}$ | Saturation at room temperature is about 3 g/L. Stir vigorously. Lowers mash pH. |
| **Magnesium sulfate** $MgSO_4 \cdot 7H_2O$ mw = 246.5 eqw $Mg^{2+}$ = 12.1 eqw $SO_4^{2-}$ = 48 | 98.6 ppm $Mg^{2+}$ 389.6 ppm $SO_4^{2-}$ | 26.0 ppm $Mg^{2+}$ 102.9 ppm $SO_4^{2-}$ | Saturation at room temperature is about 255 g/L. Lowers mash pH. |
| **Calcium chloride** $CaCl_2 \cdot 2H_2O$ mw = 147.0 eqw $Ca^{2+}$ = 20 eqw $Cl^-$ = 35.4 | 272.6 ppm $Ca^{2+}$ 482.3 ppm $Cl^-$ | 72.0 ppm $Ca^{2+}$ 127.4 ppm $Cl^-$ | Dissolves readily. Lowers mash pH. Food-grade salt may not be high purity. |
| **Magnesium chloride** $MgCl_2 \cdot 6H_2O$ mw = 203.3 eqw $Mg^{2+}$ = 12.1 eqw $Cl^-$ = 35.4 | 119.5 ppm $Mg^{2+}$ 348.7 ppm $Cl^-$ | 31.6 ppm $Mg^{2+}$ 92.1 ppm $Cl^-$ | Dissolves readily. Lowers mash pH. Food-grade salt may not be high purity. |
| **Sodium chloride** NaCl eqw $Na^+$ = 23 eqw $Cl^-$ = 35.4 | 393.4 ppm $Na^+$ 606.6 ppm $Cl^-$ | 103.9 ppm $Na^+$ 160.3 ppm $Cl^-$ | Dissolves readily. Avoid iodized salt and anti-caking agents. |

*Notes: The contributions are listed equivalently as ppm (mg/L), mEq/L, mEq/gal., or ppm as $CaCO_3$, as applicable.*

*mw = molar weight in g/mole; eqw = equivalent weight in g/Eq.*

### Reducing Alkalinity with Acid

You can also use acids to reduce alkalinity and lower mash pH. Acids reduce alkalinity by supplying protons (which are hydrogen ions, H$^+$) to convert all of the carbonate and bicarbonate ions in solution to carbonic acid, and subsequently to carbon dioxide ($CO_2$). Note that the $CO_2$ in solution must be removed from the water for the reaction to be complete. Most of the $CO_2$ will escape as gas when the water is heated and stirred. In commercial breweries, where tuns are usually closed or contained, this $CO_2$ is actively removed by agitation, bubbling with forced air or steam, or spraying. This is in order to prevent $CO_2$ accumulating in enclosed piping or tankage where it can cause severe corrosion problems.

Acid additions to reduce alkalinity are quite simple to calculate if you work in terms of milliequivalents (mEq). The total alkalinity in ppm as $CaCO_3$ is easily converted to milliequivalents per liter by dividing by calcium chloride's equivalent weight, which is 50. For example, if the total alkalinity of the water is 125 ppm as $CaCO_3$, that would equal 2.5 mEq/L. Adding 1 mEq of acid per liter would therefore reduce the total alkalinity to 1.5 mEq/L, or 75 ppm as $CaCO_3$. However, there are two aspects to consider, specifically:

1. How many milliliters of acid is 1 mEq?
2. What flavor effect does the acid have?

The answer to the first question is that the amount of acid required depends on the specific acid, and table 18.3 lists dilutions for creating 1 N solutions of several common acids. The "N" stands for *normality*, which denotes Eq/L. A 1 N solution means that 1 L of the solution supplies 1 Eq, therefore, 1 mL supplies 1 mEq. So if you needed to reduce the alkalinity of 20 L (5.3 gal.) of water by 50 ppm as $CaCO_3$ (i.e., 1 mEq/L), you would add 20 mL of your 1 N acid solution.

When considering the second question, the point is that the acid reaction will replace each equivalent of alkalinity with an equivalent of that acid's anion (the negatively charged ion, e.g., Cl$^-$). In the case of hydrochloric acid (HCl), this is one way of boosting the chloride level without adding more calcium or magnesium ions. Lactic and citric acids, however, have anions with a characteristic flavor and brewers need to consider whether those flavors will have an impact on the flavor of their beer. Choosing the acid and the final alkalinity is a matter of recipe formulation and may take a bit of trial-and-error. The method presented here reduces the alkalinity without regard to the pH. The purpose is to reduce the level of residual alkalinity, not to arrive at a specific mash pH. For more information on acid additions, see *Water: A Comprehensive Guide for Brewers,* by Palmer and Kaminski (2013).

### Pre-Boiling to Reduce Alkalinity

Boiling has been used for hundreds of years to reduce the alkalinity and hardness of water by decarbonation. Broadly, the way it works is that the rise in water temperature changes the saturation point of all the carbonate species in solution. First the dissolved $CO_2$ fizzes out of the water due to the rise in temperature. This removal of $CO_2$ unbalances the equilibrium between the bicarbonate and carbonic acid in solution, which causes conversion of bicarbonate ions to carbonic acid and aqueous $CO_2$, and in so doing consumes protons. This raises the pH. The increased pH causes some of the remaining bicarbonate ions to convert to carbonate ions. This results in saturation with respect to calcium carbonate, which precipitates. Since the formation of calcium carbonate

## TABLE 21.3—PREPARING 1 N SOLUTIONS OF COMMON ACIDS

| Acid | w/w % | Density | Molarity | mL of acid to prepare 1 L of 1 N soln | Anion contribution per mEq/L |
|---|---|---|---|---|---|
| Hydrochloric | 10 | 1.048 | 2.9 | 348 | 35.4 ppm Cl⁻ |
| | 31 | 1.18 | 12.0 | 83.5 | 35.4 ppm Cl⁻ |
| Phosphoric | 10 | 1.05 | 1.1 | 935[a] | 96 ppm $H_2PO_4^-$ |
| | 85 | 1.69 | 14.7 | 68[a] | 96 ppm $H_2PO_4^-$ |
| Lactic | 88 | 1.209 | 11.8 | 84.7 | 89 ppm lactate (~400 ppm flavor threshold) |
| Citric | (powder) | ... | ... | 96 grams in 1 L | 96 ppm citrate (~150 ppm flavor threshold) |

[a] Phosphoric acid is approximately monoprotic at mash pH.

Note: It is important to understand that the procedure is to dilute the prescribed volume up to a total volume of 1 liter. For example, 348 mL of 10% (w/w) hydrochloric acid would be poured into a volumetric flask, and water added to the flask to make exactly 1 liter.

**Take care!** Concentrated acids need to be added to a large volume of water that is already in the flask, to avoid exothermic splashing, before being topped up with additional water to the final volume.

## ACID SAFETY: WORDS OF CAUTION FOR STRONG ACIDS AND BASES

Always add acid to water and NEVER add water to acid. It sounds silly but, "Do what you ought'er, add acid to water," may help you avoid an acid splash. Do not get concentrated acid (of any kind) on your skin. Dilute acids (~10% w/w) are usually not hazardous, but will still cause irritation and can be damaging to your eyes. Wearing eye protection is always recommended.

Concentrated acids should not be handled without proper training. You should read and follow the recommendations for personal protective equipment (e.g., gloves, goggles, and apron) on the material safety data sheet (MSDS).

Finally, the acids and bases that you use to treat brewing water should be food grade. While "food grade" doesn't have a precise definition, it generally means that the substance does not contain hazardous or toxic impurities and is generally recognized as safe or suitable for human consumption in accordance with the US Food and Drug Administration. Off-the-shelf acids from the hardware or auto-parts store, for example, might contain heavy metals or other impurities. Be careful what you buy and what you use.

removes carbonate ions, this causes a further imbalance in the equilibrium, and (in accordance with Le Chatelier's principle) more bicarbonate ions convert to carbonate ions. Calcium carbonate continues to precipitate until either the calcium ion concentration or bicarbonate ion concentration is about 1 mEq/L, either 20 ppm or 61 ppm, respectively.

This last milliequivalent per liter as $CaCO_3$ does not precipitate and stays in solution. The calcium carbonate that has precipitated exists as microcrystals in suspension, which will eventually grow heavy enough to settle out. According to historical brewing texts,[6] the water would typically be boiled for a half hour to allow the $CO_2$ to be fully purged by the steam, and would then be allowed to settle overnight, leaving a white layer of calcium carbonate precipitate on the bottom of the kettle. The reduced-alkalinity water would then be decanted off the sediment for use as brewing liquor. This reaction is limited to water with moderate to high alkalinity, because it requires at least 1 mEq/L of calcium ions (20 ppm) and 1 mEq/L of bicarbonate ions (61 ppm) for the reaction to occur. In fact, unless the water contains significantly more than 1 mEq/L of each, the force driving of the chemical reaction will be low, and the method will be less effective at lowering alkalinity. The higher you can make the pH in this process, that is, the more dissolved $CO_2$ you force out of solution, the more alkalinity you can ultimately remove. This is usually accomplished by bubbling air or steam through the water to agitate it until the pH is 8.5 or more. Water can typically be decarbonated down to a total alkalinity of 50 ppm as $CaCO_3$ without too much trouble, but the concentration of calcium ions in the water is often a limiting factor. The residual calcium ion concentration after softening by boiling can be calculated with the following equation:[7]

$$[Ca^{2+}]_f = [Ca^{2+}]_i - \{\frac{[HCO_3^-]_i - [HCO_3^-]_f}{3.05}\}$$

where,

ion concentrations are denoted by square brackets, e.g, $[Ca^{2+}]$ is concentration of $Ca^{2+}$;

all of the initial (i) and final (f) concentrations are in ppm;

and the factor 3.05 accounts for the conversion between bicarbonate ion and calcium ion equivalents.

The quantity, $[HCO_3^-]_f$ is the estimate of the final bicarbonate concentration, and is assumed to be at least 61 ppm of $HCO_3^-$, which is equivalent to total alkalinity of 50 ppm as $CaCO_3$ at roughly pH 8.3. This final bicarbonate concentration of 61 ppm is based on ideal conditions. Using a more conservative value, such as 80 ppm bicarbonate, may be more realistic, allowing for conditions that are not ideal and where the reaction does not proceed to completion. A final bicarbonate concentration between 61 and 80 ppm is more typical when calcium ion concentration is not the limiting factor.

This reaction works best when the total hardness ppm as $CaCO_3$ is greater than the total alkalinity ppm as $CaCO_3$. It also works best if the permanent hardness is greater than the temporary hardness, meaning that there is plenty of calcium in solution to fuel the reaction and nearly all of the bicarbonate ions can be removed, except for that final 1 mEq/L (50 ppm as $CaCO_3$). The best way to increase the permanent-to-temporary hardness ratio is to add calcium sulfate or calcium chloride to the hot water. The salts will also act as nucleation sites and help evolve the $CO_2$ as gas.

This same reaction also happens when the water is merely heated to strike temperature. The difference is that the $CO_2$ may not come out of solution until the water is stirred. This means any calcium carbonate that does precipitate will not have time to settle out, but will remain in suspension and be carried into the mash. The reaction kinetics for calcium carbonate additions to the mash

---

[6]   For example, see Sykes and Ling (1907, 410).
[7]   Martin Brungard, "Water Knowledge," *Bru'n Water,* last updated January 8, 2015, http://sites.google.com/site/brunwater /water-knowledge.

are very slow, typically taking over two hours to change the mash pH. Boiling usually doesn't affect magnesium ion levels because magnesium carbonate is much more soluble than calcium carbonate.

For example, look at the water profile for the city of Munich:

| Ionic species | $Ca^{2+}$ | $Mg^{2+}$ | $HCO_3^-$ | $Na^+$ | $Cl^-$ | $SO_4^{2-}$ | RA |
|---|---|---|---|---|---|---|---|
| Concn. (ppm) | 77 | 17 | 295 | 4 | 8 | 18 | 177 |

The calcium and bicarbonate concentrations are high, and the residual alkalinity is comparable to Dublin, a city famous for its stouts. How did Munich become renowned for brewing pale Munich helles and amber Oktoberfest beers? One answer may be the decrease in alkalinity from pre-boiling the water. The approximate water composition after boiling would be:

| Ionic species | $Ca^{2+}$ | $Mg^{2+}$ | $HCO_3^-$ | $Na^+$ | $Cl^-$ | $SO_4^{2-}$ | RA |
|---|---|---|---|---|---|---|---|
| Concn. (ppm) | 20 | 17 | 120 | 4 | 8 | 18 | 74 |

Boiling and decanting changes the residual alkalinity of the water from 177 to 74 ppm, and this may be what enables the brewing of lighter-colored styles. Another factor may be the use of acidulated malt, as explained in the sidebar "Using Acidulated Malt."

## USING ACIDULATED MALT

Acidulated malt, or *Sauermalz*, is a base malt that has been sprayed with lactic acid produced by naturally occurring lactic acid bacteria that live on barley malt. Thus, acidulated malt meets the requirements of the Reinheitsgebot and it is how German brewers have been acidifying their mashes for a very long time. It is only used for lowering the mash pH. Several malting companies make an acid malt, but the most well-known brand is probably Weyermann® Acidulated Malt. Weyermann specifies that using their acid malt as 1% by weight of the total grain bill will lower the mash pH by 0.1 pH. The company also claims that it can be used to make up to 10% of the grain bill without imparting sour lactic flavor to the beer, although a recipe for a sour Berliner weissbier only calls for 8%. I estimate that 4%–5% may be used in any beer style to reduce the residual alkalinity and lower the mash pH without lactic character being evident in the final beer.

## Mash pH is Water Chemistry plus Malt Chemistry

The mash pH is the equilibrium between the water chemistry (the residual alkalinity) and the malt chemistry. What is malt chemistry? Well, I'm glad you asked. Every malt contains phosphates, proteins, and acids that affect the chemistry of the mash. Every malt when mashed in distilled water causes a drop in pH from the water's starting pH of 7.0 to some nominal baseline value. I would like to say that each type of malt has a characteristic baseline pH, but unfortunately this is only generally true. There is a lot of variation in baseline pH (±0.2) between different barley varieties and different maltsters for the same malt type. In base malts, the pH range is typically 5.7–6.0, averaging about 5.8. Specialty malts typically have lower baseline pH values, in the range of 4.0–5.4, with the darkest caramel malts and roast malts having the lowest. In addition, malts with similar baseline pH values can have different buffering power, meaning the capacity to resist changes in pH differs between these malts even if their baseline pH is similar.

The buffering capacity of a malt is measured as milliequivalents per kilogram of alkalinity or acidity to a target pH. Table 21.4 lists the baseline pH and buffering power for several malts to three target mash pH levels. But it must be understood that these are single data points from experiments conducted by Briess Malt & Ingredients Co. in 2015. This data is presented for information only and is not intended to represent the pH values for those malt types in general, nor to represent typical pH values for Briess' malt types, and certainly not be representative of products from other maltsters. Baseline pH and buffering capacity are not properties that any maltster specifies or controls; modification, yield, and color are much more important to the brewer. The purpose of showing the data in table 21.4 is so that you gain a general understanding of the way different malt types will affect the mash pH. Positive numbers pull mash pH up from the target, negative numbers pull the mash pH down from the target.

How do we use this sort of information? The mash pH is the equilibrium between the residual alkalinity of the water (positive or negative) and the various malts in the grain bill. To understand how this works, assume that we have chosen a target mash pH of 5.4. If we take the values reported in table 21.4, for example, the Pilsen Malt base malt, which has a baseline pH of 5.8 and buffering capacity of 9.2 mEq/kg with respect to the target pH. The base malt represents an alkaline quantity with respect to our mash target, meaning we have to overcome it's buffering capacity to get down to our target mash pH. The Pilsen Malt's total buffering capacity is 9.2 mEq/kg multiplied by its weight in the grain bill. Next, let's assume the recipe has a couple of specialty malts in the grain bill. The first is Caramel Malt 60L with a baseline pH of 4.8 and a buffering capacity of −43.4 mEq/kg with respect to the pH target of 5.4. The second is Chocolate Malt with a baseline pH of 4.7 and a buffering capacity of −39.7 mEq/kg to the target pH of 5.4. These specialty malts represent acidic quantities with respect to our mash pH, meaning we have to overcome their buffering capacities to come up to our target pH of 5.4. The buffering capacity of the brewing water is either alkaline or acidic depending on its RA—alkaline if positive, acidic if negative. Adding up these factors will give you a surplus or deficit of milliequivalents that can be adjusted to zero by adding a strong acid or base, and thus reach your target mash pH. Or to put it another way:

[Base malt alkalinity (mEq/kg) × Weight malt (kg)]
+
[Specialty malt acidity (mEq/kg) × weight malt (kg)]
+
[Water residual alkalinity (mEq/L) × volume (L)]
= 0*

* If the various buffer capacities add up to zero, it means that you will hit your target mash pH. If the number is plus or minus a couple of milliequivalents, you can add the difference with a strong acid or base to hit your mash pH target.

The point of all this is that you understand how the residual alkalinity of the brewing water and the grain bill come together to determine the mash pH. Base malts are alkaline, specialty malts are acidic, and the water can be either. We will explore the water effect more in chapter 22 when we look at famous brewing waters.

## TABLE 21.4—EXAMPLES OF BASELINE PH AND ALKALINITY/ACIDITY FOR BRIESS MALTS

| Briess malt type | Color (°L) | Baseline pH | mEq/kg of alkalinity (+) or acidity (−) to target mash pH of: | | |
|---|---|---|---|---|---|
| | | | 5.2 | 5.4 | 5.6 |
| Pilsen Malt | 1.5 | 5.8 | 14.8 | 9.2 | 4.5 |
| Brewers Malt | 1.9 | 5.6 | 11.1 | 5.0 | 0 |
| Pale Ale Malt | 2.9 | 5.6 | 9.7 | 3.4 | 0 |
| Red Wheat Malt | 2.8 | 5.8 | 18.9 | 11.1 | 4.8 |
| Goldpils® Vienna | 3.5 | 5.6 | 13.6 | 7.2 | 1.6 |
| Ashburne® Mild Malt | 4.4 | 5.5 | 9.4 | 2.6 | −1.7 |
| Bonlander® Munich Malt | 12 | 5.5 | 9.0 | 2.6 | 0 |
| Aromatic Munich Malt | 16 | 5.4 | 5.0 | 0 | −3.3 |
| Victory® Malt | 28 | 5.4 | 3.4 | −0.9 | −7.8 |
| Special Roast Malt | 40 | 5.1 | −2.2 | −13 | −20.0 |
| Caramel Malt 20L | 19 | 5.1 | −6.0 | −21.3 | −26.5 |
| Caramel Malt 40L | 40 | 4.8 | −20.7 | −30.4 | −41.2 |
| Caramel Malt 60L | 61 | 4.8 | −28.9 | −43.4 | −74.2 |
| Caramel Malt 80L | 80 | 4.7 | −32.8 | −45.1 | −58.7 |
| Caramel Malt 120L | 120 | 4.5 | −48.8 | −60.4 | −72.6 |
| Extra Special Malt | 126 | 4.6 | −51.7 | −67.5 | −85.5 |
| Roasted Barley | 292 | 4.7 | −26.6 | −38.6 | −59.2 |
| Chocolate Malt | 416 | 4.7 | −29.3 | −39.7 | −51.7 |
| Dark Chocolate Malt | 458 | 4.5 | −38.8 | −48.8 | −57.9 |
| Black Malt | 471 | 4.6 | −27.0 | −36.0 | −44.1 |
| Black Barley | 514 | 4.2 | −38.2 | −51.8 | −66.0 |

*Source: R. Hansen and J. Geurts, "Specialty malt acidity," Proceedings of MBAA Annual Conference, Jacksonville, FL, October 8–10, 2015.*

### Sparge Water Adjustment

Lowering the pH of your sparge water is usually unnecessary, provided your sparge water has sufficient calcium. The phosphates in the mash will react with the calcium and buffer the pH until the wort gravity falls below 1.012. The key is to have enough calcium ions in the water. According to Taylor (1990), a calcium ion concentration of 100–200 ppm is needed to prevent pH rise during sparging, although his data indicate that 50 ppm is sufficient to prevent excessive rise, greater than 5.6, with final total wort pH of 5.4 before the boil. This same trend was observed in a pilot batch of

IPA trialed in 2014 at the Ballast Point brewery in Little Italy, San Diego, CA, during the sparging of two mashes using only pale ale malt. The first trial batch used a calcium ion content of 54 ppm and had an RA of −44 ppm as $CaCO_3$. The wort pH rose from 5.39 for the first runnings to 5.54 for the last runnings, and had a wort pH of 5.15 after the boil. The second trial batch had a calcium ion content of 120 ppm and an RA of −99 ppm as $CaCO_3$. The wort pH did not rise during sparging, in fact, it went from 5.35 to 5.33, and the wort pH was 5.07 after the boil.

Generally, if you are building your water from distilled, or adjusting your water to have good calcium ion levels and reduced alkalinity, you will be able to sparge with the same water and not experience excessive pH rise and associated astringency in your beer. If you are not adjusting your water with brewing salts, or you are brewing with highly alkaline water, you may want to acidify it before using it for sparging. A rule of thumb when neutralizing the total alkalinity this way is to divide the total alkalinity, or bicarbonate concentration, by the equivalent weight to determine the number of acid equivalents to add. Be aware that acidifying with phosphoric acid can precipitate the calcium in your water, because the calcium ions react with phosphates in solution to form calcium phosphate. Acidifying to lower pH with phosphoric acid, such as 5.2–5.5, can prevent this. Alternatively, you can use other acids, such as hydrochloric, lactic, and citric acid, that do not have this problem. See appendix B in the *Water* book by Palmer and Kaminski (2013), for more information on this topic.

If you are mashing a dark beer with several specialty malts using an adjusted water with sufficient calcium and appropriate levels of residual alkalinity, you should be able to sparge with the adjusted water without much rise in pH. Alternatively, you could use a lower RA water (such as distilled water) to sparge with, and make up any water profile differences with brewing salt additions to the boil kettle before starting the boil.

## The Mash pH Sets Up the Beer pH

The mash pH continually decreases during the time of the mash as more calcium phosphate precipitate is formed, and as enzyme activity breaks down the malt and releases amino acids and other buffering compounds. The mash pH tends to drop by 0.2 during the course of the mash, so the wort pH in the boil kettle may be between 5.0 and 5.4. See figure 21.2 for a schematic of this process.

Wort pH also drops by about 0.3 during the boil. This is due to protein denaturing and coagulation, hop alpha-acid additions, and Maillard reactions. Hop isomerization is increased by higher wort pH, but higher wort pH is not a good way to improve hop utilization, because it results in a harsher, coarser bitterness that is not at all like the hop flavor you want.

The mash pH can affect how much the beer pH drops during fermentation as well. Proteolytic enzyme activity is favored by lower mash pH. There is a trade-off between the amino acids produced during the mash, associated yeast growth, and wort buffering. As yeast ferments the wort, it takes up amino acids and other buffering compounds as nutrients and excretes protons as part of the yeast's normal biochemical activity. With single-infusion mashes, there is typically a low amount of FAN in the wort, which allows the yeast to grow normally (i.e., not too fast or too slow) but does not give the wort a significant amount of buffering power, so the wort pH drops by about 0.5 pH during fermentation. With the addition of a typical protein rest during the mash more FAN is put into the wort, but not enough to significantly affect yeast growth; however, the extra FAN does give more buffering power and the pH drop is the least seen with this method (about 0.3). With a very long protein rest a lot of FAN is put into the wort, significantly boosting yeast growth and producing a

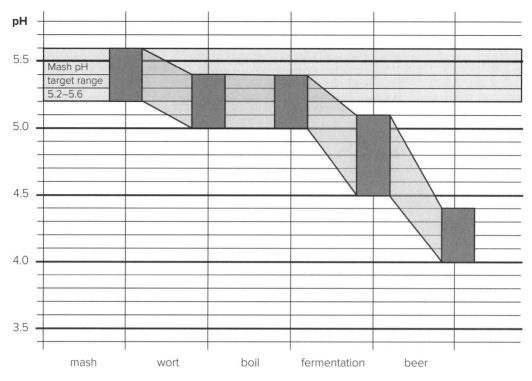

**Figure 21.2. Change in Wort pH During the Brewing Process.** A schematic of the decrease in pH throughout the brewing process.

moderate pH drop. Obviously, temperature is a bigger factor for enzymatic activity, but mash pH contributes as well (fig. 21.3).

The mash pH sets the course for the final beer pH. Table 21.5 shows four examples of how the beer pH was changed by varying the residual alkalinity across four different recipes. Fermentation does have a large moderating effect on pH, but the trend is clear.

## Beer pH Controls Beer Flavor

Beer pH can be adjusted after fermentation, but the results never seem quite as harmonious as if it had been adjusted in the mash or before fermentation. Generally, beer pH (not including sour beers) ranges between 4.0 and 4.7.

Pale beers seem to taste better with a lower beer pH, between 4.0 and 4.4. A lower beer pH makes the malt character in pale beers taste brighter and the hop character more refined. As the beer pH approaches 4.0, the character becomes sharp and crisp, as it approaches 4.4 it becomes softer.

Dark beer styles, particularly those with a high percentage of dark roasted malts, seem to taste better with a higher pH, between 4.3 and 4.7. The higher beer pH softens the malt acidity and opens up the specialty malt flavors, achieving more complexity. Dark beers with a lower pH tend to taste more one dimensional, with the flavors sharpening to a narrower range. This will not taste bad, but the flavor description tends to be summed up in one word, either roasty,

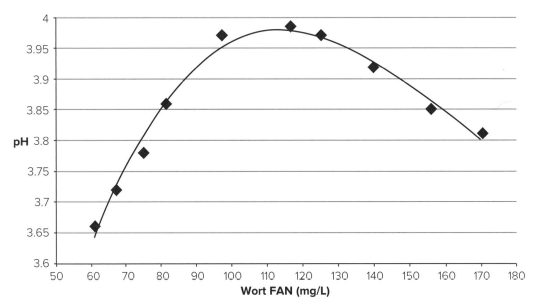

**Figure 21.3. Beer pH vs. Wort FAN.** The curve shows the change in beer pH associated with increases in the wort FAN content. (Reproduced from Taylor (1990, fig. 6) with permission.)

## TABLE 21.5—MASH PH AND BEER PH CHANGES FOLLOWING ADJUSTMENT OF RESIDUAL ALKALINITY WITH CALCIUM SALTS AND SODIUM BICARBONATE

| Reference | Beer style | RA ppm as CaCO$_3$ | Mash pH | Beer pH |
|---|---|---|---|---|
| Denver NHC 2007[a] | Pale ale | 200 | 6.1 | 4.7 |
| | Pale ale | −45 | 5.5 | 4.5 |
| | Stout | 200 | 5.4 | 4.6 |
| | Stout | −45 | 4.9 | 4.2 |
| Grand Rapids NHC 2014[b] | IPA | 140 | 5.5 | 4.6 |
| | IPA | −35 | 5.25 | 4.5 |
| 3rd Congresso Technico dos Cervejeiros Artensanais 2016[c] | Stout | 140 | 6.2 | 4.8 |
| | Stout | 14 | 5.3 | 4.1 |

[a] *John Palmer, Rick Bobbit, and Scott Jackson, "Beer Color and Residual Alkalinity: A Practical Example," Presentation at the AHA National Homebrewers Conference, Denver, June 21–23, 2007.*

[b] *John Palmer and Adam Mills, "Brewing Water Effects on Beer Flavor," Presentation at the AHA National Homebrewers Conference, Grand Rapids, MI, June 12–14, 2014.*

[c] *John Palmer, Ronaldo Dutra-Ferreira, and Malcolm Frazer, "Live Brulosophy Experiment," 3rd Congresso Técnico dos Cervejeiros Artensanais, Florianopolis, Brazil, April 22–23, 2016.*

coffee, or chocolate. Stouts and porters at the higher end of the pH range tend to taste more layered, and judges' descriptions will broaden to, for example, roasty with hints of caramel and chocolate. Of course, the flavors that appear depend on the recipe.

Every beer should have a flavor portfolio that includes:

- malt flavors and aromas;
- hop flavors and aromas;
- yeast flavors and aromas.

Every beer recipe has an ideal beer pH where all these flavors are best expressed. If you can't taste or smell every ingredient, or every ingredient favorably, then you probably haven't achieved the optimum pH for that beer. You can try adjusting the pH of the final beer with acid and base additions, but it usually works better to adjust the mash pH by 0.1–0.2 and let that change work its way through to the beer, taking the sort of approach best summed up as, "An ounce of prevention is worth a pound of cure." The brewer's art is finding that particular pH for a recipe and maintaining it, batch to batch, season to season. Achieving that particular beer pH starts with the water chemistry and the mash pH.

# Adjusting Water for Style

## Famous Brewing Waters and Their Beers

The best way to begin explaining how to adjust water for different styles is to describe two of the world's most famous beers and their brewing waters (table 22.1). The city of Pilsen in the Czech Republic was the birthplace of the Pilsener style of beer. A Pilsener is a golden, clear lager with a soft bitter finish. The water of Pilsen is very soft, free of most minerals and very low in alkalinity.

The other beer to consider is the famous stout of Ireland. The ground water in Dublin is high in bicarbonate, and it has a fair amount of dissolved calcium, but not enough to balance the bicarbonate. The result is hard, alkaline water with a lot of buffering power. The high alkalinity of the water makes it difficult to produce light pale beers that are not harsh tasting. Consequently, this region is renowned for producing outstanding dark beers. The highly roasted black barley used in making stout adds acidity to the mash. The natural acidity of these grains counteracts the alkalinity in the water, lowering the mash pH into the desired range.

The brewing waters of Pilsen and Dublin had a direct effect on the color and flavor of the beer they could best produce. Before you brew your first all-grain beer, you should get a water analysis from your local water utility (see chapter 8 for an example) and look at the mineral profile to get a better idea of the styles of beer that may work best for you. Calculate the residual alkalinity of your water. If it is high, you will probably have better success with dark beers. If it is medium, then pale ales and brown ales may work best. If it is low (or negative), then pale ales and light lagers may work best for you, and you will probably need to add alkalinity to the water to brew darker styles.

Table 22.1 lists famous brewing cities and the beers that made them famous. By looking at the city and its beer, you can gain an appreciation for how malt chemistry and water chemistry combine to create the beer. Descriptions of each city's trademark beer style follow.

## TABLE 22.1—WATER PROFILES FROM FAMOUS BREWING CITIES

| City / Style | $Ca^{2+}$ | $Mg^{2+}$ | $HCO_3^-$ | $Na^+$ | $Cl^-$ | $SO_4^{2-}$ | RA[a] |
|---|---|---|---|---|---|---|---|
| Pilsen / Pilsener | 7 | 2 | 16 | 2 | 6 | 8 | 7 |
| Dublin / Dry stout | 120 | 4 | 315 | 12 | 19 | 55 | 170 |
| Dortmund / Export lager | 230 | 15 | 235 | 40 | 130 | 330 | 20 |
| Vienna / Vienna lager | 75 | 15 | 225 | 10 | 15 | 60 | 122 |
| Munich / Dunkel | 77 | 17 | 295 | 4 | 8 | 18 | 177 |
| London / Porter | 70 | 6 | 166 | 15 | 38 | 40 | 82 |
| Edinburgh / Scottish ale | 100 | 20 | 285 | 55 | 50 | 140 | 150 |
| Burton-on-Trent / India pale ale | 275 | 40 | 270 | 25 | 35 | 610 | 1 |

Source: Martin Brungard, "Water Knowledge," Bru'n Water, last updated January 8, 2015, http://sites.google.com/site/brunwater/water-knowledge.

[a] Ion concentrations in parts per million (ppm), except RA values, which are given in "ppm as $CaCO_3$" calculated from the profile and rounded to the nearest whole number.

*Pilsen.* The water that supplies the city of Pilsen has very low hardness and alkalinity. The beer of Pilsen has a soft, rich maltiness like fresh bread. The lack of sulfate in the water provides a mellow hop bitterness that does not overpower the soft maltiness. The soft water made it easy for brewers to add calcium chloride salts to help them achieve a lower mash pH.[1]

*Dublin.* Famous for its dry stout, Dublin has the highest bicarbonate concentration of any city in the British Isles, and Irish brewers embrace it with one of the darkest, maltiest styles in the beer world. The low levels of sodium, chloride, and sulfate create an unobtrusive hop bitterness to properly balance all of the malt.

*Dortmund.* Another city famous for pale lagers, Dortmund export lager has less hop character than a Pilsener, with a more assertive malt character due to the higher level of all minerals. The balance of minerals is very similar to Vienna, but Dortmund export is bolder, drier, and lighter in color. The sodium and chloride bring out a rich roundness to the malt character.

---

[1] Dr. Ludwig Narziss, personal communication, 2014.

*Vienna.* The water of Vienna is similar to Dortmund, but lacks the calcium levels to balance the carbonates, and has less sodium and chloride for flavor. Attempts to imitate Dortmund export failed miserably until a percentage of toasted malt was added to balance the mash, and Vienna's famous red-amber lagers were born.

*Munich.* The smooth flavors of the dunkels, bocks, and Oktoberfests of the Munich region show the success of using dark malts to balance the temporary hardness in the mash. The relatively low sulfate content provides for a mellow hop bitterness that lets the malt flavor dominate. Pre-boiling to get rid of the temporary hardness lowers the residual alkalinity considerably.

*London.* The water of the upper stretches of the River Thames helped produce the famous porter beers of London. The high bicarbonate dictated the use of toasted and dark malts to lower the pH of the mash, but the water could be pre-boiled to precipitate the temporary hardness and gypsum added back to allow production of pale ales as well.

*Edinburgh.* Think of misty Scottish evenings and you think of strong Scotch ale—dark ruby highlights, a sweet malty beer with a mellow hop finish. Edinburgh's water is similar to London's but with a bit more bicarbonate and sulfate, making a beer that can embrace a heavier malt body while using less hops to achieve balance.

*Burton-on-Trent.* Compared to London, the calcium and sulfate are remarkably high in the water of Burton-on-Trent, but the hardness and alkalinity are balanced to nearly the degree of Pilsen. The high level of sulfate and low level of sodium produce an assertive, clean hop bitterness. Compared to the ales of London, Burton ales are paler, but much more bitter and drier.

## The Dogma of Virgin Water

Grapes are harvested once a year and that wine is fermented and aged until it tastes good. Vintners will ferment the grapes from a particular part of their vineyard in the hopes of expressing that location's *terroir*, the effect of the place on the wine. The wine from a particular year is called a vintage, and the mark of a good vintage is its individuality. Brewers don't brew this way. Classic beer styles may start out as a marriage of local ingredients and local water, but like any good marriage the two parts will grow into each other and the beer will evolve until the brewer is satisfied with it. (After that, God willing, it won't change. The beer from this week has to taste the same as the beer from last week, and the beer brewed next year should taste the same as well.) Brewers have been adjusting water for hundreds, if not thousands, of years in order to keep their beers consistent. These adjustments can be as simple as digging a new well closer to or further away from a river, or adding salts and acids to the brewing water.

A common misconception is that to brew a classic example of a particular style you have to return to the roots of the style and brew with the exact water that it was founded on. This is not true. Water has a profound effect on the character of beer, no question there. The issue is the dogma of virgin water—the premise that brewers did not adjust the source water to improve the character of their beer—and we know that this is false. Great beers are made, not found.

Martin Brungard gave an insightful presentation at the 2013 National Homebrewer's Conference in Philadelphia, titled "Historic Water," in which he explained how the geology of the Burton region affected the brewing of the local beer. The character of Burton's water changed significantly over the years as the city grew and more breweries started drawing on the groundwater. The city sits on sand and gravel along the River Trent, which is underlain by a Mercia mudstone layer containing high levels of gypsum and chalk, which in turn is underlain by sandstone. The result of this geology is that the local well water is a mixture of low-mineral river water and high-mineral groundwater, and the proportions of each in the water supply depend on the distance to the river and depth of the well. The permeability of the sand and gravel is much higher than that of the mudstone layer, such that, as the residents and breweries of Burton drew more water from the wells near the river, the pollution in the river got into the well water. The solution? Dig deeper wells further from the river. The result? Higher mineral concentrations than what the brewers were used to, and the need to measure parameters like "sulfate of lime" and "carbonate of lime" to ensure the consistency of their beer. There was actually a lawsuit in 1830 filed by the brewers of Burton-upon-Trent against a local public-interest group for claiming that the brewers were adulterating their beers with chemicals. The public may not have understood the reasons for acid and mineral additions to water, but the brewers certainly did. Ask a modern Burton brewer about their water, and they will tell you that they hate brewing with it: "It's too minerally!" Mitch Steele's book on IPA has more information about Burton water.[2]

To brew a great example of a classic style, you need to understand how the local water may have played its part in creating the style, but also understand what may have been done in spite of it. After all, it is the beer that we are drinking, not the water. Look at the description of the style and try to understand what the style's water requirements might be. Is the beer hoppy, malty, dark, or pale? Is the body of the beer firm or soft? Use these insights to help you brew the perfect example of the style, but don't be caged by them.

## THE PROBLEM WITH WATER QUALITY REPORTS

Unfortunately, the water reports that you find on the Internet are frequently misleading. Water quality reports are intended to show that a water source is safe to drink, not necessarily whether it is good to brew with. The ions important for brewing (calcium, chloride, and others discussed in chapters 8 and 21) are often listed at the end of the report and are probably the average of several measurements over the course of a year. This is fine if the water source is fairly constant, but can be very misleading if it changes during the year.

All natural water sources are electrically neutral— the sums of the cation charges ($Ca^{2+}$, $Mg^{2+}$, $Na^+$) and anion charges (total alkalinity, $SO_4^{2-}$, and $Cl^-$), measured as milliequivents per liter (mEq/L), should be equal (or nearly equal, if you allow for small errors). You can evaluate the composition given by a water quality report by dividing all of the ion concentrations by their equivalent weight to convert each concentration to milliequivalents per liter. It is best if the alkalinity is listed as "total alkalinity as $CaCO_3$," because if the listing is only for bicarbonate ($HCO_3^-$), you would need to calculate the total alkalinity based on the water pH. If the water pH is 8.0–8.6, then the bicarbonate is about 97% of the total alkalinity and you can use the simple conversion factor,

$$\text{Total alkalinity as } CaCO_3 = (50/61) \times [HCO_3^-] \text{ in ppm}$$

to convert between the two. But it's easier to just use the total alkalinity.

Actually, there are a lot more ions dissolved in water than just the six brewing ions, so it is not unusual for there to be a 1 mEq difference between the two sums. If the difference between the sums is greater than 1 mEq, then the cause of the discrepancy might be because the report was compiled from different locations around the city, or at different times throughout the year. If the difference is large, that is, more than 3 mEq, then that composition is unbalanced and probably not an accurate description of a naturally occurring water. A brewer trying to build that water profile would have a difficult time getting the same concentrations, but replication is not the actual goal; the goal is a good tasting beer in that style.

---

[2]    Steele (2012).

## Using Minerals to Accentuate Flavor

Let's begin by quickly summarizing the effects of the six ions that are important for brewing water:

- Calcium, magnesium, and bicarbonate (or total alkalinity) affect pH via residual alkalinity.
- Calcium, magnesium, and bicarbonate do not affect flavor but they do affect the mouthfeel of the beer, transitioning from soft to firm with increasing amounts.
- Sulfate, chloride, and sodium do not affect residual alkalinity or beer pH. Not at all.
- Sulfate directly affects beer flavor by accentuating the bitterness and hop character, and by creating a drier mouthfeel.
- Chloride and sodium directly affect beer flavor by accentuating the sweetness and fullness of the malt character.
- Too much sodium can taste minerally or metallic.

When you think about a beer, think about it as food. What kind of beer is it? Is it a sweet beer or a bitter beer? Is the body heavy or light? Is the character of the beer firm or soft? Let's look at the two ways that water profile seasoning can accentuate these characteristics. The first is the sulfate-to-chloride ratio. The second is the total amount of seasoning you add.

### Sulfate-to-Chloride Ratio

The sulfate-to-chloride ratio affects the balance of hoppiness versus maltiness in the final beer. Sulfate emphasizes the bitterness character and dries out the palate, making the beer taste drier and crisper. Chloride accentuates the maltiness of the beer, making the beer taste sweeter, fuller, and rounder.

The sulfate-to-chloride ratio is a benchmark that you can use to understand how your mineral profile and salt additions will affect the flavor of your beer, but it is not magic. You cannot fix a bad recipe or off-flavors by adjusting the seasoning. Lots of sulfate will not make a malty beer bitter, and lots of chloride will not make a bitter beer sweet. These salts can only accentuate what is already there. The following are some guidelines for using the sulfate-to-chloride ratio:

- The recommended range for sulfate is 50–150 ppm for most beers, 150–400 ppm for pale ales and IPAs.
- The recommended range for chloride is 50–150 ppm for all beer styles.
- The recommended range for the sulfate-to-chloride ratio is 5:1 to 0.5:1. IPAs can go as high as 9:1, but be careful not to oversalt the beer.
- Do not exceed 0.5:1 with excess chloride, the beer will start tasting oddly sweet, like it is underattenuated.
- Do not try to make a beer that is both hoppy and malty by maximizing both; the beer will just taste minerally. I recommend keeping the sum of sulfate and chloride below 500 ppm.
- A 3:1 ratio of 15 ppm sulfate to 5 ppm chloride does not taste the same as a 3:1 ratio of 150 ppm sulfate to 50 ppm chloride. The total amount of minerals also has an effect.

## Total Dissolved Solids

The other way that minerals season the beer is by the total dissolved solids effect, that is, the total sum of minerals in the water. Beer brewed without any minerals (i.e., distilled water) tends to taste watery. The total amount of minerals in the water gives structure to the beer flavor, making it soft, medium, or firm. I like to compare this to the difference between delicate French seasoning and heavy Cajun seasoning. A good way to describe this effect with beer is to look at three similar beer styles: Bohemian Pilsener, German Pilsner, and Dortmunder export.

Bohemian Pilsener is a rich, malty beer backed by a large, soft bitterness; no sharp edges. It has a smooth finish that is balanced between the malt and hops. German Pilsner is crisper and has a more bitter, hop-forward character, followed by a clean maltiness and dry finish. German Pilsner is a beer defined by clean edges. Dortmunder export is a robust beer that balances a rich malt character and firm, dry bitterness. It tastes like a bigger beer than the first two styles. However, the recipes for these three styles are very similar: OG about 1.050, Pilsner base malt, noble hops, and a lager yeast strain.

What makes these three styles different from each other? Well, the main difference is the brewing water in each of the cities that developed them. Look at table 22.2, which gives the mineral profile and total dissolved solids (TDS) for those minerals for each city.

### TABLE 22.2—WATER PROFILE FOR THE CITIES OF PILSEN, MUNICH, AND DORTMUND

| City | $Ca^{2+}$ | $Mg^{2+}$ | $Na^+$ | $Cl^-$ | $SO_4^{2-}$ | $HCO_3^-$ | TDS |
|---|---|---|---|---|---|---|---|
| Pilsen | 7 | 2 | 2 | 6 | 8 | 16 | 50 |
| Munich | 77 | 17 | 4 | 8 | 18 | 295 | 419 |
| Dortmund | 230 | 15 | 40 | 130 | 330 | 235 | 1000 |

Notes: Concentrations in parts per million (ppm). Munich and Dortmund profiles are after pre-boiling to eliminate alkalinity due to temporary hardness. TDS, total dissolved solids.

Source: (Palmer and Kaminski 2013, 143–4).

Looking at the TDS value for each of these water profiles, two things stand out: 1) the sulfate-to-chloride ratios are essentially 1:1, 2:1, and 3:1 for Pilsen, Munich, and Dortmund, respectively; and, more significantly to my point, is 2) there is nearly an order of magnitude increase in the TDS between each of the water profiles, from Pilsen to Munich, and then Munich to Dortmund. This is where most of the character difference in these three beers comes from—the mineral backbone of the brewing water.

The TDS value is not a brewing target, it is a zone. Bear in mind that TDS is the result of the decisions you make for salt additions. For example, you may want to add 200 ppm of sulfate to your water to punch up the hop character for an IPA, but when you look at the total mineral profile afterwards, the TDS may be over 1000 ppm, which is probably too much. Use TDS as a reference, where high values (>800 ppm) will probably taste minerally, while low values (<100 ppm) will probably be watery. Generally, you will want to be more toward the middle, but that will depend in part on your taste.

## The Brew Cube

Before we look at how to make specific water adjustments to suit specific styles, it may help to take another look at the big picture. Brewing water and beer styles go hand-in-hand, but visualizing this relationship can be very murky for some people. Thus, I invented the brew cube, basically a Rubik's® cube to help you understand how the three beer style factors of color, flavor balance, and structure, relate to the three water factors of residual alkalinity, sulfate-to-chloride ratio, and calcium level. The brew cube is illustrated in figure 22.1.

The color of the style generally indicates how acidic the grain bill is, and therefore how much (or how little) residual alkalinity is necessary to achieve the target mash pH. The overall flavor balance of the style (hoppy, balanced, or malty) is influenced by the sulfate-to-chloride ratio of the water, although the actual concentrations of these ions are more important than simply the ratio. The TDS gives structure to the beer and affects how the beer presents itself to the palate; that structure can be soft, medium, or firm.

However, TDS is not the real goal, it is a frame of reference to help you view the beer as a whole. The goal is adjusting the pH and flavor balance, and that is typically done by the addition of calcium salts. Choosing your calcium ion and residual alkalinity levels has a big impact on TDS. Therefore, the calcium ion level is a better indicator than TDS of where you want to be with the brewing water, and that is why the calcium ion level is on the cube as the driver of structure.

The purpose of the brew cube is only to help you understand the big picture of how brewing water and beer styles relate. The numbers on the cube are for frame of reference only—do not use them as targets for specific styles.

The beer style is described by color, flavor balance, and structure. The brewing water is described by residual alkalinity, sulfate-to-chloride ratio, and calcium ion concentration. To use the cube, locate the intersection of the three beer descriptors and read off the corresponding water factors. For example, if you wanted to make a pale, hoppy, and firm-bodied beer, such as American pale ale, you would see that these descriptors line up with a residual alkalinity of −100 ppm,[3] a sulfate-to-chloride ratio of 4:1, and a calcium ion concentration of 150 ppm.

However, this is not the only specific water profile that you should use, it is just an example of the type of water you could use. The high and low values for calcium ion and residual alkalinity levels given on the cube are the minimums and maximums that I recommend when adjusting water with salts alone due to potentially excessive mineral levels. The three guidelines for calcium ion concentration are 50–100, 75–125, and 100–150 ppm. They are only general guidelines. The guidelines for RA overlap as well: −100 to −25, −50 to 50, and 25 to 100 ppm. The correlation between overall beer color and acidity can be misleading due to the greater acidity of red malts versus black malts (see chapter 21). The brew cube is simply a tool to help you visualize the relationship between beer style characteristics and brewing water characteristics.

---

[3]   *Note:* when you see residual alkalinity (RA) stated as ppm in this book, this is shorthand for "ppm as $CaCO_3$." Sometimes RA is given in mEq/L, but this will always be clearly stated.

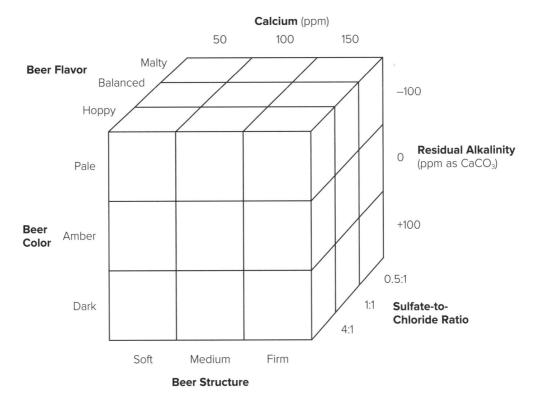

**Figure 22.1.** The brew cube describes the relationship between beer style and brewing water. If you describe the beer style by color, flavor balance, and structure, you can understand the general characteristics of the brewing water that will help you brew that style.

## Adjusting Water for Style

Let's review what we learned in the last chapter.

1. Beer is food.
2. The minerals in the water affect the beer in two ways: pH and seasoning.
3. Residual alkalinity and malt chemistry combine to determine the mash pH.
4. Mash pH sets up beer pH.
5. Beer pH controls beer flavor.

Teaching people how to adjust brewing water can be difficult. We talk a lot about the rules and guidelines for the mineral levels and tend to lose sight of the forest for the trees. The trees are the individual ion ranges and the ranges that are recommended for particular styles. The forest is composed of the mash pH, beer pH, and the resulting beer flavor. A lower mash pH leads to a lower boil pH, which affects the hop character, making it a more refined bitterness than a higher pH would. This is essential for the interplay of malt and hops in lighter beers. In order for the singular malt character to really shine, pale beers generally need a lower mash and beer pH, for example, 5.2–5.4 and 4.0–4.3, respectively.

Conversely, darker styles generally need a higher mash and beer pH, for example, 5.4–5.6 and 4.3–4.6, respectively. Darker styles usually include several specialty malts that broaden the malt character and give it complexity, and a slightly higher pH helps prevent one malt from dominating the rest.

You can adjust the mineral levels in the water as long as you don't lose sight of how these changes are going to affect the pH and how all of these adjustments suit the beer as a whole. Getting the sulfate-to-chloride ratio exactly where you want it doesn't do any good if the beer pH is out of whack, or if the beer is so minerally it's metallic. The beer will taste better with the right pH and wrong mineral levels than it will with the right mineral levels and wrong pH.

You have three tools for adjusting water to suit the style of beer. The first is adjusting pH via residual alkalinity. The second is adjusting the levels of sulfate and chloride. The third is adjusting the total dissolved solids, but this should only be addressed after adjusting the calcium ion concentration and residual alkalinity. While residual alkalinity is the biggest tool in the toolbox for adjusting water to suit a particular style, you need to remember that residual alkalinity is only a means to an end—the mash pH target is the goal.

Recommended levels for mash pH are 5.2–5.4 for pale beers, 5.3–5.5 for amber beers, and 5.4–5.6 for dark beers, but these are general guidelines. Very pale beers (straw, yellow, golden, light amber) need negative RA to help draw the mash pH down to the lower end of the mash pH range. Amber, red, and brown beers have more buffering power from the malts; these beers can accommodate 0–75 ppm RA and still be in the target mash pH range. Deep red, brown, and black beers have a lot of acidity and actually require positive RA values to prevent the mash pH from going too low.

Mash pH is also affected by the water-to-grist ratio, $Rv$ (and $R$), of the mash. When $Rv$ is higher the RA has more effect on the mash pH than when $Rv$ is lower. Think about what $Rv$ is; it is the ratio of water to malt, and therefore the ratio of water buffering capacity to malt buffering capacity. When $Rv$ is low, you may need a higher absolute value for RA (i.e., make positive RA more positive, or negative RA more negative) to drive the mash pH where you want it, but don't go overboard with it.

Any time you add brewing salts to water to build a water profile, you want to keep the final concentration of each ion within the suggested guidelines. Salt additions to change residual alkalinity can only be taken so far; they can only change the pH by one to two tenths (e.g., 5.6 to 5.4) before the mineral character becomes excessive. The grain bill has to do the rest of the work. The nomograph on the inside back cover can help you visualize how residual alkalinity and the overall color of the beer interact. Let's look at an example using the following water profile for a place known as Yourtown. *Note—all concentrations in the water profile tables that follow are in ppm unless stated otherwise:*

| City | $Ca^{2+}$ | $Mg^{2+}$ | $HCO_3^-$ | $Na^+$ | $Cl^-$ | $SO_4^{2-}$ | RA |
|------|------|------|------|------|------|------|------|
| Yourtown | 40 | 8 | 100 | 15 | 35 | 32 | 50 |

This is average water: low calcium, low sulfate and chloride, medium bicarbonate, and a moderate but positive amount of residual alkalinity (fig. 22.2). Looking at the color guide on the nomograph (*see* inside back cover), this water is best suited to an amber colored beer. You can brew with this water, but you can brew with almost any water; this water lacks purpose. Let's see what we can do with it, shall we?

## Brewing Yourtown Pale Ale

There is a wide variety of pale ale styles to choose from: some are very pale, some are maltier, some are relatively balanced between bitterness and malt, and others are very bitter indeed. Generally, the sulfate-to-chloride ratio should favor the sulfate, such as 2:1 or 4:1, although a balanced ratio of 1:1

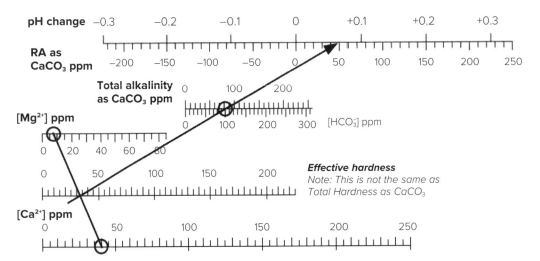

**Figure 22.2.** The RA nomograth showing the Yourtown water profile and resultant residual alkalinity.

is appropriate for some styles, such as English bitter. Generally, stronger beers can support more minerals than lighter beers. Make the total dissolved solids of the water appropriate to the character of the beer. For example, a Kölsch would probably be best with a low-to-moderate mineral profile, while a special bitter or IPA may be better with a robust mineral profile. (All of the ion contributions for the salts used in these examples can be found in table 21.2 in the previous chapter.)

*Step 1—Adjust calcium.* The first thing we should do is adjust the calcium ion ($Ca^{2+}$) concentration. The recommended range is 50–150 ppm, so let's increase the calcium ion concentration to 100 ppm, which is appropriate for a solid pale ale character. Let's use gypsum (calcium sulfate) to increase the calcium, because the additional sulfate ($SO_4^{2-}$) this contributes is also appropriate for pale ale. We need an additional 60 ppm to raise the calcium ion level to 100 ppm. Gypsum salt contributes 61.5 ppm of calcium and 147.4 ppm of sulfate when dissolved in water at a concentration of 1 g/gal. If we are treating 10 gal. of water, that would be 10 g of gypsum salt. This addition gives us a new water profile of:

| City | $Ca^{2+}$ | $Mg^{2+}$ | $HCO_3^-$ | $Na^+$ | $Cl^-$ | $SO_4^{2-}$ | RA |
|------|-----------|-----------|-----------|--------|--------|-------------|-----|
| Yourtown | 102 | 8 | 100 | 15 | 35 | 179 | 11 |

*Step 2—Review RA for your chosen style.* The brewing water now has good calcium and sulfate levels to brew a pale ale. The residual alkalinity is lower (fig. 22.3) and the full color nomograph on the inside back cover indicates that this amount of residual alkalinity would be appropriate for an amber colored beer, such as a London special bitter or East Coast American pale ale. To brew a West Coast American pale ale, which is a paler style than the East Coast variation, you would probably need to reduce the alkalinity with an acid addition to the water.

*Step 3—Adjust alkalinity with an acid addition.* Let's neutralize the bicarbonate ions ($HCO_3^-$) with an acid addition. To determine how much acid to use, convert the bicarbonate ion concentration from ppm (i.e., mg/L) to milliequivalents per liter by dividing by bicarbonate's equivalent weight of 61 mg/mEq,

$$100 \text{ mg/L} / 61 \text{ mg/mEq} = 1.64 \text{ mEq/L}.$$

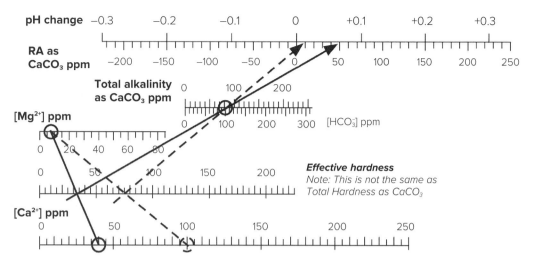

**Figure 22.3.** Notice how the calcium salt addition has raised the effective hardness and lowered the residual alkalinity compared to figure 22.2.

Using a 1N solution of lactic acid (refer to table 21.3), we would add 1.64 mL of solution for each liter of water we are treating. If we are treating 10 gal. (37.8 L), that would be 62 mL total of 1N lactic acid solution we would add. The acid solution should be added to the brewing water and stirred to mix thoroughly. The water should be stirred again later to allow the carbon dioxide that forms to fizz and vent from the water. Neutralizing the bicarbonate ions in the water gives us a new water profile with the same calcium and sulfate levels, but a lower RA (fig. 22.4), which will make for a slightly lower mash pH and beer pH that is more appropriate for a very pale beer.

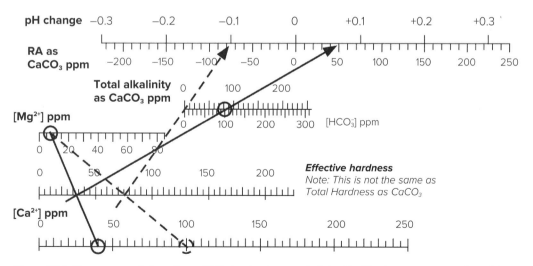

**Figure 22.4.** The nomograph from figure 22.3 showing how neutralizing the bicarbonate concentration with lactic acid has further reduced the residual alkalinity of the water. The water is now better suited to brewing a very pale colored beer than the water in figure 22.3.

| City | Ca²⁺ | Mg²⁺ | HCO₃⁻ | Na⁺ | Cl⁻ | SO₄²⁻ | RA |
|---|---|---|---|---|---|---|---|
| Yourtown | 102 | 8 | 0 | 15 | 35 | 179 | −75 |

### Brewing Yourtown Stout

A stout is a darker beer and needs more residual alkalinity to balance the acidity of the specialty malts. A malty stout should have more of a malty balance by way of a low sulfate-to-chloride ratio, either balanced at 1:1, or maltier at 0.5:1.

*Step 1—Adjust calcium.* The first thing we should do is adjust the calcium ion level. The recommended range is 50–150 ppm, so let's increase the calcium ion concentration to 50 ppm. We want to raise the calcium ion level without lowering the residual alkalinity very much. Let's use calcium chloride ($CaCl_2$) to increase the calcium so that we can raise the chloride ($Cl^-$) level as well. The initial concentration is 40 ppm, so we just need an additional 10 ppm to raise the calcium to 50 ppm. Referring again to table 21.2, calcium chloride salt contributes 72 ppm of calcium ions and 127.4 ppm of chloride when dissolved at 1 g/gal. To calculate how many grams of calcium chloride we need, we multiply the desired change in concentration by the number of gallons we need and divide by the contribution per gram:

$$(10 \text{ ppm} \times 10 \text{ gal.}) / 72 \text{ ppm/g/gal.} = 1.4 \text{ g total } CaCl_2.$$

We can now work out how much the chloride contribution will be from this calcium chloride addition:

$$(1.4 \text{ g} \times 127.4 \text{ ppm/g/gal.}) / 10 \text{ gal.} = 17.8 \text{ ppm (or about 18 ppm).}$$

This calcium chloride addition gives us a new water profile of:

| City | Ca²⁺ | Mg²⁺ | HCO₃⁻ | Na⁺ | Cl⁻ | SO₄²⁻ | RA |
|---|---|---|---|---|---|---|---|
| Yourtown | 50 | 8 | 100 | 15 | 53 | 32 | 48 |

*Step 2—Raise residual alkalinity.* Looking at the color bar on the nomograph, let's raise the residual alkalinity to 125 ppm as $CaCO_3$. Two brewing salts can be used to raise alkalinity, calcium hydroxide ($Ca(OH)_2$) and sodium bicarbonate ($NaHCO_3$). Calcium hydroxide is stronger but also raises the calcium ion concentration. Sodium bicarbonate raises the sodium ion concentration. Neither of these consequences is an issue in this case. Both the calcium and sodium ion levels will still be within the suggested ranges after additions. Let's use calcium hydroxide. Calcium hydroxide, also known as pickling lime, contributes 143 ppm calcium ions and 121 ppm hydroxide ($OH^-$) when dissolved at 1 g/gal. The equivalent weight of hydroxide is 17 mg/mEq, so each gram of calcium hydroxide contributes:

$$121 \text{ ppm} / 17 = 7.1 \text{ mEq/L of } OH^-.$$

To convert that to ppm as $CaCO_3$, multiply it by 50 mg/mEq (i.e., $7.1 \times 50 = 355$ mg/L). So each gram of calcium hydroxide added per gallon of water contributes 143 ppm of calcium ions, and 355 ppm as $CaCO_3$ alkalinity.

This means that we will have to calculate the residual alkalinity all over again. Fortunately, table 21.2 contains the ΔRA number for calcium hydroxide as well, meaning that each gram per gallon will change the residual alkalinity by 5.1 mEq/L, or 255 ppm as $CaCO_3$.

Therefore, if we are going use calcium hydroxide to raise the RA from 50 ppm to 125 ppm (i.e., by 75 ppm), we need to recalculate the total calcium ion concentration, including the addition we did in Step 1 for this stout. First we need to calculate exactly how much calcium hydroxide we will be adding, and then we can work our what the calcium ion contribution from that addition will be.

$$(75 \text{ ppm} \times 10 \text{ gal.}) / 255 \text{ ppm/g/gal.} = 2.9 \text{ g total Ca(OH)}_2 \text{ (round up to 3 g)}.$$

The current calcium ion concentration is 50 ppm from step 1. We need to add the calcium ion contribution from the calcium hydroxide to that. The calcium contribution is:

$$(3 \text{ g} \times 143 \text{ ppm/g/gal}) / 10 \text{ gal.} = 42.9 \text{ ppm Ca}^{2+} \text{ added (round up to 43)},$$

$$50 \text{ ppm} + 43 \text{ ppm} = 93 \text{ ppm Ca}^{2+}.$$

Now let's calculate the new total alkalinity value. If we convert the initial bicarbonate ion concentration to total alkalinity in ppm as $CaCO_3$, it would be:

$$(100 \text{ ppm} / 61 \text{ mg/mEq}) \times 50 \text{ mg/mEq} = 82 \text{ ppm as CaCO}_3 \text{ for HCO}_3^-.$$

The total alkalinity addition from adding the 75 ppm RA as $CaCO_3$ using calcium hydroxide is:

$$(3 \text{ g} \times 355 \text{ ppm/g/gal}) / 10 \text{ gal.} = 106.5 \text{ total alkalinity from the OH}^- \text{ (round up to 107)},$$

$$82 + 107 = 189 \text{ ppm as CaCO}_3 \text{ total alkalinity}.$$

The new water profile is now:

| City | Ca$^{2+}$ | Mg$^{2+}$ | HCO$_3^-$ | Na$^+$ | Cl$^-$ | SO$_4^{2-}$ | RA |
|------|-----------|-----------|-----------|--------|--------|-------------|-----|
| Yourtown | 93 | 8 | 189 | 15 | 53 | 32 | 125 |

*Step 3—Adjust chloride and sodium to accentuate malt.* So now we have more calcium ions (but still within the guidelines) and more residual alkalinity to compensate for the dark malts, but the chloride is not very high for a robust stout like this. The sodium ion (Na$^+$) level is still low as well. We could add pure table salt (sodium chloride) to increase the sodium and chloride levels without affecting the residual alkalinity. Let's aim for a total chloride concentration of 100 ppm. Sodium chloride adds 104 ppm sodium ions and 160 ppm chloride when dissolved at 1 g/gal. We need an additional 47 ppm to bring the chloride concentration up to 100 ppm (i.e., 100 ppm – 53 ppm), so the amount of sodium chloride to add is:

$$(47 \text{ ppm} \times 10 \text{ gal.}) / 166 \text{ ppm/g/gal} = 2.8 \text{ g NaCl total (or 3 g)}.$$

The sodium ion contribution would be:

$$(3 \text{ g} \times 104 \text{ ppm/g/gal.}) / 10 \text{ gal.} = 31 \text{ ppm Na}^+.$$

The new water profile is now:

| City | Ca$^{2+}$ | Mg$^{2+}$ | HCO$_3^-$ | Na$^+$ | Cl$^-$ | SO$_4^{2-}$ | RA |
|------|-----------|-----------|-----------|--------|--------|-------------|-----|
| Yourtown / Stout | 93 | 8 | 195 | 46 | 100 | 32 | 125 |

That is a good water profile for a stout. Granted, the sulfate-to-chloride ratio is now pretty much 1:3 (in other words, 0.32:1), which exceeds the guideline of not going lower than 0.5:1, but 100 ppm chloride is comfortably below the recommended 150 ppm limit, and 32 ppm sulfate is not that far from 50 ppm, so it should be fine. It's OK to brew outside the box as long as you know where the box is and why it's there.

### Brewing Yourtown Pilsener

Invariably, every brewer wants to make a Pilsener style beer at some point. It is a difficult style to brew well because there is not a lot of flavor for any brewing flaws to hide behind. A long-standing myth about Czech Pilsener is that it is made with very soft water, which is true; but what is not mentioned was that the water was then adjusted with brewing salts to improve mash and fermentation performance, although probably at low levels. So, when adjusting Yourtown's water, we will target 50 ppm of calcium and low levels of the rest of the ions.

*Step 1—Reduce residual alkalinity.* Yourtown water has 100 ppm bicarbonate, or a total alkalinity of 82 ppm as $CaCO_3$. We need to reduce this. The best way to do that is by diluting the water with distilled water and then adding calcium chloride to bring the calcium level up. Diluting the Yourtown water 1:1 with distilled water will cut the ion concentrations in half, becoming:

| City | $Ca^{2+}$ | $Mg^{2+}$ | $HCO_3^-$ | $Na^+$ | $Cl^-$ | $SO_4^{2-}$ | RA |
|---|---|---|---|---|---|---|---|
| Yourtown | 20 | 4 | 41 | 8 | 18 | 16 | 25 |

(An alternative of course would be to build new brewing water from distilled water. While that is certainly a valid option, it is not as good an example of adjusting water as the way we are going to do it. If we were going to build the water from distilled, then using 1 g/gal. of calcium chloride would give concentrations of 72 ppm calcium ion and 127 ppm chloride, which is within the general guidelines and very easy to do.)

*Step 2—Raise calcium ion concentration.* Let's calculate how much calcium chloride is needed to bring the calcium concentration to 50 ppm. The difference is 30 ppm, so:

$$(30 \text{ ppm} \times 10 \text{ gal.}) / 72 \text{ ppm/g/gal} = 4.17 \text{ g } CaCl_2 \text{ (or 4 g)}.$$

The new chloride contribution is therefore:

$$(4 \text{ g} \times 127.4 \text{ ppm/g/gal.}) / 10 \text{ gal.} = 50.96 \text{ ppm } Cl^- \text{ (or about 51 ppm) added}.$$

This addition of calcium chloride gives us a new water profile of:

| City | $Ca^{2+}$ | $Mg^{2+}$ | $HCO_3^-$ | $Na^+$ | $Cl^-$ | $SO_4^{2-}$ | RA |
|---|---|---|---|---|---|---|---|
| Yourtown / Pilsener | 50 | 4 | 41 | 15 | 69 | 32 | 3 |

The residual alkalinity of 3 ppm as $CaCO_3$ is essentially zero, meaning that this water is very similar to distilled water for its effect on mash pH, but it has more minerals for flavor. This water will not bring the mash pH down towards 5.2, however, which is where the mash pH for this sort of style should probably be. There are two ways of achieving this target mash pH. One way would be to use acidulated malt. The other way would be to acidify the brewing water or the mash, either by adding acid to the water to neutralize the remaining alkalinity (41 ppm $HCO_3$), or by adding drops of acid to the mash itself and monitoring the results with a pH meter until the target pH is reached.

## Summary

Just as with seasoning in cooking, there is no single correct answer for the quantity and type of brewing salts (i.e., what water profile) a great beer should receive. The answer will largely depend on the other ingredients in the beer and your own tastes. Do not blindly follow a water report for a brewing city; look at the profile and characterize it in terms of what it means to the flavor of the beer. Think of the beer style as you would any other type of food and think of mash pH, wort pH, beer pH, and mineral content as the seasoning for your food. Build the water to suit the beer, and not the other way around.

# Section III
## Recipes, Experimenting, and Troubleshooting

# Some of My Favorite Beer Styles and Recipes

## A Description of Style

There are so many styles of beer, it's hard to know where to begin. There is a lot more to a style than just whether it's light or dark. Each beer style has a characteristic taste, imparted by its ingredients and brewing method. Change any single item and you have probably hopped into another style category (no pun intended). Each country, each geographic region, even each town, can have its own style of beer. In fact, you may be starting to realize by now that many beer styles originate from local brewing conditions. Access to ingredients, the local water profile, and the climate—all of these elements combine to dictate the character of the beer that the brewer can best produce. Your success and satisfaction as a homebrewer is going to depend on understanding these factors, and then using your knowledge and skill to brew beyond them.

The first place to start when defining a style is the fermentation. Next is the recipe and brewing method. Each malt and grain has a unique flavor that it contributes to the beer. The hop varieties play a part in defining the style too. One of the differences between English pale ale and American pale ale is the difference in flavor between English and American hops. Even the same variety of hop grown in different regions will produce a different character. Fuggles grown in the USA has an American character compared to the original British variety. In fact, the same hop variety grown in two different fields will have a different character. Regional differences for a variety are generally small but still noticeable.

"Beerspace" can be visualized by plotting the type of maltiness (bready, dark bread crust, caramel, cocoa, and coffee) versus the hoppy to malty balance of the beer. The trouble with actually putting this to paper is that it's a three-dimensional space with the progression of the different malty flavors on two of the axes and then the bitterness, or hoppy to malty balance, of the beer on the third. Suffice to say that you can have any combination of malty flavors with different degrees of balance between the malt and hops.

Figure 24.1 helps to visually illustrate the similarities and differences between beer styles based on the ratio of the hop bitterness to the original gravity. Note how so many of the styles overlap there in the middle and yet these are distinct styles; this is due to the multitude of combinations of malty flavors and fermentation character. The chart is based on the averages of the OG, IBU, and %ABV as listed in the Beer Judge Certification Program (BJCP) *2015 Style Guidelines*. Further description of the particular styles and their attributes can be found at bjcp.org.

To really understand styles (and probably find better recipes for them) I urge you to read other books, such as: *Brewing Classic Styles* by Jamil Zainasheff and me; *Modern Homebrew Recipes* by Gordon Strong; *Radical Brewing* and *Tasting Beer*, both by Randy Mosher; *Brewing with Wheat* and *Brew Like a Monk*, both by Stan Hieronymus; *Experimental Homebrewing* by Drew Beechum and Denny Conn; and *Home Brew Recipe Bible* by Chris Colby. This list is by no means exhaustive—there are a multitude of books written by many notable authors.

Finally, it is very important to realize that styles constantly change. Reading Ron Pattinson's book, *The Home Brewer's Guide to Vintage Beer*, quickly shows how fluid the world of beer is, as you look at the recipes for a style or class of beer across several breweries over several decades. Brewers continually have to adapt to changes in the availability of raw materials, changes in taxation, changes in customer preference, and innovation in ingredients and methods. For example, it has only been ten years since the previous edition of *How To Brew*, and yet in that time the quantity and variety of IPAs has quite figuratively exploded! But you also see these same trends for other styles throughout history. The more things change, the more they stay the same.

## Notes on Recipes

Generally, each recipe in this chapter will be presented in two versions, an extract-with-steeped-grain recipe (a.k.a. Palmer Brewing Method) and an all-grain recipe. The character of each version of a recipe is intended to be same, or nearly the same, but sometimes the defining ingredients do not really work in the extract brewing method. Base malts like Pilsner and Munich, and kilned malts like biscuit or smoked malt, require a mash to convert their starches to fermentable sugars.

Remember: mashing is basically steeping at specific temperatures to convert unfermentable starch to fermentable sugar using the enzymes in the base malts. If the malts in the recipe do not have enough diastatic power for good conversion, you can add a packet of Palmer's Instamash® brewing enzymes to the kettle to facilitate it. Look for this product at your local brewshop.

### Batch Size and Boil Gravity

The recipes in this chapter are built around two brewing methods: a) malt extract and steeping grain brewing using a 3 gal. (11.4 L) boil in a 5 gal. (20 L) kettle; or b) all-grain brewing using a full-volume wort boil in an 8 gal. (30 L) pot. The partial boil version has been calculated to be very similar to the all-grain version in terms of boil gravity and melanoidin development so that the two methods should produce the same beer.

The goal is for you to have five gallons (19 L) of clear beer to bottle or keg. Therefore, a full-volume boil starts out with about seven gallons (27 L) of wort at a lower gravity than the target OG. The wort will increase in gravity during the boil. There should be about six gallons (23 L) of wort at the target OG to go into the fermentor after boiling and decanting off the trub. Fermenting six gallons should yield about five gallons of beer after decanting or racking off the trub and yeast cake. Your exact volumes and wort losses may differ depending on your own methods and equipment, but the basis for all of these calculations has been given in previous chapters, so it should be easy to adapt any recipe to your system. Don't drive yourself crazy trying to hit the exact numbers given in the recipe; close enough is good enough. The important thing is to brew a tasty beer.

### Hop Schedules

I have attempted to keep the hop schedules the same for both versions of a recipe by manipulating the Wort A and Wort B ingredients so that the boil gravity is the same for both the extract partial boil and all-grain full-volume boil methods. I wanted to avoid any confusion with different hop quantities between different versions of the same recipe, but sometimes you will need to add a little bit more of the main bittering hop (e.g., at 60 minutes) for the all-grain option to get the same IBU after the boil.

In fact, you may need to recalculate your hop additions anyway, because the alpha acid percentage of each hop variety varies from year to year. Or, you may need to substitute a different hop variety due to availability. See the IBU calculations in chapter 5 for help in recalculating your hop quantities. But remember, this is cooking, not chemistry class; close enough is good enough. The goal is a tasty beer.

### Extract and Steeping Grain Version

These recipes use the Palmer Brewing Method (i.e., partial boil, Worts A and B) discussed in chapters 1 and 9. You will start with three gallons (11.4 L) of water in the kettle. The Wort A ingredients will increase the volume to between 3.25 and 4 gal. (12.3–15 L). Removing the steeping grains will take away about 0.5 qt. per pound of grain (1 L/kg). The volume will decrease by about 0.5 gal. (1.9 L) during a one-hour boil. The Wort B ingredients will raise that volume by about 0.25–0.4 gal. (0.95–1.5 L) so that, when added to three gallons (11.4 L) of water in the fermentor, the total volume will be about six gallons (23 L) at the recipe gravity. Don't worry if the numbers are a little bit off, it is not that important.

The extract recipes can also be used with a full-volume boil by combining Wort A and Wort B into six gallons (23 L) of water for the boil. The volume of the ingredients will bring the volume of the boil to about seven gallons (27 L).

## All-Grain Version

The all-grain version (i.e, the mashing option) for the recipes assumes an extract efficiency of 75% (see table 18.5) and seven gallons (27 L) of wort being collected and boiled to yield six gallons (23 L) into the fermentor at the target OG, to produce the same five gallons (19 L) of clear beer. Depending on your equipment and brewing efficiency, you may want to adjust the recipe amounts to compensate.

I usually mash with a water-to-grist ratio of 2 qt./lb. (4 L/kg) and use a traditional continuous sparge technique. The grain bills for the recipes are based on those methods. To use other techniques, such as BIAB, see the lautering efficiency tables in chapter 19 for help in choosing your mashing/lautering method and water-to-grist ratio.

## Mash Schedule

The temperature target given for all single infusion mashes is 153°F (67°C) because this is the generally accepted upper limit for starch gelatinization in barley. Although 149°F (65°C) is generally recommended for best fermentability, it assumes that complete gelatinization has already occurred. I recommend that you adjust your strike water to achieve 153°F (67°C) for single temperature infusion mashes, the reason being that you will be better assured of complete starch gelatinization, and if the mash cools a few degrees during the mash your fermentability will only be enhanced. If you wish to improve the fermentability of your wort over that of a single temperature infusion mash, you can use a two-step infusion for both beta- and alpha-amylase rests as described in chapter 17. Several recipes include a decoction mashing step, and those recipes use an initial mash temperature of 150°F (65°C) because the mash will be heated to the alpha-amylase rest temperature later in the schedule. The primary purpose of the decoction step is melanoidin formation, not to solubilize and convert residual starch, although that happens too.

## Recommended Water Profile

Each recipe includes a general recommendation for the brewing water as described by the brew cube in chapter 22. A more specific water profile for the style is given below that. These recommendations are most appropriate for brewing the all-grain version of the recipe. Remember that these recommendations are guidelines and the beer will not be horrible if you don't adhere to them. (However, the beer will usually be better if you do.)

Generally, the brewing water for the extract and steeping grain recipes should be either distilled or low-mineral water. The malt extract generally contains all of the other minerals necessary for the style. However, you may want to add additional sulfate or chloride salts to the water to enhance the beer flavor. If so, use the sulfate and chloride levels under the "Recommended water profile (ppm)" part for guidance, keeping in mind that there may already be a fair amount of each in the malt extract.

## Yeast Strain and Pitching Rate

I am leaving the specific yeast strain and form to your choice. The number of yeast companies and strains has grown tremendously since the previous edition and it is not reasonable to single out one over another, especially since my choice may not be the best for you. An individual yeast strain is like a horse in that you and it need to adapt to each other and your brewing environment for best performance.

The yeast pitching rate given for each recipe is based on the typical pitching rates given in chapter 7, "Yeast Management," multiplied by the nominal fermentation volume of 6 gal. (23 L). As such, it really isn't a rate, but instead the actual pitch in billions of cells. A pitch of 200 billion cells would be roughly equal to two yeast packages, depending on the specific manufacturer and form.

## THE ALE STYLES

## Wheat Beer

Wheat has nearly as long a tradition in brewing as barley. Wheat does not have a husk and is therefore trickier to mash and lauter than barley. For this reason you will rarely see 100% wheat beers; most wheat beers are 60%–70% wheat with the remainder being Pilsner malt. Wheat is not as sweet and rich tasting as barley, and I find that adding 10%–30% wheat to many beer styles can help dry out the character of the beer and make it taste lighter and less filling, but without sacrificing mouthfeel or head retention, unlike what happens when using sugar adjuncts.

In my opinion, there are four main styles of wheat beer: German *weizen*, German *sauerweiss*, Belgian *witbier*, and American wheat. Some would argue that American wheat is not a true style, that it's simply a mongrel of the other wheat styles with its head shaved, but American wheat beers were around before Prohibition and are a recognized style today, so there. But first, let's talk about German wheat beers.

### German Wheat Beers

German *weizenbier* and *hefeweizen* were the most popular of the wheat styles in the latter half of the twentieth century and are still very popular today. German wheat beer is characterized as moderate strength, low bitterness beer dominated by a fruity, spicy fermentation character. Weizenbier can be clear or hazy, but most people are more familiar with the hazy hefeweizen form. The haze comes from suspended yeast.

Decoction mashing is traditional for German wheat beers, and even though modern malts are highly modified and don't need it for modification and better yield, many home and professional breweries still swear by it. The consensus is that it adds that little something extra to the quality of the malt and fermentation character.

German wheat beers are best when they are fresh out of the fermentor; the signature clove and banana flavor and aroma will fade within a month or so. It's not that they go stale faster than other styles, but they do become less interesting.

The classic Bavarian *weissbier* has an OG range of 1.044–1.052, bitterness of 8–15 IBU, and a pale gold color of 2–6 SRM (4–12 EBC).

## Klasse Bavarian Weissbier

**Original gravity:** 1.049
**Final gravity:** 1.011
**IBU:** 11

**SRM (EBC):** 3 (6)
**ABV:** 5%

### Version: Extract and Steeping Grain

| Wort A | Gravity points | |
|---|---|---|
| 3 lb. (1.4 kg) wheat DME | 45 | |
| Boil gravity for 3 gal. (11.4 L) | 1.045 | |
| **Hop schedule** | **Boil time (min.)** | **IBUs** |
| 0.5 oz. (15 g) Mandarina Bavaria 9% AA | 30 | 11 |
| **Wort B (add at knockout)** | **Gravity points** | |
| 3 lb. (1.4 kg) wheat DME | 45 | |
| **Yeast strain** | **Pitch (billions of cells)** | **Fermentation temp.** |
| German wheat beer | 200 | 62°F (17°C) |

### Version: All-Grain

| Grain bill | Gravity points | | | | |
|---|---|---|---|---|---|
| 6 lb. (2.7 kg) wheat malt | 25 | | | | |
| 4 lb. (1.8 kg) Pilsner malt | 16 | | | | |
| 0.33 lb. (150 g) rice hulls | 0 | | | | |
| Boil gravity for 7 gal. (27 L) | 1.041 | | | | |
| **Mash schedule** | **Rest temp.** | | **Rest time (min.)** | | |
| Conversion rest – Infusion | 153°F (67°C) | | 30 | | |
| Single decoction (see fig. 17.3) | (boil) | | 20 | | |
| **Hop schedule** | **Boil time (min.)** | | **IBUs** | | |
| 0.5 oz. (15 g) Mandarina Bavaria 9% AA | 30 | | 11 | | |
| **Yeast strain** | **Pitch (billions of cells)** | | **Fermentation temp.** | | |
| German wheat beer | 200 | | 62°F (17°C) | | |
| **Recommended Water Profile (ppm)** | | | | **Brew cube:** Pale, Balanced, Soft | |
| Ca | Mg | Total alk. | $SO_4$ | Cl | RA |
| 50–100 | 10 | 0–50 | 0–50 | 50–100 | −50–0 |

### American Wheat

Wheat beer became extinct with Prohibition in the United States, and has only been revived in the last couple of decades. Today's American wheat beer is loosely modeled after German weizen, but is made with American ale yeast and not the specialized German weizenbier yeasts with their spicy, clove-like character. The noble-type hops are traditional, but citrusy American varieties also work very well. Wheat beers are usually pale gold, but *dunkel* (dark), *bock* (strong), and *dunkel weizenbock* are common variations.

The American wheat ale style varies between a German weizen made with California ale yeast and an American pale ale made out of wheat. In fact, the style guidelines for American wheat are nearly identical to those for American blonde ale, except for the emphasis on wheat instead of barley. They are typically pale straw to light amber, with only a dash of caramel malt for additional color, if at all. Most American wheat ales are only mildly bitter, between 15–30 IBUs, and have a drier, less sweet character than American blonde ale due to wheat as the primary malt instead of barley.

If you want an easy drinking, thirst-satisfying beer, look no further than American wheat beer. The OG ranges from 1.040–1.055, FG from 1.008–1.013, and the color from 3–6 SRM (6–12 EBC).

## Three Weisse Guys—American Wheat

**Original gravity:** 1.049  
**Final gravity:** 1.011  
**IBU:** 25  

**SRM (EBC):** 3 (6)  
**ABV:** 5%  

| Version: Extract and Steeping Grain | | |
|---|---|---|
| **Wort A** | **Gravity points** | |
| 3 lb. (1.4 kg) wheat DME | 45 | |
| Boil gravity for 3 gal. (11.4 L) | 1.045 | |
| **Hop schedule** | **Boil time (min.)** | **IBUs** |
| 0.5 oz. (15 g) Sterling 7% AA | 60 | 12 |
| 0.75 oz. (15 g) Liberty 4% AA | 30 | 10 |
| 1 oz. (15 g) Liberty 4% AA | Steep 15 | 3 |
| **Wort B (add at knockout)** | **Gravity points** | |
| 3 lb. (1.4 kg) wheat DME | 45 | |
| **Yeast strain** | **Pitch (billions of cells)** | **Fermentation temp.** |
| California ale yeast | 200 | 65°F (18°C) |

### Version: All-Grain

| Grain bill | Gravity points | |
|---|---|---|
| 6 lb. (2.7 kg) wheat malt | 25 | |
| 4 lb. (1.8 kg) Pilsner malt | 16 | |
| 0.33 lb. (150 g) rice hulls | 0 | |
| Boil gravity for 7 gal. (27 L) | 1.041 | |

| Mash schedule | Rest temp. | Rest time (min.) |
|---|---|---|
| Conversion rest – Infusion | 153°F (65°C) | 60 |

| Hop schedule | Boil time (min.) | IBUs |
|---|---|---|
| 0.5 oz. (15 g) Sterling 7% AA | 60 | 12 |
| 0.75 oz. (15 g) Liberty 4% AA | 30 | 10 |
| 1 oz. (15 g) Liberty 4% AA | Steep 15 | 3 |

| Yeast strain | Pitch (billions of cells) | Fermentation temp. |
|---|---|---|
| California ale yeast | 200 | 65°F (18°C) |

| Recommended Water Profile (ppm) | | | Brew cube: Pale, Balanced, Medium | | |
|---|---|---|---|---|---|
| Ca | Mg | Total alk. | SO$_4$ | Cl | RA |
| 50–100 | 10 | 0–50 | 50–100 | 50–100 | –50–0 |

## Witbier

Wheat beers also work well with spices and fruit. Witbier is a very old style of beer from Belgium that was revived by Pierre Celis in the 1960s. Historically it was said to be a sour beer, and I have included a sour version in chapter 14, "Brewing Sour Beers." American wit, while currently not a recognized BJCP style, is similar to the classic Belgian witbier, but without the sour character. For a great recipe for an unsour Belgian witbier, see Jamil's recipe for "Wittebrew" in *Brewing Classic Styles* (Zainasheff and Palmer 2007). The following recipe is a clone recipe for Coronado Brewing Co.'s Orange Avenue Wit, which is more of a modern American witbier. This beer doesn't have the starch haze of a traditional wit, but uses honey and orange peel and is very light and refreshing.

## Casual Wit—Fruit and Spice Specialty

**Original gravity:** 1.046      **SRM (EBC):** 5 (10)
**Final gravity:** 1.008      **ABV:** 5%
**IBU:** 10

### Version: Extract and Steeping Grain

| Wort A | Gravity points |
|---|---|
| 1.8 lb. (0.8 kg) wheat DME | 27 |
| 0.5 lb. (225 g) Pilsner DME | 7 |
| 0.75 lb. (340 g) CaraVienne malt (20°L) – Steeped | 5 |
| Boil gravity for 3 gal. (11.4 L) | 1.039 |

| Hop schedule | Boil time (min.) | IBUs |
|---|---|---|
| 0.36 oz. (10 g) Northern Brewer 9% AA | 60 | 10 |
| **Wort B (add at knockout)** | **Gravity points** | |
| 3 lb. (1.4 kg) pale DME | 42 | |
| 0.5 lb. (225 g) honey | 4 | |
| ~1.75 oz. (50 g) orange peel (Steep 15 minutes) | | |
| ~0.5 oz. (15 g) ground coriander (Steep 15 minutes) | | |
| **Yeast strain** | **Pitch (billions of cells)** | **Fermentation temp.** |
| American ale | 200 | 65°F (18°C) |

### Version: All-Grain

| Grain bill | Gravity points | |
|---|---|---|
| 7 lb. (3.2 kg) Pilsner malt | 27.5 | |
| 2 lb. (0.9 kg) wheat malt | 8 | |
| 0.75 lb. (340 g) CaraVienne (20°L) | 2 | |
| 0.5 lb. (225 g) honey (Steep 15 minutes) | (3) | |
| ~1.75 oz. (50 g) orange peel (Steep 15 minutes) | | |
| ~0.5 oz. (15 g) ground coriander (Steep 15 minutes) | | |
| Boil gravity for 7 gal. (27 L) | 1.038 | |
| **Mash schedule** | **Rest temp.** | **Rest time (min.)** |
| Conversion rest – Infusion | 153°F (67°C) | 60 |
| **Hop schedule** | **Boil time (min.)** | **IBUs** |
| 0.36 oz. (10 g) Northern Brewer 9% AA | 60 | 10 |
| **Yeast strain** | **Pitch (billions of cells)** | **Fermentation temp.** |
| American ale | 200 | 65°F (18°C) |

| Recommended Water Profile (ppm) | | | | Brew cube: Pale, Malty, Soft | |
|---|---|---|---|---|---|
| **Ca** | **Mg** | **Total alk.** | **SO$_4$** | **Cl** | **RA** |
| 50–100 | 10 | 0–50 | 0–50 | 50–100 | –100–0 |

*Notes*

1. *Any bittering hop will work, the goal is 10 IBUs of clean bitterness.*

2. *Wash two medium sized oranges and use a vegetable or apple peeler to carefully remove the outer rind. This should be about 50 g total. Do not peel/shave too deeply; the peel should be thin, bright orange without white pith.*

3. *After the boil, heat off; stir in the honey, then add the orange peel and coriander, and allow them to hot steep for 15 minutes before chilling. The orange peel can be removed before fermentation, but doesn't have to be.*

## Saison

*Saison* is a very new style, created in the 1950's from the ideals of what an old Belgian farmhouse ale should have been like. The primary ingredient is Pilsner malt, although other grains are common ingredients as well, such as wheat, maize, oats, rye, and spelt. Sugar or honey is often used to help dry out the beer. The yeast character plays a dominant role in the beer; it should be fruity and spicy, but not the banana and clove character that is common in German and Belgian yeasts. The saison yeast character is more peppery and has citrus or pear. The hop character generally takes a back seat to the yeast character, although some interplay of floral, fruity, and/or herbal aroma is common. The overall character of the beer is a dry, grainy malt character, firm bitterness and mineral structure, with lots of aromatics coming from the yeast. Saison is a good base for adding spices or fruit, but these are variations on the style; normally these flavors and aromas come from the saison yeast (which may have its origins as a red wine yeast). The guidelines for the style are OG 1.048–1.065, FG 1.002–1.008, 20–35 IBU, and 5–14 SRM (10–28 EBC), although darker and stronger versions are common as well.

## Battre L'oie—Saison

**Original gravity:** 1.049  **SRM (EBC):** 3 (6)
**Final gravity:** 1.007  **ABV:** 5.6%
**IBU:** 27

| Version: Extract and Steeping Grain | | |
|---|---|---|
| **Wort A** | **Gravity points** | |
| 3 lb. (1.4 kg) wheat DME | 45 | |
| Boil gravity for 3 gal. (11.4 L) | 1.045 | |
| **Hop schedule** | **Boil time (min.)** | **IBUs** |
| 1 oz. (30 g) Aramis 7% AA | 60 | 22 |
| 1 oz. (30 g) Barbe Rouge 8.5% AA | Steep 15 | 5 |
| **Wort B (add at knockout)** | **Gravity points** | |
| 3 lb. (1.4 kg) wheat DME | 45 | |
| **Yeast strain** | **Pitch (billions of cells)** | **Fermentation temp.** |
| Belgian saison | 200 | 62°F (17°C) |

| Version: All-Grain | |
|---|---|
| **Grain bill** | **Gravity points** |
| 6.6 lb. (3 kg) Pilsner malt | 26 |
| 1.1 lb. (500 g) Vienna malt | 4 |
| 1.1 lb. (500 g) wheat malt | 5 |
| 1.1 lb. (500 g) white table sugar | 7 |
| Boil gravity for 7 gal. (27 L) | 1.042 |

| Mash schedule | Rest temp. | | Rest time (min.) | |
|---|---|---|---|---|
| Conversion rest – Infusion | 153°F (67°C) | | 60 | |
| **Hop schedule** | **Boil time (min.)** | | **IBUs** | |
| 1 oz. (30 g) Aramis 7% AA | 60 | | 22 | |
| 1 oz. (30 g) Barbe Rouge 8.5% AA | Steep 15 | | 5 | |
| **Yeast strain** | **Pitch (billions of cells)** | | **Fermentation temp.** | |
| Belgian saison | 200 | | 62°F (17°C) | |
| **Recommended Water Profile (ppm)** | | | **Brew cube:** Pale, Balanced, Medium | |
| **Ca** | **Mg** | **Total alk.** | **SO$_4$** | **Cl** | **RA** |
| 75–125 | 10 | 0–50 | 100–150 | 100–150 | –100–0 |

## Pale Ale

Pale ales are common worldwide; every country and every region often has its own substyle. They are easy to brew, refreshing to drink, and similar around the world, despite some differences. Generally, they are made from a simple recipe of pale base malt, a readily available adjunct or specialty malt, local hops, and a local yeast. "Pale" is a relative term and it originally meant "pale as compared to dark;" most early malts and beers were darker due to the typical wood fire kilning technology of previous millennia.

Truly pale malts are a relatively recent invention of about 500 years ago, with the use of coke (a clean burning derivative of coal) for kilning. For a better history of pale ales I will refer you to Ron Pattinson and his book, *The Home Brewer's Guide to Vintage Beer*. Pale ale was exported to the far reaches of the British Empire in the eighteenth and nineteenth centuries, and many of its modern variations were born in that time. Today there are English, Scottish, Irish, Belgian, American, Australian, and even German versions of pale ale. Modern pale ales can range from golden to deep amber and are typically a balanced sort of beer, in other words, there should be equal prominence between the malt and the hops. Sometimes the hops will be forward with a malty finish and sometimes the malt will be forward with a hoppy finish, but mainly in balance. A pale ale that is too hoppy is an IPA, and a pale ale that is too malty is an amber or brown ale. Modern pale ales are intended to be session beers, generally in the range of 3.5%–5.5% ABV, although there will be the occasional 6%. The top fermenting ale yeast and warm fermentation temperatures give pale ales a subtle fruitiness compared to lagers. Pale ales are best served cool, not cold, about 45–55°F (7–12°C), to allow the fruit and malty notes to emerge.

The OG for ordinary bitter, the smallest beer, is 1.030–1.039, with 25–35 IBU, and an amber color of 8–14 SRM (16–28 EBC). Best bitter has a step higher OG at 1.040–1.048, and 25–40 IBU. Strong bitter steps up again to 1.048–1.060 OG and 30–50 IBUs, and the color can be a bit darker as well, up to 18 SRM (36 EBC).

### English Pale Ale (Bitter)

Modern English pale ales include the bitters and what have variously been called London ale and Burton ale in recent years (e.g., Bass Ale), although style gurus now classify these as strong bitters. English bitters tend to have more malt character than American pale ales, and the malt character tends to be a bit toastier and sweeter. Cask ale is a subset of English pale ale that is not really different in recipe, but in how it is served. Drinkers of megabrewery beers in the US would probably describe them as flat. The beer is brewed to a low final gravity, yielding a dry finish with only a low level of residual sweetness that does not mask the hop finish.

## Crouchback's Strong Bitter

**Original gravity:** 1.055  
**Final gravity:** 1.012  
**IBU:** 36  

**SRM (EBC):** 8 (16)  
**ABV:** 5.8%  

### Version: Extract and Steeping Grain

| Wort A | Gravity points | |
|---|---|---|
| 2 lb. (0.9 kg) pale DME | 30 | |
| 0.66 lb. (300 g) corn syrup solids | 9 | |
| 0.5 lb. (225 g) biscuit malt | 3 | |
| 0.25 lb. (115 g) caramel 40°L – Steeped | 1 | |
| 0.25 lb. (115 g) caramel 80°L – Steeped | 0.5 | |
| Boil gravity for 3 gal. (11.4 L) | 1.045 | |
| **Hop schedule** | **Boil time (min.)** | **IBUs** |
| 1 oz. (30 g) Challenger 7.5% AA | 60 | 24 |
| 1 oz. (30 g) EK Goldings 5% AA | 30 | 12 |
| 1 oz. (30 g) EK Goldings 5% AA | 0 | 0 |
| **Wort B (add at knockout)** | **Gravity points** | |
| 4 lb. (1.8 kg) pale DME | 56 | |
| **Yeast strain** | **Pitch (billions of cells)** | **Fermentation temp.** |
| British ale | 200 | 67°F (19°C) |

### Version: All-Grain

| Grain bill | Gravity points |
|---|---|
| 10 lb. (4.5 kg) pale ale malt | 40 |
| 1 lb. (450 g) flaked corn (maize) | 4 |
| 0.5 lb. (225 g) biscuit malt | 1.5 |
| 0.25 lb. (115 g) caramel 40°L | 1 |
| 0.25 lb. (115 g) caramel 80°L | <1 |
| Boil gravity for 7 gal. (27 L) | 1.047 |

| Mash schedule | Rest temp. | Rest time (min.) |
|---|---|---|
| Conversion rest – Infusion | 153°F (67°C) | 60 |

| Hop schedule | Boil time (min.) | IBUs |
|---|---|---|
| 1 oz. (30 g) Challenger 7.5% AA | 60 | 24 |
| 1 oz. (30 g) EK Goldings 5% AA | 30 | 12 |
| 1 oz. (30 g) EK Goldings 5% AA | 0 | 0 |

| Yeast strain | Pitch (billions of cells) | Fermentation temp. |
|---|---|---|
| British ale | 200 | 67°F (19°C) |

| Recommended Water Profile (ppm) | | | Brew cube: Pale, Hoppy, Firm | | |
|---|---|---|---|---|---|
| Ca | Mg | Total alk. | SO$_4$ | Cl | RA |
| 75–125 | 10 | 0–50 | 150–300 | 50–100 | –100–0 |

## Scottish Ales

Scottish ales were very similar to English pale ale for much of the nineteenth century, but after World War II they were defined as being maltier and less hoppy. They also had a reputation for having a peat smoke character, but diligent research has shown this to be categorically untrue; it probably was the work of some American travel agency. Today, Scottish pale ales are similar to English bitters in strength, but generally have a slightly richer, slighter sweeter caramel malt character. The beers are still very well attenuated and dry, but the absence of late hop additions to the overall hop character lets the malt take center stage. While the Scottish version of barleywine (wee heavy) is definitely sweet and heavy, the pale ales are, dare I say, thrifty in their richness and are one of the most quaffable styles in existence.

The strengths of Scottish ales follows the three tiers of English pale ales, with light, heavy, and export strengths of 1.030–1.035, 1.035–1.040, and 1.040–1.060 OG, respectively. The bitterness for light and heavy are 10–20 IBU, and for export is 15–30 IBU. The color for the three versions are all pretty similar, being in the 10–20 SRM (20–40 EBC) range.

## Thistle Do Well Scottish Export

**Original gravity:** 1.049  
**Final gravity:** 1.011  
**IBU:** 24  

**SRM (EBC):** 11 (22)  
**ABV:** 4.8%  

| Version: Extract and Steeping Grain | |
|---|---|
| **Wort A** | **Gravity points** |
| 2 lb. (0.9 kg) pale DME | 30 |
| 1 lb. (450 g) caramel 80°L | 5 |
| 0.55 lb. (250 g) Briess Carabrown® (55°L) | 2 |
| 0.28 lb. (125 g) corn syrup solids | 4 |
| Boil gravity for 3 gal. (11.4 L) | 1.041 |

| Hop schedule | Boil time (min.) | IBUs |
|---|---|---|
| 1 oz. (30 g) Challenger (7.5%) | 60 | 24 |
| **Wort B (add at knockout)** | **Gravity points** | |
| 3 lb. (1.4 kg) pale DME | 45 | |
| **Yeast strain** | **Pitch (billions of cells)** | **Fermentation temp.** |
| Scottish ale | 200 | 65°F (18°C) |

### Version: All-Grain

| Grain bill | Gravity points | |
|---|---|---|
| 9 lb. (4.08 kg) pale ale malt | 36 | |
| 0.5 lb. (225 g) flaked corn (maize) | 2 | |
| 1 lb. (450 g) caramel 80°L | 2 | |
| 0.55 lb. (250 g) Briess Carabrown® (55°L) | 2 | |
| Boil gravity for 7 gal. (27 L) | 1.042 | |
| **Mash schedule** | **Rest temp.** | **Rest time (min.)** |
| Conversion rest – Infusion | 153°F (67°C) | 60 |
| **Hop schedule** | **Boil time (min.)** | **IBUs** |
| 1 oz. (30 g) Challenger (7.5%) | 60 | 24 |
| **Yeast strain** | **Pitch (billions of cells)** | **Fermentation temp.** |
| Scottish ale | 200 | 65°F (18°C) |
| **Recommended Water Profile (ppm)** | **Brew cube:** Amber, Malty, Medium | |

| Ca | Mg | Total alk. | SO₄ | Cl | RA |
|---|---|---|---|---|---|
| 75–125 | 10 | 75–125 | 50–100 | 100–150 | 0–50 |

## Irish Red Ale

It may seem to be a bit of a stretch to say that Irish red ale is a pale ale, but that is the role it fills. In fact, the flavor of an Irish red ale is very similar to the English and Scottish ales, but differentiated by a grainy malt flavor as opposed to the biscuit of the English or the caramel of the Scottish. The deep red color comes from the use of about 3% by weight roasted malt in the grain bill, enough to give deep red highlights but without contributing to the flavor of the beer as seen in brown ales, porters, or stouts. The use of flaked barley gives a lot of mouthfeel to the beer, similar to a stout but without the coffeelike flavors. The Briess Carabrown is an amber-type malt and one of my favorites. It gives a distinct toasted bread crust flavor to the beer, which perfectly rounds out this style.

The Irish were content with one version of their pale ale, it having a range of 1.036–1.046 OG, 18–28 IBU, and a deep red color of 9–14 SRM (18–28 EBC).

## Fuil Croi—Irish Red Ale

**Original gravity:** 1.042  
**Final gravity:** 1.011  
**IBU:** 26

**SRM (EBC):** 12 (24)  
**ABV:** 4.1%

### Version: Extract and Steeping Grain

| Wort A | Gravity points |
|---|---|
| 1.5 lb. (0.68 kg) pale ale DME | 22.5 |
| 1 lb. (450 g) flaked barley – Instamash® | 7.5 |
| 0.33 lb. (150 g) Briess Carabrown® (55°L) – Steeped | 1.5 |
| 0.33 lb. (150 g) roast barley (300°L) – Steeped | 2 |
| Boil gravity for 3 gal. (11.4 L) | 1.033 |

| Hop schedule | Boil time (min.) | IBUs |
|---|---|---|
| 0.9 oz. (25 g) UK Phoenix 9% AA | 60 | 26 |

| Wort B (add at knockout) | Gravity points |
|---|---|
| 3 lb. (1.4 kg) pale DME | 42 |

| Yeast strain | Pitch (billions of cells) | Fermentation temp. |
|---|---|---|
| Irish ale | 200 | 65°F (18°C) |

### Version: All-Grain

| Grain bill | Gravity points |
|---|---|
| 7.5 lb. (3.4 kg) pale ale malt | 30 |
| 1 lb. (450 g) flaked barley | 3.5 |
| 0.33 lb. (150 g) Briess Carabrown® (55°L) – Steeped | 1 |
| 0.33 lb. (150 g) roast barley (300°L) – Steeped | 1.5 |
| Boil gravity for 7 gal. (27 L) | 1.036 |

| Mash schedule | Rest temp. | Rest time (min.) |
|---|---|---|
| Conversion rest – Infusion | 153°F (67°C) | 60 |

| Hop schedule | Boil time (min.) | IBUs |
|---|---|---|
| 0.9 oz. (25 g) UK Phoenix 9% AA | 60 | 26 |

| Yeast strain | Pitch (billions of cells) | Fermentation temp. |
|---|---|---|
| Irish ale | 200 | 65°F (18°C) |

| Recommended Water Profile (ppm) | | | Brew cube: Amber, Malty, Medium | | |
|---|---|---|---|---|---|
| Ca | Mg | Total alk. | $SO_4$ | Cl | RA |
| 75–125 | 10 | 50–100 | 50–100 | 100–150 | 0–50 |

### Belgian Pale Ale

Belgian pale ale takes the idea of English pale, but uses Continental ingredients to bring the flavor home. Pilsner, Vienna, and Munich malts combine to give a flavor exactly like that of bread fresh from the oven. Triskel and Aramis are two relatively new hop varieties from the Alsace region that are both traditional in their basis, yet new in their higher alpha acid and aromatic oil levels. The Belgian ale yeast gives the beer a complex spiciness from the phenolics and esters to create a marvel in every glass.

The OG for Belgian pale ranges from 1.048–1.054, the bitterness from 20–30 IBU, and it has a gold to amber color of 8–14 SRM (16–28 EBC).

## Marvel Banale—Belgian Pale Ale

**Original gravity:** 1.052　　　　　　　**SRM (EBC):** 12 (24)
**Final gravity:** 1.011　　　　　　　　**ABV:** 5.5%
**IBU:** 27

### Version: Extract and Steeping Grain

| Wort A | Gravity points | |
|---|---|---|
| 1.5 lb. (680 g) Briess Vienna LME | 18 | |
| 1.1 lb. (500 g) Munich DME | 16.5 | |
| 1 lb. (450 g) CaraVienne malt – Steeped | 7 | |
| 0.5 lb. (225 g) biscuit malt – Steeped | 3 | |
| 0.25 lb. (115 g) Special "B" malt (140°L) – Steeped | 1 | |
| Boil gravity for 3 gal. (11.4 L) | 1.046 | |
| **Hop schedule** | **Boil time (min.)** | **IBUs** |
| 1 oz. (30 g) French Triskel 8% AA | 30 | 25 |
| 1 oz. (30 g) French Aramis 8% AA | Steep 15 | 2 |
| **Wort B (add at knockout)** | **Gravity points** | |
| 3.3 lb. (1.5 kg) pale DME | 42 | |
| **Yeast strain** | **Pitch (billions of cells)** | **Fermentation temp.** |
| Belgian ale | 200 | 70°F (21°C) |

### Version: All-Grain

| Grain bill | Gravity points |
|---|---|
| 7 lb. (3.2 kg) pale ale malt | 28 |
| 2 lb. (0.9 kg) Vienna malt | 7 |
| 1 lb. (450 g) Munich malt | 4 |
| 1 lb. (450 g) CaraVienne (20°L) | 4 |
| 0.5 lb. (225 g) biscuit malt | 1 |
| 0.25 lb. (115 g) Special "B" malt (140°L) | 1 |
| Boil gravity for 7 gal. (27 L) | 1.045 |

| Mash schedule | Rest temp. | Rest time (min.) |
|---|---|---|
| Conversion rest – Infusion | 153°F (67°C) | 60 |
| **Hop schedule** | **Boil time (min.)** | **IBUs** |
| 1 oz. (30 g) French Triskel 8% AA | 30 | 25 |
| 1 oz. (30 g) French Aramis 8% AA | Steep 15 | 2 |
| **Yeast strain** | **Pitch (billions of cells)** | **Fermentation temp.** |
| Belgian ale | 200 | 70°F (21°C) |

| **Recommended Water Profile (ppm)** | | | **Brew cube:** Amber, Balanced, Soft | | |
|---|---|---|---|---|---|
| **Ca** | **Mg** | **Total alk.** | **SO$_4$** | **Cl** | **RA** |
| 50–100 | 10 | 100–125 | 75–100 | 75–100 | 25–75 |

## Australian Sparkling Ale

Australian Sparkling ale is named for the effervescence of the style, not so much the clarity, because it is one of the few styles where the yeast sediment is often swirled into suspension before consumption; lots of good vitamins in there. The ale is all pale malt with only a small amount of pale crystal malt to help build a golden color. The flavor is bready backed by a firm bitterness. English hop varieties and American Cluster hops were traditional, although Pride of Ringwood largely replaced those in the middle of the twentieth century. The aroma of the beer should be equally balanced between malt, hop, and yeast character; at once bready, herbal, earthy, and fruity.

The guidelines for Australian sparkling ale are similar to those for best bitter, having an OG of 1.038–1.050, a bitterness of 20–35 IBU, but a paler golden color of 4–7 SRM (8–14 EBC).

## Fair Drop—Australian Sparkling Ale

**Original gravity:** 1.045
**Final gravity:** 1.008
**IBU:** 30

**SRM (EBC):** 5 (10)
**ABV:** 5%

| *Version: Extract and Steeping Grain* | | |
|---|---|---|
| **Wort A** | **Gravity points** | |
| 2.5 lb. (1.13 kg) Pilsner DME | 37 | |
| 0.5 lb. (225 g) CaraVienne malt (20°L) – Steeped | 3 | |
| Boil gravity for 3 gal. (11.4 L) | 1.040 | |
| **Hop schedule** | **Boil time (min.)** | **IBUs** |
| 0.8 oz. (23 g) Pride of Ringwood 9% AA | 60 | 24 |
| 0.5 oz. (15 g) New Zealand Wakatu 7.5% AA | 15 | 6 |
| **Wort B (add at knockout)** | **Gravity points** | |
| 3 lb. (1.4 kg) Pilsner DME | 42 | |
| **Yeast strain** | **Pitch (billions of cells)** | **Fermentation temp.** |
| Australian or English ale | 150 | 65°F (18°C) |

## Version: All-Grain

| Grain bill | Gravity points | |
|---|---|---|
| 9.5 lb. (3.2 kg) Pilsner malt | 38 | |
| 0.5 lb. (340 g) CaraVienne 20°L malt | 1 | |
| Boil gravity for 7 gal. (27 L) | 1.039 | |
| **Mash schedule** | **Rest temp.** | **Rest time (min.)** |
| Conversion rest – Infusion | 153°F (67°C) | 60 |
| **Hop schedule** | **Boil time (min.)** | **IBUs** |
| 0.8 oz. (23 g) Pride of Ringwood 9% AA | 60 | 24 |
| 0.5 oz. (15 g) New Zealand Wakatu 7.5% AA | 15 | 6 |
| **Yeast strain** | **Pitch (billions of cells)** | **Fermentation temp.** |
| Australian or English ale | 150 | 65°F (18°C) |

| Recommended Water Profile (ppm) | | | Brew cube: Pale, Balanced, Medium | | |
|---|---|---|---|---|---|
| Ca | Mg | Total alk. | SO$_4$ | Cl | RA |
| 75–125 | 20 | 0–50 | 100–150 | 100–150 | –100–0 |

## Kölsch

German *Kölsch* is the softest of the pale ales, generally having bitterness of 18–30 IBU and an OG of 1.044–1.050. While Germany is normally known for its lagers, the beers of Cologne (*Köln*) have long been renowned for their brilliant clarity and perfect balance of soft malt and hop character. They are brewed as ales, but then cold conditioned to give them great clarity. This is easily one of the most popular styles of homebrewers around the world, but often a difficult style to do well. However, perfect or not, it still is a very enjoyable beer. I have always thought of Kölsch as the "California" beer of Germany, thus the name of the recipe.

The guidelines for Kölsch are 1.044–1.050 OG, bitterness of 18–30 IBU, and a pale gold color of 3.5–5 SRM (7–10 EBC).

## Surfin' Vogel—Kölsch

**Original gravity:** 1.046
**Final gravity:** 1.011
**IBU:** 28

**SRM (EBC):** 4 (8)
**ABV:** 4.6%

## Version: Extract and Steeping Grain

| Wort A | Gravity points |
|---|---|
| 2.5 lb. (1 kg) Briess Vienna LME | 30 |
| 0.66 lb. (300 g) Pilsner DME | 9 |
| Boil gravity for 3 gal. (11.4 L) | 1.039 |

| Hop schedule | Boil time (min.) | IBUs |
|---|---|---|
| 1.25 oz. (35 g) German Tradition 6% AA | 60 | 24 |
| 1.1 oz. (30 g) German Opal 6% AA | Steep 15 | 4 |
| **Wort B (add at knockout)** | **Gravity points** | |
| 3.3 lb. (1.5 kg) Pilsner DME | 46 | |
| **Yeast strain** | **Pitch (billions of cells)** | **Fermentation temp.** |
| German Kölsch | 200 | 65°F (18°C) |

### Version: All-Grain

| Grain bill* | Gravity points |
|---|---|
| 7 lb. (1.6 kg) Pilsner malt | 27.5 |
| 3.3 lb. (450 g) Vienna malt | 12.5 |
| Boil gravity for 7 gal. (27 L) | 1.040 |
| * This recipe may require some lactic acid or acidulated malt to help hit a mash pH of 5.2. | |

| Mash schedule | Rest temp. | Rest time (min.) |
|---|---|---|
| Conversion rest – Infusion | 153°F (67°C) | 60 |
| **Hop schedule** | **Boil time (min.)** | **IBUs** |
| 1.25 oz. (35 g) German Tradition 6% AA | 60 | 24 |
| 1.1 oz. (30 g) German Opal 6% AA | Steep 15 | 4 |
| **Yeast strain** | **Pitch (billions of cells)** | **Fermentation temp.** |
| German Kölsch | 200 | 65°F (18°C) |

| Recommended Water Profile (ppm) | | | | Brew cube: Pale, Balanced, Soft | |
|---|---|---|---|---|---|
| **Ca** | **Mg** | **Total alk.** | **SO$_4$** | **Cl** | **RA** |
| 50–100 | 10 | 0–50 | 50–100 | 50–100 | –75–0 |

## American Pale Ale

American pale ale is a hoppier adaptation of English pale ale. The American ale yeast strains generally produce less esters than comparable English ale yeasts, and thus American pale ales have a less fruity taste than their counterparts from the British Isles. American pale ales also have more of their hop character coming from aroma and flavor additions than bittering additions. The color ranges from gold to dark amber, and the flavor often has a roundness from the use of crystal malt, although typically less than in an English pale ale.

By definition, American pale ales use American hop varieties, while English pale ales use English varieties, although these days everyone seems to be using whatever hop variety they like. It almost seems like it is easier to find a classic bitter in the US and a classic American pale in the UK! American pale ales typically have an OG of 1.045–1.060, an FG of 1.010–0.015, and 30–50 IBUs. Stylistically, it is important for the bitterness units to be less than the gravity units, in a BU:GU ratio of about 3:4 (or between 0.7–0.8); otherwise the beer will be more similar to other styles such as blonde ale or IPA.

## Lady Liberty—American Pale Ale

**Original gravity:** 1.047
**Final gravity:** 1.011
**IBU:** 36

**SRM (EBC):** 6 (12)
**ABV:** 4.8%

| *Version: Extract and Steeping Grain* | | |
|---|---|---|
| **Wort A** | **Gravity points** | |
| 1.5 lb. (680 g) pale DME | 30 | |
| 1 lb. (450 g) Munich DME | 15 | |
| 0.5 lb. (225 g) caramel 40°L – Steeped | 3 | |
| Boil gravity for 3 gal. (11.4 L) | 1.048 | |
| **Hop schedule** | **Boil time (min.)** | **IBUs** |
| 0.75 oz. (23 g) Centennial 11% AA | 30 | 22 |
| 1 oz. (30 g) Cascade 7% AA | 15 | 10 |
| 1 oz. (30 g) Cascade 7% AA | Steep 15 | 4 |
| **Wort B (add at knockout)** | **Gravity points** | |
| 3 lb. (1.4 kg) pale DME | 42 | |
| **Yeast strain** | **Pitch (billions of cells)** | **Fermentation temp.** |
| American ale | 200 | 65°F (18°C) |

| *Version: All-Grain* | | |
|---|---|---|
| **Grain bill** | **Gravity points** | |
| 9 lb. (4.5 kg) pale ale malt | 36 | |
| 1 lb. (450 g) Munich malt | 3.5 | |
| 0.5 lb. (115 g) caramel 40°L | 2 | |
| Boil gravity for 7 gal. (27 L) | 1.042 | |
| **Mash schedule** | **Rest temp.** | **Rest time (min.)** |
| Conversion rest – Infusion | 153°F (67°C) | 60 |
| **Hop schedule** | **Boil time (min.)** | **IBUs** |
| 0.75 oz. (23 g) Centennial 11% AA | 30 | 22 |
| 1 oz. (30 g) Cascade 7% AA | 15 | 10 |
| 1 oz. (30 g) Cascade 7% AA | Steep 15 | 4 |
| **Yeast strain** | **Pitch (billions of cells)** | **Fermentation temp.** |
| American ale | 200 | 65°F (18°C) |

| **Recommended Water Profile (ppm)** | | | **Brew cube:** Pale, Hoppy, Medium | | |
|---|---|---|---|---|---|
| **Ca** | **Mg** | **Total alk.** | **SO$_4$** | **Cl** | **RA** |
| 75–125 | 10 | 0–50 | 150–300 | 50–100 | –100–0 |

*See notes for Lady Liberty on following page.*

*Notes*

*To make a rye pale ale, substitute 3 lb. of Briess CBW Rye Malt Extract for both Wort A and B, (i.e., 6 lb. [2.7 kg] total). Everything else is unchanged.*

*To make a Session IPA, double the 15 min. boil and hop steep additions to 2 oz. (60 g) each for a total of about 50 IBU. Everything else is unchanged.*

## Blonde Ale

Blonde ales are very similar to pale ales except that the hops have been dialed back to create a beer that is drinkable as a light lager but much easier to brew. The bitterness for blonde ales typically ranges from 15–25 IBU (compared to 25–25 IBU for pale ales). Blonde ales generally do not contain any specialty malts except for the lightest of crystals or Carapils® (10°L). American blonde ales are often the most approachable beer for the new craft beer drinker, typically having more malt character than light lager or cream ale, but half the hops of a typical pale ale. To brew an American blonde ale from the following Argentinian recipe, change the Pilsner malt to pale ale or American two-row base malt, and the hops to American varieties. Everything else is the same.

### Argentinian Pampas Golden Ale

The Argentines are dedicated homebrewers and I have greatly enjoyed my trips there over the years. I should say they are no more dedicated than the other homebrewers of South America, but they have distinguished themselves by defining and documenting their own style of blonde ale, the *Dorada Pampeana*, or golden Pampas ale. The basic recipe was created in 2001 in the San Telmo district of Buenos Aires by Marcelo Cerdán. The beer is made from Pilsner-style malt, grown and malted in Buenos Aires, and hopped with Argentine hops to create a beer of 1.042–1.054 OG and 15–22 IBU. The yeast is typically a dried American ale yeast, although that is broadening as the brewing market in Argentina grows. Kölsch yeast is also a popular choice.

## Nuestro Pan Diaro—Dorada Pampeana

**Original gravity:** 1.045  
**Final gravity:** 1.008  
**IBU:** 18  

**SRM (EBC):** 5 (10)  
**ABV:** 4.9%  

| Version: Extract and Steeping Grain | | |
|---|---|---|
| **Wort A** | **Gravity points** | |
| 2.5 lb. (1.13 kg) Pilsner DME | 37 | |
| 0.5 lb. (225 g) caramel 20°L malt – Steeped | 3 | |
| Boil gravity for 3 gal. (11.4 L) | 1.040 | |
| **Hop schedule** | **Boil time (min.)** | **IBUs** |
| 0.5 oz. (15 g) Argentine Cascade 9% AA | 60 | 12 |
| 0.5 oz. (15 g) New Zealand Wakatu 7.5% AA | 15 | 6 |

| Wort B (add at knockout) | Gravity points | |
|---|---|---|
| 3 lb. (1.4 kg) Pilsner DME | 42 | |
| **Yeast strain** | **Pitch (billions of cells)** | **Fermentation temp.** |
| American ale | 200 | 70°F (21°C) |

### Version: All-Grain

| Grain bill | Gravity points | |
|---|---|---|
| 9.5 lb. (3.2 kg) Pilsner malt | 38 | |
| 0.5 lb. (340 g) crystal 20°L malt | 1 | |
| Boil gravity for 7 gal. (27 L) | 1.039 | |
| **Mash schedule** | **Rest temp.** | **Rest time (min.)** |
| Conversion rest – Infusion | 150°F (65°C) | 60 |
| **Hop schedule** | **Boil time (min.)** | **IBUs** |
| 0.5 oz. (15 g) Argentine Cascade 9% AA | 60 | 12 |
| 0.5 oz. (15 g) New Zealand Wakatu 7.5% AA | 15 | 6 |
| **Yeast strain** | **Pitch (billions of cells)** | **Fermentation temp.** |
| American ale | 200 | 70°F (21°C) |
| **Recommended Water Profile (ppm)** | | **Brew cube:** Pale, Malty, Medium |

| Ca | Mg | Total alk. | SO$_4$ | Cl | RA |
|---|---|---|---|---|---|
| 75–125 | 10 | 0–50 | 50–100 | 100–150 | –100 to –50 |

## Amber Ale

Part of the American ale style spectrum, amber ales bridge between pale and brown ales by adding body and sweetness, and by shifting the beer's balance away from the hops to the malt. Amber ales are sweeter than brown ales, but will have more hop flavor and aroma dancing on top—high malt sweetness balanced by lots of hops. Amber ales have become one of my favorite beers. I like the balance of these beers—they are very hearty and satisfying.

The guidelines for American amber ale run from 1.045–1.060 OG, 25–40 IBU, and a copper color of 10–17 SRM (20–34 EBC). This is my clone of Red Nectar Ale.

### Big Basin Amber—American Amber Ale

**Original gravity:** 1.055
**Final gravity:** 1.014
**IBU:** 40

**SRM (EBC):** 15 (30)
**ABV:** 5.5%

### Version: Extract and Steeping Grain

| Wort A | Gravity points |
|---|---|
| 2.5 lb. (1.13 kg) pale ale DME | 37 |
| 1 lb. (450 g) crystal 40°L malt – Steeped | 6 |

| 1 lb. (450 g) crystal 80°L malt – Steeped | 5 | |
|---|---|---|
| Boil gravity for 3 gal. (11.4 L) | 1.048 | |
| **Hop schedule** | **Boil time (min.)** | **IBUs** |
| 1 oz. (30 g) Nugget 12% AA | 60 | 35.5 |
| 0.5 oz. (15 g) Amarillo 9% AA | Steep 15 | 3 |
| 0.5 oz. (15 g) East Kent Goldings 5% AA | Steep 15 | 1.5 |
| **Wort B (add at knockout)** | **Gravity points** | |
| 3.75 lb. (1.7 kg) Pilsner DME | 52.5 | |
| **Yeast strain** | **Pitch (billions of cells)** | **Fermentation temp.** |
| California ale | 230 | 65°F (18°C) |

### Version: All-Grain

| **Grain bill** | **Gravity points** | |
|---|---|---|
| 11 lb. (5 kg) pale ale malt | 44 | |
| 1 lb. (450 g) crystal 40°L malt | 2.5 | |
| 1 lb. (450 g) crystal 80°L malt | 2 | |
| Boil gravity for 7 gal. (27 L) | 1.048 | |
| **Mash schedule** | **Rest temp.** | **Rest time (min.)** |
| Conversion rest – Infusion | 153°F (67°C) | 60 |
| **Hop schedule** | **Boil time (min.)** | **IBUs** |
| 1 oz. (30 g) Nugget 12% AA | 60 | 35.5 |
| 0.5 oz. (15 g) Amarillo 9% AA | Steep 15 | 3 |
| 0.5 oz. (15 g) East Kent Goldings 5% AA | Steep 15 | 1.5 |
| **Yeast strain** | **Pitch (billions of cells)** | **Fermentation temp.** |
| California ale | 230 | 65°F (18°C) |

| **Recommended Water Profile (ppm)** | | | | **Brew cube:** Amber, Malty, Soft | |
|---|---|---|---|---|---|
| **Ca** | **Mg** | **Total alk.** | **SO$_4$** | **Cl** | **RA** |
| 50–100 | 10 | 50–100 | 50–100 | 75–100 | 25–75 |

## India Pale Ale (IPA)

According to popular mythos, the IPA style arose from the months-long sea journey to India, during which the beer conditioned with hops in the barrel. Extra hops were added to help prevent spoilage during the long voyage. This conditioning time mellowed the hop bitterness to a degree and imparted a wealth of hop aroma to the beer. For the real history of IPA, see Mitch Steele's book, *IPA* (Steele, 2012). It details the history of genuine India pale ale, and gives many, many recipes for classic examples of the style, from historic to modern day, including black, white, session, and double IPAs.

The world of IPAs has erupted, and now every hop-forward beer of any color is being marketed as some type of IPA. It is worth noting that the BJCP in the recent *2015 Style Guidelines* has moved away from calling these beers "India Pale Ales" and is only referring to them as "IPA," because these modern derivatives were never made or drank in India. The BJCP now has IPA substyles for Belgian, White, Red, Rye, Brown, and Black.

Let's review the main characteristics of these beers. . . (*dramatic pause*). . . Hops! All of these beers have a dominating hop balance in both aroma and flavor. The malt and fermentation character will always take a back seat to the hop character in an IPA, but these characters play a very important role in defining the substyle, and their interplay will determine whether the beer itself is any good or not.

Every beer is a symphony of flavors and aromas, and if a blonde ale is the nearest thing to a string quartet, an IPA is a big brass marching band. And the best marching bands include woodwinds and drums to create dynamism and complexity. The *Monty Python's Flying Circus* theme song—otherwise known as *The Liberty Bell March* by John P. Sousa—comes to mind. The malt and fermentation characters need to play this supporting role in IPA; they provide contrast and balance to the hops with grainy malt flavors, sweetness, and esters. The best IPAs, like music, contain other characters to create a better composition.

I will outline several of the current styles but will not provide recipes for them all. Recipes are a dime a dozen anyway, and therefore I will just include recipes for English IPA, American IPA, double IPA, brown IPA, and black IPA, which are five of my favorites.

Belgian IPA is an adaptation of American IPA using Belgian yeast strains, or to look at it from the other side of the fence, an adaptation of Belgian golden strong ale or *tripel* with American hopping rates. The high hop bitterness is offset by a higher alcohol content (which makes it taste sweeter) and the hop aroma is enhanced with fruity esters and spicy phenols from the Belgian yeast strains. Many different hop varieties may be used, including American, European, and Pacific varieties, but it is important that the hops share a common element with the fermentation character. The spicy and fruity elements of more recent hop varieties tend to work better than the more sulfurous and piney character (i.e., dank garlic and onion) in older hop varieties that can clash with the phenolic yeast character. Belgian IPAs tend to be golden to amber in color, with an OG of 1.058–080, FG of 1.008–1.016, and 50–100 IBUs.

White IPAs are a cross between the Belgian wit style and American IPA. The Belgian witbier yeast character and the use of spices, such as coriander and chamomile, enhance the hop character. The goal with white IPA is to strike a cooperative balance between the traditional witbier character and the higher bitterness of an IPA. The result is a more assertive witbier—drier, fruitier, and spicier, with a more robust bitterness. Again, it is important to avoid the danker hop varieties that would clash with the light body and spiciness of the witbier style. The substyle is characterized by a pale gold or hazy amber color, OG of 1.056–1.065, FG of 1.010–1.016, and 40–70 IBUs.

Rye IPAs, or RIPAs, add just that little something extra to American IPA to make them more interesting. Generally, about 15%–20% rye malt is substituted for the same amount of pale ale malt. Crystal malts are used sparingly, just as in American IPA. The hops are still the boss. The rye malt adds a pleasant, light grainy spiciness to the malt character, and the overall effect is a slightly drier, more complex beer. American ale yeast should be used to avoid confusion with the Belgian IPA substyle. The color of rye IPA is similar to American IPA, but can be a slightly darker golden-to-reddish amber; however, it should not be sweeter. The OG of RIPA is 1.056–1.075, FG 1.008–1.014, and 50–75 IBUs. To brew a RIPA, simply use the American IPA recipe and substitute in some rye malt for the base malt.

Red IPA is the evolution of American amber ale to higher hoppiness. A red IPA balances the higher bitterness with more malt sweetness, but not to the point where you lose the essential balance of an IPA. In other words, a red IPA is not an imperial red ale; the malt character still only plays a supporting role to the hop character, but the tone of that character is sweeter and more caramel or toffee, rather than golden bread or light toast. The color of a red IPA is deep amber red to brown with ruby highlights. The OG is 1.056–1.070, FG 1.008–1.016, and 40–70 IBU. The hops should be American or Pacific varieties; fruitier English hop varieties are also acceptable.

## English IPA

Good British pale ale malt, British hops, and English ale yeast. Keep Calm and Add More Hops! The style guidelines for English IPA are 1.050–1.075 OG, 40–60 IBU, and amber to copper color of 6–14 SRM (12–28 EBC).

## Victory & Chaos—English IPA

**Original gravity:** 1.060
**Final gravity:** 1.012
**IBU:** 58

**SRM (EBC):** 7 (14)
**ABV:** 6.5%

| *Version: Extract and Steeping Grain* | | |
| --- | --- | --- |
| **Wort A** | **Gravity points** | |
| 2.5 lb. (1.13 kg) pale ale DME | 38 | |
| 1 lb. (450 g) Munich DME | 14 | |
| 1 lb. (450 g) crystal 20°L malt – Steeped | 3 | |
| Boil gravity for 3 gal. (11.4 L) | 1.055 | |
| **Hop schedule** | **Boil time (min.)** | **IBUs** |
| 1 oz. (30 g) Northdown 8% AA | 60 | 22 |
| 1 oz. (30 g) First Gold 7.5% AA | 30 | 16 |
| 1 oz. (30 g) East Kent Goldings 5% AA | 30 | 11 |
| 1 oz. (30 g) First Gold 7.5% AA | Steep 15 | 4 |
| 1 oz. (30 g) East Kent Goldings 5% AA | Steep 15 | 3 |
| 1 oz. (30 g) First Gold 7.5% AA | Dry Hop | (1) |
| 1 oz. (30 g) East Kent Goldings 5% AA | Dry Hop | (1) |
| **Wort B (add at knockout)** | **Gravity points** | |
| 3.4 lb. (1.5 kg) Pilsner DME | 48 | |
| 0.56 lb. (250 g) corn syrup solids | 8 | |
| **Yeast strain** | **Pitch (billions of cells)** | **Fermentation temp.** |
| English ale | 250 | 65°F (18°C) |

### Version: All-Grain

| Grain bill | Gravity points | |
|---|---|---|
| 11 lb. (5 kg) pale ale malt | 44 | |
| 1 lb. (450 g) Munich 10°L malt | 3.5 | |
| 0.5 lb. (225 g) crystal 20°L malt | 1.5 | |
| 1 lb. (450 g) flaked corn (maize) | 3.5 | |
| Boil gravity for 7 gal. (27 L) | 1.052 | |

| Mash schedule | Rest temp. | Rest time (min.) |
|---|---|---|
| Conversion rest – Infusion | 150°F (65°C) | 60 |

| Hop schedule | Boil time (min.) | IBUs |
|---|---|---|
| 1 oz. (30 g) Northdown 8% AA | 60 | 22 |
| 1 oz. (30 g) First Gold 7.5% AA | 30 | 16 |
| 1 oz. (30 g) East Kent Goldings 5% AA | 30 | 11 |
| 1 oz. (30 g) First Gold 7.5% AA | Steep 15 | 4 |
| 1 oz. (30 g) East Kent Goldings 5% AA | Steep 15 | 3 |
| 1 oz. (30 g) First Gold 7.5% AA | Dry Hop | (1) |
| 1 oz. (30 g) East Kent Goldings 5% AA | Dry Hop | (1) |

| Yeast strain | Pitch (billions of cells) | Fermentation temp. |
|---|---|---|
| English ale | 250 | 65°F (18°C) |

| Recommended Water Profile (ppm) | | | | Brew cube: Pale, Hoppy, Medium | |
|---|---|---|---|---|---|
| Ca | Mg | Total alk. | SO$_4$ | Cl | RA |
| 100–150 | 10 | 50–100 | 200–400 | 50–100 | –100–0 |

## American IPA

American IPA is all about hops; typically American "C" hops such as, Chinook, Centennial and Cascade, but also the new Pacific varieties from New Zealand and Australia with their tropical fruit aromas. There are two kinds of American IPA, West Coast and not. West Coast IPA is primarily a pale ale malt beer so that the hop flavors are unimpeded. East Coast IPA tends to be maltier, with more emphasis on caramel malt and Munich malt character, but not to excess. If West Coast IPA is the clarion call of a trumpet, then East Coast is a trumpet, French horn, and baritone. The Rushmore IPA recipe below is a West Coast IPA with just enough caramel and Munich malt to add complexity. East Coast IPA would add some biscuit malt, in addition to the use of a darker caramel instead of the 40°L, to create a bit more malt balance with the hop character. Northeast IPA (the New England region) goes for a very juicy mouthfeel by substituting wheat and oats into the grain bill in place of the caramel and Munich, and moving all the hops to late additions (i.e., hop steep) to retain as much hop oil as possible. The northeast beer tends to be hazy due to all of the hop polyphenols and resin.

In general, the guidelines for American IPA are 1.056–70 OG, 40–70 IBU, and pale to copper color range of 6–14 SRM (12–28 EBC). In my experience, West Coast versions tend to run to 4–8 SRM (8–16 EBC).

## Rushmore—American IPA

**Original gravity:** 1.059
**Final gravity:** 1.015
**IBU:** 70-ish

**SRM (EBC):** 6 (12)
**ABV:** 6%

### Version: *Extract and Steeping Grain*

| Wort A | Gravity points | |
|---|---|---|
| 2.2 lb. (1 kg) pale DME | 33 | |
| 1 lb. (450 g) Munich DME | 15 | |
| 0.5 lb. (225 g) caramel 40°L – Steeped | 3 | |
| Boil gravity for 3 gal. (11.4 L) | 1.051 | |

| Hop schedule | Boil time (min.) | IBUs |
|---|---|---|
| 1 oz. (30 g) Nugget 13% AA | 60 | 38 |
| 0.5 oz. (15 g) Cascade 6% AA | 15 | 4 |
| 0.5 oz. (15 g) Amarillo 10% AA | 15 | 7 |
| 0.5 oz. (15 g) Centennial 10.5% AA | 15 | 7.5 |
| 0.5 oz. (15 g) Cascade 6% AA | Steep 30 | 3 |
| 0.5 oz. (15 g) Amarillo 10% AA | Steep 30 | 4.5 |
| 0.5 oz. (15 g) Centennial 10.5% AA | Steep 30 | 4.5 |
| 0.5 oz. (15 g) Cascade 6% AA | Dry | (1) |
| 0.5 oz. (15 g) Amarillo 10% AA | Dry | (1) |
| 0.5 oz. (15 g) Centennial 10.5% AA | Dry | (1) |

| Wort B (add at knockout) | Gravity points | |
|---|---|---|
| 4 lb. (2.3 kg) pale DME | 60 | |

| Yeast strain | Pitch (billions of cells) | Fermentation temp. |
|---|---|---|
| American ale | 250 | 65°F (18°C) |

### Version: *All-Grain*

| Grain bill | Gravity points | |
|---|---|---|
| 11 lb. (4.5 kg) pale ale malt | 44 | |
| 1 lb. (450 g) Munich malt | 4 | |
| 0.5 lb. (225 g) caramel 40°L | 2 | |
| Boil gravity for 7 gal. (27 L) | 1.050 | |

| Mash schedule | Rest temp. | Rest time (min.) |
|---|---|---|
| Conversion rest – Infusion | 153°F (67°C) | 60 |

| Hop schedule | Boil time (min.) | IBUs |
|---|---|---|
| 1 oz. (30 g) Nugget 13% AA | 60 | 38 |
| 0.5 oz. (15 g) Cascade 6% AA | 15 | 4 |
| 0.5 oz. (15 g) Amarillo 10% AA | 15 | 7 |
| 0.5 oz. (15 g) Centennial 10.5% AA | 15 | 7.5 |
| 0.5 oz. (15 g) Cascade 6% AA | Steep 30 | 3 |
| 0.5 oz. (15 g) Amarillo 10% AA | Steep 30 | 4.5 |
| 0.5 oz. (15 g) Centennial 10.5% AA | Steep 30 | 4.5 |
| 0.5 oz. (15 g) Cascade 6% AA | Dry | (1) |
| 0.5 oz. (15 g) Amarillo 10% AA | Dry | (1) |
| 0.5 oz. (15 g) Centennial 10.5% AA | Dry | (1) |

| Yeast strain | Pitch (billions of cells) | Fermentation temp. |
|---|---|---|
| American ale | 250 | 65°F (18°C) |

| Recommended Water Profile (ppm) | | | Brew cube: Pale, Hoppy, Medium | | |
|---|---|---|---|---|---|
| Ca | Mg | Total alk. | SO$_4$ | Cl | RA |
| 100–150 | 10 | 0–50 | 200–400 | 50–100 | –100–0 |

## Brown IPA

In the 2006 edition of *How To Brew* I wrote:

> *The hoppy brown ales, which can be nutty also, arose from the US homebrew scene when hop-crazy homebrewers decided that most brown ales were just too wimpy. But, American brown ales should not be brown IPAs! They should be malt-dominated beers with a toasted malt character, and the hops should be riding the crest of the wave of the beer's flavor. The hops should not be a tsunami. (p. 221)*

Well, apparently the challenge was accepted, and now we have a brown IPA category. In fact, hoppy brown ales were a thing in Texas during the 1980s and 90s, and were called Texas brown ales due to the number of Texas homebrewers brewing them (such as Houston's Foam Rangers club). Brown IPAs are a logical extension from American IPA in the same way that American brown ale differs from American pale and amber ales. The goal is a hop-forward beer backed by a dry brown ale character. These are eminently drinkable beers, with an OG of 1.056–1.070, FG of 1.008–1.016, and 40–70 IBUs. They are, of course, brown. American or English hop varieties seem to work best.

## Terra Firma—Brown IPA

**Original gravity:** 1.061
**Final gravity:** 1.012
**IBU:** 63

**SRM (EBC):** 15 (30)
**ABV:** 6.6%

### Version: Extract and Steeping Grain

| Wort A | Gravity points |
|---|---|
| 1.75 lb. (0.8 kg) pale ale DME | 26.5 |
| 2 lb. (0.9 kg) Briess Carabrown® malt (55°L) – Steeped | 7.5 |
| 1.15 lb. corn syrup solids | 16 |
| 0.5 lb. (225 g) caramel 80°L malt – Steeped | 3 |
| Boil gravity for 3 gal. (11.4 L) | 1.053 |

| Hop schedule | Boil time (min.) | IBUs |
|---|---|---|
| 0.5 oz. (15 g) Bravo 15% AA | 60 | 21 |
| 1 oz. (30 g) Bravo 15% AA | 15 | 21 |
| 1 oz. (30 g) Delta 6% AA | 15 | 8 |
| 1 oz. (30 g) Bravo 15% AA | Steep 15 | 8 |
| 1 oz. (30 g) Delta 6% AA | Steep 15 | 3 |
| 1 oz. (30 g) Bravo 15% AA | Dry Hop | (1) |
| 1 oz. (30 g) Delta 6% AA | Dry Hop | (1) |

| Wort B (add at knockout) | Gravity points |
|---|---|
| 4 lb. (1.8 kg) pale ale DME | 60 |

| Yeast strain | Pitch (billions of cells) | Fermentation temp. |
|---|---|---|
| American ale | 275 | 65°F (18°C) |

### Version: All-Grain

| Grain bill | Gravity points |
|---|---|
| 9 lb. (4.1 kg) pale ale malt | 36 |
| 2 lb. (0.9 kg) Briess Carabrown® malt (55°L) | 7 |
| 2 lb. (0.9 kg) flaked maize | 8.5 |
| 0.5 lb. (225 g) caramel 80°L malt – Steeped | 1 |
| Boil gravity for 7 gal. (27 L) | 1.052 |

| Mash schedule | Rest temp. | Rest time (min.) |
|---|---|---|
| Conversion rest – Infusion | 153°F (67°C) | 60 |

| Hop schedule | Boil time (min.) | IBUs |
|---|---|---|
| 0.5 oz. (15 g) Bravo 15% AA | 60 | 21 |
| 1 oz. (30 g) Bravo 15% AA | 15 | 21 |
| 1 oz. (30 g) Delta 6% AA | 15 | 8 |
| 1 oz. (30 g) Bravo 15% AA | Steep 15 | 8 |
| 1 oz. (30 g) Delta 6% AA | Steep 15 | 3 |
| 1 oz. (30 g) Bravo 15% AA | Dry Hop | (1) |
| 1 oz. (30 g) Delta 6% AA | Dry Hop | (1) |

| Yeast strain | Pitch (billions of cells) | Fermentation temp. |
|---|---|---|
| American ale | 275 | 65°F (18°C) |

| Recommended Water Profile (ppm) | | | Brew cube: Amber, Hoppy, Medium | | |
|---|---|---|---|---|---|
| Ca | Mg | Total alk. | SO$_4$ | Cl | RA |
| 75–125 | 20 | 100–150 | 150–250 | 100–150 | 25–75 |

## Black IPA

Black IPAs are not stout IPAs; instead they are a typical American IPA with the malt character of a *schwarzbier*; black, dry, but not dominated by roast malt. Instead, the roast malt and crystal malt characters are just strong enough to round out the base malt character, resulting in a hoppy beer that tastes fuller and better balanced than its pale cousins. Debittered black malts are best; these are smooth black ales without astringency, and even less grainy or biscuit flavor than in a brown IPA. Any hop variety can be used, but it is important that the hop bitterness and malt bitterness don't compound each other; avoid the more sulfurous hop varieties. The color of black IPAs ranges from deep brown to black with ruby highlights. This beer should not really be opaque; if it is, I would bet that the specialty malts have exceeded 10% of the grain bill. In fact, the roast malt additions should be 5% or less of the grain bill. The OG of black IPA ranges from 1.050–1.085, FG 1.010–1.016, and the bitterness 50–90 IBUs.

## Glorious Abyss—Black IPA

**Original gravity:** 1.070  
**Final gravity:** 1.017  
**IBU:** 75

**SRM (EBC):** 21 (42)  
**ABV:** 7.7%

| *Version: Extract and Steeping Grain* | |
|---|---|
| Wort A | Gravity points |
| 3.5 lb. (1.6 kg) pale DME | 52.5 |
| 0.5 lb. (225 g) caramel 60°L – Steeped | 3 |
| 0.5 lb. (225 g) Briess Black Prinz® wheat malt – Steeped | 5 |
| Boil gravity for 3 gal. (11.4 L) | 1.060 |

| Hop schedule | Boil time (min.) | IBUs |
|---|---|---|
| 1 oz. (30 g) Nugget 13% AA | 60 | 34 |
| 1 oz. (30 g) Citra 13% AA | 15 | 17 |
| 1 oz. (30 g) Simcoe 12% AA | 15 | 15 |
| 1 oz. (30 g) Cascade 7% AA | Steep 15 | 3 |
| 1 oz. (30 g) Amarillo 10% AA | Steep 15 | 5 |
| 1 oz. (30 g) Cascade 7% AA | Dry | (1) |
| 1 oz. (30 g) Amarillo 10% AA | Dry | (1) |

| Wort B (add at knockout) | Gravity points |
|---|---|
| 2.5 lb. (1.13 kg) pale DME | 45 |
| 2 lb. (0.9 kg) wheat DME | 30 |

| Yeast strain | Pitch (billions of cells) | Fermentation temp. |
|---|---|---|
| American ale | 300 | 67°F (19°C) |

## Version: All-Grain

| Grain bill | Gravity points |
|---|---|
| 12 lb. (5.4 kg) pale ale malt | 48 |
| 2.5 lb. (1.13 kg) wheat malt | 10 |
| 0.5 lb. (225 g) caramel 60°L | 1 |
| 0.5 lb. (225 g) Briess Black Prinz® wheat malt | 2 |
| Boil gravity for 7 gal. (27 L) | 1.061 |

| Mash schedule | Rest temp. | Rest time (min.) |
|---|---|---|
| Conversion rest – Infusion | 150°F (65°C) | 60 |

| Hop schedule | Boil time (min.) | IBUs |
|---|---|---|
| 1 oz. (30 g) Nugget 13% AA | 60 | 34 |
| 1 oz. (30 g) Citra 13% AA | 15 | 17 |
| 1 oz. (30 g) Simcoe 12% AA | 15 | 15 |
| 1 oz. (30 g) Cascade 7% AA | Steep 15 | 3 |
| 1 oz. (30 g) Amarillo 10% AA | Steep 15 | 5 |
| 1 oz. (30 g) Cascade 7% AA | Dry | (1) |
| 1 oz. (30 g) Amarillo 10% AA | Dry | (1) |

| Yeast strain | Pitch (billions of cells) | Fermentation temp. |
|---|---|---|
| American ale | 300 | 67°F (19°C) |

| Recommended Water Profile (ppm) | | | Brew cube: Amber, Hoppy, Medium | | |
|---|---|---|---|---|---|
| Ca | Mg | Total alk. | SO$_4$ | Cl | RA |
| 100–150 | 20 | 50–100 | 200–400 | 50–100 | 0–50 |

## American Strong Ale

American strong ale is another of my favorite styles. Although its name has changed over the years from such appellations as "stock ale" and "imperial red," it is essentially the big brother to American amber ale, being very malty and rich with a balancing (but not overweening) hop bitterness and aroma. The malt character tends to dominate, or at least balance, the hop character; it is but a short step (or hop) to red IPA from here. In fact, American strong ale tends to be a bigger and more bitter beer than red IPA; the difference is emphasis on malt flavor versus hop flavor.

The malt flavors of American strong ale come from the caramel and biscuit malts. Although most examples are amber colored, roast malts can be used sparingly to add a deeper red color. There shouldn't be any chocolate or coffee flavors from roast malts, because those tend to transform the beer into a Baltic porter, although much too bitter for that style.

The following recipe is my version of a popular American beer that seems to straddle the line between American strong ale and red IPA.

### Confident Bastard—American Strong Ale

**Original gravity:** 1.070  
**Final gravity:** 1.016  
**IBU:** 75

**SRM (EBC):** 18 (36)  
**ABV:** 7.5%

| Version: Extract and Steeping Grain | | |
|---|---|---|
| **Wort A** | **Gravity points** | |
| 3.5 lb. (1.6 kg) pale ale DME | 53 | |
| 0.5 lb. (225 g) Briess Victory® malt – Steeped | 3 | |
| 1 lb. (450 g) Weyermann Caraaroma® malt – Steeped | 3 | |
| Boil gravity for 3 gal. (11.4 L) | 1.059 | |
| **Hop schedule** | **Boil time (min.)** | **IBUs** |
| 1 oz. (30 g) Chinook 12% AA | 60 | 32 |
| 1 oz. (30 g) Centennial 10.5% AA | 15 | 14 |
| 1 oz. (30 g) Cascade 6% AA | 15 | 8 |
| 1 oz. (30 g) Chinook 12% AA | Steep 30 | 10 |
| 1 oz. (30 g) Centennial 10.5% AA | Steep 30 | 8.5 |
| 1 oz. (30 g) Chinook 12% AA | Dry Hop | (1) |
| 1 oz. (30 g) Cascade 6% AA | Dry Hop | (1) |
| 1 oz. (30 g) Centennial 10.5% AA | Dry Hop | (1) |
| **Wort B (add at knockout)** | **Gravity points** | |
| 4.5 lb. (2.04 kg) pale ale DME | 67.5 | |
| **Yeast strain** | **Pitch (billions of cells)** | **Fermentation temp.** |
| English ale | 300 | 65°F (18°C) |

| Version: All-Grain | | |
| --- | --- | --- |
| **Grain bill** | **Gravity points** | |
| 13.77 lb. (6.25 kg) pale ale malt | 55 | |
| 0.5 lb. (225 g) Briess Victory® malt | 1.5 | |
| 1 lb. (450 g) Weyermann Caraaroma® malt 150°L | 3.5 | |
| Boil gravity for 7 gal. (27 L) | 1.060 | |
| **Mash schedule** | **Rest temp.** | **Rest time (min.)** |
| Conversion rest – Infusion | 150°F (65°C) | 60 |
| **Hop schedule** | **Boil time (min.)** | **IBUs** |
| 1 oz. (30 g) Chinook 12% AA | 60 | 32 |
| 1 oz. (30 g) Centennial 10.5% AA | 15 | 14 |
| 1 oz. (30 g) Cascade 6% AA | 15 | 8 |
| 1 oz. (30 g) Chinook 12% AA | Steep 30 | 10 |
| 1 oz. (30 g) Centennial 10.5% AA | Steep 30 | 8.5 |
| 1 oz. (30 g) Chinook 12% AA | Dry Hop | (1) |
| 1 oz. (30 g) Cascade 6% AA | Dry Hop | (1) |
| 1 oz. (30 g) Centennial 10.5% AA | Dry Hop | (1) |
| **Yeast strain** | **Pitch (billions of cells)** | **Fermentation temp.** |
| English ale | 300 | 65°F (18°C) |
| **Recommended Water Profile (ppm)** | **Brew cube:** Amber, Balanced, Medium | |

| Ca | Mg | Total alk. | SO$_4$ | Cl | RA |
| --- | --- | --- | --- | --- | --- |
| 75–125 | 20 | 75–125 | 100–200 | 100–150 | 0–50 |

## Brown Ale

Surprisingly, brown ales are a relatively recent style, only becoming common in the twentieth century. Historically, they were similar to the brown malt porters of the eighteenth and nineteenth century but less hopped. Brown ales actually disappeared for about a century when brown malt fell out of favor as a base malt. Today there are many variations of brown ale: mild, sweet, nutty, and hoppy.

Brown ales as a class have grown to bridge the gap between pale ales and porters. There are two types according to the BJCP guidelines: English and American. The English version tends to be softer and maltier, whereas the American version tends to be drier and more balanced between the malt and hops. Contrary to popular myth, there are no nuts or nut extracts in typical brown ales; toasted malts give the beer a nutlike flavor and nut-brown color. The recipe below makes use of a particularly nice malt from Briess Malt & Ingredients Co. in Wisconsin; it has a very toasty, bread crust and graham cracker flavor without harshness. The recipe is named for the Tittabawasee River that runs through my hometown of Midland, MI. The recipe can be made more English by adding more crystal malt, or more American by changing up the hops and increasing the bitterness. I prefer this version, which is right down the middle. The dark chocolate malt is also one of my favorites, but I use it here mainly for color (I use more of it in my porter recipe).

The style guidelines for brown ales range from 1.040–1.060 OG, 20–30 IBU, and all colors of brown, from copper to nearly black, 12–35 SRM (24–70 EBC).

## Tittabawasee Brown—Brown Ale

**Original gravity:** 1.050
**Final gravity:** 1.013
**IBU:** 26

**SRM (EBC):** 18 (36)
**ABV:** 5%

### Version: Extract and Steeping Grain

| Wort A | Gravity points | |
|---|---|---|
| 2 lb. (0.9 kg) pale ale DME | 30 | |
| 2 lb. (450 g) Briess Carabrown® malt 55°L – Steeped | 7 | |
| 0.5 lb. (225 g) caramel 80°L malt – Steeped | 3 | |
| 0.15 lb. (70 g) Briess Dark Chocolate malt 400°L – Steeped | 1.5 | |
| Boil gravity for 3 gal. (11.4 L) | 1.042 | |
| **Hop schedule** | **Boil time (min.)** | **IBUs** |
| 0.5 oz. (15 g) Nugget 12% AA | 60 | 18.5 |
| 0.5 oz. (15 g) East Kent Goldings 5% AA | 15 | 4.5 |
| 1 oz. (30 g) East Kent Goldings 5% AA | Steep 15 | 3 |
| **Wort B (add at knockout)** | **Gravity points** | |
| 3.3 lb. (1.5 kg) pale ale DME | 67.5 | |
| **Yeast strain** | **Pitch (billions of cells)** | **Fermentation temp.** |
| English ale | 200 | 65°F (18°C) |

### Version: All-Grain

| Grain bill | Gravity points | |
|---|---|---|
| 9.5 lb. (4.3 kg) pale ale malt | 38 | |
| 2 lb. (450 g) Briess Carabrown® malt 55°L | 3.5 | |
| 0.5 lb. (225 g) caramel 80°L malt – Steeped | 1 | |
| 0.15 lb. (70 g) Briess Dark Chocolate malt 400°L | 0.5 | |
| Boil gravity for 7 gal. (27 L) | 1.043 | |
| **Mash schedule** | **Rest temp.** | **Rest time (min.)** |
| Conversion rest – Infusion | 150°F (65°C) | 60 |
| **Hop schedule** | **Boil time (min.)** | **IBUs** |
| 0.5 oz. (15 g) Nugget 12% AA | 60 | 18.5 |
| 0.5 oz. (15 g) East Kent Goldings 5% AA | 15 | 4.5 |
| 1 oz. (30 g) East Kent Goldings 5% AA | Steep 15 | 3 |
| **Yeast strain** | **Pitch (billions of cells)** | **Fermentation temp.** |
| English ale | 200 | 65°F (18°C) |

| Recommended Water Profile (ppm) | | | Brew cube: Amber, Balanced, Medium | | |
|---|---|---|---|---|---|
| Ca | Mg | Total alk. | SO$_4$ | Cl | RA |
| 75–125 | 20 | 75–125 | 100–150 | 100–150 | 0–50 |

## Porter

Historically, porters preceded stouts and had a much different character than today. Porter was the first beer mass produced on an industrial scale, and was matured in swimming pool-sized wooden vats for about six months before serving. The dominant flavor came from the use of highly kilned brown malt, which was used as the base malt. The long aging time was necessary for the brown malt's rough flavors (like tree bark) to mellow. What starts out as harshly bitter-malt beer became much smoother as the tannins settled out.

Today, porter is a dark ale with a malty flavor and a roasted finish. I would compare the flavor of porter to stout as the difference between strong black tea and coffee. Similarly, I would compare the flavor of brown ale to porter as the difference between light tea and strong tea. A brown ale is toasty; a porter is toasty, but darker and more intense. The defining character of a porter is chocolate malt, which gives the beer a cocoa flavor bridging between toast and roast. Porters are balanced, but should have some residual sweetness to round out the malt character. A porter is richer and has a higher gravity than brown ale.

Similar to the distinction between brown ales, there are two types as defined by the BJCP *2015 Style Guidelines*, English porter and American porter. English porter is softer and sweeter than American porter, which will generally have a stronger roast component and higher bittering. The OG for English porter ranges from 1.040 to 1.052, while American porter is stronger at 1.050–1.070. English porter has 18–35 IBU, whereas the bitterness is increased accordingly for the stronger American porter at 25–50 IBU. The color range overlaps, 20–30 SRM (40–60 EBC) for English porter and 20–40 SRM for American. If the roast character is increased with roast malts other than chocolate malt, a porter turns into a stout.

A third type, Baltic porter, is basically an imperial porter, where the essential character of a porter is doubled in the grain bill. Porters should be fairly well attenuated (dry), though sweet ("brown") porters are popular too. A porter should be lighter in both body and color when compared to stout. In addition, it should not be opaque, a porter should have a deep ruby red glow when held up to the light.

British, Irish, and American yeast strains are good choices for porters and stouts.

## Palmer's Porter—Porter

**Original gravity:** 1.055  
**Final gravity:** 1.016  
**IBU:** 40

**SRM (EBC):** 18 (36)  
**ABV:** 5.3%

| *Version: Extract and Steeping Grain* | |
|---|---|
| **Wort A** | **Gravity points** |
| 2 lb. (0.9 kg) pale ale DME | 24 |

| 1 lb. (450 g) caramel 80°L malt | 4 | |
|---|---|---|
| 1 lb. (450 g) Briess Carabrown® malt 55°L – Steep | 3 | |
| 1 lb. (450 g) Briess Dark Chocolate malt 400°L – Steep | 7 | |
| 0.33 lb. (150 g) Briess Black Prinz® malt 500°L – Steep | 3 | |
| Boil gravity for 3 gal. (11.4 L) | 1.041 | |
| **Hop schedule** | **Boil time (min.)** | **IBUs** |
| 1 oz. (30 g) Centennial 10.5% AA | 60 | 32 |
| 0.5 oz. (15 g) Willamette 5% AA | 15 | 5 |
| 1 oz. (30 g) East Kent Goldings 5% AA | Steep 15 | 3 |
| **Wort B (add at knockout)** | **Gravity points** | |
| 3.75 lb. (1.7 kg) pale ale DME | 53 | |
| **Yeast strain** | **Pitch (billions of cells)** | **Fermentation temp.** |
| English ale | 350 | 65°F (18°C) |

## Version: All-Grain

| **Grain bill** | **Gravity points** | |
|---|---|---|
| 9.5 lb. (4.3 kg) pale ale malt | 38 | |
| 1 lb. (450 g) caramel 80°L malt | 2 | |
| 1 lb. (450 g) Briess Carabrown® malt 55°L | 1.5 | |
| 1 lb. (450 g) Briess Dark Chocolate malt 400°L | 3.5 | |
| 0.33 lb. (150 g) Briess Black Prinz® malt 500°L | 1.5 | |
| Boil gravity for 7 gal. (27 L) | 1.047 | |
| **Mash schedule** | **Rest temp.** | **Rest time (min.)** |
| Conversion rest – Infusion | 150°F (65°C) | 60 |
| **Hop schedule** | **Boil time (min.)** | **IBUs** |
| 1 oz. (30 g) Centennial 10.5% AA | 60 | 32 |
| 0.5 oz. (15 g) Willamette 5% AA | 15 | 5 |
| 1 oz. (30 g) East Kent Goldings 5% AA | Steep 15 | 3 |
| **Yeast strain** | **Pitch (billions of cells)** | **Fermentation temp.** |
| English ale | 350 | 65°F (18°C) |

| **Recommended Water Profile (ppm)** | | | **Brew cube:** Dark, Balanced, Medium | | |
|---|---|---|---|---|---|
| Ca | Mg | Total alk. | $SO_4$ | Cl | RA |
| 75–125 | 20 | 100–150 | 100–150 | 100–150 | 50–100 |

## Stout

Arguably one of the most popular styles among homebrewers, stouts vary a lot in flavor, degree of roastiness, and body. There are dry stouts, sweet stouts, export stouts, oatmeal stouts, coffee stouts, and more besides. The one defining characteristic of a stout is the use of highly roasted malts and/or unmalted roast barley. The most popular, Guinness® Extra Stout, is the defining example of Irish dry stout and uses only pale malt, unmalted roast barley, and flaked barley; no crystal malt is used. English stouts tend to be of the sweet stout style and will include chocolate and crystal malts. Some English stouts do not use any black malt or roast barley at all, getting their color from amber malt, dark crystal and chocolate malt. Dry stouts are session beers, with an OG of 1.036–1.040 and bitterness of 25–45 IBU. The color is dark brown to black at 25–40 SRM (50–80 EBC). Export stouts are brewed to a very high gravity, 1.075–1.100 OG, with a huge complexity of flavors, being sweet and tarry, fruity and quite bitter. Oatmeal stout is my favorite, being a sweet stout with the smooth silkiness of oatmeal added in. Coffee stouts are another homebrew favorite, because the taste of coffee perfectly complements the roast character of a stout.

The following recipe is more typical of the sweeter stouts of Cork, Ireland, but not as sweet as an English or sweet stout.

## Mill Run Stout

**Original gravity:** 1.040
**Final gravity:** 1.012
**IBU:** 25

**SRM (EBC):** 30 (60)
**ABV:** 3.7%

| *Version: Extract and Steeping Grain* | | |
|---|---|---|
| **Wort A** | **Gravity points** | |
| 1 lb. (450 g) pale ale DME | 14 | |
| 1 lb. (450 g) roast barley 300°L – Steep | 8 | |
| 1 lb. (450 g) flaked barley – Instamash® | 8 | |
| 0.44 lb. (200 g) Briess Dark Chocolate malt 400°L – Steep | 3.5 | |
| 0.44 lb. (200 g) crystal 80°L – Steep | 2 | |
| Boil gravity for 3 gal. (11.4 L) | 1.036 | |
| **Hop schedule** | **Boil time (min.)** | **IBUs** |
| 1 oz. (30 g) East Kent Goldings 5% AA | 60 | 18 |
| 0.5 oz. (15 g) East Kent Goldings 5% AA | 30 | 7 |
| **Wort B (add at knockout)** | **Gravity points** | |
| 2.75 lb. (1.25 kg) pale ale DME | 38.5 | |
| **Yeast strain** | **Pitch (billions of cells)** | **Fermentation temp.** |
| English or Irish ale | 175 | 65°F (18°C) |

### Version: All-Grain

| Grain bill | Gravity points | |
|---|---|---|
| 6.25 lb. (4.3 kg) pale ale malt | 25 | |
| 1 lb. (450 g) roast barley (300°L) – Steep | 4 | |
| 1 lb. (450 g) flaked barley – Instamash® | 3.5 | |
| 0.44 lb. (200 g) Briess Dark Chocolate malt 400°L – Steep | 1.5 | |
| 0.44 lb. (200 g) crystal 80°L – Steep | 1 | |
| Boil gravity for 7 gal. (27 L) | 1.035 | |
| **Mash schedule** | **Rest temp.** | **Rest time (min.)** |
| Conversion rest – Infusion | 150°F (65°C) | 60 |
| **Hop schedule** | **Boil time (min.)** | **IBUs** |
| 1 oz. (30 g) East Kent Goldings 5% AA | 60 | 18 |
| 0.5 oz. (15 g) East Kent Goldings 5% AA | 30 | 7 |
| **Yeast strain** | **Pitch (billions of cells)** | **Fermentation temp.** |
| English or Irish ale | 175 | 65°F (18°C) |

| Recommended Water Profile (ppm) | | | Brew cube: Dark, Malty, Medium | | |
|---|---|---|---|---|---|
| **Ca** | **Mg** | **Total alk.** | **SO₄** | **Cl** | **RA** |
| 75–125 | 20 | 100–150 | 50–100 | 100–150 | 50–100 |

*Note: SO₄ is rendered as $SO_4$.*

***Variations***

***Dry stout option:*** *Eliminate the crystal and chocolate malt. Add same weight of base malt to maintain same OG.*

***Oatmeal stout:*** *Add one pound (450 g) of instant rolled oats and use Palmer's Instamash® mash enzymes to convert them in the steep. There really is no difference in flavor between instant and old-fashioned oats. Instant oats can be added directly to the mash/steep, whereas the other grades would need to be cooked first before adding to the brew. Original gravity 1.056–1.075, and 30–50 IBU.*

***Coffee stout:*** *This is an easy variation to any stout recipe. Cold steeping carries more aroma into the beer, while hot steeping carries more flavor. To cold steep, place ¼–½ lb. (115–225 g) of freshly ground coffee into 1–2 qt. (0.95–1.9 L) of water in the refrigerator for 24 hours. Pour that cold coffee through a paper coffee filter and add it to your beer after the boil. To hot steep, place the coffee inside a paper filter and pour hot wort into the filter after the boil. Add the hot coffee to the wort and then chill as usual.*

## Imperial Stout

This stout recipe was developed for the 2016 Congresso Técnico dos Cervejeiros Artensanais in Florianopolis, Brazil. The smoked malt is just for accent, a touch of smoke to add to the malt complexity of dark chocolate and coffee. The style guidelines for imperial stout are 1.075–1.115 OG, 50–90 IBUs, and black in color.

## Namorada Bela Stout—Imperial Stout

**Original gravity:** 1.090    **SRM (EBC):** 50 (100)
**Final gravity:** 1.023-ish    **ABV:** 9.5%
**IBU:** 70

| Version: *Extract and Steeping Grain* | |
|---|---|
| **Wort A** | **Gravity points** |
| 3 lb. (1.4 kg) pale ale DME | 31.5 |
| 2 lb. (900 g) Briess Roast Barley 300°L | 12 |
| 1.1 lb. (500 g) flaked barley – Instamash® | 6 |
| 1.1 lb. (500 g) Briess Dark Chocolate malt 400°L | 7.5 |
| 1.1 lb. (500 g) Briess Apple Smoked malt | 7 |
| 1.1 lb. (500 g) Weyermann Caraaroma® 130°L | 6.5 |
| Boil gravity for 4 gal. (15.2 L) | 1.071 |

| Hop schedule | Boil time (min.) | IBUs |
|---|---|---|
| 2 oz. (60 g) UK Phoenix 10% AA | 60 | 48 |
| 1 oz. (30 g) UK Phoenix 10% AA | 30 | 18 |
| 1 oz. (30 g) UK Phoenix 10% AA | Steep 15 | 5 |

| Wort B (add at knockout) | Gravity points |
|---|---|
| 6 lb. (2.75 kg) pale ale DME | 84 |

| Yeast strain | Pitch (billions of cells) | Fermentation temp. |
|---|---|---|
| English or Irish ale | 375 | 65°F (18°C) |

| Version: *All-Grain* | |
|---|---|
| **Grain bill** | **Gravity points** |
| 13.25 lb. (6 kg) pale ale malt | 53 |
| 2 lb. (900 g) Briess Roast Barley 300°L | 7.5 |
| 1.1 lb. (500 g) flaked barley | 3.5 |
| 1.1 lb. (500 g) Briess Dark Chocolate malt 400°L | 4 |
| 1.1 lb. (500 g) Briess Apple Smoked malt | 3.5 |
| 1.1 lb. (500 g) Weyermann Caraaroma® 130°L | 3.5 |
| Boil gravity for 7 gal. (27 L) | 1.076 |

| Mash schedule | Rest temp. | Rest time (min.) |
|---|---|---|
| Conversion rest – Infusion | 150°F (65°C) | 60 |

| Hop schedule | Boil time (min.) | IBUs |
|---|---|---|
| 2 oz. (60 g) UK Phoenix 10% AA | 60 | 48 |
| 1 oz. (30 g) UK Phoenix 10% AA | 30 | 18 |
| 1 oz. (30 g) UK Phoenix 10% AA | Steep 15 | 5 |

| Yeast strain | | Pitch (billions of cells) | Fermentation temp. |
|---|---|---|---|
| English or Irish ale | | 375 | 65°F (18°C) |
| **Recommended Water Profile (ppm)** | | **Brew cube:** Dark, Malty, Medium | |

| Ca | Mg | Total alk. | SO₄ | Cl | RA |
|---|---|---|---|---|---|
| 75–125 | 20 | 100–150 | 50–100 | 100–150 | 50–100 |

*Notes*

*You will need a bigger kettle due to the large amount of steeping grains in this recipe. You will probably need an 8 gal. (30 L) kettle and will need 4 gal. (15.2 L) of water for the steeping, due to the amount that the wet grain will carry away. The use of the flaked barley will require the use of extra mash enzymes, such as Palmer's Instamash or White Labs Opti-mash.*

## Barleywine

Barleywine is the drink of the gods, the intellectual gods anyway. Few beverages can equal the complexity of flavors that a properly aged barleywine has: malt, fruit, spice, and warmth from the high level of alcohol (9%–14%). Barleywine as a style has been around for several hundred years, but Bass coined the name in 1903.

Barleywines tend to require the use of malt extracts to help achieve the high gravities that are their hallmark. Barleywine usually consist solely of pale and crystal malts to avoid masking the flavor with roasted malts. The color of barleywine ranges from deep gold to ruby red. Wheat and rye malts can be used for "accent," counterbalancing the heavy maltiness of the barley. A barleywine is meant to be sipped in front of the fire on a cold winter's night, providing the fuel for philosophical thoughts on science and the wonders of metallurgy.

My recipe is atypical in its high usage of wheat, but I had the Rubicon Brewing Company's Winter Wheat Warmer in Sacramento and liked it. And yes, I read the Tolkien books as a seventh grader (1977), and developed this recipe and the name in 1994—long before the movie came out. It seemed like a draught for the toughest orcs.

## Fighting Urak Hai—Barleywine

**Original gravity:** 1.098
**Final gravity:** 1.025-ish
**IBU:** 98

**SRM (EBC):** 22 (44)
**ABV:** 10.5%

| *Version: Extract and Steeping Grain* | |
|---|---|
| **Wort A** | **Gravity points** |
| 5 lb. (2.27 kg) wheat DME | 70 |
| 0.5 lb. (225 g) Briess Dark Chocolate malt 400°L | 4 |
| 0.5 lb. (225 g) Weyermann Caraaroma 130°L | 4 |
| Boil gravity for 3 gal. (11.4 L) | 1.078 |

| Hop schedule | Boil time (min.) | IBUs |
|---|---|---|
| 2 oz. (60 g) Glacier 12% AA | 60 | 50 |
| 1.5 oz. (45 g) Glacier 12% AA | 30 | 30 |
| 1 oz. (30 g) Glacier 12% AA | 15 | 13 |
| 1 oz. (30 g) Glacier 12% AA | Steep 15 | 5 |
| **Wort B (add at knockout)** | **Gravity points** | |
| 7.3 lb. (3.3 kg) pale ale DME | 102 | |
| **Yeast strain** | **Pitch (billions of cells)** | **Fermentation temp.** |
| American ale | 400 | 65°F (18°C) |

### Version: All-Grain

| Grain bill | Gravity points | |
|---|---|---|
| 13.25 lb. (6 kg) pale ale malt | 53 | |
| 6.9 lb. (3.1 kg) wheat malt | 28 | |
| 0.5 lb. (225 g) Briess Dark Chocolate malt 400°L | 2 | |
| 0.5 lb. (225 g) Weyermann Caraaroma 130°L | 1.5 | |
| Boil gravity for 7 gal (27 l) | 1.084 | |
| **Mash schedule** | **Rest temp.** | **Rest time (min.)** |
| Conversion rest – Infusion | 150°F (65°C) | 60 |
| **Hop schedule** | **Boil time (min.)** | **IBUs** |
| 2 oz. (60 g) Glacier 12% AA | 60 | 50 |
| 1.5 oz. (45 g) Glacier 12% AA | 30 | 30 |
| 1 oz. (30 g) Glacier 12% AA | 15 | 13 |
| 1 oz. (30 g) Glacier 12% AA | Steep 15 | 5 |
| **Yeast strain** | **Pitch (billions of cells)** | **Fermentation temp.** |
| American ale | 400 | 65°F (18°C) |

| Recommended Water Profile (ppm) | | | Brew cube: Amber, Malty, Medium | | |
|---|---|---|---|---|---|
| Ca | Mg | Total alk. | $SO_4$ | Cl | RA |
| 75–125 | 20 | 50–100 | 50–100 | 100–150 | 0–50 |

## THE LAGER STYLES

### Pilsener

Beer as the world knew it changed dramatically in 1842 when a brewery in the town of Pilsen produced the first light golden lager. Until that time, beers had been rather dark, varying from amber ("pale"), to deep brown or black. Today Pilsner Urquell® is that same beer, "the original of Pilsen." The original Pilsener beer is a hoppy, dry beer of 1.045 OG. The Pilsener style is imitated more than any other and interpretations run from the light flowery lagers of Germany to the maltier, more herbal versions of the Netherlands, to the increasingly tasteless varieties of light and dry from the United States and Japan. Most of these are broadly in the Pilsener style, but lack the assertive noble hop bitterness and flavor of the original.

Brewing a true Pilsener can be fairly difficult, especially from an all-grain point of view. Pilsen has very soft water, the next closest thing to distilled water, and the relatively high baseline pH of lightly kilned lager malt makes achieving a proper mash pH difficult. In 2014, at the joint ASBC/MBAA conference in Chicago, I was able to ask Dr. Ludwig Narziss how the Pilsen brewers achieved good conversion with such soft water (remember, the pH scale and measurement technique weren't invented until 1909). "Oh, they used salts," he replied. "Burtonization was high technology at the time and everyone was doing it." It is not known how much brewing salt they added, but adding at least 50 ppm of calcium is a good place to start.

When brewing an all-grain Pilsener, it is best to use distilled or de-ionized water and build it up with calcium chloride and magnesium sulfate to improve the mash conditions and prevent tannin astringency.

## Plzenske Pivo—Pilsener Lager

**Original gravity:** 1.046  
**Final gravity:** 1.011  
**IBU:** 40

**SRM (EBC):** 4 (8)  
**ABV:** 4.5%

### Version: Extract and Steeping Grain

| Wort A | Gravity points | |
| --- | --- | --- |
| 3 lb. (1.36 kg) Pilsner DME | 42 | |
| Boil gravity for 3 gal. (11.4 L) | 1.042 | |

| Hop schedule | Boil time (min.) | IBUs |
| --- | --- | --- |
| 1 oz. (30 g) German Tradition 6% AA | 60 | 19 |
| 1.5 oz. (45 g) German Select 5% AA | 30 | 18 |
| 2 oz. (60 g) Saaz 4% AA | Steep 15 | 3 |

| Wort B (add at knockout) | Gravity points | |
| --- | --- | --- |
| 3 lb. (1.36 kg) Pilsner DME | 42 | |

| Yeast strain | Pitch (billions of cells) | Fermentation temp. |
| --- | --- | --- |
| Czech lager | 380 | 52°F (11°C) |

### Version: All-Grain

| Grain bill | Gravity points |
| --- | --- |
| 10 lb. (4.5 kg) Pilsner malt | 39 |
| Boil gravity for 7 gal. (27 L) | 1.039 |

| Mash schedule | Rest temp. | Rest time (min.) |
|---|---|---|
| Conversion rest – Infusion | 150°F (65°C) | 30 |
| Single decoction (see fig. 17.3) | (boil) | 20 |
| **Hop schedule** | **Boil time (min.)** | **IBUs** |
| 1 oz. (30 g) German Tradition 6% AA | 60 | 19 |
| 1.5 oz. (45 g) German Select 5% AA | 30 | 18 |
| 2 oz. (60 g) Saaz 4% AA | Steep 15 | 3 |
| **Yeast strain** | **Pitch (billions of cells)** | **Fermentation temp.** |
| Czech lager | 380 | 52°F (11°C) |

| Recommended Water Profile (ppm) | | | Brew cube: Pale, Balanced, Soft | | |
|---|---|---|---|---|---|
| Ca | Mg | Total alk. | SO$_4$ | Cl | RA |
| 50–100 | 10 | 0–50 | 50–100 | 50–100 | −75–0 |

## Munich Helles

Not every pale lager is a Pilsner, and many of today's pale lager beers are actually based on Munich *helles*, which was, in fact, developed to compete with pale Pilsner-type beer. Previously, the most typical beer of Munich was a *dunkel*, made entirely with Munich malt. A helles has a soft, grainy malt flavor and a soft, clean finish. The balance of the beer is toward the malt, but only lightly. The beer is like a waltz, the malt leads but the hops are in perfect step. The hop character plays a supporting role and should be floral, lightly spicy, and/or herbal. Overall the character is balanced, but the malt always leads the dance. The style guidelines for helles are an OG of 1.044–1.048, 16–22 IBU, and color of 3–5 SRM (6–10 EBC).

## Schönes Mädel—Munich Helles

**Original gravity:** 1.046  
**Final gravity:** 1.011  
**IBU:** 22  

**SRM (EBC):** 4 (8)  
**ABV:** 4.6%  

| *Version: Extract and Steeping Grain* | | |
|---|---|---|
| **Wort A** | **Gravity points** | |
| 2.2 lb. (1 kg) Pilsner DME | 31 | |
| 1 lb. (450 g) Carahell® 10°L – Steeped | 8 | |
| Boil gravity for 3 gal. (11.4 L) | 1.042 | |
| **Hop schedule** | **Boil time (min.)** | **IBUs** |
| 1 oz. (30 g) German Tradition 6% AA | 60 | 19 |
| 1 oz. (30 g) Saaz 4% AA | Steep 15 | 3 |
| **Wort B (add at knockout)** | **Gravity points** | |
| 3.3 lb. (1.5 kg) Pilsner DME | 46 | |

| Yeast strain | Pitch (billions of cells) | Fermentation temp. |
|---|---|---|
| German lager | 380 | 52°F (11°C) |

### Version: All-Grain

| Grain bill | Gravity points | |
|---|---|---|
| 9 lb. (4 kg) Pilsner malt | 35.5 | |
| 1 lb. (450 g) Carahell® malt 10°L | 3.5 | |
| Boil gravity for 7 gal. (27 L) | 1.039 | |

| Mash schedule | Rest temp. | Rest time (min.) |
|---|---|---|
| Conversion rest – Infusion | 150°F (65°C) | 60 |

| Hop schedule | Boil time (min.) | IBUs |
|---|---|---|
| 1 oz. (30 g) German Tradition 6% AA | 60 | 19 |
| 1 oz. (30 g) Saaz 4% AA | Steep 15 | 3 |

| Yeast strain | Pitch (billions of cells) | Fermentation temp. |
|---|---|---|
| German lager | 380 | 52°F (11°C) |

| Recommended Water Profile (ppm) | | | | Brew cube: Pale, Malty, Soft | |
|---|---|---|---|---|---|
| Ca | Mg | Total alk. | SO$_4$ | Cl | RA |
| 50–100 | 10 | 0–50 | 0–50 | 50–100 | –50–0 |

*Note*

*This recipe may require some lactic acid or acidulated malt to help hit a mash pH of 5.2.*

## Dortmunder Export (Helles Exportbier)

Historically, Dortmunder and other export-strength beers were brewed less bitter than Pilsner beer, with the difference being that the brewing water of the region had more permanent hardness than Munich; water that was actually more similar to Burton-upon-Trent than any other region in Europe. The result was a pale lager with a firm mineral structure, making the beer more similar to pale ale than Pilsner. The result was a lager with a robust malt flavor and light clean bitterness, as compared to the softer structure and bitterness perception of the Pilsner. The current guidelines for *Exportbier* are an OG of 1.048–1.056, 20–30 IBU, and a color of 4–7 SRM (8–14 EBC). The current bitterness guideline for Czech premium lager is 30–45 IBU, and for German Pilsner is 22–40 IBU.

## Spieltag – Dortmunder Export

**Original gravity:** 1.056
**Final gravity:** 1.012
**IBU:** 38

**SRM (EBC):** 4 (8)
**ABV:** 5.9%

### Version: Extract and Steeping Grain

| Wort A | Gravity points | |
|---|---|---|
| 1.66 lb. (750 g) Pilsner DME | 23 | |
| 1.66 lb. (750 g) Munich DME | 23 | |
| Boil gravity for 3 gal. (11.4 L) | 1.046 | |
| **Hop schedule** | **Boil time (min.)** | **IBUs** |
| 1.5 oz. (45 g) Mandarina Bavaria 8% AA | 60 | 35 |
| 1 oz. (30 g) German Opal 6% AA | Steep 15 | 3 |
| **Wort B (add at knockout)** | **Gravity points** | |
| 4 lb. (1.8 kg) Pilsner DME | 56 | |
| **Yeast strain** | **Pitch (billions of cells)** | **Fermentation temp.** |
| German lager | 470 | 52°F (11°C) |

### Version: All-Grain

| Grain bill | Gravity points | |
|---|---|---|
| 11 lb. (5 kg) Pilsner malt | 43 | |
| 1.4 lb. (635 g) Munich malt | 5.5 | |
| Boil gravity for 7 gal. (27 L) | 1.049 | |
| **Mash schedule** | **Rest temp.** | **Rest time (min.)** |
| Conversion rest – Infusion | 150°F (65°C) | 60 |
| **Hop schedule** | **Boil time (min.)** | **IBUs** |
| 1.5 oz. (45 g) Mandarina Bavaria 8% AA | 60 | 35 |
| 1 oz. (30 g) German Opal 6% AA | Steep 15 | 3 |
| **Yeast strain** | **Pitch (billions of cells)** | **Fermentation temp.** |
| German lager | 470 | 52°F (11°C) |

| Recommended Water Profile (ppm) | | | | Brew cube: Pale, Balanced, Firm | |
|---|---|---|---|---|---|
| **Ca** | **Mg** | **Total alk.** | **SO$_4$** | **Cl** | **RA** |
| 100–150 | 10 | 0–50 | 50–150 | 50–150 | –100–0 |

## Classic American Pilsner

Around the turn of the century in the United States the Pilsner style was very popular, but with a typically American difference. That difference was corn (maize). It's only natural in the largest corn growing region in the world that some would wind up in beer as a fermentable. In addition, six-row barley was the most common variety available, but its higher protein levels (12%–14%) made it difficult to brew with. Adding corn (8%–9% protein) to the mash helped dilute the total protein levels and added some flavor complexity as well. Unfortunately, Prohibition and higher brewing costs that came afterward helped to increase the use of corn and rice in American Pilsner-style beers to the point of blandness.

The beer of our grandfathers was a delicious, malty sweet beer with a balanced hoppiness. The brewing water should be low in sulfates. Today, there are only a few commercially produced examples that adequately represent this beer that started the lager revolution in the United States. The strength of this beer was typically between 1.045 and 1.050 OG, with a hopping of 25–40 IBUs. The style had become lighter by the time of Prohibition and afterward tended to have an average gravity of 1.040–1.044 OG, with a correspondingly lower hopping rate of 20–30 IBUs. Classic American Pilsner is best brewed with a mash using flaked maize, or using the cereal mash method in chapter 17 with corn grits. I have also successfully brewed this style using pale malt extract and brewer's corn syrup, which has a high percentage of maltose (~50%) and more closely mimics the sugar profile of barley wort. High-fructose corn syrup and refined corn sugar just don't have the same corn character. Corn grits are not hominy grits like you see in the grocery store. Brewer's corn grits most closely resembles cornmeal, although grocery store cornmeal is often fortified with vitamins and minerals. Definitely avoid any grits or cornmeal that is fortified with iron.

## Your Father's Mustache—Pre-Prohibition Lager

**Original gravity:** 1.056  
**Final gravity:** 1.014  
**IBU:** 35  

**SRM (EBC):** 3 (6)  
**ABV:** 5.5%  

| *Version: Extract and Steeping Grain* | | |
|---|---|---|
| **Wort A** | **Gravity points** | |
| 2 lb. (0.9 kg) Pilsner DME | 28 | |
| 2 lb. (1.36 kg) flaked maize – Instamash® | 18 | |
| Boil gravity for 3 gal. (11.4 L) | 1.046 | |
| **Hop schedule** | **Boil time (min.)** | **IBUs** |
| 1 oz. (30 g) Cluster 7% AA | 60 | 22 |
| 1 oz. (30 g) Saaz 4% AA | 30 | 10 |
| 1 oz. (30 g) Saaz 4% AA | Steep 15 | 3 |
| **Wort B (add at knockout)** | **Gravity points** | |
| 3 lb. (1.36 kg) Pilsner DME | 42 | |
| **Yeast strain** | **Pitch (billions of cells)** | **Fermentation temp.** |
| German lager | 470 | 52°F (11°C) |

| Version: All-Grain | | |
|---|---|---|
| **Grain bill** | **Gravity points** | |
| 10 lb. (4.5 kg) Pilsner malt | 40 | |
| 2 lb. (1.36 kg) flaked maize | 8 | |
| Boil gravity for 7 gal. (27 L) | 1.048 | |
| **Mash schedule** | **Rest temp.** | **Rest time (min.)** |
| Conversion rest – Infusion | 150°F (65°C) | 30 |
| Single decoction (see fig. 17.3) | (boil) | 20 |
| **Hop schedule** | **Boil time (min.)** | **IBUs** |
| 1 oz. (30 g) Cluster 7% AA | 60 | 22 |
| 1 oz. (30 g) Saaz 4% AA | 30 | 10 |
| 1 oz. (30 g) Saaz 4% AA | Steep 15 | 3 |
| **Yeast strain** | **Pitch (billions of cells)** | **Fermentation temp.** |
| German lager | 470 | 52°F (11°C) |
| **Recommended Water Profile (ppm)** | **Brew cube:** Pale, Balanced, Soft | |

| Ca | Mg | Total alk. | SO$_4$ | Cl | RA |
|---|---|---|---|---|---|
| 50 100 | 10 | 0 50 | 50 100 | 50 100 | 75–0 |

*Note*

*If you are using corn grits or polenta, you need to cook this cereal first.*

*Recipe contributed by my good friend Jeff Renner*

## Bock

Bock beer is an old style, dating from the late sixteenth century. The style grew out of the then world-famous beer of Einbeck. It was a strong beer brewed using one-third wheat and two-thirds barley, with a pale color, crisp taste, and a hint of acidity (the acidity was a carryover from the sour wheat beers of the day). It was brewed as an ale, but was stored cold for extended periods. Einbecker beer was widely exported and was the envy of the region.

For years, the nobles of Munich tried to imitate the strong northern beer in their breweries with limited success. Finally, in 1612, the brewmaster of Einbeck was persuaded to go south and work on producing a strong beer for Munich. The result was a strong beer interpretation of the Munich *braunbier*, a rich, malty brown ale. The classic Munich bock beer is a lager with an assertive malt character, a warmth from the higher alcohol level, and only enough hop bitterness to just balance the sweetness of the malt. Bock and its big monastic brother, *doppelbock*, should not have any fusel alcohol character nor any of the fruitiness of ales.

The style guidelines for traditional dunkel bocks are an OG of 1.064–1.072, 20–27 IBU, and a rich brown color of 14–22 SRM (28–44 EBC).

## Copper Country Bock—Dunkel Bock

**Original gravity:** 1.068  **SRM (EBC):** 20 (40)
**Final gravity:** 1.014  **ABV:** 7.4%
**IBU:** 27

### Version: Extract and Steeping Grain

| Wort A | Gravity points | |
|---|---|---|
| 3.5 lb. (1.6 kg) Munich DME | 53 | |
| 0.5 lb. (225 g) Briess Midnight Wheat malt – Steeped | 4 | |
| Boil gravity for 3 gal. (11.4 L) | 1.057 | |
| **Hop schedule** | **Boil time (min.)** | **IBUs** |
| 0.8 oz. (28 g) German Magnum 12% AA | 60 | 25 |
| **Wort B (add at knockout)** | **Gravity points** | |
| 4.5 lb. (2 kg) Munich DME | 68 | |
| **Yeast strain** | **Pitch (billions of cells)** | **Fermentation temp.** |
| German lager | 575 | 52°F (11°C) |

### Version: All-Grain

| Grain bill | Gravity points | |
|---|---|---|
| 7.5 lb. (3.4 kg) Pilsner malt | 29.5 | |
| 7.5 lb. (3.4 kg) Munich malt | 28 | |
| 0.5 lb. (225 g) Briess Midnight Wheat malt | 1.5 | |
| Boil gravity for 7 gal. (27 L) | 1.059 | |
| **Mash schedule** | **Rest temp.** | **Rest time (min.)** |
| Conversion rest – Infusion | 150°F (65°C) | 30 |
| Single decoction (see fig. 17.3) | (boil) | 20 |
| **Hop schedule** | **Boil time (min.)** | **IBUs** |
| 0.8 oz. (28 g) German Magnum 12% AA | 60 | 25 |
| **Yeast strain** | **Pitch (billions of cells)** | **Fermentation temp.** |
| German lager | 575 | 52°F (11°C) |

| Recommended Water Profile (ppm) | | | Brew cube: Amber, Malty, Medium | | |
|---|---|---|---|---|---|
| **Ca** | **Mg** | **Total alk.** | **SO$_4$** | **Cl** | **RA** |
| 75–125 | 10 | 100–150 | 50–100 | 100–150 | 0–50 |

## Vienna

The Vienna style of lager was developed in the mid-1800s in the city of Vienna, Austria. It grew from the Märzen and Oktoberfest styles of Bavaria, but was influenced by the rise of the Pilsener style of Bohemia. Attempts to imitate the Pilsen style had resulted in harsh beers due to the differences in brewing water between the two regions. The local water was higher in carbonates than that of Bohemia (in what is now the Czech Republic). The sweet amber lager now known as Vienna was the result of the efforts of Vienna's brewers to produce a lighter beer. It became immensely popular and was copied in other brewing countries.

There was a lot of immigration from central Europe to Texas and Mexico at that time, and of course the people brought their beer and brewing techniques with them. The hot climate was abysmal for lager brewing though, and commercial offerings were poorly regarded. Fortunately, by the late 1800s, refrigeration became commercially viable and variations of Old World-style lagers became very popular. The principle variation of the Vienna style in the New World is the Graf-style Vienna, named after the Mexican brewer (Santiago Graf) who developed it. It incorporated a small percentage of heavily roast malt to compensate for the more alkaline water of the region, giving it a deep amber color with hints of red.

Many people think a Mexican lager is a pale yellow beer that you stick a slice of lime into, but this recipe is what I think of. The OG for Vienna ranges from 1.048 to 1.055, and the bitterness from 18 to 30 IBU. The warm amber color of Vienna is typically in the 9–15 SRM (18–30 EBC) range, although it can be darker at 25 SRM (50 EBC) in the Mexican style.

## Corazon y Alma—Vienna Lager

**Original gravity:** 1.052  **SRM (EBC):** 14 (28)
**Final gravity:** 1.013  **ABV:** 5.3%
**IBU:** 27

| Version: Extract and Steeping Grain | | |
| --- | --- | --- |
| **Wort A** | **Gravity points** | |
| 3.3 lb. (1.6 kg) Vienna LME | 40 | |
| 3.5 oz. (100 g) Briess Midnight Wheat malt – Steeped | 2 | |
| Boil gravity for 3 gal. (11.4 L) | 1.042 | |
| **Hop schedule** | **Boil time (min.)** | **IBUs** |
| 1 oz. (30 g) Mandarina Bavaria 8% AA | 60 | 27 |
| **Wort B (add at knockout)** | **Gravity points** | |
| 4.5 lb. (2 kg) Vienna LME | 54 | |
| **Yeast strain** | **Pitch (billions of cells)** | **Fermentation temp.** |
| Mexican or German lager | 350 | 65°F (18°C) |

| Version: All-Grain | |
| --- | --- |
| **Grain bill** | **Gravity points** |
| 11 lb. (5 kg) Vienna malt | 43 |
| 3.5 oz. (100 g) Briess Midnight Wheat malt | 1 |
| Boil gravity for 7 gal. (27 L) | 1.044 |

| Mash schedule | Rest temp. | Rest time (min.) |
|---|---|---|
| Conversion rest – Infusion | 150°F (65°C) | 30 |
| Single decoction (see fig. 17.3) | (boil) | 20 |

| Hop schedule | Boil time (min.) | IBUs |
|---|---|---|
| 1 oz. (30 g) Mandarina Bavaria 8% AA | 60 | 27 |

| Yeast strain | Pitch (billions of cells) | Fermentation temp. |
|---|---|---|
| Mexican or German lager | 350 | 65°F (18°C) |

| Recommended Water Profile (ppm) | | | Brew cube: Amber, Malty, Medium | | |
|---|---|---|---|---|---|
| Ca | Mg | Total alk. | SO$_4$ | Cl | RA |
| 75–125 | 10 | 100–150 | 50–100 | 100–150 | 0–50 |

## Oktoberfest

The Märzen and Oktoberfest beers were part of the basis of the Vienna style. Whereas the Vienna was intended to be the everyday premium drinking beer, the Oktoberfest was made for festivals. The original festival celebrated the royal wedding of Crown Prince Ludwig I and Princess Therese in 1810, and they have been celebrating ever since (some beers are worth it). Examples of this rich amber style vary quite a bit, from being soft and malty, malty and dry, to malty and balanced, to malty and bitter. Be that as it may, the hallmark of the Oktoberfest/Märzen style is a rich maltiness but high attenuation to make it less filling. If you plan to polka for 12 hours straight, then this is your beer.

## Klang Freudenfest—Oktoberfest Lager

**Original gravity:** 1.055  
**Final gravity:** 1.013  
**IBU:** 27  

**SRM (EBC):** 9 (18)  
**ABV:** 5.6%  

| *Version: Extract and Steeping Grain* | | |
|---|---|---|
| **Wort A** | **Gravity points** | |
| 3.3 lb. (1.5 kg) Vienna LME | 40 | |
| 1 lb. (450 g) aromatic Munich 20°L malt – Steeped | 2 | |
| 0.33 lb. (150 g) CaraVienne – Steeped | 2 | |
| Boil gravity for 3 gal. (11.4 L) | 1.042 | |
| **Hop schedule** | **Boil time (min.)** | **IBUs** |
| 1 oz. (30 g) German Tradition 6% AA | 60 | 18 |
| 1 oz. (30 g) German Tettnanger 4% AA | 20 | 7 |
| **Wort B (add at knockout)** | **Gravity points** | |
| 3.5 lb. (1.6 kg) Munich DME | 49 | |
| **Yeast strain** | **Pitch (billions of cells)** | **Fermentation temp.** |
| German lager | 465 | 52°F (11°C) |

| Version: All-Grain | |
|---|---|
| **Grain bill** | **Gravity points** |
| 3.5 lb. (1.6 kg) Pilsner malt | 14 |
| 3.5 lb. (1.6 kg) Munich malt | 14 |
| 4 lb. (1.8 kg) Vienna malt | 15 |
| 1 lb. (450 g) aromatic Munich 20°L malt | 4 |
| 0.33 lb. (150 g) CaraVienne malt | 1 |
| Boil gravity for 7 gal. (27 L) | 1.048 |

| **Mash schedule** | **Rest temp.** | **Rest time (min.)** |
|---|---|---|
| Conversion rest – Infusion | 150°F (65°C) | 30 |
| Single decoction (see fig. 17.3) | (boil) | 20 |

| **Hop schedule** | **Boil time (min.)** | **IBUs** |
|---|---|---|
| 1 oz. (30 g) German Tradition 6% AA | 60 | 18 |
| 1 oz. (30 g) German Tettnanger 4% AA | 20 | 7 |

| **Yeast strain** | **Pitch (billions of cells)** | **Fermentation temp.** |
|---|---|---|
| German lager | 465 | 52°F (11°C) |

| **Recommended Water Profile (ppm)** | | | **Brew cube:** Amber, Malty, Medium | | |
|---|---|---|---|---|---|
| **Ca** | **Mg** | **Total alk.** | **SO$_4$** | **Cl** | **RA** |
| 75–125 | 10 | 100–150 | 50–100 | 100–150 | 0–50 |

## Summary

So there you have it—the *Reader's Digest* version of some of the classic beer styles of the world. There are many, many more. If all this talk of different malts and tastes has made you thirsty, zip on down to your local bottle shop and bring back some samples for research and development. Don't be shy—how else can you decide what you want to brew next?

# Developing Your Own Recipes

Now it's time to take off the training wheels and strike out on your own. You have read about many of the various beer styles of the world and you should now have a better idea of the kind of beer you like best and want to brew. Homebrewing is all about brewing your own beer. Recipes are a convenient starting point until you have honed your brewing skills and gained familiarity with the ingredients. Do you need a recipe to make a sandwich? Of course not! You may start out by buying a particular kind of sandwich at a sandwich shop, but soon you will be buying the meat and cheese at the store, cutting back on the mayo a little, giving it a shot of Tabasco, using real mustard instead of that yellow stuff and voila—you have made your own sandwich just the way you like it. Brewing your own beer is the same process.

However, don't forget to keep a notebook of everything that you do. It would be tragic to brew your best beer ever and then be unable to remember how you made it!

This chapter will present more guidelines for using ingredients to attain a desired characteristic. You want more body, more maltiness, a different hop profile, less alcohol? Each of these can be accomplished and this chapter will show you how.

## Recipe Basics

Recipe design is easy and can be a lot of fun. Pull together the information on yeast strains, hops, and malts, and start defining the kind of tastes and characters you are looking for in a beer. Do you want it malty, hoppy, big-bodied, or dry? Choose a style that is close to your dream beer and decide what you want to change about it. Change just one or two things at a time so you will better understand the result. Make sure you understand the signature flavors of the style before you start adding or changing lots of things, otherwise you will probably end up with a beer that just tastes weird. You cannot achieve complexity without balance.

To help get your creative juices flowing, here is a rough approximation of basic recipes for the common ale styles assuming a 5 gal. (19 L) batch:

- pale ale—base malt plus a half-pound (225 g) of caramel malt
- amber ale—pale ale plus another half-pound (225 g) of dark caramel malt
- brown ale—pale ale plus a half-pound (225 g) of chocolate malt
- porter—amber ale plus a half-pound (225 g) of chocolate malt
- stout—porter plus a half-pound (225 g) of roast barley

Yes, those recipes are crude, but I want you to realize how little effort it takes to produce a different beer. When adding a new malt to a recipe, start out with a half-pound or less (≤225 g) for a 5 gal. (19 L) batch. Brew the recipe and then adjust up or down depending on your tastes. Try commercial beers that are available in each style, and use the recipes and guidelines in this book to develop a feel for the flavors the different ingredients contribute.

Read recipes listed in brewing magazines, even if they are all-grain and you are not a grain home-brewer. By reading an all-grain recipe and the descriptions of the malts they are using, you will gain a feel for what that beer would taste like and you will get an idea of the proportions to use. For example, if you look at five different recipes for a regular 5 gal. (19 L) batch of amber ale, you will probably notice that no one uses more than 1 lb. (450 g) of any one crystal malt—all things are good in moderation. If you see an all-grain recipe that sounds good, but aren't ready to brew all-grain, use the principles given in chapter 4 to duplicate the recipe using extract and specialty grains for steeping. You may need to use a partial mash for some recipes, but most can be reasonably duplicated without.

The first thing you need to understand when designing your own recipes is that the base malt is the main ingredient, just like the bread in a sandwich. The other ingredients all play a supportive role, and should not overwhelm the main ingredient. Just like you don't want an unbalanced sandwich, you don't want one aspect to overwhelm the composition of the beer. The whole should be greater than the sum of its parts.

The second thing to understand is that beer styles are like sandwich styles; if you start making a bologna sandwich but use turkey instead, well, that's a turkey sandwich, isn't it? In other words, signature ingredients, such as roasted barley, are associated with specific styles, such as stout. You can use a little roasted barley in another recipe, such as a brown ale, but if you use too much, it's now a stout.

The proportions of the ingredients are very important in a good recipe. The base malt is usually 80%–90% of the grain bill. A signature specialty malt can be 10% of the grain bill, but any other malts should be 5% or less. Now, granted, some porters and stouts will often have five different malts, but simpler is usually better. Too many flavors will start competing with one another and you achieve muddiness instead of complexity. Again, think of a sandwich. A good sandwich will have, at most, three main flavors playing off one another, and one of them is the bread. In beer, you have the base malt, the hop character, and your specialty malts. You can have one specialty malt to provide a signature character for the style, but the other specialty malts should only be complementary accents.

The signature specialty malt in pale styles, such as pale ale and Pilsner, is actually the base malt. You don't want to crowd it with lots of other specialty malts; instead, you will use specialty malts only as complimentary accents, such as a little bit of Munich for more rich bread character, a little bit of light caramel for enhanced sweetness, or some flaked barley or flaked wheat for more body. As you move into the darker styles, such as brown ales, dunkels, bocks, and porters, the accent specialty malts are used to complement the signature specialty malt, which itself will be a highly toasted malt, a roasted malt, or (rarely) a dark caramel. The complementary malts should share a character of the signature malt, but add a little more variety to the signature flavor, such as adding a bit of smoked malt to the chocolate malt in a porter.

The yeast strain has a big impact on beer flavor. Take any ale recipe and change the ale yeast strain to a lager strain and you have a lager recipe (though not necessarily an example of a particular lager style). Look at information about yeast strains and determine what flavors different strains would give to the recipe. Use the calculations in chapters 4, 5, and 15 to estimate the gravity and IBUs of the beer. Plan the FG for your beer and decide what factors you will manipulate to achieve it, such as ingredients, mash schedule, yeast strain, and fermentation temperature. As the brewer, you have almost infinite control over the end result. Don't be afraid to experiment.

Don't get hung up on particular hop varieties for recipes, hops are very easy to substitute. If you are trying to brew a recipe to style but can't find the hop that it calls for, then substitute a similar hop from the same region (see table 5.3). Try to choose a hop that has similar percentages of alpha acids and aromatic oils, and adjust the quantity and boil time to get the same IBU contribution. The results should be very similar to what's intended in the original recipe.

Lastly, keep in mind the balance between bitterness and maltiness in different styles. Maltiness is a complex term that has several facets. First it means malt flavor, the flavors of bread and bread crust that you can smell and taste in fresh malt. Second, maltiness is directly proportional, at least in part, to the strength of the OG. So maltiness can be represented by OG when plotting a chart of flavor balance, as in figure 24.1. It shows how many of the most common styles relate to one another in terms of bitterness and gravity. Most of the smaller styles (lower ABV) group very closely in the 1.040–1.060 OG versus 20–40 IBUs range, at a ratio of 2:4 (or 1:2). American IPA and double IPA are the only styles you see exceeding the 4:4 line (or 1:1).

## SMASH and the Single Beer

The acronym SMASH stands for "single malt and single hop," and it is a good way to focus on the signature flavors of hops and malts. You typically brew a style the way you normally would, but with only a single base malt, no specialty malts, and only a single hop variety used for all hop additions. While SMASH can produce a good beer, and is a great way to teach yourself to recognize specific hop varieties, it generally doesn't produce an interesting beer.

A good way to explore new ideas and new recipes is a method my friend Drew Beechum calls "brewing on the ones." The idea is to intentionally restrict yourself to one ingredient per category, so one base malt, one specialty malt, one adjunct (if using), one hop variety for the boil, one hop variety for dry hopping, and so on. It is similar to SMASH, but provides for a more practical level of complexity in a beer. This is because brewing on the ones allows you to round out the beer with

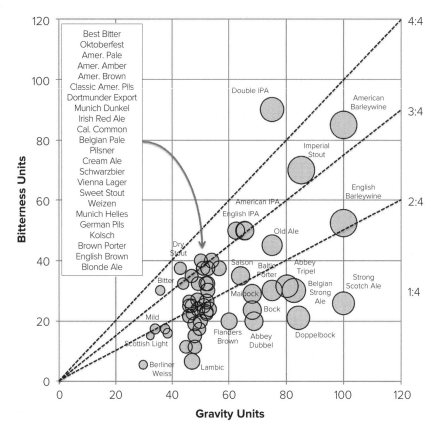

**Figure 24.1. OG and %ABV vs Bitterness for Common Beer Styles.** This chart shows the mean IBUs versus the mean OG for different beer styles according to the BJCP Style Guidelines. A lot of styles are not shown on this chart, because so many styles overlap in the 20–40 IBU and 1.040–1.060 OG region. The size of the circle indicates the relative % ABV, with the largest representing about 12% ABV and the smallest about 3% ABV. This chart can give you an idea of other beer styles to taste and brew for yourself.

a couple of extra ingredients, while maintaining the necessary discretion until you gain sufficient experience to experiment outside the box.

## Increasing the Body

Very often, brewers say that they like a beer but wish it had more body. What exactly is "more body"? Is it a physically heavier, denser beer? Does it denote more viscosity, and more flavor? It can mean all those things.

Proteins contribute most of the mouthfeel and body to a beer. Compare an oatmeal stout to a regular stout and you will immediately notice the difference. Brewers refer to these mouthfeel-enhancing proteins as "medium-sized proteins." During the protein rest, proteolytic enzymes break down large proteins into medium-sized proteins, while peptidases cleave the ends of proteins into small peptides and individual amino acids. High-protein malts and adjuncts, like wheat and oatmeal, can substantially increase the body of the beer.

Increasing the total amount of fermentables in the recipe will increase the body, but will increase the alcohol as well. Unfermentable sugars can enhance the body without increasing the alcohol.

Dextrins and other unfermentable sugars are long-chain sugars that cannot be broken down by the mash enzymes or consumed by the yeast. However, the effect of these sugars on mouthfeel is fairly limited, and some brewers suspect that dextrins are a leading cause of "beer farts," which result when these unfermentable carbohydrates are finally broken down by bacteria in the large intestine.

Dark caramel and roasted malts, such as 80°L and 120°L crystal malt, Special "B," chocolate malt, and roast barley, contain a high proportion of unfermentable sugars due to Maillard reactions. The total soluble extract (percent by weight) of these malts is close to that of base malt, but just because it's soluble does not mean it is fermentable. These sugars are only partially fermentable and contribute both a residual sweetness and higher FG to the finished beer that can be perceived as body.

In addition to using the methods above, grain brewers can add dextrin malt, caramel malt, unmalted barley, or oatmeal. Grain brewing lends more flexibility in fine-tuning the wort than extract brewing.

## Changing Flavors

What if you want a beer that tastes maltier? A bigger, more robust malt flavor is usually achieved by adding more malt or malt extract to the recipe. A 1.050 OG beer is maltier than a 1.035 OG beer. If you add more extract, be sure to increase the bittering hops proportionally to keep the beer balanced. This brings up another way to enhance the maltiness of a beer, which is to cut back on the flavor and aroma hop additions by 5–10 IBU. You can keep the total hop bitterness and balance the same by adding more bittering hops at the beginning of the boil, but, by cutting back on the middle and late hop additions, the malt flavors and aromas will be more dominant.

But what if you don't want the increased alcohol level that comes with an increase in gravity? The solution will depend on what flavor profile you are trying to achieve. If you want a toastier malt flavor, substitute a pound or two of a toasted malt (e.g., Vienna, Munich, or amber) in place of some of the base malt to help produce the malty aromas of German bock and Oktoberfest. Change the type of caramel malt you are using if you want to change the sweet malt character of the beer. You can add carastan malt, or 20°L or 40°L caramel malt, in place of a 60°L–80°L caramel malt to produce a lighter, honey-like sweetness, or make use of the bittersweet 120°L caramel and Special "B."

If the flavor profile tastes a bit flat, or you want to add some complexity to the beer, you can try substituting smaller amounts of a variety of specialty malts in place of a larger single specialty malt addition, while keeping the same OG. For instance, if a recipe calls for 0.5 lb. (225 g) of crystal 60°L malt, try using 0.25 lb. (115 g) each of crystal 40°L and crystal 80°L. For another example, if a recipe calls for 0.5 lb. of chocolate malt, try using just 0.25 lb. of chocolate malt and adding another 0.25 lb. of toasted malt or Caramunich. This will give you more flavors for the same strength beer.

The different growing regions around the world for the hop and malt varieties can also have a significant effect on the character of the beer. Sometimes these differences can be quite dramatic, and it's a sign of a really good brewer to utilize this. Tasting the ingredients before brewing with them, either by chewing or making a tea from them, is a useful way to teach yourself these differences.

Yeast has a strong effect on beer flavor, and not just by directly contributing yeasty flavors or aromas. The degree of attenuation, the ester profile, and the resultant pH of the beer are all affected by the yeast strain and fermentation conditions, and can accentuate the hops, or the malt character. Some yeast strains add fruity, woody, or earthy notes, or can impart a whiff of spicy phenol. Brewers make the wort, but yeast make the beer.

The best brewers use every little trick: the right malt, the right hops, technique, yeast, temperature, and more, all working together to deliver a unified and memorable experience to the drinker.

---

### EXPLORING NEW INGREDIENTS WITH TEA

What is the best way to learn the flavors and aromas of new ingredients? Make a tea from it!

To make one pint of wort (about half a liter) dissolve 2 oz. (60 g) of dry malt extract into 16 fl. oz. (0.5 L) of water. Add two hop pellets. Stir to break up the pellets. You can heat the wort in the microwave for a minute, if you wish to mimic more of a boiled hop character than a dry hop character. This hop tea will give you a very good idea of the aroma and flavor of the hop.

---

## Sugars Used in Brewing

Brewing beer is all about working with sugars—glucose, fructose, sucrose, maltose, and all the rest. If you are like me, you want someone to quickly explain it without getting too technical. I will do so, but please be patient because I need to lay some groundwork and describe the different building blocks. It won't take long, and once you understand what everything's made of it becomes a lot easier to explain what it means to brewing.

Sugars (technical name: *saccharides*) are made up of groups of three or more carbon atoms variously bonded to oxygen and hydrogen. The common sugars, like glucose and fructose, consist of a group of six carbon atoms and are therefore called hexoses. A *monosaccharide* is a single sugar group, a *disaccharide* is composed of two sugar groups joined together, a *trisaccharide* is three sugar groups joined together, and so on. Generally, saccharides consisting of many sugar groups joined together are called *polysaccharides*. Polysaccharides can consist of many thousands of sugar groups joined together in long, often multi-branched, chains.

The most common type of sugar is the monosaccharide glucose, (a.k.a. dextrose or corn sugar). Other monosaccharides relevant to brewing are fructose and galactose. Compositionally, glucose, fructose, and galactose are all the same, but they are isomers of each other. What this means is that, although their basic chemical formula is the same ($C_6H_{12}O_6$), the different structural arrangement of each sugar gives them different properties. For instance, fructose (a.k.a. fruit sugar) is an isomer of glucose, but tastes sweeter than glucose.

In general, monosaccharides are sweeter tasting than the other saccharides, although sucrose (a disaccharide) is sweeter than glucose. In descending order of sweetness: fructose is sweeter than sucrose, which is sweeter than glucose, which is sweeter than maltose (disaccharide), which is sweeter than maltotriose (trisaccharide). Not that it really matters in fermentation, I just thought you would like to know.

Lots of different sugars can be used in brewing, but it really begs the question—why would we want to use anything other than the sugars that come naturally from the barley? Well, there are a few reasons:

- To raise the alcohol level without increasing the body of the beer.
- To reduce the body of the beer while maintaining the alcohol level.
- To add some interesting flavors.
- To prime the beer for carbonation.

The first two reasons are two sides of the same coin, of course, but they illustrate two different ways that refined-sugar adjuncts can be used to create different styles of beer. Belgian strong golden ale has an OG of 1.065–1.080 and uses partially refined sugar syrup to achieve a brilliantly clear, light-bodied beer with a high ABV. American light lager recipes have an OG of 1.035–1.050 and use high-maltose corn syrup or rice syrup solids to obtain a very light-bodied beer of average ABV that is perfect for a hot day with the lawn mower.

Different sugars have different flavors. The monosaccharides are usually considered to not have a definable flavor other than "sweet," although the soda beverage industry seems to promote cane sugar over both beet sugar and high-fructose corn syrup for use in soft drinks. But other natural sugars, like honey and maple syrup, and processed sugars, like molasses, have characteristic flavors that make a nice accent for a beer. This is really what homebrewing is all about—taking a standard beer style and dressing it up for your own tastes. You can make a maple syrup porter, or a honey-raspberry wheat beer, or an imperial Russian stout with hints of rum and treacle. The possibilities are myriad. On the other hand, I have coined a phrase that will serve you well in your experimentation with sugars, "Discretion is the better part of flavor." A beer with 20% molasses is going to taste like fermented molasses, not beer.

And let's not forget the topic of priming, which is the addition of 2–3 gravity points of fermentable sugar to carbonate the batch. Most folks do their brewing at the boiling stage—they don't want to change their flavor profile at the priming stage, they simply want an unobtrusive sugar to carbonate the beer. Other brewers actually look at this last stage of fermentation with an ulterior motive, wanting to add character with this final step. Whatever your goal, you can select one of several sugars to accomplish it (table 24.1).

## Pure Glucose

The most common example of a simple brewing sugar is the corn sugar commonly used for priming. Corn sugar is about 92% solids and 8% moisture, with the solid portion consisting of about 99% glucose. Corn sugar is highly refined and does not contain any corn character. Brewers seeking a corn-like character for a classic American Pilsner need to cook and mash corn grits as part of an all-grain recipe. It is known that a relatively high proportion of glucose in a wort (>15%–20%) will inhibit the fermentation of maltose. Fermentation can also be impaired, or become stuck, if the yeast is underpitched or if the wort is lacking in FAN and other nutrients.

## High-Maltose Syrups and Solids

There are corn syrups, rice syrups, and syrup solids available that are produced to have high maltose levels, making them more similar to wort than other refined food syrups. Maltose is a disaccharide made from two glucose units joined together. High-maltose corn syrups can be produced from corn in an actual mash or simply by using prepackaged enzymes on corn starch (this process is discussed in more detail in the sidebar on high-fructose corn syrup). The maltose corn syrup usually doesn't have much, if any, corn character, but this type of product is suitable for brewers wanting to produce a light lager similar to Coors or Miller products.

Rice syrup solids are another maltose product, and there are different types. For example, one high-glucose type is about 75% sugars (50% glucose, 25% maltose), 20% other carbohydrates (dextrins), and 5% moisture. A high-maltose rice syrup solid would typically contain only 5% glucose, the rest being made up of 45% maltose and 45% other carbohydrates, and the same 5% moisture content. Extract brewers seeking to brew a Budweiser clone can use rice syrup solids and will obtain nearly the same malt character as the real thing (assuming that all the other brewing variables are also the same).

## HIGH-FRUCTOSE CORN SYRUP

There is a lot of hullaballoo over high-fructose corn syrup. Similar to wort, corn syrup is made by the enzymatic hydrolysis of starch. The difference between the manufacture of corn syrup and barley wort is the source of the enzymes. For corn syrup, the alpha-amylase and alpha-glucosidase enzymes are produced from bacteria and fungi; beta-amylase is not present. Relatively small amounts of hydrochloric acid and sodium hydroxide are used to adjust the pH of the corn mash to the optimum range for these enzymes. This process produces glucose. A third (and more expensive) enzyme, glucose isomerase, converts some of the glucose to its isomer, fructose. Fructose is the sweetest of the monosaccharides.

The goal of the food industry was to produce a cheaper substitute for sucrose (i.e., cane and beet sugar). Pure glucose syrup is easy to produce, but doesn't taste as sweet or have the same body as sucrose. Converting some of the glucose to fructose mimics the sweetness level of sucrose, and this glucose-fructose mixture (i.e., invert sugar) has the added benefit of better solubility than sucrose, because the invert sugar won't crystallize at high concentrations. Sucrose is a disaccharide of glucose and fructose, and has a ratio of 50:50 of these two mono-saccharides. Honey has a ratio of about 40:60 glucose to fructose, making it sweeter than table sugar, but the particular ratio can vary by location, flowers, and bees. There are two main kinds of high-fructose corn syrup used in the United States, HFCS-55 and HFCS-42, corresponding to 55% and 42% fructose, respectively. High-fructose corn syrup gives more sweetness per calorie than sucrose, but costs 40% less to produce than sucrose.

The bottom line, though, is that there is really no reason to use high-fructose corn syrup in brewing. It is effectively the same as using honey, which has more flavor.

### Sucrose-Type Sugars

Pure sucrose is the reference standard for all fermentable sugars, because it contributes 100% of its weight as fermentable extract and has zero percent moisture. Sucrose dissolved at 1 lb./gal. will form a solution that has a gravity of 1.046, or, in other words, an extract yield of 46 PPG (which equates to 384 PKL; see also chapter 13).

Lots of different brewing products are made from sucrose or the semi-refined byproducts of sucrose. Both sugar cane and sugar beet are used to make table sugar, and this refined product is indistinguishable whether it comes from cane or beet. However, when considering byproducts, you do not get useful brewing sugars from sugar beet, only sugar cane. Molasses is a common byproduct of sugar cane and is added to refined cane sugar to make brown sugar. The fermentation of molasses produces rum-like notes and sweet flavors, but there may be sharp, harsh notes as well. Brown sugar, which only contains a small amount of molasses, will only contribute a light rummy flavor. Partially refined cane sugars, like piloncillo or panela, demerara sugar, turbinado sugar, and muscovado, have better flavors than molasses, which often contains more impurities from the refining process. Here's a quote from Sykes and Ling:

> The impurities of the raw sugar derived from the beetroot are of a nauseous character, whilst those of sugarcane sugar have an agreeable, full, and luscious flavor; and so much is this the case, that the impure cane sugars are more valuable for brewing purposes than the refined, since the raw variety yield to the beer their luscious flavor. (Sykes and Ling 1907, p. 94)

Invert sugar syrups, such as Lyle's Golden Syrup, are made from sucrose that has been hydro-lyzed to separate the glucose and fructose. This has two effects: one, it makes the sugar more syrupy and less likely to crystallize; and two, it makes it sweeter. Invert sugar syrup is like artificial honey without the characteristic honey flavors. Golden syrup-type products tend to be a bit salty tasting due to the acid–base reactions that occur during manufacture. Treacle is partially inverted molasses combined with other syrups. The flavor contributions from treacle can be strong, so it is best to use it in heavier bodied beers like English strong ales, porters, and sweet stout. A recommended starting point for strongly flavored syrups is 8 fl. oz. (250 mL) per 5 gal. (19 L) batch.

Belgian candi syrups are a class of adjunct that originated in England in the early to mid-1700s, due to the high cost of sugar. These syrups were created from the remnants left in the kettle of boiling sugar into hard candy; the kettles were cleaned by pouring hot water onto the hot residue stuck to the bottom of the kettle and the resulting candi syrups were either incorporated into the next batch or stored for aftermarket sale. This syrup was money. Eventually sugar prices dropped with the advent of large sugar plantations in the New World, and the subsequent rise in sugar use and candy production led to greater availability of these syrups.

The candi syrups darkened due to Maillard reactions and developed a variety of unique flavors depending their origin and the length of time spent cooking in the kettle. These syrups became very popular coloring and flavoring adjuncts for British brewers, and this brewing practice spread to Belgium by the late 1800s. Today, these candi syrups are specifically produced for brewing and are available in several colors (degrees Lovibond), such as 5°L, 45°L, 90°L, 180°L, and 240°L. The lightest colors have flavors reminiscent of honey, light citrus, and golden raisins. The medium colors (45°L and 90°L) have caramel, toffee, vanilla, chocolate, and stone fruit flavors. The darkest colors have dark caramel, dark chocolate, coffee, and stone fruit flavors. Candi syrups are typically used at a rate of 1–3 lb. (0.5–1.4 kg) per 5 gal. (19 L) batch, depending on the style being made, and typically contribute about 32 PPG (267 PKL).

## Maple Syrup

The sap from maple trees typically contains about 2% sucrose. Maple syrup is standardized at a minimum of 66 degrees Brix,[1] and is typically composed of ≥95% sucrose. Grade B maple syrup can contain 6% invert sugar, while Grade A Light Amber will contain less than 1%. You will get more maple flavor from the Grade B syrup. The characteristic maple flavors tend to be lost during primary fermentation, so to retain as much flavor as possible it is recommended that the syrup is added after primary fermentation is over. This practice will also help the beer to ferment more completely, because it will not trigger inhibition of maltose fermentation as discussed in chapter 6. For a noticeable maple flavor, 1 gal. (3.8 L) of Grade B syrup is recommended per 5 gal. (19 L) batch.

## Honey

Honey is about 18% water by weight. The other 82% is carbohydrates. Ninety-five percent of those carbohydrates are fermentable, typically consisting of 38% fructose, 31% glucose, 8% various disac-charides, and 4% unfermentable dextrins by weight. Honey contains wild yeast strains and bacte-ria, but its low water content keeps these microorganisms dormant. Honey also contains amylase enzyme, which can break down unfermentable sugars in the wort into highly fermentable sugars

---

[1]     Degrees Brix gives the sugar content of a solution by percent weight. It measures the solution density by refractometry, and the Brix scale is roughly equivalent to the Plato scale.

## TABLE 24.1—COMMON BREWING SUGARS

| Sugar | PPG (PKL) | Fermentabilty | Constituents | Comments |
|---|---|---|---|---|
| Corn sugar | 42 (351) | 100% | Glucose (~8% moisture) | Can be used for priming, or as a wort component to increase the alcohol while lightening the body of the beer. |
| Rice syrup solids | 42 (351) | Varies by grade, high-glucose grade is ~80% | Glucose, maltose, other (~10% moisture) | Can be used for priming, or as a wort component to increase the alcohol while lightening the body of the beer. |
| Table sugar | 46 (384) | 100% | Sucrose | Can be used for priming, or as a wort component to increase the alcohol while lightening the body of the beer. |
| Lyles Golden Syrup | 38 (317) | 100% | Glucose, fructose (18% water) | An invert sugar that has been broken down to fructose and glucose. A bit salty due to acids and bases used during processing. |
| Belgian candi syrup | 32 (267) | Varies ~50%–80% | Sucrose, melanoidins | Variety of color and flavor products. Key ingredient for many Belgian ale styles. |
| Molasses / Treacle | 36 (300) | Varies, ~50%–70% | Sucrose, invert sugars, dextrins | Composition varies, therefore degree of fermentability varies, probably around 50%–70%. Can cause rumlike or winelike flavors. |
| Lactose | 46 (384) | … | Lactose (<1% moisture) | Lactose is an unfermentable sugar—a disaccharide of glucose and galactose. Used in milk stouts to impart a smooth, sweet flavor. |
| Honey | 38 (317) | 95% | Fructose, glucose, sucrose (~18% water) | Honey is a mixture of sugars high in fructose. Will impart honey-like flavors that depend on nectar source. |
| Maple syrup | 31 (259) | 100% | Sucrose, fructose, glucose (~34% water) | Maple syrup is mostly sucrose. Grade B syrup will provide more maple flavor than Grade A syrup. |
| Maltodextrin (powder) | 42 (351) | … | Dextrins (5% moisture) | Maltodextrin powder contains a small amount of maltose, but is mostly unfermentable. Adds mouthfeel and some body. |

Notes: The extract yield of sucrose-based brewing syrups can be estimated by multiplying the percent solids (i.e., the inverse of percent moisture) or degrees Brix by the reference standard for sucrose, which is 46 PPG. The extract of powdered sugars can be estimated by subtracting the percent moisture and multiplying the remainder by 46 PPG.

like maltose and sucrose. For these reasons, honey should be pasteurized before adding it to the fermentor. The National Honey Board (www.nhb.org) recommends that honey be pasteurized for 30 min. at 176°F (80°C), then cooled and diluted to the wort gravity. To retain the most honey flavor, and ensure best fermentation performance, the honey should be added to the fermentor after primary fermentation.

The NHB recommends the following percentages (by weight of total fermentables in the wort) when brewing with honey:

3%–10%     This gives a subtle honey flavor in most light ales and lagers.

11%–30%    This allows a distinct honey flavor note to develop. Stronger hop flavors, caramelized or roasted malts, spices, or other ingredients should be considered when formulating the recipe to balance the strong honey flavors at this level.

30%–66%    The flavor of honey will dominate the beer. This range is associated with braggot, which is considered by the BJCP Style Guidelines[2] to have a maximum honey-to-malt ratio of 2:1.

>66%       Any brew with more than 66% honey is considered to be a form of mead, according to the BJCP.

## Toasting Your Own Malt

As a homebrewer, you should feel free to experiment in your kitchen with malts. Oven-toasted base malt adds nutty and toasty flavors to the beer, which is a nice addition for brown ales, porters, bocks, and Oktoberfest beers. It is easy to toast-your-own, and the toasted grain can be used by both steeping and mashing. If steeped, the malt will contribute a high proportion of unconverted starch to the wort and the beer will be hazy, but a nice nutty, toasted flavor will be evident in the final beer. There are several combinations of time and temperature that can be used in producing these malts (table 24.2).

The principal reaction that takes place when you toast malt is the browning of starches and proteins, known as Maillard reactions. As the starches and proteins brown, various flavor and color compounds are produced. The color compounds are called "melanoidins," which can improve the flavor stability of beer by slowing oxidation and staling reactions as the beer ages.

Since the Maillard reactions are influenced by the wetness of the grain, water can be used in conjunction with the toasting process to produce different flavors in the malt. Soaking the uncrushed malt in water for an hour will provide the water necessary to optimize the Maillard browning reactions. Toasting wet malt will produce more of a caramel flavor due to partial starch conversion taking place caused by the heat. Toasting dry grain will produce more of a toast or Grape-Nuts® cereal flavor that is perfect for nut-brown ales.

The toasted malt should be stored in a paper bag for two weeks prior to use. This will allow time for the harsher aromatics to escape. Commercial toasted malts are often aged for six weeks before sale. This aging is more important for the highly toasted malts, those that are toasted for more than 30 min. (dry) or 60 min. (wet).

2    Beer Judge Certification Program, *2015 Beer Style Guidelines,* http://www.bjcp.org/docs/2015_Guidelines_Beer.pdf

## TABLE 24.2—GRAIN TOASTING TIMES AND TEMPERATURES

| Oven temp. | Dry/Wet | Time | Flavors |
|---|---|---|---|
| 275°F (140°C) | Dry | 60 min. | Light nutty taste and aroma. |
| 350°F (180°C) | Dry | 15 min. | Light nutty taste and aroma. |
| 350°F (180°C) | Dry | 30 min. | Toasty, Grape-Nuts® cereal flavor. |
| 350°F (180°C) | Dry | 60 min. | More roasted flavor, very similar to commercial brown malt. Can be harsh. |
| 350°F (180°C) | Wet | 60 min. | Light sweet toasty flavor. |
| 350°F (180°C) | Wet | 90 min. | Toasty, malty, slightly sweet. |
| 350°F (180°C) | Wet | 120 min. | Strong toasted/roasted flavor similar to brown malt, but slightly sweet. |

## Discretion Is the Better Part of Flavor

There comes a time in every homebrewer's development when they look at an item (e.g., maple syrup, molasses, Cheerios®, chile peppers, potatoes, pumpkins, loquat fruit, ginger root, spruce tips, heather, licorice, stale bread, mismatched socks) and say, "Hey, I could ferment that!" While many of the mentioned items will indeed work in the fermentor (socks work well for dry-hopping), it is easy to get carried away and make something that no one really wants to drink a second glass of. I thought I would like spiced holiday beer—I didn't. I thought I would like a molasses porter—I didn't. I thought I would like loquat wheat beer—four hours peeling and seeding three bags of those little bastards for something I couldn't even taste!

Experimentation is fine and dandy, but be forewarned that you may not like the result. Refined sugars, like molasses, candy sugar, honey, and maple syrup, can taste wonderful in the right proportion, and that proportion should be strictly as an accent to the beer. Also, keep firmly in mind that you are brewing beer and not a liqueur. Refined sugars often generate fusel alcohols, which can have solventlike flavors. If you want to try a new fermentable or two in a recipe, go ahead, but use a small amount so that it doesn't dominate the flavor. I feel hypocritical telling you to hold back after first saying to spread your wings and develop your own recipes. But I don't want you to spend a lot of time making a batch that is undrinkable. Just because it can be done, doesn't mean it should be done.

Okay, enough said.

In the next chapter, chapter 25, I will lead you through common problems and their causes and define some of the most common off-flavors.

# Is My Beer Ruined? 25

"Is my beer ruined?!" This phrase has to be the most frequently asked question by new homebrewers, and usually the answer is, "No." Depending on the cause, the beer may end up with an odd flavor or aroma, but you will still be able to drink it and chalk it up as another lesson along the road to brewing that perfect beer. Although a lot can potentially go wrong with a batch, most problems arise from just a couple of root causes.

Let's review our Top Five Priorities for Brewing Great Beer:
1. Sanitation
2. Fermentation temperature control
3. Proper yeast management
4. The boil
5. The recipe

This list is prioritized from highest to lowest, that is to say, if you make a mistake in a higher priority, you can't fix it by doing a lower priority correctly. Let's also add a sixth priority to this list—the ingredients. The quality of the ingredients are obviously important to the quality of the beer, but the quality of ingredients will not make much difference if you screw up the brewing process. If you have good ingredients, but a poor recipe, then the beer will not be very good. If you have a good recipe, but you didn't cook it right (the boil), then it won't be very good. If you made a good yeast starter, aerated well, and fermented at the proper temperature, but forgot to clean the pickup tube on the fermentor from the previous batch, it won't be very good. Lastly, if everything else was done right, low quality ingredients such as old malt or high alkalinity water may be a problem. Let's examine some common fermentation problems and their possible causes.

## Common Problems with Fermentation

### Problem: I added the yeast two days ago and nothing is happening.

*Cause 1: Too cold.* Lack of fermentation could be due to several things. The fermentation conditions may be too cold for otherwise healthy yeast to be active. Ale yeast strains vary in their temperature sensitivity, but tend to go dormant below 60°F (16°C). If the yeast was rehydrated in really warm water, for example, 105°F (41°C), and then pitched to a much cooler wort of 65°F (18°C), the large difference in temperature can thermally shock the yeast and cause a longer lag time as they adjust. You can pitch colder yeast and yeast starters into warmer worts, but not the other way around.

*Cure:* The temperature of the yeast or yeast starter should be within 5°F (3°C) of the wort. If you have already pitched the yeast and are experiencing a long lag time (i.e., the adaptation phase is >48 hours), try warming the fermentor by 5°F (3°C); it may make all the difference.

*Cause 2: Poor yeast management.* When a batch is not fermenting, the most common problem is with the yeast. If dry yeast has been properly packaged and stored it should be fully viable for up to two years. Check the manufacturing or "best by" date on the package. If the yeast is too old, or has been subjected to poor storage conditions, it may be mostly dead. The viability of liquid yeast can also be gauged by its color; it is creamy white when fresh and darkens to gray or brown as it ages. Old yeast needs to be revitalized before use by growing it to an adequate pitching rate with a starter.

Yeast needs to be treated with care and be given the proper growing conditions. Dry yeast is dehydrated—the cells are parched and in no condition to start work. They need some nice warm water for rehydration, some time to do some stretching, maybe an appetizer, and then they will be ready to tackle a full wort. If the dry yeast is just sprinkled onto the surface of the wort, some of the yeast cells will be up to the challenge, but most won't. Give your yeast the best chance to make good beer for you.

*Cure:* Rehydration of yeast in plain water is strongly recommended, because of how osmosis works. Dried yeast cells contain adequate sugar reserves to get them going, so there is no need for sugar in the rehydration process. However, in wort with a high concentration of dissolved sugar, yeast cells cannot draw the water across their cell membranes to hydrate their own nutritional reserves. The water is instead locked up in the wort, hydrating the sugars in solution. Dry pitching does work, but it is not best practice.

Likewise, a liquid yeast culture also needs its breakfast routine. It has been kept in a refrigerator, so needs to be warmed and fed before there will be enough active yeast cells to do the job properly. There are a lot more yeast cells in a dry yeast packet than in a liquid packet. The liquid packet yeast needs to be grown in a starter to produce enough cells to take on the job of a full five-gallon wort. Both liquid and dry yeast cultures will have a lag time from when they are pitched until they start fermenting in earnest. Aeration, the process of dissolving oxygen into the wort, provides the yeast cells with the oxygen they need to faciliate their growth and make enough yeast cells to do the job properly.

*Cause 3: Mechanical issues.* If the airlock is not bubbling, it may be due to a poor seal between the lid and the bucket, or between the airlock and the lid. Fermentation may be taking place, but the $CO_2$ is not coming out through the airlock.

  *Cure:* This is not a real problem and it probably won't affect the batch. Fix the seal or replace the appropriate part.

### Problem: I added the yeast yesterday and it bubbled all day, but it's slowing down or has stopped today.

*Cause 1: Poor yeast management.* Yeast that is poorly prepared, whether from not being rehydrated, low pitching rate (i.e., no starter used beforehand), or lack of aeration, will often fail to finish the job.

  *Cure:* Pitch new yeast. Don't be overly concerned about delaying the fermentation; if your cleaning and sanitation were adequate, a day or two spent preparing new yeast will not hurt the beer.

*Cause 2: Too cold.* Temperature is a major factor for fermentation performance. If the temperature of the room where the fermentor is located cools down, even by only 5°F (3°C) overnight, then yeast activity can be slowed dramatically.

  *Cure:* Always strive to keep the fermentation temperature constant, the yeast will thank you for it.

*Cause 3: Too warm.* Instead of too cold, the other side of the coin could be that the temperature was too warm, for example, 75°F (24°C), and the yeast got the job done ahead of schedule. This often happens when a lot of yeast is pitched, where the primary fermentation can be complete within 48 hours. This is not necessarily a good thing, as fermentations above 70°F (21°C) tend to produce a lot of esters and phenolics that just don't taste right. Fermentation happened, but not as well as it might have.

  *Cure:* Always strive to keep the fermentation temperature within the recommended temperature range, the yeast will thank you for it.

### Problem: The last batch (did that) but this batch is (doing this).

*Cause: It is a different fermentation.* Every fermentation is unique, even if you are brewing the same recipe with the same ingredients. If you change a single thing you will have brewed a different beer. Sometimes the difference in a beer is small and sometimes it is profound. If you are brewing identical recipes at the identical temperatures, then a difference in fermentation vigor or length may be due to differences in yeast health, aeration, or other factors. Only if something like odor or taste is drastically different should you worry.

  *Cure:* Be patient. Wait for the fermentation to finish and taste it; the beer may be just fine. Specific off-flavors and their causes are discussed in the next section.

### Problem: The airlock is clogged with gunk.

*Cause: Vigorous fermentation.* Sometimes fermentations are so vigorous that the kräusen is forced into the airlock. Pressure can build up in the fermentor if the airlock gets plugged, and you may end up spraying brown yeast and hop resins on the ceiling.

*Cure:* The best solution to this problem is to switch to a blowoff hose. Fit a large diameter hose (e.g., 1", or 2.5 cm) into the opening of the bucket or carboy and run it down to a bucket of water.

### Problem: White/brown/green stuff is floating/growing/moving.

*Cause 1: Normal fermentation.* The first time you look inside your fermentor, you will be treated to an amazing sight. There will be whitish yellow-brown foam on top of the wort, containing greenish areas of hops and resins. This is perfectly normal. Even if the kräusen appears slightly slimy, it is probably normal; several yeast strains have a very wet looking kräusen.

*Cure:* Relax, don't worry.

*Cause 2: Contamination.* The beer may be contaminated with bacteria. Bacterial contamination will often generate a gelatinous layer on top of the beer, called a pellicle. Pellicles are generally whitish, or a yellow-brown-ivory color, or at least they should be. (Other colors could be mold.) They can have a powdery or slimy appearance, and can be smooth or textured. Search the Internet for example pictures. Sometimes a contamination will produce ropy gelatinous strands in the beer that resemble rice noodles (but much softer). The beer may smell acidic, phenolic, or buttery, depending on the type of bacteria that is contaminating your beer. Chapter 14, "Brewing Sour Beers," has more information.

*Cure:* As a general rule, if *you* did not *intentionally* sour the beer with a commercial, or otherwise known, bacterial culture, then you should probably dump it. You could keep it and hope it turns out to be a good sour, but that may take several weeks, months, or a year, and not every sour beer is a good sour beer. Clean and sanitize your equipment thoroughly, and clean and sanitize the room(s) where you prepare and ferment the wort. Old plastic items should generally be replaced with new ones if contamination is a recurring problem.

*Cause 3: Mold.* The beer may be contaminated with mold. Mold can be blueish, greenish, or black in color. It can also be hairy. Most mold and bacterial contamination in beer is non-hazardous, but there are a couple of types that do pose a health risk. For example, black mold can cause respiratory distress in many people.

*Cure:* When in doubt, dump it out. Better safe than sorry. Clean and sanitize your equipment thoroughly, and clean and sanitize the room(s) where you prepare and ferment the wort. Sometimes it is easier to replace plastic equipment than to thoroughly clean it. Old plastic items should generally be replaced with new ones if contamination is a recurring problem.

### Problem: It smells like vinegar.

*Cause: Bacterial contamination.* If bacteria have contaminated your beer, chances are the vinegar smell probably is vinegar. *Acetobacter* (vinegar producing bacteria) and *Lactobacillus* (lactic acid producing bacteria) are common contaminants in breweries. Sometimes the contaminated beer will smell like cider vinegar, other times like malt vinegar. It will depend on which type of bacteria is living in your wort. *Acetobacter* may form a pellicle. *Acetobacter* is an aerobe, so its presence indicates that oxygen is getting to your beer.

*Cure:* As a general rule, if *you* did not *intentionally* sour the beer with a commercial, or otherwise known, bacterial culture, then you should probably dump it. You could keep it and hope it turns out to be a good sour, but that may take several weeks, months, or a year, and not every sour beer is a good sour beer. If you do decide to keep the batch and hope for the best, *Brettanomyces* can be pitched to clean up the viscosity.

### Problem: It smells like extra-butter microwave popcorn.

*Cause 1: Diacetyl.* Diacetyl is a vicinal diketone that is created by oxidation of fermentation byproducts produced early in fermentation by the yeast. The yeast should clean up the diacetyl toward the end of fermentation, but if the fermentation was weak, hurried, or incomplete, then diacetyl can occur in the beer. These causes are all yeast related, and are often due to cooling temperatures toward the end of fermentation.

*Cure:* Pitch sufficient amounts of fresh, healthy yeast and conduct a diacetyl rest toward the end of fermentation. Diacetyl is further discussed later on in this chapter, and also in the lager chapter (chapter 11).

*Cause 2: Bacterial contamination. Pediococcus* is a common souring bacteria that produces a lot of diacetyl early in the fermentation. It can also cause the beer to be cloudy with haze. A pellicle may form.

*Cure:* As a general rule, if *you* did not *intentionally* sour the beer with a commercial, or otherwise known, bacterial culture, then you should probably dump it. You could keep it and hope it turns out to be a good sour, but that may take several weeks, months, or a year, and not every sour beer is a good sour beer. *Brettanomyces* can be pitched to clean up the diacetyl and viscosity.

### Problem: It smells funky, like cloves or a barnyard.

*Cause: Wild yeast contamination.* There are many different species of yeast in addition to the *Saccharomyces* species that make up brewer's yeast. Like *Saccharomyces*, species of *Brettanomyces* can readily metabolize wort sugars. The aroma and flavor from a *Brettanomyces* fermentation is different from that of *Saccharomyces*, and tends to consist of many phenol compounds. In some beers (e.g., Belgian ales), these phenol compounds can be an integral part of the style and give a wonderful complexity. However, when unintended and uncontrolled, these compounds tend to clash with the base beer that has become contaminated. The spectrum of flavors is quite broad, but the predominant ones are typically described as horse blanket, barnyard, or leather. In some cases, pineapple or tropical fruit esters can be present.

*Cure:* As a general rule, if *you* did not *intentionally* pitch the beer with a commercial, or otherwise known, *Brettanomyces* culture, then you should probably dump it. A "Brett" beer is not a sour beer, although it can be; instead, it is most often a funky tasting beer. Some people like Brett beers but many do not. You may want to pitch a sour beer bacterial culture to clean up the funk a bit.

### Problem: It smells rotten, or like rotten eggs.

*Cause 1: Bacterial contamination.* Bacterial contamination can also produce sulfury odors. If you are not brewing a lager beer, then this odor is a good sign that you have a bacteria contaminating your beer. Sometimes the odor may be like butyric acid, that is to say, it smells like vomit.

*Cure:* As a general rule, if *you* did not *intentionally* sour the beer with a commercial, or otherwise known, bacterial culture, then you should probably dump it. You could keep it and hope it turns out to be a good sour, but that may take several weeks, months, or a year, and not every sour beer is a good sour beer.

*Cause 2: Yeast strain.* Many lager yeast strains produce noticeable amounts of hydrogen sulfide (the rotten egg odor) during fermentation. The smell will dissipate after fermentation and will not affect the flavor of the beer.

*Cure:* A vigorous fermentation should scrub those odors from the beer and the final beer should not taste like it smells during fermentation. Warming the fermentation by a few degrees can invigorate the yeast, producing more $CO_2$ and help to flush those odors.

### Problem: It's been one or two weeks and it's still bubbling.

First, check the fermentation temperature; is it in the right range? If so, then take a gravity reading and see if it is between the OG and anticipated FG (a typical FG falls between 1.008 and 1.014). If so, then there is probably nothing wrong—see Cause 1 below. If the specific gravity is less than the anticipated FG, then you may have a bacteria or wild yeast contaminant. Taste the gravity sample for any off-flavors.

*Cause 1: Cool temperatures.* A beer that has been steadily fermenting (bubbling in the airlock) for a long time (more than a week for ales, more than two weeks for lagers) may not have anything wrong with it. It is often due to the fermentation being a bit too cool, so the yeast is working slower than normal. A slow but normal fermentation should display one bubble in the airlock every few seconds, with an abrupt decline in activity as the beer reaches final gravity.

*Cure:* This condition is not a problem.

*Cause 2: Gusher contamination.* Sustained bubbling is often due to "gusher-type" contamination. This type of contamination can occur at any time and is due to the presence of wild yeast or bacteria that eat the normally unfermentable sugars, like dextrins. The result is a beer that keeps bubbling (one or two bubbles per minute) until all of the carbohydrates are fermented, leaving a beer that has no body and very little taste. If it occurs at bottling time, the beer will overcarbonate and will fizz like soda pop, fountaining out of the bottle.

If the beer seems to be bubbling too long, check the gravity with a hydrometer. Use a siphon or turkey baster to withdraw a sample from the fermentor when checking the gravity. If the gravity is still high, above 1.020, then it is probably due to lower than optimum temperature or low yeast pitching rate. If the gravity is below 1.010 and you still see bubbling in the airlock at the rate of a couple per minute, then your beer may be contaminated with wild yeast or bacteria. The beer will not be worth drinking, either due to the total lack of flavor or because unpleasant phenols will be present.

*Cure:* Improve your sanitation next time.

### Problem: The fermentation seems to have stopped but the hydrometer reads high.

*Cause 1: Too cold.* This situation is commonly referred to as a "stuck fermentation" and it can have a couple of causes. The most common cause is low temperature. As mentioned at the start of this section, a significant drop in temperature can cause the yeast to go dormant and yeast cells will start to settle to the bottom.

*Cure:* Moving the fermentor to a warmer room and swirling the fermentor to rouse the yeast will often fix the problem. You should see more bubbling in the airlock, and the gravity should be closer to your target after a few days.

*Cause 2: Weak or underpitched yeast.* The other most common cause is underpitching. Referring back to the previous discussion of yeast preparation above, weak yeast, or low volumes of healthy yeast, will often not be up to the task of fermenting high-gravity wort. This problem is most common with higher gravity beers, where the OG is greater than 1.075.

*Cure:* Add an actively fermenting yeast starter to the beer. Making yeast starters is described in chapter 7.

*Cause 3: Low fermentability.* Another common cause is low fermentability of the wort itself, which can be caused by high levels of specialty malts or dextrinous adjuncts. It can also be caused by high mash temperatures; just a few degrees can make a big difference in fermentability.

*Cure:* Caramel malts and roast malts typically have low fermentability. Estimate your FG by assuming about 50% attenuation from those malts versus 75% attenuation from base and kilned malts. Conducting a forced fermentation (i.e., one dried yeast packet to 1 L wort at warm temperatures) would show you what your maximum attenuation for that wort would be.

Mash temperature can also be the culprit. A mash temperature of 156–160°F (69–71°C) will convert well, but will have a higher proportion of unfermentable dextrin sugars than if the mash had been between 149–155°F (65–68°C). You may want to invest in a higher quality thermometer that has a calibration certificate.

## Common Problems After Fermentation

### Problem: It won't carbonate.
*Cause: Needs more time.* Time, temperature and yeast strain all combine to form a government committee with a charter to determine the range of times when they can expect to be 90% finished with the Carbonation & Residual Attenuation Project. This committee works best without distractions—the meetings should be held in a warm room. If the committee was given enough budget (priming sugar), then it should arrive at a consensus in about two weeks. If the committee don't get its act together within a month, then it's time to rattle the cage and shake things up a bit.

*Cure:* If the temperature is too cool in the room, moving the bottles to a warmer room may do the trick. The yeast may have settled out prematurely, so the bottles need to be swirled to rouse the yeast and get it back into suspension. Beers that have sat in the fermentor for a long time may not have enough viable yeast left to do the job of carbonation. Fresh yeast may need to be added at bottling.

### Problem: The bottles are overcarbonated.
*Cause 1: Too much sugar.* You used too much priming sugar.

*Cure:* Vent and re-cap all of the bottles. Venting and re-capping may have to be done several times, as it simply lets out gas from the head space, not what is dissolved in the beer. It might be a good idea to uncap and then cover each bottle opening with little squares of aluminum foil, and let them stand for several minutes or hours before re-capping. The important thing is not to use too much sugar the next time.

I recall one story I read on the rec.crafts.brewing forum where a brewer recounted how both he and his partner had each added ¾ cup of priming sugar to the batch, thinking that the other one had not. By venting and re-capping all the remaining bottles after the initial explosions, they thought they had saved the batch. Then a massive storm front swept through and the corresponding drop in barometric pressure caused the rest of the bottles to explode. Be careful!

*Cause 2: Bottled too soon.* You bottled before fermentation was complete.

*Cure:* Vent and re-cap all of the bottles. Next time be more aware of the fermentation and make sure it is complete before you bottle. Think about your recipe and make sure you know what its FG should be. Make sure that you pitch enough healthy yeast for a strong fermentation.

*Cause 3: Unmixed priming sugar.* Sometimes the priming sugar solution doesn't get mixed very well with the beer and you get uneven carbonation across your bottles.

*Cure:* You can't really fix this batch, so be more careful on the next one. Stir or swirl the priming sugar and beer as you rack the next batch.

*Cause 4: Wild yeast contamination.* A "gusher bug" has gotten into the beer. Gusher bugs (i.e., *Brettanomyces* and other wild yeast strains) are a real problem, as they will keep on fermenting all the sugars until there is nothing left but fizzy, bitter, alcoholic water. The real danger with over-carbonation is exploding bottles. Bottle grenades can be very dangerous both from flying glass and from glass slivers left in the carpet.

*Cure:* Refrigerate the bottles and drink them while there is still some flavor left.

### Problem: The (finished) beer is hazy or cloudy.

*Cause 1: Chill haze.* Chill haze is the number one cause of cloudy homebrew. It is partly caused by an insufficient boil (too weak or short), resulting in the haze-active proteins not coagulating.

*Cure:* The boil should be vigorous and you should see a fair amount of protein flocs floating around in the wort. Chilling the wort quickly down to room temperature (or lower) can help reduce haze by improving the cold break.

*Cause 2: Starch.* If you made an all-grain beer and had incomplete conversion, or steeped malt for an extract batch that actually should have been mashed, then you probably have residual starches in the beer that will cause cloudiness.

*Cure:* Check your malts to determine if they need to be steeped or mashed. Verify the mash temperature with a calibrated thermometer and monitor the temperature to make sure that it stays in the right range. Add an infusion of hot water or do a decoction to bring the temperature back up if it is too low. Use an iodine test at the end of the mash to check for residual starch.

*Cause 3: Yeast strain.* Some yeast strains that have low flocculation, such as German hefeweizen yeast, will cause the beer to be cloudy.

*Cure:* Give the beer more time to settle, or use a more flocculant yeast strain if you want the beer to clear faster. Low calcium ion levels in the brewing water (<50 ppm) can also lead to flocculation and haze problems. Add calcium sulfate or calcium chloride salts to the brewing water next time.

In all cases, beer cloudiness can be combated by adding fining agents (e.g., isinglass or gelatin) after fermentation. Irish moss can be added towards the end of the boil to enhance clarity. See appendix C for more info.

## Common Off-Flavors and Aromas

There are many flavors that contribute to the overall character of a beer. Some of these flavors have been described as malty, fruity, or bitter. However, when it's time to figure out why a beer tastes bad, we need to be more specific. In this section, we discuss different off-flavors and aromas and what could cause each.

### Acetaldehyde

Acetaldehyde is an intermediate compound in the formation of alcohol that is reduced by the yeast to ethanol toward the end of fermentation. Some yeast strains produce more than others do, but generally the presence of acetaldehyde in beer indicates that the fermentation was stressed or incomplete. The aroma of acetaldehyde can be like green (unripe) apples, cut pumpkin, or the smell of a freshly painted room. The flavor tends to be cidery, or like pumpkin or unripe apple.

Residual acetaldehyde can be produced by fermenting a beer too aggressively. This can be caused by overpitching and over-aerating; and also by starting the fermentation too warm and allowing the fermentor to cool as it progresses. A high proportion of simple sugars, such as glucose, fructose, and sucrose, present early in fermentation promote acetaldehyde formation. However, repeatedly feeding the fermentation with new fermentables also promotes acetaldehyde accumulation, so strike a balance. The depletion of the yeast's own glycogen reserves by leaving a starter on a stir plate for too long (>3 days), storing a yeast starter for more than a few days prior to pitching, or extended fermentations with multiple sugar additions, have all been shown to increase acetaldehyde production and impair its uptake later in fermentation. The yeast must be active and in good health going into the maturation phase of fermentation to effectively clean up the acetaldehyde.

To reduce the likelihood of acetaldehyde in your beer:
- Do not underpitch or overpitch.
- Do not under-aerate or over-aerate.
- Do not pitch warm and allow the fermentation temperature to cool during fermentation.
- If using simple sugars, add them after the first day (24–36 hours) of fermentation.

To clean up acetaldehyde in the fermentor:
- Use a diacetyl rest toward the end of fermentation.
- Do not rush the fermentation; give it time.
- Rouse the yeast to keep it suspended (if necessary).
- Use a less flocculant yeast strain (if necessary).

## Alcoholic

An alcoholic flavor is a sharp, sometimes spicy flavor that can be mild and pleasant or hot and bothersome. When an alcohol taste detracts from a beer's flavor it can usually be traced to one of two causes. The first problem is often hot fermentation temperatures. At temperatures above 80°F (27°C), the yeast can produce too much of the higher weight fusel alcohols, which have lower taste thresholds than ethanol. These alcohols taste harsh to the tongue, not as bad as cheap tequila, but bad nonetheless.

The second cause is generally related to high-gravity worts and a corresponding low pitching rate, which stresses the yeast. Be sure to pitch a healthy, properly fed, and adequately sized yeast starter to properly aerated wort to ensure an ideal fermentation.

To reduce the amount of fusel alcohols produced during fermentation:
- Do not stress the yeast.
- Do not overdo the aeration and yeast nutrient supplements.
- Pitch at the intended fermentation temperature; do not pitch warm.
- Ferment at the lower end of the recommended temperature range for the yeast.
- Do not add large amounts of sucrose or other refined sugars to the wort.

## Astringency

Astringency differs from bitterness by having a puckering quality, like sucking on a tea bag. It is dry, somewhat powdery, and is usually due to excess polyphenols (tannins). Astringency is basically haze forming on your tongue; the polyphenols react with proteins in your mouth and coat your tongue. Excess polyphenols can come from over-hopping with low-alpha-acid hops or excessive dry hopping, steeping grains too hot, or oversparging the mash. In other words, high hops, high temperatures, or high pH. Tannin extraction due to high mash pH is a common problem for pale beers brewed with high-alkalinity water. An astringent character in dark beers is more likely due to high temperature than high pH, but pH can still be a cause. Excessively low pH can make dark beers taste acrid, which is not the same as being astringent, but excessive alkalinity can release tannins from dark malts just as easily as from pale malts.

## Cidery

Cidery flavors can have several causes, but is usually due to acetaldehyde, as characterized by a green-apple flavor. Acetaldehyde is a common fermentation byproduct, and different yeasts will produce different levels of it depending on the recipe and temperature. Cidery flavors are often produced by the use of a high proportion of simple sugars, such as glucose and sucrose. Oxidation of acetaldehyde can produce acetic acid, which also contributes to the overall cidery character. See the "Acetaldehyde" subsection above.

## Diacetyl

Diacetyl is a vicinal diketone that is produced by a purely chemical reaction from precursors that are excreted by the yeast early in fermentation. It is most often described as a butter or butterscotch flavor. Smell a bag of extra-butter microwave popcorn for a good example. Diacetyl is desired to a small degree in many beer styles to round out the flavor, but it can easily become overwhelming. Diacetyl can be the result of the normal fermentation process, or the result of a bacterial infection (e.g., *Pediococcus*). Ideally, diacetyl precursors are produced early in the fermentation cycle, which are then oxidized into diacetyl during fermentation and subsequently broken down by the yeast toward the end of fermentation. A brew that experiences a long lag time, usually due to weak yeast or insufficient aeration, will produce lots of diacetyl precursors before the main fermentation begins. A diacetyl rest, where the beer temperature is raised by 5°F (3°C) toward the end of fermentation, can invigorate the yeast and help it break down the diacetyl that has formed. A beer that is fermented warm and fast and rushed through the fermentor may leave lots of diacetyl precursors behind in the beer, which will then manifest as diacetyl in packaging.

To reduce diacetyl in your beers:
- Do not stress the yeast.
- Do not rush the fermentation.
- Do not overoxygenate your wort, and minimize oxygen exposure after fermentation starts.

To clean up diacetyl in the fermentor:
- Increase the temperature towards the end of fermentation (i.e., use a diacetyl rest).
- Keep the beer on the yeast (i.e., don't bottle too soon)
- Rouse the yeast to keep it suspended (if necessary).
- Use a less flocculent yeast strain (if necessary).

## Dimethyl Sulfide / Cooked Vegetable Flavors

Dimethyl sulfide (DMS) is common in many light lagers and is considered to be part of the character in small amounts, just like diacetyl is in ales. However, DMS is considered an off-flavor in most other styles. In pale beers, DMS can have a creamed corn aroma and flavor, whereas in dark beers the character is more tomatolike. Dimethyl sulfide is produced in the wort during the boil by the chemical reduction of another compound, S-methylmethionine (SMM), which is itself produced during malting. S-methylmethionine is volatile, and more will vaporize the longer the malt is kilned at the end of the malting process. Pale ale malt is kilned longer and hotter than lager malt, which explains why DMS is more prevalent in pale lagers.

Dimethyl sulfide is continuously produced in the wort from SMM while it is hot and is usually removed by volatilization during the boil. If the boil is weak (a simmer), the DMS will not be scrubbed from the wort. Longer boils are recommended for beers using Pilsner malt (as opposed to pale ale base malt) to help remove all of the SMM and DMS. Any SMM remaining in the wort can react to produce DMS later in the beer. Although DMS production is promoted by temperatures above 140°F (60°C), it will also be formed at cooler temperatures, so chilling the wort below 140°F after the boil is not the solution—the key is to boil long enough to get rid of the precursor. Dimethyl sulfide is not cleaned up by the yeast during maturation.

When caused by bacterial contamination, DMS has a more rancid character, more liked cooked cabbage than creamed corn. Bacterial contamination is usually the result of poor sanitation. Repitching the yeast from an infected batch of beer will perpetuate the problem.

## Estery / Fruity

Ales are supposed to be slightly fruity, and Belgian and German wheat beers are expected to have a small amount of banana ester character (e.g., isoamyl acetate), but sometimes a beer comes along that could flag down a troop of monkeys. Ethyl acetate, an ester with the aroma of nail polish remover, is also a common problem. Esters are produced by the yeast, and different yeast strains will produce different amounts and types of ester compounds. Most esters in beer are produced from ethanol and only a small percentage are made from fusel alcohol precursors. In general, stressing the yeast produces more esters. Yeast can be stressed by low pitching rates, high fermentation temperatures, low fermentation temperatures, and high-gravity worts; all these conditions will result in more ester formation.

Ester formation is complicated, but it involves the molecule acetyl-CoA, which is normally used to convert various nutrients into cellular building blocks, such as fatty acids and sterols. As yeast cells grow and divide, they utilize acetyl-CoA to facilitate metabolism and growth. Ester formation occurs when acetyl-CoA is no longer needed for further cell growth, which happens right after the exponential growth phase. Therefore, ester formation is promoted by high aeration, high levels of nutrients, low pitching rates, and high yeast growth. Ester formation is not promoted by high pitching rates and low aeration. It is also not promoted when there are sufficient sterols and fatty acids for growth supplied in the trub. In other words, a turbid wort supplies nutrients that the yeast would otherwise have to synthesize for themselves using acetyl-CoA; therefore, less acetyl-CoA is needed, and less esters are subsequently formed.

Ester formation is inhibited by hydrostatic pressure, which increases the solubility of $CO_2$ and inhibits yeast metabolism. These kind of pressures are found at the bottom of tall 100-barrel cylindroconical fermentors. This is obviously not a viable method for controlling esters for homebrewers—I mention it because it is one reason that commercial brewers tend to ferment their beers a few degrees warmer than homebrewers. The tall tanks and higher levels of dissolved $CO_2$ means a warmer fermentation is needed to obtain a similar ester profile to that seen in under normal low pressure conditions.

To reduce excessive ester formation in your beer, reduce those factors that promote excessive yeast growth or yeast stress:
- Use sufficient aeration.
- Ferment cooler but not cold.
- Use an appropriate pitching rate.
- Do not generate high amounts of FAN in the wort.
- Do not separate all the trub from the wort prior to fermentation.

## Grassy

Flavors reminiscent of fresh-cut grass occasionally occur and are most often linked to poorly stored ingredients. Poorly stored malt can pick up moisture and develop musty smells or grassy odors. These odors are often aldehydes, including acetaldehyde. Hops are another source of these "green" flavors. If the hops are poorly stored, or not properly dried prior to storage, polyphenol compounds will become evident in the beer, and these off-flavors can vary from grassy to grainy to astringent.

## Husky / Grainy

Husky and grainy flavors are similar to the grassy flavors described above. Grainy can also include astringency from the grain husks. These flavors are more evident in all-grain beers and are due to poor grain steeping or sparging practices. Follow the same procedures recommended to prevent astringency to correct the problem (see the "Astringency" section above).

Highly toasted malts can also contribute grainy flavors. If you are making your own toasted malts, allow them to age at least two weeks after crushing so the harsher aromatic compounds can dissipate. Cold conditioning the beer for a month or two will often help these harsh compounds settle out along with the yeast. Clarifying your beer with gelatin may also help.

## Medicinal

Chlorophenols result from reactions involving chlorine-based sanitizers (bleach) and phenol compounds. Chlorophenols have very low taste thresholds, and the flavors are often described as medicinal or "Band-Aid like," or they can be highly phenolic (i.e., spicy like cloves), or smell like hot plastic. The phenols are produced by the yeast, and these then react with chlorine or iodine from leftover sanitizer. Items sanitized with chlorine and iodine-based sanitizers need to drain and dry completely before use to avoid these kinds of off-flavors. Do not exceed recommended concentrations when preparing sanitizing solutions.

## Meaty

Meaty, brothy, and umami flavors (reminiscent of ham soup) are typically caused by yeast autolysis, in which the yeast cells die and spill their guts into the beer. This is why you will see yeast extract or hydrolyzed yeast as an ingredient in bouillon cubes. Autolysis is caused by starvation of the yeast and is more likely to happen with old yeast packages, poor fermentation conditions, or extended fermentations (i.e., lasting months). Racking the beer away from the yeast cake prior to a long maturation period is the best defense.

## Metallic

Metallic flavors can be caused by old malt, by adding too much brewing salt for water treatment, or by the presence of iron or manganese from well water. Shiny new aluminum pots will sometimes turn black when boiling water due to chlorine and carbonates in the water, which may cause off-flavors the first time but is not hazardous. Aluminum pots usually won't cause metallic flavors unless the brewing water is highly alkaline (pH >9), which is rare.

## Moldy

Molds are quickly recognized by their smell and taste. Black bread molds and mildew can grow in both wort and beer. Contamination from mold spores is likely if the wort or beer is exposed to musty or damp areas after the boil or early in fermentation.

## Oxidized

Oxidation is the most common problem affecting all beers. Oxygen exposure after fermentation and at packaging is the primary cause. Storage temperature is the second factor. Once carbonated, beer should be stored cold for longest shelf life. The oxidation of fatty acids produces *trans*-2-nonenal, which has a cardboard taste and the aroma of old paper. See the discussion of oxygen and the wort in chapter 6.

## Soapy

Soapy flavors can be caused by not rinsing your beer glass very well, but they can also be caused by chemical reactions in the wort. Soapy flavors can result from the breakdown of fatty acids in the trub and by autolysis of the yeast. Soap is, by definition, the salt of a fatty acid; so you are literally tasting soap.

## Solventlike

Solventlike flavors includes alcohol and ester flavors, but are harsher to the tongue. These flavors often result from a combination of high fermentation temperatures and oxidation. The solvents in some plastics, such as rigid PVC, can leach out as a result of high temperatures. Do not use common PVC irrigation tubing for brewing. Make sure your plastics are food-grade!

## Skunky

Skunky or cat-musk aromas in beer are caused by photochemical reactions of isomerized hop compounds. The wavelengths of light that cause the skunky smell are in the blue and ultraviolet range. Brown glass bottles effectively screen out these wavelengths, but green bottles do not. Skunky flavors will result if the beer is left in direct sunlight, or stored under fluorescent lights (as happens in supermarkets). Beers that use pre-isomerized hop extract or very little flavoring hop additions will be fairly immune to damage from ultraviolet light.

## Sweaty / Goaty

I often encounter sweaty or goaty aromas and flavors at restaurants that don't clean their beer lines regularly. It could be caused by mold, *Brettanomyces*, or *Pediococcus* contamination. It is always a sanitation issue.

## Yeastlike

The cause of a yeastlike flavor is pretty easy to understand. If the beer is too young and the yeast has not had time to settle out, it will have a yeastlike taste. Watch your pouring method too, keep the yeast layer on the bottom of the bottle.

If the yeast is unhealthy and begins autolyzing, this will release compounds that may be initially described as yeastlike aromas, but may smell soapy or meaty as the beer ages.

# Section IV

## Appendices

# Appendix A

## Using Hydrometers
## and Refractometers

## Using Hydrometers

A hydrometer measures the specific gravity of a solution by how high the hydrometer floats in the solution. Specific gravity is a measure of relative density, and a hydrometer measures the density of a solution relative to water. Water has a specific gravity of 1.000. Hydrometer readings are always quoted to the standard temperature, which is 59°F (15°C). The density of a liquid is dependent on temperature, so hydrometer readings are adjusted to state the gravity of the liquid at the standard temperature (see table A.1 at the end of this appendix).

A hydrometer can be used to monitor fermentation because it lets you know the degree of attenuation. Attenuation is the degree of conversion of sugar to ethanol by the yeast. A typical wort starts with an original gravity (OG) of 1.035–1.060, and typically ferments to a final gravity (FG) of 1.005–1.015. Champagnes and meads can have final gravities less than 1.000 because these drinks contain a higher percentage of ethanol, which has a specific gravity of 0.794.

Knowing what wort gravity is and why it should be measured makes the hydrometer a useful tool in the hands of a homebrewer. Beer recipes usually list the OG and FG to better describe the beer to the reader. The rule of thumb is that the FG should be about one-quarter of the OG, which corresponds to an apparent attenuation of 75%. For example, a typical beer OG of 1.040 should finish on about 1.010, but a couple of points either way is not unusual.

It needs to be emphasized that the stated FG of a recipe is not the goal. The goal is to make a good tasting beer. The homebrewer should only be concerned about a high hydrometer reading

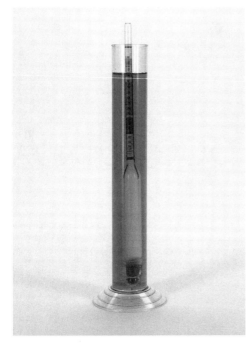

**Figure A.1.** A hydrometer is a useful tool for brewers to measure the specific gravity of a liquid.

when primary fermentation has apparently ended and the reading is about ⅓–½ of the OG, instead of the nominal one-quarter. Proper yeast preparation should prevent this problem. Low FG measurements may also be a concern if they are significantly lower than the target; an FG ≥4 gravity points lower than intended may indicate the action of a wild yeast that is fermenting the usually unfermentable dextrins in the beer that give it flavor and body.

Beginning brewers often make the mistake of checking the gravity too frequently. Every time you open the fermentor you are risking contamination from airborne microbes. Check the gravity when you are ready to pitch the yeast, then leave it alone until the bubbling in the airlock stops. Checking the gravity in between these times will not change anything except to possibly contaminate your beer. Also, always remove a sample of the wort to test it. Don't stick the hydrometer into the whole batch. Use a sanitized siphon or wine thief (basically like a turkey baster) to withdraw a sample of the wort to a hydrometer jar (any tall, narrow jar) and float the hydrometer in that. There is less chance of contamination and you can drink the sample to see how the fermentation is coming along. It should taste like beer, although the flavor may be a bit yeastlike.

Remember to always quote the standardized value when recording or discussing a specific gravity reading. The hydrometer temperature correction table is shown in table A.1. An example: If the wort temperature is 108°F, and the measured gravity of the sample is 1.042, the delta G value (ΔG) would be between 0.0077 and 0.0081. We can round off the average to the third decimal place, which gives us 0.008. Add this 0.008 to the measured 1.042 value, which yields 1.050 as the standardized reading.

## Using Refractometers

The benefit of using a handheld refractometer to measure wort gravity rather than a hydrometer is that it only takes a couple of drops of wort versus filling a hydrometer jar, and you can get the reading from a refractomer in seconds, rather than having to wait for the wort in the hydrometer jar to cool or go looking up the temperature correction factor. A refractometer is less sensitive to temperature due to the very small sample size, and very handy to have at the boil kettle to check your extraction when you are all-grain brewing. The only problem is that refractometers do not directly measure wort density or specific gravity.

Refractometers measure the refraction of light passing through a solution. The denser the solution, the slower the light will travel through it and the more the light will be refracted. Handheld refractometers are calibrated with respect to the density of a sucrose solution at 68°F (20°C). The scale in the viewing window of a refractometer is scaled in degrees Brix (°Bx), which is roughly equivalent to degrees Plato (°P). The only problem is that beer wort is not made of sucrose, it is made up of several different

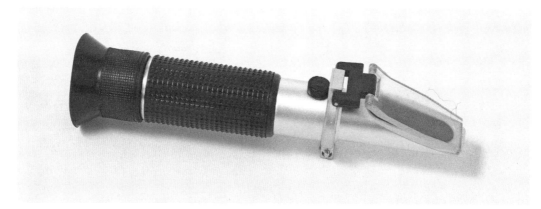

**Figure A.2.** A refractometer is used to quickly measure the specific gravity of a liquid using a small sample size.

sugars and therefore has a slightly different refraction index than a pure sucrose solution. There are industry standard equations that allow you to convert between the various scales with a reasonable degree of accuracy. One such equation to convert the refractometer reading between °Bx and °P is:

$$\text{Wort (°P)} = \frac{\text{Refractometer reading (°Bx)}}{1.04}$$

So, for example, if your wort's refractometer reading is 12°Bx that is more precisely 11.5°P.

Okay, what does the Plato scale measure? Well, like degrees Brix, degrees Plato defines the percentage weight of sucrose in solution. So, a wort of 10°P is 10% by weight sucrose. The specific gravity of the same sucrose solution is approximately four times the degrees Plato, so 10°P is a specific gravity of 1.040. But this approximation becomes less accurate after you exceed 13°P. Fortunately, the American Society of Brewing Chemists (ASBC) has published an official conversion table, and I have reproduced some of that table here (table A.2). Degrees Plato for specific gravities over 1.084 are estimated from ASBC published conversion equations, and are given in italics.

## TABLE A.1—HYDROMETER READING TEMPERATURE CORRECTIONS

| °F | °C | ΔG | °F | °C | ΔG |
|------|------|----------|-------|------|----------|
| 32.0 | 0 | −0.0007 | 77.0 | 25 | +0.0021 |
| 33.8 | 1 | −0.0008 | 78.8 | 26 | +0.0023 |
| 35.6 | 2 | −0.0008 | 80.6 | 27 | +0.0026 |
| 37.4 | 3 | −0.0009 | 82.4 | 28 | +0.0029 |
| 39.2 | 4 | −0.0009 | 84.2 | 29 | +0.0032 |
| 41.0 | 5 | −0.0009 | 86.0 | 30 | +0.0035 |
| 42.8 | 6 | −0.0008 | 87.8 | 31 | +0.0038 |
| 44.6 | 7 | −0.0008 | 89.6 | 32 | +0.0041 |
| 46.4 | 8 | −0.0007 | 91.4 | 33 | +0.0044 |
| 48.2 | 9 | −0.0007 | 93.2 | 34 | +0.0047 |
| 50.0 | 10 | −0.0006 | 95.0 | 35 | +0.0051 |
| 51.8 | 11 | −0.0005 | 96.8 | 36 | +0.0054 |
| 53.6 | 12 | −0.0004 | 98.6 | 37 | +0.0058 |
| 55.4 | 13 | −0.0003 | 100.4 | 38 | +0.0061 |
| 57.2 | 14 | −0.0001 | 102.2 | 39 | +0.0065 |
| 59.0 | 15 | 0 | 104.0 | 40 | +0.0069 |
| 60.8 | 16 | +0.0002 | 105.8 | 41 | +0.0073 |
| 62.6 | 17 | +0.0003 | 107.6 | 42 | +0.0077 |
| 64.4 | 18 | +0.0005 | 109.4 | 43 | +0.0081 |
| 66.2 | 19 | +0.0007 | 111.2 | 44 | +0.0085 |
| 68.0 | 20 | +0.0009 | 113.0 | 45 | +0.0089 |
| 69.8 | 21 | +0.0011 | 114.8 | 46 | +0.0093 |
| 71.6 | 22 | +0.0013 | 116.6 | 47 | +0.0097 |
| 73.4 | 23 | +0.0016 | 118.4 | 48 | +0.0102 |
| 75.2 | 24 | +0.0018 | 120.2 | 49 | +0.0106 |

Notes: Hydrometer readings are standardized to 59°F (15°C) because the density of a liquid changes with temperature. When discussing specific gravity of wort and beer with other brewers, always quote the standardized value. Measure the specific gravity of your wort, take the temperature, and add the correction value (ΔG) given in the table.

## TABLE A.2—ASBC PUBLISHED SPECIFIC GRAVITY TO DEGREES PLATO (°P) CONVERSIONS

| Specific gravity | °P | Specific gravity | °P |
|---|---|---|---|
| 1.008 | 2.0 | 1.060 | 14.7 |
| 1.010 | 2.6 | 1.062 | 15.2 |
| 1.012 | 3.1 | 1.064 | 15.7 |
| 1.014 | 3.6 | 1.066 | 16.1 |
| 1.016 | 4.1 | 1.068 | 16.6 |
| 1.018 | 4.6 | 1.070 | 17.0 |
| 1.020 | 5.1 | 1.072 | 17.5 |
| 1.022 | 5.6 | 1.074 | 18.0 |
| 1.024 | 6.1 | 1.076 | 18.4 |
| 1.026 | 6.6 | 1.078 | 18.9 |
| 1.028 | 7.1 | 1.080 | 19.3 |
| 1.030 | 7.5 | 1.082 | 19.8 |
| 1.032 | 8.0 | 1.084 | 20.2 |
| 1.034 | 8.5 | 1.087 | *21* |
| 1.036 | 9.0 | 1.092 | *22* |
| 1.038 | 9.5 | 1.096 | *23* |
| 1.040 | 10.0 | 1.101 | *24* |
| 1.042 | 10.5 | 1.106 | *25* |
| 1.044 | 11.0 | 1.110 | *26* |
| 1.046 | 11.4 | 1.115 | *27* |
| 1.048 | 11.9 | 1.120 | *28* |
| 1.050 | 12.4 | 1.129 | *29* |
| 1.052 | 12.9 | 1.134 | *30* |
| 1.054 | 13.3 | 1.139 | *31* |
| 1.056 | 13.8 | 1.144 | *32* |
| 1.058 | 14.3 | 1.149 | *33* |

Notes: *The Plato and Brix scales are roughly equivalent and are the preferred units of large-scale brewers when measuring and reporting gravity. Values in italics are estimated from ASBC published conversion equations.*

Source: *ASBC's* Laboratory Methods for Craft Brewers *(Crumplen 1997).*

# Appendix B

## Beer Color

Ah, the many wonderful colors of beer! The pale straw of Belgian wit, the rich gold of Pilsner, the burnished copper hue of special bitter, the rich mahogany of brown ale, the ruby-black highlights of porter, and the pre-dawn darkness of stout—the very sight of these colors whets our imagination for the characteristic flavors of each style. Beer color comes from the malts used; different types of malt have different characteristic colors, and these colors become expressed in the wort. Malt extract is just concentrated wort, and the color of the extract will depend on the malts that were mashed to make it. In addition, there are other factors in the brewing process that can also influence the color, such as caramelization during the boil.

When beers are judged in a competition, beer color is often the first thing a judge checks when determining how well a beer has been brewed to style. About an inch of beer will be poured into one of the clear plastic judging cups, the beer will be swirled to dispel any bubbles on the sides, then held up to the light to gauge the color against a color guide. Dark beers are often examined with a flashlight shone from behind the beer to determine the clarity and its effect on the color. We can reasonably predict the final color of our beer by calculating the color contribution of each malt, malt extract, and adjunct that we use in our recipe. All malts are analyzed for color during production, and in the case of specialty malts, are produced to a specific color range. We can use the color ratings provided by the manufacturers to determine whether our recipe will meet the range for the intended style. Typical color ratings for several malts, malt extracts, and adjuncts are given in tables B.1 and B.2.

## TABLE B.1—SPECIFIED COLORS FOR UNHOPPED MALT EXTRACTS[a]

| Extract type | Coopers | Muntons[b] |
|---|---|---|
| Wheat LME | 4.5°L | <5.0°L |
| Extra Light LME | ... | 2.0–3.5°L |
| Light DME | 3.0°L | 3.5–6.0°L |
| Light LME | 3.5°L | 4.0–6.0°L |
| Amber DME | ... | 12–22°L |
| Amber LME | 16°L | 8–10°L |
| Dark DME | ... | 22–35°L |
| Dark LME | 66°L | 25–30°L |

°L, degrees Lovibond; DME, dried malt extract; LME, liquid malt extract.

[a] Information taken from manufacturers' websites.

[b] Converted from EBC.

## TABLE B.2—TYPICAL COLOR RATINGS OF SELECTED COMMON MALTS AND ADJUNCTS

| Malt / Adjunct | SRM rating |
|---|---|
| Two-row lager malt | 1.5°L |
| Wheat malt | 2°L |
| Pale ale malt | 3°L |
| Vienna malt | 4°L |
| Munich malt | 10°L |
| Biscuit malt | 25°L |
| Crystal 40 | 40°L |
| Crystal 60 | 60°L |
| Crystal 120 | 120°L |
| Chocolate malt | 350°L |
| Black "patent" malt | 500°L |
| Flaked barley | 1.5°L |
| Flaked corn | 1.0°L |
| Flaked rice | 1.0°L |
| Flaked rye | 2.0°L |
| Flaked wheat | 2.0°L |
| Torrified wheat | 1.5°L |
| Maltodextrin powder | zero |
| Dextrose, glucose, sucrose, and fructose | zero |

°L, degrees Lovibond; SRM, Standard Reference Method.

## The Basis of Color Rating

Historically, the color of beer and brewing malts has been rated in degrees Lovibond (°L). This system was created in 1883 by J.W. Lovibond, and consisted of glass slides of various shades that could be combined to produce a range of colors. A standard sample of beer or wort would be compared to combinations of these slides to determine the rating. Malt color is determined by conducting a Congress mash (a standardized method) with the malt and measuring the color of the resulting wort. This system was later modified to the Series 52 Lovibond scale, which consisted of individual slides or solutions for specific Lovibond ratings, but the system still suffered from inconsistency due to fading, mislabeling, and human error.

In 1950, the American Society of Brewing Chemists (ASBC) adopted the use of optical spectrophotometers to measure the absorbance through a standard-sized sample at a specific wavelength of light (430 nanometers). A darker wort or beer absorbs more light, and yields a higher measurement. This method allowed for consistent measurement of samples and so the Standard Reference Method (SRM) for determining color was born. The SRM scale was originally set up to approximate the Series 52 Lovibond scale and the two scales can be considered to be nearly identical for most of their range. However, the resolution of a spectrophotometer diminishes greatly as the wort or beer darkens and very little light can penetrate the sample to reach the detector. To accommodate a dark wort or beer, the sample is diluted and the measurement subsequently scaled to give an undiluted value. Unfortunately, dilutions have long been known to be non-linear for beers made from highly colored malts.

When provided with consistent, precise references, the human eye can distinguish very narrow color differences, because the range of wavelengths of visible light coming from a sample and entering the eye is so much greater when compared with the information conveyed by a single wavelength to a spectrophotometer. Although there is less variation in a single wavelength measurement, there is a corresponding loss in range. For this reason, the Series 52 Lovibond scale is still in use today, in the form of precision visual comparators, and is most often used to determine the rating of dark or roasted malts. The use of comparators is most prevalent in the malting industry, which is why the color of malt is discussed in terms of °L, while the color of beer is discussed in terms of the SRM scale, even though they are calibrated to a common standard (absorbance at 430 nm).

Prior to 1990, the European Brewery Convention (EBC) used a different wavelength for measuring absorbance, and conversion between the EBC and SRM scale was an approximation. Today, the EBC scale uses the same wavelength as the ASBC for measuring beer color (i.e., 430 nanometers), but the EBC uses a smaller diameter sample glass. An EBC rating for a malt or beer is not equivalent to its Lovibond rating. The current EBC scale gives a rating for beer color that is roughly twice the SRM rating. The actual conversion factor between the two methods is 1.97, but to argue whether an Irish stout with an EBC rating of 90 has an SRM rating of 45 or 45.6 is pointless.

Color swatches to illustrate seven levels of beer color are shown on the inside front cover of this book. These color swatches for SRM colors were taken from Promash Brewing Software, v.1.8,[1] and are representative of beer that has been poured to a depth of about 1.5" in a typical 6 fl. oz. clear plastic judging cup, swirled to de-gas, and held up to good lighting against a white background.

The main constituents of color in malts are the melanoidins produced by Maillard browning reactions. Browning reactions between sugars and amino acids occur whenever food is heated, for example, when you make toast. Different heating methods with different sugars and amino acids will produce different colors, ranging from amber, to red, to brown, and black. Thus, the wide spectrum of beer colors is due to the variety of germination and kilning procedures used in malt production. The final color of a beer can be estimated from a recipe by adding up the contribution of melanoidins in the form of malt color units (MCU). An MCU is like an alpha-acid unit (AAU) used for calculating hop bitterness (given in IBU). The color rating of the malt (in °L) is multiplied by the weight of malt used in the recipe to generate the MCU, just like the weight of a hop addition is multiplied by its alpha-acid rating to give the AAU.

To estimate the SRM rating of a beer, the MCUs are divided by the recipe volume and multiplied by a constant that is like the percent utilization in the IBU calculation. For light-colored beers (yellow/gold/light amber) the relationship between SRM and MCU is approximately 1-to-1.

As a brief example, let's take 8 lb. of two-row lager malt (2°L) and 2 lb. of Vienna malt (4°L) being used for a 5 gal. batch:

$$\text{Estimated color} = [(8 \times 2) + (2 \times 4)] / 5$$

$$= 4.8 \text{ SRM (round up to 5 SRM)}.$$

[1]   Jeffrey Donovan, "ProMash Brewing Software," v. 1.8, Sausalito Brewing Co., Santa Barbara, CA, 2003, http://www.promash.com. The ProMash website landing page has been a simple holding message since 2015. It is not clear at this writing if and when the full website will be made available again.

## TABLE B.3—PROPOSED MODELS FOR BEER COLOR AS A FUNCTION OF MCU

| SRM = MCU | (traditional) |
|---|---|
| SRM = (0.3 × MCU) + 4.7 | R. Mosher[a] |
| SRM = (0.2 × MCU) + 8.4 | R. Daniels[b] |
| SRM = 1.49 × MCU$^{0.69}$ | D. Morey[c] |

[a] *Randy Mosher,* The Brewer's Companion *(Seattle: Alephenalia Publications, 1994).*

[b] *Ray Daniels,* Designing Great Beers *(Boulder: Brewer's Publications, 1996).*

[c] *Dan Morey, "Approximating °SRM Beer Color," http://www.morebeer.com /brewingtechniques/beerslaw/morey.html.*

Unfortunately, this simple model does not work when the total MCUs exceed 15 (fig. B.1). Linear models have been proposed by Mosher and Daniels, but data for the full spectrum of beer color may be better fit by an exponential curve, such as the one described in the equation by Morey (table B.3).

The fault of the linear models proposed by Mosher and Daniels (table B.3) is that there is a lower limit for beer color at 4.7 and 8.4 SRM, respectively. Obviously, there are beer styles, such as Belgian wit, Pilsner, and American light lager, that are lighter than these limits. Morey's exponential equation fits the data better, because the function is nearly equal to MCU at low MCU values, but as the MCUs go to higher values (corresponding to brown ales, porters, and stout), the actual color diverges from the SRM = MCU line and increases at a lower rate as depicted by the linear models of Mosher and Daniels. A beer with an MCU of 200 compared to one with an MCU of 100 is still just "very dark," instead of being twice as dark as the SRM = MCU model would have you believe. Even expert beer judges cannot discern a difference between color for values greater than 40 SRM.

Another aspect of beer color that needs to be mentioned is "hue." Different beers with the same SRM rating can actually be different hues, because the measurement is based on the absorbance of a single wavelength of (blue-violet) light. And actually, it isn't the absorbance of the light that's measured, but how much of the light that gets through the sample to the detector. The human eye sees all the visible wavelengths, and will perceive other colors transmitted through or reflected from the sample that the spectrophotometer detector will not. This drawback in the current ASBC method was noted in a brewing study at UC-Davis.[2] In this study, four beers, consisting of two lagers, one pale ale, and one stout, were diluted to the same rating (3.5–3.6 SRM). A group of 31 people were presented with ten pairings of these diluted beers and asked to determine if they were the same or different. The results clearly showed that the panelists could correctly determine a difference in color between different beers, except in the case of the two lagers, which were perceived as being the same. The original color of the undiluted all-malt lager was 8 SRM, and the undiluted color of the other lager, containing cereal adjuncts, was 4 SRM; by contrast, the pale ale was 25 SRM. When analyzed, there was virtually no difference (<1 SRM) between the ASBC method and a Series 52 Lovibond Comparator (Tintometer Ltd., Salisbury, UK) in the undiluted color results for each of the two lagers and the pale ale. The stout was the exception, with the comparator measuring 115°L, while the spectrophotometer measured 86 SRM. This difference illustrates the drawback of the ASBC method for determining the color of very dark malts and beers.

---

[2]  Smythe and Bamforth (2000).

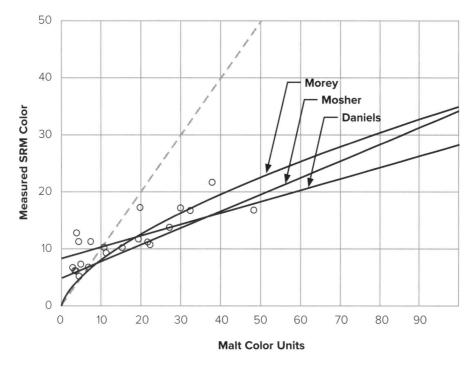

**Figure B.1. SRM Malt Color as a Function of Malt Color Units** A comparison of four models for final beer color. The dashed line is a reference line for SRM = MCU. Lines for each of the suggested color models are shown compared to the actual data. Analysis of commercial as well as homebrewed beer has shown that the measured color tapers off, even as the malt color contributions increase dramatically. Data taken from Daniels (1995).

## Other Factors Determining Color

The color of the malt is not the only factor that determines a beer's final color. Other variables, such as boil time, heating method, hopping rate, yeast flocculation, clarity, and oxidation, will each affect the absorbance of light and the perceived color. Long boil times over high heat will promote the Maillard reactions between sugars and proteins that darken the wort. The oxidation of polyphenols (tannins) from grain husks or hop cones also contributes to wort darkening. Wort that has been oversparged or heavily hopped (like an IPA) will have a greater propensity for darkening as it ages. Wort will also lighten during the boil and subsequent chilling as proteins combine with polyphenols to form the hot and cold break. Other color carrying compounds will settle out during fermentation as the yeast flocculates. And, of course, the beer's overall clarity will affect the degree to which light is absorbed and therefore its perceived color. For instance, cloudy beers appear darker, and will measure darker if the beer is not clarified by centrifuge or filtering.

These other factors that affect beer color are diverse and significant enough that the actual color could be vary ±20% from the calculated SRM value from Morey. So, that being the case, a simplified exponential equation of SRM = $1.5 \times MCU^{0.7}$ is just as valid as the derived values of Morey's equation. My purpose in stating this isn't to propose a new model, but instead to point out the inherent limits of any model for beer color. None of the three models discussed above (fig. B.1) is necessarily any more correct than another, although Morey's may be more forgiving for very light beer styles. I hope this caveat will prevent the more technically minded reader from trying to calculate color to the fourth decimal place. A nomograph for beer color based on the Morey model is given in Figure B.2.

## Estimating Beer Color

To plan the color of your recipe, calculate the MCU values for each of your malts and adjuncts, and then apply the result to one of the color models. As discussed above, remember that the MCU is found by multiplying the malt's Lovibond rating by the weight of malt (in pounds) that you are going to use, and then dividing by the recipe volume (in gallons). Let's look at some examples.

**BJCP Style Guidelines for California Common Beer:**
OG = 1.044–1.055
FG = 1.011–1.014
IBU = 35–45
Color = 8–14 SRM

**No.4 Shay Steam—California Common Beer 6**
Recipe OG = 1.048

| Recipe malts | Color rating | MCU |
|---|---|---|
| 6.0 lb. Light LME | 5°L | 6 |
| 0.75 lb. crystal 40L malt | 40°L | 6 |
| 0.25 lb. maltodextrin powder | 0°L | 0 |

**Calculated color: (Mosher, Daniels, Morey)**

| Total MCU | Mosher | Daniels | Morey |
|---|---|---|---|
| 12 | 8 | 11 | 8 |

*Comment:* In this example recipe for California common beer (above), the malts yield an SRM color rating within the BJCP guidelines for all three color models. The brewer can be confident that the entry would not be marked down for color.

**BJCP Style Guidelines for Brown Porter:**
OG = 1.040–1.050
FG = 1.008–1.014
IBU = 20–30
Color = 20–35 SRM

**BJCP Style Guidelines for Robust Porter:**
OG = 1.050–1.065
FG = 1.012–1.016
IBU = 25–45
Color = 30+ SRM

**Port O' Palmer—Porter**
Recipe OG = 1.048

| Recipe Malts | Color Rating | MCUs |
|---|---|---|
| 6.0 lb. Light LME | 5°L | 6 |
| 0.5 lb. crystal 60L malt | 60°L | 6 |
| 0.5 lb. chocolate malt | 350°L | 35 |
| 0.25 lb. black "patent" malt | 500°L | 25 |

**Calculated color: (Mosher, Daniels, Morey)**

| Total MCU | Mosher | Daniels | Morey |
|---|---|---|---|
| 72 | 26 | 23 | 28 |

*Comment:* In this example porter recipe (above), the malts yield an SRM color rating and an OG within the BJCP guidelines for Brown Porter, although the use of black "patent" malt adds some roast character that is more appropriate to the Robust Porter category. In this case, the brewer can use color modeling to adjust the recipe to firmly place it in the Robust Porter if they wish. Both the OG and the SRM rating would need to be increased by a few points. There are many options to do this; here are three.

In the first option, add 1.5 lb. of Dark DME. This will add about 12 points to both the OG and the MCU total, giving color estimates of 30, 25, and 32 SRM for Mosher, Daniels, and Morey, respectively. The drawback to this approach is that the OG has increased significantly (to 1.060) without changing the color very much.

For the second option, increase the chocolate malt to 1 lb. This will change the MCU total to 107, but without changing the OG all that much, giving color estimates of 37, 30, and 37 SRM. Though you will still need to increase the gravity to ensure it makes the Robust Porter category.

For the third option, increase the Light LME to 7 lb., and increase the chocolate malt to 1 lb. The MCU total is almost the same as that reached using the method in the previous option, keeping the total color almost exactly the same. However, the extra pound of extract increases the OG to 1.055, which is respectable for a Robust Porter.

## Summary

I hope this has given you a good understanding of how beer color is measured and how it can be estimated to help your brewing. It is also important to remember that final beer color is driven by many factors from all parts of the brewing process, and that these are not factored into the color models. You will need to examine your equipment, your processes, and your beers to determine which model works best for you—just as you would in the case of hops and IBU calculations. These tools are not the end, they are the means to an end, and the proof is in the beer.

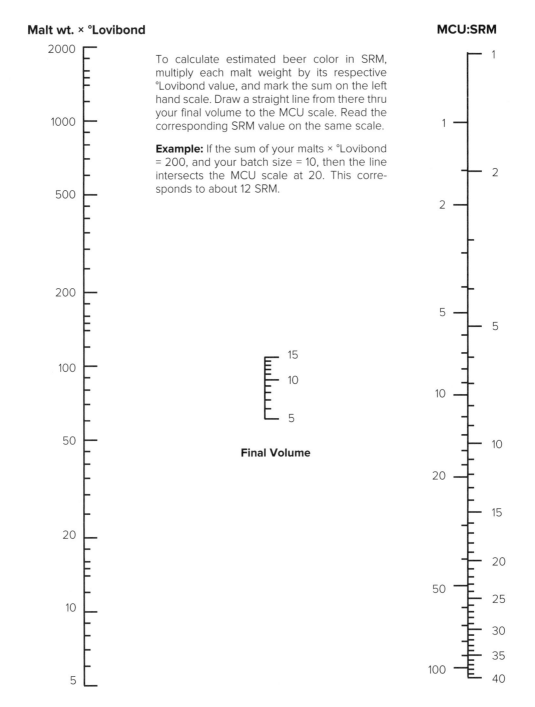

**Figure B.2.** Nomograph for calculating SRM from malt color units (MCUs). To use, multiply weight of the grain (pounds) by it's color rating in °Lovibond, and mark the number on the Malt wt. × °Lovibond scale. Draw a line from there through the batch size number (e.g., 5 gallons) to the MCU scale. This is the MCU contribution for that particular grain. You will need to add up the MCUs for all the malts in the grain bill, and mark that total on the MCU scale to read off the calculated SRM. This nomograph utilizes the Morey color model.

**Author's Note**
This work was originally published as a feature article in *Brew Your Own*, May/June 2003, 28–33. (Used with permission.)

# Appendix C

## Beer Clarity

Last month you planned and brewed what should be your best beer yet, and today is the first pour. This time, your recipe used a new brand of malt extract, or a new specialty grain or adjunct, and you also added some dry hops, or fruit or spices, to the secondary fermentor to give it that extra flavor you were looking for. The guys at the homebrew shop had tried pulling your leg when you told them the recipe, suggesting that you add seaweed to the boil, and put fish guts and gelatin in the fermentor. What a bunch of kidders, you didn't believe a word of it!

And now you pour your best beer and, and . . . it's cloudy! What happened?!

There are several possible causes of cloudy, or hazy, beer. Maybe it's simply that your yeast has not flocculated (settled out) yet. You could try cold conditioning the beer for a few days to see if that helps. You might have wild yeast or bacterial contamination; it could be haze due to unconverted or insoluble starch, or fruit pectin; or it could be a protein-polyphenol haze. How can you tell? What should you do about it? Contamination is an issue unto itself and needs to be evaluated as such. Hazes due to starches represent a food source to many wild yeasts and bacteria, and can cause flatulence as the starches are broken down in your gut. Fruit pectin haze can be combated by the use of pectic enzymes (pectinase) or by changing how you prepare your fruit for the fermentor.

Finally, proteins and polyphenols from the malt and hops can combine to form both temporary and permanent hazes, although these can often be mitigated by fining agents that will clarify the beer.

Those of us used to seeing crystal clear American light lagers may assume that this is how clear every beer should be. Wrong! In fact, the low-protein, high-adjunct beers of the US are some of the clearest beers in the world, if not the clearest. The ingredients have a lot to do with it, but the other half of the equation is the filtering capability that large commercial breweries have available to them. Filtering systems for homebrewers are available, but require that you keg the beer and force carbonate, because the yeast is filtered out too.

But other than aesthetics, why should you care about cloudy beer? You can't taste haze . . . or can you? Haze can often be an indicator of another problem, such as bacterial contamination. Bacteria often cause clouding of the beer and characteristic off-flavors. For instance, *Pediococcus damnosus* is a commonly feared brewery contaminant that generates high amounts of diacetyl. In addition to the tartness of lactic acid, *Lactobacillus* can produce a variety of flavors, some of which are pleasant, as in lambic beers. Other strains of lactobacilli will produce excessive amounts of diacetyl, much like *Pediococcus*. A third type of haze-causing bacteria are coliforms, and these bacteria will often produce vegetal off-flavors, reminiscent of parsnips and old celery. Hazes due to bacterial contamination will most likely develop in the bottle after fermentation, and the sudden appearance of haze can be an indicator that something has gone wrong.

## What Is Beer Haze?

What is beer haze? Beer haze, including chill haze, is a combination of haze-active proteins and haze-active polyphenols, which complex together via hydrogen bonding to create large molecules suspended in the beer that create a visible haze. Hydrogen bonding is strongest at colder temperatures. At warm temperatures, the component molecules are vibrating too much for the weak hydrogen bonds to hold the protein-polyphenol complex together. That's why chill haze behaves the way it does. With time, the protein-polyphenol complexes can oxidize and polymerize into permanent haze, but it all starts with haze-active protein and haze-active polyphenols.

Why do we say, "haze-active"? What does that mean? Basically it refers to the type and size of the proteins and polyphenols, both of which are large, complex molecules that come in many forms. Haze-active proteins are of the same size class as foam-active proteins, but haze-active proteins happen to be rich in the amino acid proline, which appears to be the site where haze-active polyphenols can attach.[1] Haze-active proteins can be broken down into non-haze-active proteins by proteolytic enzymes, but those enzymes also break down the foam-active proteins, which is obviously a problem.

### Why Do We Care about Beer Haze?

Millions of dollars are spent annually researching and combating beer haze. Why? Because in addition to the aesthetic appeal of a clear beer, the polyphenols that contribute to haze are part of a chemical equilibrium that contributes to oxidative staling reactions. You may be wondering just what a polyphenol is. You may have heard of them in terms of being an off-flavor in beer—compounds that have spicy, plastic, or medicinal flavors. No, actually those compounds that have characteristic off-flavors are *phenols*. Polyphenols are polymers of phenol compounds. You may have heard that polyphenols are tannins. Actually it's the other way around—tannins are a type of large

---

[1]    Siebert and Lynn (1997).

polyphenol molecule. And you have probably heard that oversparging or having the wrong mash pH will leach tannins into your wort. This is indeed true, because tannins (and other polyphenols) will be extracted from the malt husks and papery hop cones under such conditions.

There will always be some level of polyphenols in the beer, but it's like complaining about having sand in the desert. Unless there is a sandstorm, you just accept it and work around it. If you think of phenols as being like Lego® blocks, you will get an idea of the possible variations in size and how small polyphenols can link up to form large polyphenols, including tannins. The most common manifestation of protein-polyphenol haze is "chill haze," which is formed by small polyphenols cross-linking with haze-active proteins. These complexes are insoluble when the beer is chilled, but don't have enough mass to settle out effectively and so remain suspended in the beer. These chill haze complexes break up when the beer is warmed to room temperature.

Larger polyphenols form larger protein-polyphenol complexes that can settle out as hot and cold break material, while the smaller polyphenols are carried over into the final beer. As mentioned earlier, these small polyphenols can grow by polymerization, especially in the presence of oxygen. If a beer with chill haze was poorly handled during bottling, the oxygen can cause the chill haze to become permanent haze.

It is interesting to note that, while searching abstracts in the professional brewing journals, haze seems to have become a bigger problem since the late 1990s. While this increase could simply be attributed to growth in the craft brewing industry and a tighter focus on quality, a better explanation might be that there is also more awareness and control of oxidation in the wort production process. A reduction in wort oxidation will result in less polymerization of the smaller polyphenols, such that less polyphenols and tannins are precipitated in the break material during boiling and cooling. Thus, more polyphenols survive into the packaged beer where they contribute to chill haze. In other words, a beer produced 50 years ago with little regard to wort oxidation before fermentation (not aeration for yeast growth) may have been more prone to staling and had a shorter shelf life, but it was probably clearer than comparable beers today.

## Fixing Haze in the Recipe

To reduce the chances of haze forming in your beer, you can take steps during your brewing process to try to reduce the haze-active protein levels, reduce the polyphenols, or a bit of both. You can make these reductions by tweaking the recipe or by using clarifiers and finings. Each option has its pros and cons. To reduce the level of haze-active proteins and polyphenols in the recipe, you can change from an all-malt recipe to one that uses a percentage of low-protein adjunct, such as corn, rice, or refined sugar. This approach is exemplified by American light lager, Belgian tripel and Belgian golden strong ale. Using wheat, or wheat extract, in a recipe to reduce polyphenols (because wheat doesn't have a husk) can be a double-edged sword. At low levels, 5%–12% of the grain bill, the high protein levels in wheat can cause extensive haze, but as the percentage of wheat increases to 40% the total polyphenol levels are substantially decreased and the beer is very clear.

Hops are another source of polyphenols. Many brewers swear by the exclusive use of low-alpha-acid aroma hops for bittering, justly claiming a more refined hop character in the beer. The downside to this is a greater proportion (up to four times) of hop cone material in the wort, and the large amount of polyphenols that will be extracted from this material during the boil. I brew an American wheat extract beer that tends to be hazy due to the wheat gluten, but last time I brewed it I switched from using Nugget

(12% AA) as my bittering hop to using all Liberty (3.5% AA). That batch had a superb hop character that was as rich as royalty, and a creamy head that needed a spoon to clean the glass, but it was hazier than previous batches.

It has been shown that 70% of malt polyphenols can survive the hot and cold break, while only 20% of hop polyphenols do.[2] The message here for reducing haze-active polyphenols and proteins is to achieve a good hot break and cold break to minimize the excess protein that will bind with the polyphenols to cause haze. Dry hopping after the boil will introduce more hop polyphenols, which is why you often see hazy IPAs.

If you are an all-grain brewer, your malts and the way you mash and sparge can affect your polyphenol levels too. New barley varieties have been developed to have less polyphenol than current varieties, and while test batches have been promising for reduced haze, these new malts haven't really caught on in the market. Your sparging method can also affect the total level of polyphenols. While the first runnings generate the highest concentration of the small polyphenols, the last runnings of a continuous sparge contain the highest proportion of tannin-type polyphenols extracted from the husks. This is due to the rise in mash pH as the buffering power of the malt acids is rinsed away from the grain bed. This problem can be mitigated by either calcium additions to or acidification of the sparge water (see chapter 21). Using a batch sparge (where the final runnings typically don't fall below 1.020) or a no-sparge technique (where there is no rinsing), can also minimize pH rise and prevent excessive tannin extraction into your wort.

## Fixing Haze with Clarifiers

Now we come to the seaweed, fish guts, and gelatin. You can add clarifiers to your wort and beer that will bond with haze-forming molecules and allow them to settle to the bottom. Irish moss, isinglass, and gelatin are the most common clarifiers used by homebrewers. Polyvinylpolypyrrolidone and silica gel products are most commonly used by commercial brewers, because they cost less in bulk and are generally more effective than carrageen and collagen. However, they are not digestible and need to be physically separated from the beer before packaging. Proline-specific enzymes are also now available, and these have proven to be very effective at eliminating haze while preserving beer foam.

### Irish Moss

Irish moss, or carrageen moss, is a type of red seaweed containing a certain class of long-chain polysaccharides called carrageenans, which preferentially attracts and binds to large proteins. Irish moss is the only clarifier that you add to your boil. All other clarifiers are added after fermentation. Irish moss is added during the last 5–20 min. of the boil, where it greatly enhances the clumping and precipitation of proteins that would otherwise contribute to haze and staling reactions.

In the past, it was generally accepted that haze-active proteins were different from the proteins responsible for head retention (i.e., foam-active proteins). However, more recent studies[3] have shown that the two groups of proteins are similar enough that any attempt to eliminate haze-forming proteins using either enzymes or non-specific protein-absorbing additives (e.g., bentonite) will also affect the head retention and body of the beer. What this means to you is that either adding a protein rest to your mash schedule or enzyme clarifiers to your wort is probably not a good idea.

---

[2]     McMurrough, *et al.* (1985).
[3]     Ishibashi, *et al.* (1996) and Bamforth (1999).

In addition, misuse of the right clarifiers can also be trouble. If too much Irish moss is used in the boil, not only can the proteins responsible for head retention be affected, but it could also reduce the free amino nitrogen (FAN) that the yeast need for nutrition. For this reason, Irish moss is not recommended for use with malt extract or adjunct worts.

Irish moss is commonly available as dry flakes that are rehydrated before use. A typical dose is 1 teaspoon of flakes for 5 gal. of boil volume (125 mg/L). Another form of Irish moss is a product called Whirlfloc® from Australia, consisting of a large tablet that you simply drop into your wort. Each tablet is good for a 5 gal. (19 L) batch. People have reported excellent results using Whirlfloc.

## Isinglass

The other popular fining agent is isinglass, commonly used in English cask ales. Composed almost entirely of the protein collagen, isinglass is obtained by cleaning and drying the swim bladders of the sturgeon, cod, hake, and other fishes. It is an excellent clarifier for yeast and haze, but expensive given its source.

Isinglass is sold as dehydrated powder to be used at a dosage of 30–60 mg/L, but it is most commonly available for homebrewers as a ready-to-use liquid. To use, add isinglass to the fermentor after fermentation has finished, or to the bottling bucket when you add your priming sugar solution. Do not attempt to heat up isinglass, because it is easily denatured. Two ounces of the liquid product will treat five gallons of beer (3 mL/L). Isinglass is considered to be better than gelatin for use in cask conditioned ales, because it settles readily at cask serving temperatures after being disturbed. Gelatin typically requires coolers temperatures to resettle after movement.

*(Seriously, you have to wonder whose idea this was. "Igor, go get me some fish guts to add to the beer. What! Nothing fresh? Well just scrape some of that dried stuff off the cutting board there. . .")*

## Gelatin

Gelatin is a byproduct of the collagen extraction process from cow hooves and pigskin. It is not as effective as isinglass at settling the yeast mass, needing about three times as much to do the same job, but it is less expensive. The recommended dosage rate is about 0.2–0.4 g/L (or 0.75–1.5 g/gal.), although brewers have reported good results with dosage as low as 0.08 g/L (0.3 g/gal.). As with any fining agent, the exact dosage depends on the yeast and haze content of the particular beer. Too much of any fining agent can cause residual haze and a condition commonly referred to as "fluffy bottoms," where the sediment is easily disturbed and pours out with the beer.

Gelatin can be added to the beer in the fermentor before racking to the bottling bucket or keg, or it can be added directly to the keg. The clarifying action of gelatin does not depend on the gelling action that it is known for. Instead, gelatin acts as a bridge between haze active proteins and polyphenols, attracting and hydrogen bonding them together into clumps with enough mass to settle to the bottom. The beer should be cold, between 35°F and 45°F (2–7°C), in order to strengthen the hydrogen bonds and facilitate clumping.

To prepare gelatin, it only needs to be thoroughly dissolved in a small amount of water. Powdered gelatin acts a lot like dry malt extract when it is added to water. It will hydrate without clumping in cold water, but needs to be heated to about 150°F (65°C) to fully dissolve. Adding powdered gelatin directly to hot or boiling water tends to result in clumps that take a lot of stirring to dissolve. Once dissolved, the hot solution of gelatin can be poured directly into chilled beer and typically clarifies the beer within 24 hours. Swirling the beer to distribute the gelatin solution will help but is not usually necessary. Best results are obtained when the clarified beer is racked away from the settled haze and trub before bottling or kegging. There is typically enough yeast in suspension to support priming and natural carbonation, although adding a small pitch of fresh yeast will speed the process.

### Polyvinylpolypyrrolidone / Polyclar®

Polyvinylpolypyrrolidone (PVPP), also known as crospovidone, is a micronized white powder with a high surface-area-to-volume ratio that readily adsorbs polyphenols, including tannins. The necessary contact time is only a few hours. Commercially, PVPP is the most popular clarifier and stabilizer (a common brand is Polyclar®). For homebrewing, about 6–10 g per 5 gal. (19 L) is added after fermentation but prior to bottling. The powder is commonly combined with cooled boiled water to form a slurry that is then added gently to the fermentor. The PVPP slurry needs to be mixed thoroughly with the beer and allowed to settle out, which should take less than a day. Then the beer should be carefully racked off the sediment and bottled or kegged. This material is not approved by the FDA for ingestion. Commercial breweries remove PVPP by filtration.

### Silica Gel

Silica hydrogels and xerogels are the other half of the one-two punch that commercial brewers use to control haze and improve shelf life. Where PVPP works to bind polyphenols, silica gel binds to proteins. In fact, silica gel binds preferentially to haze-active proteins, because chemically it reacts with the same proline sites that polyphenols do. Like PVPP, silica gel is used at the same rate of 6–10 g per 5 gal. (19 L), and added using the same procedure. Silica gel and PVPP work synergistically to reduce haze more than each would alone. A combined product called Polyclar Plus™ is available to commercial breweries; at time of writing I don't know if it has been packaged for use at a homebrewing scale. Silica gel is not approved by the FDA for ingestion. Commercial breweries remove it by filtration. Allowing the material to settle and carefully racking away from the sediment should be sufficient.

### Proline-Specific Enzymes

Proline-specific endoproteases, such as Clarity Ferm from Whitelabs, act by cleaving the haze-active proteins where a proline amino acid is accessible in the protein chain. This cleavage reduces them to smaller, non-haze-active proteins. The proline sites will still bond with haze-active polyphenols, but the size of the complex can't grow to form a haze. This type of enzyme has the bonus benefit of breaking up the proteins that form gluten. Industry studies have shown that beer treated with this enzyme measured less than 20 ppm gluten based on the R5 Mendez Competitive ELISA assay.[4] However, even though the current benchmark for considering a food to be gluten-free is <20 ppm gluten, there is enough variation among people with gluten-related allergies that marketing a beer as gluten-free may be impossible from a legal liability standpoint. For more information, see appendix I, "The Trouble with Producing Gluten-Free Beer."

The dosage rate for Clarity Ferm is listed as being 12 mL per barrel for both gluten and haze reduction. This translates to 0.4 mL/gal. (about 0.11 mL/L). The enzyme must be stored cold (39–46°F [4–8°C]) to ensure viability, but has the convenience of being able to be added to the fermentor at pitching time.

---

[4] ELISA stands for enzyme-linked immunosorbent assay, and is a detection technique that is highly specific due to the use of antibodies.

## Summary

Beer haze can have many possible causes, but a hazy beer that still tastes good is probably suffering from protein-polyphenol haze. Haze is usually treatable by the use of different ingredients, including malt and hop varieties, as well as by additives like clarifiers and finings (table C.1). I hope this discussion will help you understand how these hazes form and how to best address the cause and solution when making your own homebrews.

### TABLE C.1—CLARIFIER SUMMARY TABLE

| Clarifier | Purpose | Amount | Comments |
|---|---|---|---|
| Irish moss | Protein coagulant | 1 teaspoon/5 gal. (125 mg/L) | A good clarifier for almost all worts, though not recommended for high-adjunct or extract-based worts. |
| Whirlfloc | Protein coagulant | 1 tablet/5 gal. (19 L) | A good clarifier for almost all worts, though not recommended for high-adjunct or extract-based worts. |
| Isinglass | Yeast and haze flocculent | 30–60 mg/L, or 2 fl. oz./5 gal, (3 mL/L) in liquid preparation | Most effective for settling yeast. Will also settle some protein haze. |
| Gelatin | Yeast and haze flocculent | 0.3–0.6 g/gal. (80–160 mg/L) | Generally, the most economical and effective clarifier for homebrewing. Suggested dosage is 0.75–1.5 g/gal. (0.2–0.4 g/L).[a] Typically clarifies in 1 day. |
| Polyvinylpoly-pyrrolidone (PVPP) / Polyclar® | Polyphenol binder | 6–10 g/5 gal. (0.30–0.52 g/L) | A non-aerated slurry should be mixed into the beer before bottling and allowed to settle out. Typically clarifies in 1 day. |
| Silica gel | Haze- active protein binder | 6–10 g/5 gal. (0.30–0.52 g/L) | A non-aerated slurry should be mixed into the beer before bottling and allowed to settle out. Typically clarifies in 1 day. |
| Proline-specific endoproteases (e.g., Clarity Ferm) | Haze- active protein reducer | 12 mL/bbl., or 0.4 mL/gal. (0.1 mL/L) | A proteolytic enzyme that breaks up haze-active protein molecules (typically hordeins), reducing both haze and gluten content. Added to fermentor at pitching. |

[a] *Dosage based on Siebert and Lynn (1997).*

## Author's Note

Portions of this work were originally published as a feature article in *Zymurgy*, vol. 26, September–October, 2003. Gelatin and Clarity Ferm notes revised 2015.

# Appendix D

## Building Wort Chillers

The purpose of wort chillers is to cool the wort quickly after the boil, which helps minimize the risk of bacterial contamination as it cools. There are three basic types of wort chiller: immersion, counterflow, and plate. An immersion chiller works by circulating cold water through a large coil of copper tubing and submersing the coil in the hot wort. A counterflow chiller works by running the hot wort through the copper tubing while cold water runs outside in the opposite direction. Here in the United States, the basic material for both these types of chiller is ⅜" (10 mm) outer diameter (OD) soft copper tubing. Half-inch (13 mm) OD tubing works better for large-scale immersion chilling, but ⅜" OD works better for counterflow, and works well for immersion chilling too. Do not use less than ⅜" because the restricted water flow impairs cooling efficiency.

A plate chiller is a more efficient heat exchanger, and consists of thin copper and stainless steel plates brazed together in a stack to provide the most cooling surface area in the smallest possible package. Plate chillers cannot be made at home but are readily available commercially (as are the immersion and counterflow chillers). The best thing about wort chillers is that they prevent you from having to lift a heavy pot of hot wort.

There are several factors that determine how effectively and efficiently a wort chiller works. An ideal chiller drops the wort temperature by the largest amount in the shortest period of time using

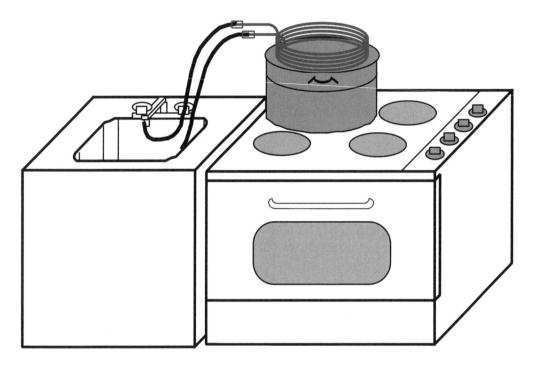

**Figure D.1.** Immersion chilling on the stove.

the least amount of cooling water. It's the ol' "faster, better, cheaper" criteria that engineers wrestle with all the time. Guess what? The best you can do is two out of three. You can have fast and cheap but not better, or better and cheaper but not faster, or faster and better but not cheaper. It's the same with chilling—the more water you can put through the chiller, the more you will cool the wort. The amount of water is essentially your cost, although the length of copper tubing is definitely a major cost as well. The longer the copper tubing, the more surface area you have for cooling water to act on the wort. The final factor is the temperature difference between the cooling water and the wort. The colder the cooling water, the more effectively it will chill the wort.

## Immersion Chillers

Immersion chillers are the simplest to build and maintain. To make one, simply coil 25–50 ft. (8–15 m) of soft copper tubing around a pot, or other cylindrical object, that fits well inside your brewing pot. Spring-like tube benders can be used to prevent kinks from forming during bending. Be sure to bring both ends of the tube up high enough to clear the top of your boiling pot. Attach compression-to-pipe-thread fittings to the tubing ends. Then attach a pipe-thread-to-standard garden hose fitting. This is the easiest way to run water through the chiller without it leaking. The cold water IN fitting should connect to the top coil and the hot water OUT should be coming from the bottom coil for best chilling performance. Several examples of immersion chillers are shown in figure D.2.

The primary performance factors for immersion chillers are chiller surface area and water flow rate. The more water you push through the chiller, the more heat you can remove from the wort. Therefore, a larger diameter tube is better than a smaller one. A longer, larger coil will also help, but

**Figure D.2.** Here are commercial examples of two immersion chillers (top), a plate chiller (center), and a counter-flow chiller (right).

only up to a point. If the coil length is too long, the exiting cooling water will be the same temperature as the wort for much of the coil's length, and that is wasted capacity. The answer is to increase the flow rate of the water so that the cooling water in the chiller is always cooler than the wort. However, faster and better is not cheaper; you will use a lot of water that way. I have found that a 50 ft. (15 m) coil of ½" (13 mm) OD copper tubing works very well for 10 gal. (38 L) batches, and 25 ft. (8 m) of ⅜" OD copper tubing works fine for 5 gal. (19 L) batches.

The advantages of an immersion chiller are that it is easily sanitized by being placed in the boil, and the wort will be cooled before it is poured into the fermentor. This allows you to separate the wort from the cold break. Make sure the chiller is clean (but not necessarily shiny) before you put it into the wort. Placing the chiller in the hot wort soon after the heat is turned off will ensure it is thoroughly sanitized as long as you wait 10 seconds before starting the cooling water. If you are conducting a hop steep, or whirlpool hopping, you can immerse the chiller, but do not turn on the cooling water until you are done with the hop steep.

A good way to improve the performance of an immersion chiller is to stir the wort gently during chilling so that hot wort is continually moving against the cold coils. You can also move the chiller up and down or use it to stir with, which will do much the same thing. Keep the wort moving and it will chill more quickly.

Working with cool wort is much safer than hot wort. The cool wort can be poured into the fermentor with vigorous splashing for aeration without having to worry about oxidation damage. The wort can also be poured through a strainer to keep the spent hops and much of the break material out of the fermentor.

## Counterflow Chillers

Counterflow chillers are almost the opposite of immersion chillers, because instead of immersing a small chiller in a large wort you are flowing a small amount of wort through a relatively larger chiller. The drawbacks of this method are that the cold break is carried into the fermentor, it's harder to keep the inside of the chiller clean, and it's hard to keep hops and hot break material in the kettle from clogging the intake. A pot scrubby made from copper or stainless steel and attached to the end of the racking cane will help.

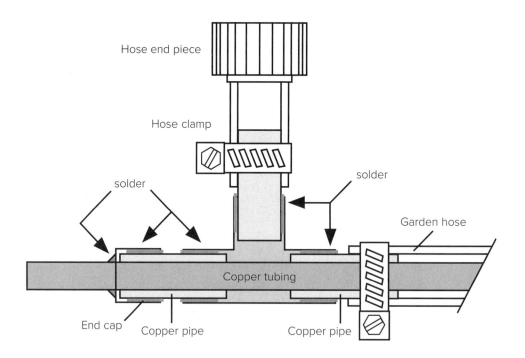

**Figure D.3.** Suggested counterflow wort chiller design.

The increased efficiency of a counterflow chiller lets you use a shorter length of tubing to achieve the same amount of wort cooling. The tube-within-a-tube configuration can be coiled into a convenient roll. The hot side of the chiller, where the racking tube intake is, needs to be copper or some other heat resistant material. Plastic racking canes tend to melt from the heat of the pot when the hot wort is siphoned into the chiller. Counterflow chillers work best with a spigot mounted on the side of the boil kettle, which negates the need to siphon the wort.

## TABLE D.1—COOLING WATER TO HOT WORT RATIOS BY FLOW AREA[a]

| Hose ID inch/mm | Copper tube OD[b] inch/mm | Water flow area (sq.in.) | Wort flow area (sq.in.) | Flow ratio |
|---|---|---|---|---|
| 0.5/13 | 0.375/10 | 0.086 | 0.076 | 1.13 |
| 0.625/16 | 0.375/10 | 0.196 | 0.076 | 2.58 |
| 0.75/19 | 0.375v10 | 0.331 | 0.076 | 4.36 |
| 0.625/16 | 0.5/13 | 0.110 | 0.149 | 0.74 |
| 0.75/19 | 0.5/13 | 0.245 | 0.149 | 1.64 |

*ID, internal diameter; OD outer diameter.*

[a] *If you are outside the USA, check your local hardware store for standard tube, hose, and pipe sizes, and select your parts accordingly. The important thing is to calculate the flow ratio and choose for highest cooling efficiency.*

[b] *The copper tubing wall thickness is 0.032 in. for both sizes in USA.*

The ratios in table D.1 show that a 12.5 mm (½") internal diameter (ID) garden hose (a typical size in the UK and Australia) with 10 mm (⅜") OD tubing has a smaller ratio of cooling water flow area to wort flow area than the larger 19 mm (¾") ID hose. In the USA, the typical garden hose is ⅝" ID (16 mm), and the flow ratio when using ⅜" tubing is significantly higher. Choose a combination that has a higher ratio of water flow to wort flow for better chilling efficiency.

To test cooling capacity, a typical US counterflow chiller was built and tested, consisting of 25 ft. of ⅝" ID hose and ⅜" OD copper tubing. This combination was able to chill 5.3 gal. (20 L) of boiling water to 70°F (21°C) in about 10 min. using 63°F (17°C) cooling water. The exit temperature of the cooling water was about 90°F (32°C). Remember, it is a counterflow chiller, so the warmer exit water is being presented to the boiling hot wort, and the colder source water is cooling the exiting wort. This arrangement maximizes the temperature difference, making the chiller more efficient than if it was the other way around.

The other two factors that greatly influence counterflow chiller performance are chiller length and flow rates. A longer chiller will allow the wort temperature to approach more closely the cooling water temperature. A higher cooling water flow rate will also bring the wort temperature closer to the water temperature, as will a slower wort flow. However, you don't want to use too much water (unless you are using it to fill your swimming pool), and you don't want to take forever to fill your boil kettle, so building a longer chiller may be a better option if your cooling water temperature is high (e.g., >77°F [25°C]). Table D.2 summarizes these effects.

## TABLE D.2—FACTORS IMPROVING COUNTERFLOW CHILLER PERFORMANCE

| Factor | Change | Effect |
|---|---|---|
| Water-to-wort flow ratio | Increase | More cooling |
| Chiller length | Increase | More cooling |
| Water flow rate | Increase | More cooling |
| Wort flow rate | Decrease | More cooling |
| Cooling water temperature | Decrease | More cooling |

Figure D.3 and figure D.4 show one example for building the counterflow fittings and assembling the copper tubing inside the garden hose using standard US parts. If you are outside the US, choose your part sizes accordingly. The parts are common ½" ID rigid copper tube, an end cap, and T sweat-type fittings (fig. D.4 Step 1). The parts are soldered together using lead-free silver solder and a propane torch. The ends of the garden hose are cut off and reattached via the tube clamps to the T fittings. The 3/8" OD soft copper tubing (for the wort) exits the fitting assembly through a 3/8" diameter hole in the end cap. The opening for the tubing is sealed with a fillet joint soldered around the hole.

**USA parts list:**
- 25 ft. (8 m) of ⅜" (10 mm) copper tubing coil
- 25 ft. (8 m) of ⅝" (16 mm) fitted garden hose
- 10 in. (24 cm) of ½" (13 mm) ID, ⅝" OD (16 mm) copper pipe
- 2 of ½" copper pipe T fittings
- 2 of ½" copper pipe cap fittings
- 4 of stainless steel hose clamps
- 2 of ⅜" compression to ½" male pipe thread adapters (or equivalent)
- silver-tin plumbing solder and flux

### Assembling the Chiller
**Step 1.** Cut the 10-inch (24 cm) piece of ½" (13 mm) ID copper pipe into six 1.5-inch (4 cm) stubs.

**Step 2.** Apply flux inside the T fittings and to one end of each stub and solder the T fittings as shown.

**Step 3.** Drill a ⅜" hole in each end cap. Be precise and use a pilot hole or a step-drill bit for best results. You want to have a close fit between the cap and copper coil tubing, as shown.

**Step 4.** Cut off 4–6 in. (10–15 cm) from both ends of the garden hose and pour half a bottle of dishwashing soap down the hose. Leave the plastic cap on the copper coil tubing (or cover the end with tape) to prevent soap from getting in the tubing, because it will interfere with soldering later.

Gently uncoil the copper tubing and feed it into the hose. Do not straighten the tubing, but widen the coils to about a six or seven feet diameter (~2 m) as you insert it into the hose. This will better enable you to recoil the chiller when it is complete. Allow 4–6 in. (10–15 cm) of tubing to protrude from both ends of the garden hose before cutting.

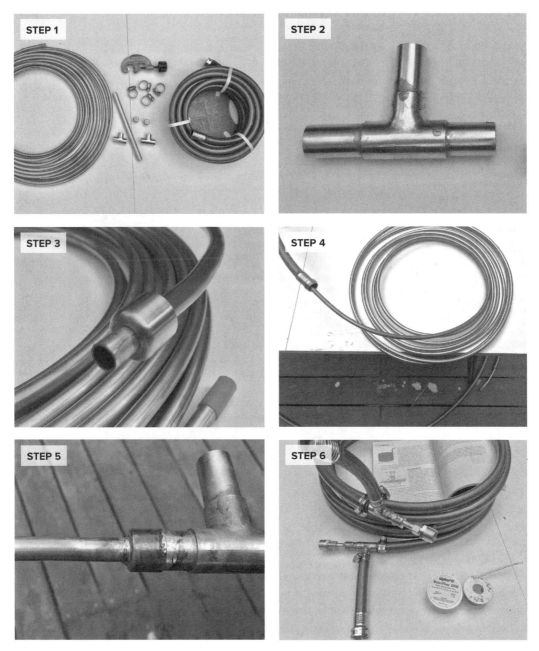

**Figure D.4.** Building a counterflow chiller using copper tubing inside a garden hose. See main text for details of each step.

**Step 5.** Slide on a hose clamp and T assembly to both ends of the chiller. Insert the stub of the T fully into the hose, but do not clamp it tight yet. Wait until after soldering. Apply flux to the entire inside of the end cap and slide it over the tube and fully onto the other stub. Apply flux to the outside of the cap where the tubing exits. Heat the cap with your propane torch. When hot, apply the solder to the interfaces of the cap, stub, and tubing and allow the molten solder to wick into each joint as shown. Do not overheat or you will melt the garden hose. It doesn't have to be pretty, it just needs to seal the joint.

**Step 6.** Trim the ends of the tubing to the desired length and affix the compression fittings. These will allow you to attach your wort hoses to the chiller. Gently recoil the chiller to a tighter diameter. Attach the garden hose fittings to the T fittings with the remaining hose clamps, and you are done! Be sure to flush the soap out of the garden hose before storing it.

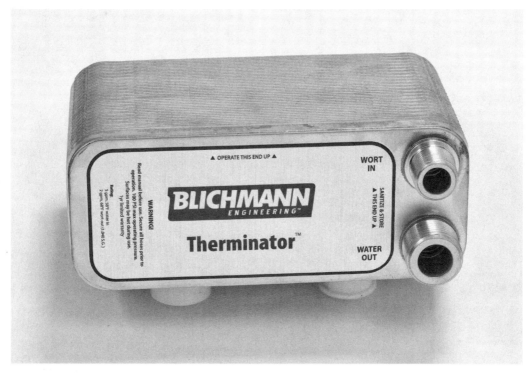

**Figure D.5.** Plate chiller with threaded fittings.

## Plate Chillers

Plate chillers (fig. D.5) are the most efficient chilling option and are very similar to what professional brewers have used daily for years. A homebrew plate chiller can chill 10 gal. to within a few degrees of the water temperature within five minutes. Plate chillers are made from thin sheets of brazed stainless steel and typically require a pump to push the wort through. The chief drawback of plate chillers is that they are easily clogged by hops and trub. I recommend using them in conjunction with a hop back or a fine filter in the kettle to prevent clogging.

Trub tends to build up in plate chillers if they are not cleaned thoroughly after each use, and that can cause ongoing contamination problems. Soaking in a bucket of PBW or other non-corrosive cleaner will loosen buildup. Always flush the wort section thoroughly with clean water, backwards and forwards, after each use and drain them as well as you are able. Do not store them wet or full of sanitizer, as this will promote corrosion.

To sanitize before use, run boiling water or a sanitizer solution through the wort section for a short time.

# Appendix E

## Lauter Tun Design
## for Draining

On any given day, there is at least one person standing in front of the brass fittings at the hardware store, trying to figure out what he needs to build his mash tun. Building a mash tun from a cooler is inexpensive and the easiest way to start all-grain brewing. You can use either a rectangular chest cooler or a cylindrical beverage cooler (fig. E.1).

As was discussed earlier in Chapter 19, "Clearing the Wort Out (Lautering)," there are two types of lautering, draining and rinsing. Draining lautering methods simply drain the existing wort and don't require any uniformity of flow through the grainbed. On the other hand, rinsing the grainbed via the continuous sparging method requires uniform flow to insure that all the grist is equally well-rinsed of wort (see appendix F). This appendix will discuss how to build a mash-lauter tun for the draining lautering methods (i.e., the batch sparging and no-sparge techniques). The BIAB method does not require a lauter tun and this appendix doesn't apply to it.

**Figure E.1.** Examples of both cylindrical (round or circular) and rectangular coolers or ice chests. Choose a cooler that provides for a good grain bed depth of between 4–12 inches (10–30 cm).

## Choosing a Cooler

The original home lautering system was probably the bucket-in-a-bucket false bottom championed by Charlie Papazian in the first edition (1984) of *The Complete Joy of Homebrewing*. This setup is effective and very cheap to assemble. Using two food-grade 5 gal. buckets, the inner bucket is drilled with lots of small holes to form a false bottom that holds the grain and allows the liquid to run off; the sweet wort passes into the outer bucket and is drawn off through a hole in the side.

Picnic coolers (also called cool boxes, or "eskies" after the Esky® brand) offer a few advantages over buckets, adding both simplicity and efficiency. A cooler's built-in insulation provides better mash temperature stability than a bucket can provide. Their size also allows mashing and lautering in the same vessel. Thus, all-grain brewing is as simple as pouring the grain into the cooler, adding hot water, waiting an hour, and then draining the sweet wort.

The shape of the cooler determines your grain bed depth. In general, deeper is better. If the grain bed is wide and shallow (<4 in. [<10 cm]), it won't filter efficiently and your wort will be cloudy with debris. However, if the grain bed is too deep it is more likely to compact from lautering too fast and stop flowing. My advice is to pick your cooler based on your average batch size. Don't pick one larger than you really need and think that a larger one will give you more flexibility for future batches. If you pick one that is too large for the majority of your batches, the grain bed depth for your usual batch size will be too shallow and it won't hold the heat as well. The minimum grain bed depth should be about 4 in. (10 cm). The maximum should be about 16 in. (40 cm), although deeper will work if you are careful.

A 10 gal. (38 L) cylindrical beverage cooler with either a false bottom or manifold works well for both 5 and 10 gallon (19 and 38 liter) batches. The rectangular ice chest coolers also work well, and are commonly sized at 20, 24, 34, and 48 quarts (roughly equivalent in liters), which offers a good choice for any batch size. Many coolers have drain spigots that can be removed to make it easy to drain the wort via a bulkhead fitting. If you are using a cooler that does not have a drainage opening or spigot, lautering works just as well if you come over the side with a vinyl hose and siphon the wort out (fig. E.3). You should use a stopcock or clamp to regulate the flow, and as long as you keep air bubbles out of the line, it will work great.

Everything you need to build a mash-lauter tun is readily available at a hardware store. In the US, you should find that the total investment for the cooler and all the parts to convert it into a mash-lauter tun is usually less than $50.

---

## ESTIMATING GRAIN BED DEPTH

In order to estimate the typical grain bed depth in a cooler, you need to know the dimensions of the cooler and the OG of your typical batch.

Here is how you calculate grain bed depth:

**1.** Calculate the total mash volume.

$Vm = G \times (Rv + 0.38)$  For pounds and quarts

$Vm = G \times (Rv + 0.8)$  For kilograms and liters

where:

$Vm$ = total volume (quarts or liters)

$G$ = (dry) weight of grain (pounds or kilograms)

$Rv$ = water to grain ratio of the mash (qt./lb. or L/kg)

**2.** Convert the mash volume to cubic inches or cubic centimeters. There are 57.75 cubic inches in a quart, and 1,000 cubic centimeters in a liter.

US standard units: quarts × 57.75 cu. in./qt. = cu. in.

Metric units: liters × 1000 = cc

**3.** Divide the mash volume by the floor area of the tun to get the resultant depth.

---

## Rinsing versus Draining—a Recap

Traditionally, commercial brewers have sparged their mashes using water sprinklers, or rotating sparge arms, and grain rakes to rinse the mash uniformly for the best extraction. They use large lauter tuns, anywhere from 10–20 ft. (3–6 m) in diameter, because they are lautering tons of grain. The only way to add water uniformly is to use large sprinkler devices to distribute over the whole tun. It's different on the homebrewing scale. Our mash and lauter tuns are barely 2 ft. (60 cm) wide; it is quite easy to simply drop in a hose and have the water uniformly spread out across the grain bed. A sparge arm or diffuser plate can help, but it is not strictly needed.

To get the most uniform flow through all parts of the grain bed, the mash must be kept fully hydrated with free water above the grist to help prevent settling and compaction. Lautering was typically an hour-long process of monitoring the flow rates and the run-off gravity to assure good extraction. I spent a year conducting fluid flow experiments with ground-up corncobs and food coloring, and enlisted the aid of two hydrologists and an astrophysicist, in order to gain a thorough understanding of how to optimize extraction of wort via continuous sparging. That work is collected in the appendix F.

Meanwhile, other homebrewers, whom were not so concerned with optimizing efficiency, simply drained their mashes using a slotted pipe manifold or stainless steel screen, dumped in another batch of water, drained again, and got on with their day. Frankly, I was disturbed by this flagrant disregard for technology; sure, it was easy, and took less time, but where was the fun in that? Eventually, I realized that there was room for intellectual discussion and elegance of design in batch sparging, and I joined the bandwagon.

Draining the wort, rather than rinsing the grain bed, changes the design requirements. In the steady state flow conditions of continuous sparging, you want the flow rate to be the same at every point in the grain bed, and a false bottom is the best solution for realizing this goal. The problem with this solution is that a uniformly high outflow rate can compact the grain bed uniformly into an impenetrable layer that results in a stuck sparge. High performance comes with a high-risk price tag. If you are simply draining the wort, you don't care whether it is drained from over here or over there, as long as it drains. Extraction uniformity is achieved with stirring in the next batch of sparge water. The drainage points could be entirely along one side of the tun—it doesn't matter so long as you can drain it.

In batch sparging, if the grain bed is drained from a single point with a high flow rate, the grain will quickly compact around that point and flow will cease. The more you distribute that collection point, the lower the effective flow rate will be at any of those distributed points. This is the benefit of a slotted pipe or long screen—any decrease in flow at one point can be alleviated by an increase in flow at another point. Actually, false bottoms operate the same way, but the difference is that they are more uniform and symmetrical so that a high flow rate at one point is a high flow rate everywhere. Slotted pipes are not efficient enough to have that problem. Thus, slotted pipes and screens work better for draining the wort than false bottoms at high flow rates. Nevertheless, fluid dynamics states that the initial wort flow be slow, to avoid compacting the bed and causing a stuck sparge. Start slow and gradually increase the rate.

Batch sparging also works fine with false bottoms, you just need to be aware that it if you throw the valve wide open, you run a high risk of compacting the bed. Start draining slowly and you won't have any problems.

### Siphon or Bulkhead Fitting?

You have two options for actually getting the wort out of the tun: you can use a bulkhead fitting, or you can siphon the wort out. Many coolers have drain spigots that can be removed to make it easy to drain the wort via a bulkhead fitting. If your cooler does not have a drainage opening or spigot, you can buy a hole saw for your drill that will easily cut a nice hole in the cooler. Bulkhead fittings are available from several suppliers, or you can make your own. A bulkhead fitting is a short section of fully threaded pipe with two flat washers, two rubber washers, and two nuts for sealing around the hole that the pipe passes through. A hose barb and vinyl tubing can be used to connect to the lautering device on the inside, and a ball valve or a hose barb is connected to the outside. Another suggested design using off-the-shelf fittings is detailed in figure E.2.

With the siphon method, vinyl tubing connects directly to the lautering device and just comes out over the side of the cooler. During the mash, the tubing can be coiled inside the tun with the lid on to help retain the heat. Both methods work well, though the bulkhead fitting looks spiffier, and it is hard to siphon all the wort out. Whichever method you choose, you will also need a proper valve to regulate the flow rate. Ball valves are readily available in brass, chrome-plated brass, or stainless steel. Plastic stopcocks are an inexpensive option and work nicely in-line with the siphoning method (see fig. E.3).

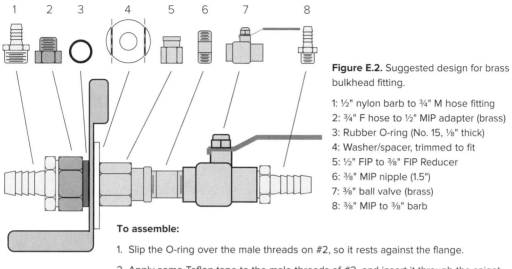

**Figure E.2.** Suggested design for brass bulkhead fitting.

1: ½" nylon barb to ¾" M hose fitting
2: ¾" F hose to ½" MIP adapter (brass)
3: Rubber O-ring (No. 15, ⅛" thick)
4: Washer/spacer, trimmed to fit
5: ½" FIP to ⅜" FIP Reducer
6: ⅜" MIP nipple (1.5")
7: ⅜" ball valve (brass)
8: ⅜" MIP to ⅜" barb

**To assemble:**

1. Slip the O-ring over the male threads on #2, so it rests against the flange.

2. Apply some Teflon tape to the male threads of #2, and insert it through the spigot hole from the inside of the cooler.

3. Slip the spacer over the threads and hand tighten #5 to make a good seal.

4. Assemble the rest of the parts in the sequence shown.

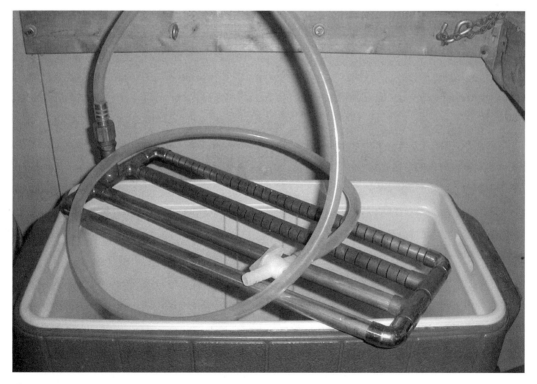

**Figure E.3.** This cooler doesn't have a drainage port and therefore is using a siphon hose connected to the manifold to siphon the wort out.

## False Bottom, Pipe Manifold, or Screen?

There are several options for lautering devices available. Here is a list of pros and cons for each.

### False Bottoms

*Pros:*

- Pre-fabricated false bottoms for cylindrical beverage coolers are readily available from several suppliers and are easy to assemble.
- False bottoms are always more uniform than manifolds when continuous sparging—near 100%.

*Cons:*

- False bottoms are tedious to fabricate yourself and difficult to fit to rectangular coolers. They should fit closely around the edges of the tun to prevent gaps that can allow sparge water to bypass the grain bed and reduce the yield.
- False bottoms are more prone to stuck sparges when the lauter flow is too fast, because they will compact the grain bed uniformly.

**Figure E.4.** This is a picture of a perforated false bottom designed for cylindrical beverage coolers. A drainage hose would connect to the hose barb in the center.

## Manifolds

*Pros:*

- Copper pipe manifolds are easy to build and fit to any size cooler.
- Stuck sparges are rare with manifolds, because the grain bed will not compact uniformly.
- Highly efficient configurations are easily built.

*Cons:*

- The efficiency of manifolds depends on the pipe spacing and grain bed depth.
- The grain bed is not lautered below the manifold, so the pipe slots should face down and be as close to the bottom of the tun as possible.
- Many slots to cut.

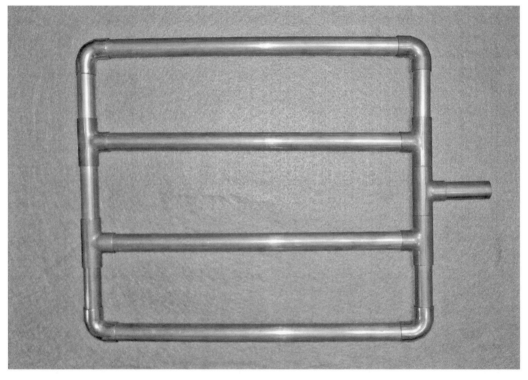

**Figure E.5.** A manifold is a distributed drainage device that can easily be made with copper or CPVC water pipe. Regular rigid PVC pipe should not be used because it will contribute plastic and solvent flavors and aromas to the beer.

### Stainless Steel Screens and Braids

*Pros:*
- No slots to cut.
- Screens clear very quickly during the lautering recirculation step.
- Stuck sparges are rare with screens and braids, because the grain bed will not compact uniformly.
- Pre-fabricated screens and braid assemblies are available from several suppliers.

*Cons:*
- Cutting the braid free from a rubber hose is a bit of work.
- Half-inch diameter, light gauge steel braids can collapse from the weight of the mash and make lautering difficult. One-inch hose braids won't collapse.
- Not well-suited for continuous sparging.

**Figure E.6.** Stainless steel mesh screens or braids can be purchased or made yourself from off-the-shelf materials.

## Building a Copper Pipe Manifold

A manifold can be made of either soft or rigid copper tubing. Choose a form to suit your cooler and design. In a round cooler, the best shape is a circle divided into quadrants, although an inscribed square works nearly as well. In a rectangular cooler, the best shape is rectangular with several legs to adequately cover the floor area.

Copper sweat fittings can be used to join the legs together. The fittings don't need to be soldered, simply crimping the ends slightly with a pair of pliers will provide the friction to hold the assembly together. When you cut the ½" copper water pipe lengths to fit the cooler, don't forget to take the assembled elbow and T fitting lengths into account.

Use a standard hacksaw blade to cut the slots into the ½" copper water pipe—they don't need to be any narrower. The slots in the pipes should only be cut halfway through and don't need to be closer than a quarter-inch apart. Even a half-inch apart is fine. The slots should face down—wort that is physically below the slots will not defy gravity and flow upward.

A wide variety of off-the-shelf brass fittings, such as hose barbs and compression fittings, can be used to connect the manifold to a siphon or bulkhead.

## Building a Stainless Steel Braided Ring

The stainless steel braid from hot water hoses makes good lautering screens. They are a bit of work to cut and disassemble, however, so here is one suggestion for making one.

### Parts List:
- 24" × 1" diameter water heater connecter
- ⅝" compression Tee brass fitting (with included ferrules)
- Two 1" lengths of ½" diameter copper tubing

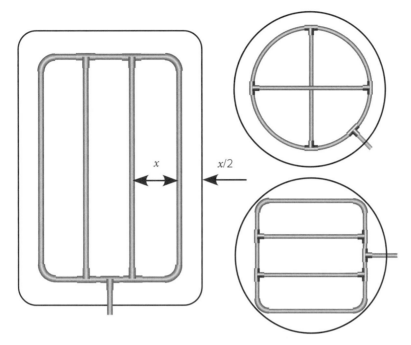

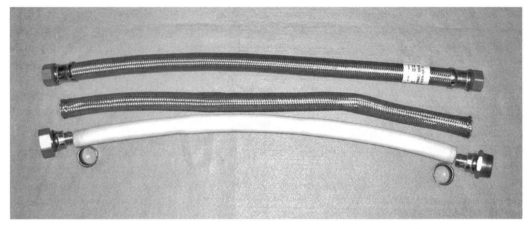

**Figure E.7.** Manifolds should be shaped to uniformly fit the cooler and distribute the drainage points. The pipes should be spaced at half the inter-pipe distance to the wall, to ensure uniform flow throughout the grain bed to the manifold.

**Figure E.8.** This picture shows a common hot water heater hose before (top) and after the stainless steel outer braid has been removed.

## Procedure:

1. Clamp one of the end fittings in a shop vise.
2. Cut all the way through the metal sleeve that binds the hose braid to the end fittings with a hacksaw. This way the ends are evenly trimmed and won't fray.
3. Pull the hose off the fitting in the vise.
4. Clamp the other end and repeat.

5. Now axially compress the braid to work it loose and slide it off the hose (see fig. E.8).
6. Pull or compress the ends of the braid to make it narrow and slide one of the copper tubing pieces onto each end of the braid. Let the end of the braid extend about ⅛" beyond the tubing.
7. Slide a compression Tee nut over each end of the braid onto the copper pipe.
8. Insert a ferrule into each end of the braid (see fig. E.9).
9. Slide the assembly snugly into the Tee and tighten the nut so that it crimps down on the copper pipe.
10. Repeat step 9 for the other side of the Tee, and now you have a braided ring manifold that won't come apart in the mash. This ring can now be connected to a bulkhead or siphon like the other systems (see fig. E.10)

A ring that divides the area equally in half (half area is inside/outside ring) is nearly as effective as a false bottom is for uniformity of sparge flow during continuous sparging. Flow uniformity doesn't really matter for batch sparging, but it's nice to have. See appendix F for more information on uniform fluid flow during lautering.

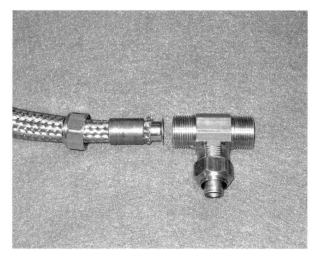

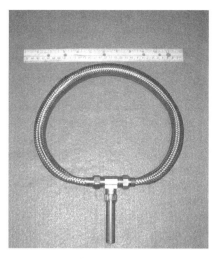

**Figure E.9.** Assemble the copper tube, compression nut and brass ferrule onto the steel braid as shown. When assembled and tightened, the nut will be attached to the tube, and the braid will be securely attached to the T fitting but can be easily disassembled for cleaning.

**Figure E.10.** The braid diameter can be sized to fit any round cooler. For uniformity, the diameter of the ring should be 0.707 times the diameter of the tun.

## Design Examples

### Design Option 1—Cylindrical Cooler with False Bottom

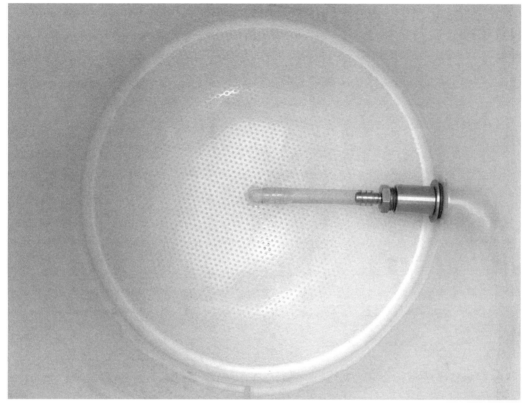

**Figure E.11.** A cylindrical beverage cooler with a false bottom.

*Works for:* Continuous sparging, batch sparging, no-sparge
*Degree of difficulty:* Easy

A cylindrical cooler with false bottom system is probably the easiest to assemble, because it is readily available off the shelf. This system can lauter any beer by any method—the only caveat is that you need to watch your initial flow rate so that you don't compact the grain bed and get a stuck sparge.

## Design Option 2—Cylindrical Cooler with Manifold

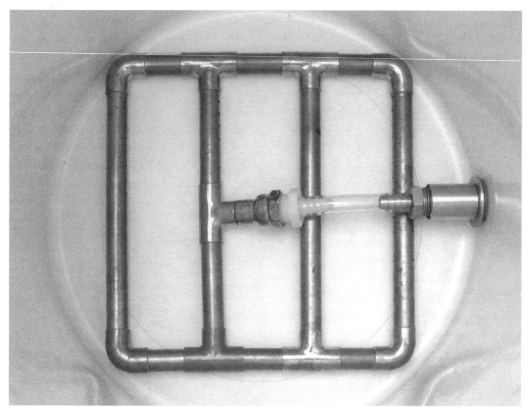

**Figure E.12.** A cylindrical beverage cooler fitted with a square manifold.

*Works for:* Continuous sparging, batch sparging, no-sparge
*Degree of difficulty:* Strenuous

The cylindrical cooler with manifold design can be optimized for distribution and uniformity, allowing flexibility for any sparging method. Soft copper tubing and compression fittings can be used to make a ring. The advantage of the manifold over the false bottom in this case is the reduced risk of a stuck sparge caused by uniform compaction.

## Design Option 3—Rectangular Cooler with Manifold

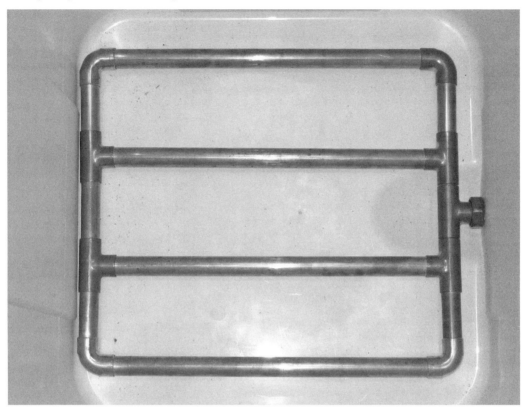

**Figure E.13.** A rectangular cooler fitted with a rectangular manifold.

*Works for:* Continuous sparging, batch sparging, no-sparge
*Degree of difficulty:* Moderate

The rectangular cooler with manifold design can be optimized for distribution and uniformity, allowing flexibility for any sparging method. The slotted pipes require some time to make, but the resulting manifold is very durable. Many brewers prefer the larger, wider rectangular coolers for ease of stirring and water additions.

## Design Option 4—Cylindrical Cooler with Braided Ring

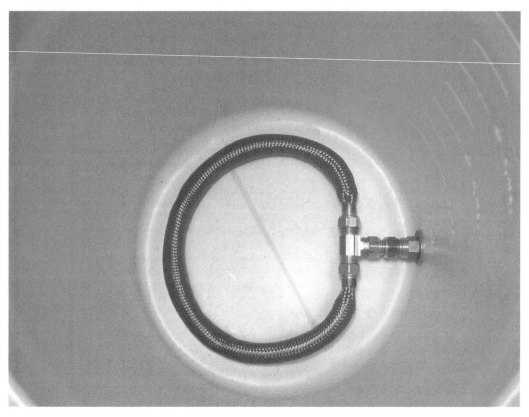

**Figure E.14.** A cylindrical beverage cooler fitted with a stainless steel braid ring.

*Works for:* Batch sparging, no-sparge, continuous sparging
*Degree of difficulty:* Moderate

A circular braided ring in a cylindrical cooler that divides the area evenly inside and outside the ring is actually more uniform than a false bottom of the same size. The other advantage of a ring over the false bottom here is the reduced risk of a stuck sparge from uniform compaction.

## Design Option 5—Rectangular Cooler with Single Braid Tube

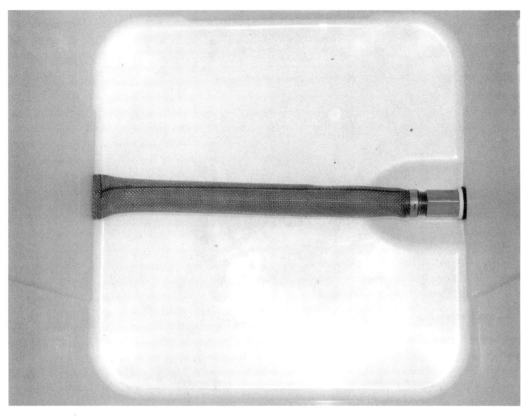

**Figure E.15.** A rectangular cooler with a single straight stainless steel braid drain tube.

*Works for:* Batch sparging, no-sparge
*Degree of difficulty:* Easy

A single screen or manifold pipe a rectangular cooler configuration doesn't adequately cover the floor area to work well for continuous sparging, but will still work fine for the draining methods. Stainless steel braids clear quickly during the recirculation step. Many homebrewers prefer the larger, wider rectangular coolers for ease of stirring and water additions. The only problem is that the braid tends to crush flat and it is hard to hold down on the bottom. Stainless steel springs are often placed inside the braid to help it hold its shape and maintain good flow.

## Design Option 6—Cylindrical Cooler with a T-Screen

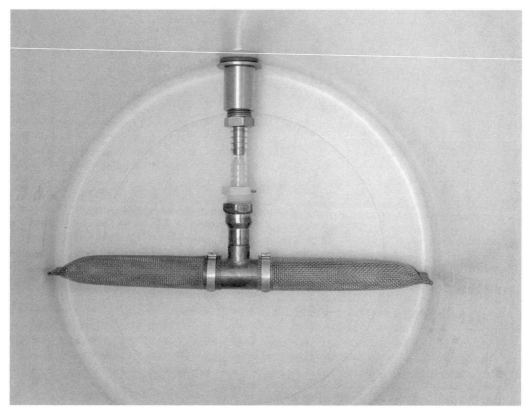

**Figure E.16.** A cylindrical cooler with a T-screen.

*Works for:* Batch sparging, no-sparge
*Degree of difficulty:* Easy

Cylindrical and rectangular coolers with a T-screen or manifold work fine for batch sparging and no-sparge brewing. They tend to be rigid and movement is not a problem. These sorts of drains do not work for continuous sparging because the collection area is not uniform. Uniformity is discussed in more detail in appendix F.

# Appendix F

## Lauter Tun Design for Continuous Sparging

Continuous sparging is all about understanding steady state flow. We need to understand how the sparge water is going to flow through the grain bed so we can predict where the grain will be rinsed of sugar and where it won't. Before I lead you through all the theory and explanations, let me cut to the chase and tell you how it works best:

- A false bottom works best, but a large multi-pipe manifold works almost as well.
- Regulate the flow with a valve to achieve a slow flow rate—about 1 qt. or 1 L per minute—to prevent compaction of the grain bed and channeling.
- Maintain an inch (couple of centimeters) of water over the grain bed during the lauter to ensure fluidity and free flow.

To explain why false bottoms work best, I have to turn to fluid dynamics and some mathematical models that a friend put together for me. I spent a year conducting fluid flow experiments with ground-up corncobs and food coloring in an aquarium in order to understand how slotted pipes worked under continuous sparging. Those experiments demonstrated that the flow converged to the pipe and that all points along a slotted pipe drained at the same rate, regardless of distance to the drain (see fig. F.1).

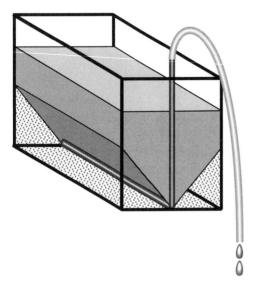

**Figure F.1.** Depiction of fluid flow to a single pipe. The flow converges, leaving unsparged zones in the corners.

The goal in the continuous sparging process is to rinse all of the grain particles in the tun of all their sugar. To do this we need to focus on two things:

- Keep the grain bed completely saturated with water.
- Make sure that the fluid flow through the grain bed to the drain is slow and uniform.

By keeping the grain bed covered with at least an inch of water, the grain bed is in a fluid state and not subject to compaction by gravity. Each particle is free to move and the liquid is free to move around it. Settling of the grain bed due to loss of fluidity leads to preferential flow and poor extraction.

Continuous sparging depends on being able to rinse the sugars from the grain, and a big part of this process is diffusion. If you rinse quickly you will end up with mostly sparge water in your boiler, because the sugar won't have time to diffuse into the sparge water as it flows past.

## Fluid Mechanics

To extract the wort from all regions of the grain bed, there must be fluid flow from all regions to the drain. In a perfect world, the wort would separate easily from the grain and we could simply drain the grain bed and be done. However, it is not a perfect world, and we must rinse, or sparge, the grain bed to get most of the sugars, and some sugar is still left behind. If some regions of the grain bed are far from the drain, and experience only 50% of the sparge flow, then only 50% of the wort from those regions will make it to the drain. Fluid mechanics allows us to model the flow rates for all regions of the grain bed and determine how well a grain bed is rinsed. These differences can be quantified, enabling us to compare different lauter tun configurations.

To illustrate, let's look at a cross section of a 10-inch wide by 8-inch deep lauter tun that uses a single pipe manifold (fig. F.2). For every unit volume of water that rinses the grain bed, grain at the top of the grain bed will experience "unit" flow, that is to say, 100% flow. As the flow moves deeper into the grain bed, it must converge to the single drain. This means that the region immediately above the drain can experience ten times the unit flow, while a region off to the side will only experience one-tenth of unit flow. The vector flow plot for a lauter tun of the same size that uses two pipes demonstrates the same behavior (fig. F.3), although the convergence effect is less. Figure F.4 shows the flow rate distribution for the single pipe manifold lauter tun. For purposes of illustration, unit flow (the big white upper area) is drawn within the bounds of ±10% of actual unit flow, and lines for 50%, 90%, 110%, and 200% of unit flow are shown. Figures F.2 and F.4 convey the same idea, but figure F.4 lets us quantify the percentages of flow for this grain bed. The histogram (fig. F.5) constructed from this data summarizes the flow distribution and we can use the histogram to measure two aspects of lautering performance—efficiency and uniformity. For comparison, flow distribution and associated histogram for a full-sized false bottom are illustrated in figures F.6 and F.7. The differences will be explored more fully in the next couple of sections as we look at the concepts of lauter efficiency and uniformity.

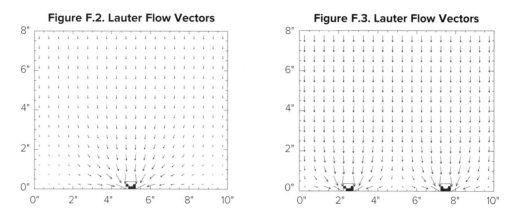

**Figure F.2 and F.3.** These diagrams show the flow vectors in a lauter tun consisting of a single pipe (F.2) and double pipes (F.3). The size of the arrows indicates the relative speed of flow. Note how the flows converge to the pipes, leaving low flow areas in the corners. This same behavior has been observed when flowing food coloring dye through a grain bed in a glass aquarium.

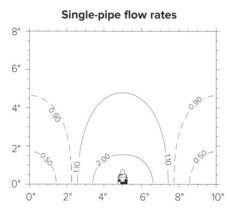

**Figure F.4.** Flow rate distribution for the same single-pipe lauter tun shown in figure F.2. Darcy's Law allows us to quantify the flow velocity at any point within the lauter tun. The lines show the boundaries for 50%, 90%, 110%, and 200% of unit flow velocity. The area above the 0.90 and 1.10 flow lines is a region of 100% uniform flow.

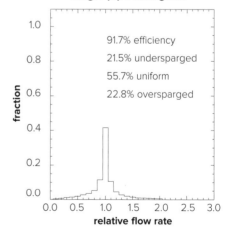

**Figure F.5.** Histogram showing the relative amounts of different percentages of unit flow as described by figure F.4.

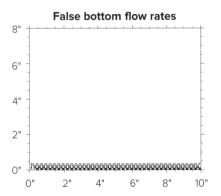

**Figure F.6.** Flow rate distribution for a lauter tun with a full-sized false bottom. As in figure F.4, the lines show the boundaries for 50%, 90%, 110%, and 200% of unit flow, but the convergence zone is so small that you can't pick them out. This false bottom model represents ⅛" holes on ¼" centers.

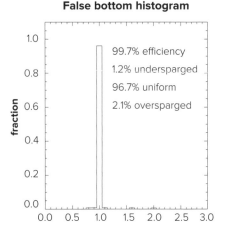

**Figure F.7.** Histogram showing the relative amounts of different percentages of unit flow as described by figure F.6. As expected, the histogram shows that the vast majority of the grainbed lies within the big white uniform region above the convergence zone.

## A NUMERICAL MODEL FOR FLUID FLOW

For fluid flow through porous material, Darcy's Law states that flow velocity is proportional to pressure variations and inversely proportional to the flow resistance. The resistance (or, inversely, the permeability) to flow is determined by the media (i.e., the porous material), which, in our case, is the grist. By combining this velocity-pressure relationship with a statement that water is conserved (i.e., water is not created or destroyed), we can form a numerical model that describes the fluid flow throughout the tun.

Darcy's law: $q = (-K/\mu)\cdot\nabla p$

where $p$ is the pressure (actually the velocity potential), $\mu$ is the absolute viscosity (or shear), $K$ is the permeability, and $q$ is the Darcean velocity (bulk flow velocity).

Conservation of water: $\nabla\cdot q = 0$

Combination: $\nabla^2\cdot p = 0$ (Laplace's equation)

Assuming $K$ and $\mu$ are constant everywhere, this equation for Darcy's law holds true everywhere except the top of the tun, where water is added, and the pipe slot, where liquor flows out of the tun. The tun walls are rigid (they can support any pressure), and nothing flows through the walls.

## Lauter Efficiency

Earlier we stated that 50% of the unit flow rate would only extract 50% of the sugar. However, we cannot say that a 200% flow rate will extract 200% of the sugar. If we assume that 100% of the unit flow rate extracts 100% of the sugar, then there is no more sugar to extract; higher flow rates do not extract anything further, except possibly tannins. If we add up all of the extraction rates from the different flow regions of the grain bed, we can determine the efficiency percentage for that configuration.

For example, if a single pipe manifold system lautered 5% of the grain bed at 40% of unit flow, 10% at 60%, 15% at 80%, and 70% of the grain bed at ≥100% of unit flow, the efficiency of that tun would be calculated as 90%:

$$(5 \times 40) + (10 \times 60) + (15 \times 80) + (70 \times 100) = 90\% \text{ efficiency.}$$

A "perfect" false bottom would lauter the entire grain bed with 100% of unit flow, because every region would have equal access to the drain, and would be 100% efficient. The computer model that applies the numerical model for fluid flow (see sidebar) estimates a real false bottom (⅛" holes on ¼" centers) to be 99.7% efficient (fig. F.7).

### Lauter Uniformity

While efficiency gives a measure of the extract quantity, the uniformity gives a measure of its quality. To discuss uniformity, we look at three percentages of flow: flow less than 90%, flow between 90% and 110%, and flow greater than 110%. With these three percentages, we can compare different configurations that have similar efficiency, and determine if one configuration is more uniform than another. Regions of the grain bed with flow values between 90% and 110% are considered uniformly sparged, with values less than 90% being undersparged, and values over 110% oversparged. Generally, the percentage of oversparging is roughly the same as the percentage of undersparging for any one configuration.

Returning to our single pipe manifold example, let's look at the histogram in figure F.5. From the histogram we can determine that only 56% of the grain bed is uniformly sparged, with 21% being undersparged, and 23% oversparged. This means 23% of the grain bed is subject to tannin extraction. But these percentages can be adjusted dramatically by tweaking a few variables.

## Factors Affecting Flow

My thanks to Brian Kern—an astrophysicist at Caltech and a homebrewer—who co-developed this material with me.

The same computer model mentioned above was used to analyze 5,184 configurations of lauter tun and manifold in order to determine the primary factors for flow efficiency and uniformity. In descending order of influence, the factors are:

- inter-pipe spacing,
- wall spacing,
- grain bed depth.

The analysis also determined that the pipe slots should always face down, being as close to the bottom as possible, because wort is not collected from below the manifold.

### Inter-Pipe Spacing

By increasing the number of pipes across the width of the tun, you are effectively decreasing the inter-pipe spacing. Interestingly, analysis of the models (figures F.8–F.11) shows a nearly linear relationship between pipe spacing and both efficiency and uniformity, which peaks at a center-to-center pipe spacing of four times the pipe diameter. For a half-inch pipe, maximum efficiency and uniformity occur at a center-to-center spacing of two inches. Although optimum, it is not necessary for the pipes to be that close; the relationship between spacing and efficiency/uniformity starts to flatten out at three inches, or six times the pipe diameter. As can be seen in table F.1, only 1%–2% gains are realized by decreasing the pipe spacing from 3 to 2 in., although when the grain bed is shallow (<4 in.), the differences approach 5%.

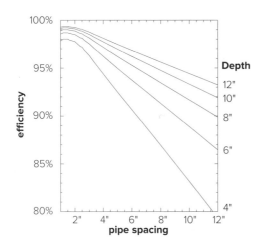

**Figure F.8.** Lautering efficiency as a function of pipe spacing and grain bed depth.

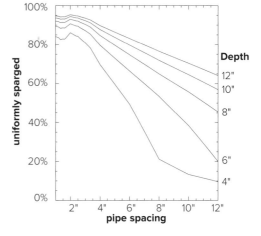

**Figure F.9.** Lauter flow uniformity as a function of pipe spacing and grain bed depth.

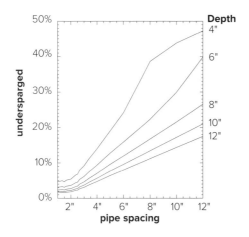

**Figure F.10.** Lauter flow under 90% as a function of pipe spacing and grain bed depth.

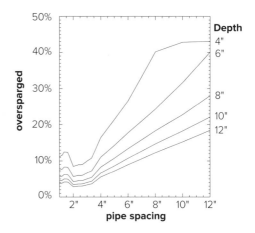

**Figure F.11.** Lauter flow over 110% as a function of pipe spacing and grain bed depth.

## TABLE F.1—EFFECT OF INTER-PIPE SPACING IN 10"W × 8"H (25 × 20 CM) GRAIN BED

| No. of pipes | Center-to-center spacing | Efficiency | Undersparged | Uniformly sparged | Oversparged |
|---|---|---|---|---|---|
| 1 | 10.00 | 91.7% | 21.5% | 55.7% | 22.8% |
| 2 | 5.00 | 96.2% | 9.6% | 79.7% | 10.7% |
| 3 | 3.33 | 97.8% | 5.5% | 89.0% | 5.5% |
| 4 | 2.50 | 98.6% | 3.4% | 92.1% | 4.5% |
| 5 | 2.00 | 98.9% | 2.7% | 93.0% | 4.3% |

## Wall Spacing

The next most significant factor is the spacing of the pipes with respect to the walls of the tun. There are three ways to do this (fig. F.12):

- Edge spacing—the two outermost pipes are placed flush against the walls and any other pipes are spaced evenly between them.
- Even spacing—the spacing between the outer pipes and the walls is the same as the inter-pipe spacing.
- Balanced spacing—the spacing between the outer pipes and the walls is half of the inter-pipe spacing.

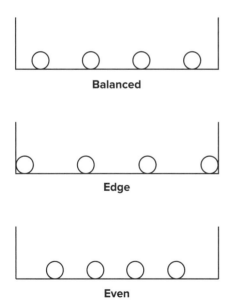

**Balanced**

**Edge**

**Even**

**Figure F.12.** Schematic showing three approaches to spacing of the pipes with respect to the walls of the lauter tun.

As you can see in table F.2, balanced spacing is the most efficient. Balanced spacing places the wall at half of the inter-pipe spacing so that flow velocity is symmetrical around every pipe in the manifold, and the manifold draws as uniformly as possible from the grain bed. Another way of looking at this variable is to say that, for a given tun width, balanced spacing covers the most area with the closest inter-pipe spacing using the least number of pipes. This is most significant for large inter-pipe spacings; at closer spacings the uniformity difference between balanced and edge spacing is smaller (5% or less). But with edge spacing, you need to be aware of the propensity for preferential flow down the walls to the drain. This phenomenon is often referred to as "channeling."

Fluid mechanics describes a "boundary effect," in which the flow resistance decreases at the wall, due to a lack of interlocking particles as a function of particle size (i.e., the wall is not grist, so there is more open area for flow at the wall). The boundary layer for crushed malt is about ⅛" (3 mm) wide. Likewise, if the edges of a false bottom do not conform to the tun walls, the flow will divert into the gaps. These low-resistance paths can result in a significant percentage of the sparge water bypassing the grain bed, which decreases the yield from each volume of wort collected. This means that balanced spacing is preferable to edge spacing for manifolds, and false bottoms should be fitted closely to the tun to minimize the effect.

### TABLE F.2—EFFECT OF WALL SPACING IN A 10"W × 8"H (25 × 20 CM) GRAIN BED

| No. of pipes | Wall spacing | Efficiency | Undersparged | Uniformly sparged | Oversparged |
|---|---|---|---|---|---|
| 2 | Balanced | 96.2% | 9.6% | 79.7% | 10.7% |
| 2 | Even | 94.5% | 14.1% | 66.6% | 19.3% |
| 2 | Edge | 92.0% | 20.4% | 56.9% | 22.7% |
| 3 | Balanced | 97.8% | 5.5% | 89.0% | 5.5% |
| 3 | Edge | 96.4% | 9.0% | 80.3% | 10.7% |
| 3 | Even | 96.2% | 9.7% | 76.3% | 14.0% |
| 4 | Balanced | 98.6% | 3.4% | 92.1% | 4.5% |
| 4 | Edge | 98.0% | 4.9% | 89.9% | 5.2% |
| 4 | Even | 97.1% | 7.1% | 84.3% | 8.6% |
| 5 | Balanced | 98.9% | 2.7% | 93.0% | 4.3% |
| 5 | Edge | 98.6% | 3.4% | 92.0% | 4.6% |
| 5 | Even | 97.7% | 5.3% | 87.6% | 7.1% |

### Grain Bed Depth

The depth of the grain bed is the final significant factor affecting flow—not the total depth of the grain and sparge water, only the depth of the grain itself. For both false bottoms and manifolds, the amount of flow convergence depends only on the drain size and spacing. The size of the convergence zone does not change significantly with depth (pressure). The ratio of under-flow, uniform flow, and over-flow within the convergence zone are nearly constant, and the size (height) of the convergence zone is nearly constant. In the case of false bottoms, the drain features are quite small, so the convergence zone is narrow (less than a half inch in our model). But the drain features of manifolds are larger and more spread out, so the convergence zone is large and affects a larger proportion of the mash.

In other words, increasing the grain bed depth increases the proportion of the grain bed that is outside the convergence zone, which increases the proportion of uniform flow, which increases the extraction efficiency as a whole (table F.3). Thus, the efficiency of false bottoms (small convergence zones) are not significantly affected by grain bed depth, while manifolds (large convergence zones) are, although you can minimize the effect with a manifold by decreasing the inter-pipe spacing (by increasing the number of pipes) to reduce the height of the convergence zone.

## TABLE F.3—EFFECT OF GRAIN BED DEPTH FOR A 10" (25 CM) WIDE GRAIN BED

| No. of pipes | Depth | Efficiency | Undersparged | Uniformly sparged | Oversparged |
|---|---|---|---|---|---|
| 1 | 4" (10 cm) | 83.2% | 43.8% | 13.4% | 42.8% |
| 1 | 6" (15 cm) | 88.9% | 29.9% | 38.4% | 31.7% |
| 1 | 8" (20 cm) | 91.7% | 21.5% | 55.7% | 22.8% |
| 1 | 10" (25 cm) | 93.3% | 17.2% | 64.5% | 18.3% |
| 1 | 12" (30 cm) | 94.4% | 14.3% | 70.4% | 15.3% |
| 1 | 24" (61 cm) | 97.2% | 7.2% | 85.1% | 7.7% |
| 1 | 48" (122 cm) | 98.6% | 3.6% | 92.5% | 3.8% |
| 5 | 4" (10 cm) | 97.8% | 5.4% | 86.1% | 8.5% |
| 5 | 6" (15 cm) | 98.5% | 3.6% | 90.7% | 5.7% |
| 5 | 8" (20 cm) | 98.9% | 2.7% | 93.0% | 4.3% |
| 5 | 10" (25 cm) | 99.1% | 2.2% | 94.4% | 3.4% |
| 5 | 12" (30 cm) | 99.2% | 1.8% | 95.3% | 2.9% |
| 5 | 24" (61 cm) | 99.6% | 0.9% | 97.7% | 1.4% |
| 5 | 48" (122 cm) | 99.8% | 0.5% | 98.8% | 0.7% |

For example: if you had only one pipe in a 10 in. wide tun with an 8 in. deep grain bed, the convergence zone is about 3.5 in. deep, and the percentage of uniform flow is 55.7%. If the grain bed is 48 in. deep, the convergence zone is still about 3.5 in., but the percentage of uniform flow is now 92.5%. With five pipes, the zone height is 0.5 in. and 90% of the flow is uniform at 8 in. depth. When you build a manifold lautering system, both the pipe spacing and wall spacing affect the actual size of the convergence zone, and the grain bed depth affects its relative size. To get the best performance from a manifold system you should optimize all three factors.

The four plots in figures F.8 and F.11 summarize the analysis of all the numerical models for inter-pipe spacing, wall spacing, and grain bed depth, for ½" diameter pipes. Each plot shows the behavior of the stated quantity as a function of center-to-center pipe spacing and as a function of grain bed depth. The relationships are nearly linear except at close pipe spacings and shallow grain bed depths.

## Designing Pipe Manifolds for Continuous Sparging

To summarize:

- The manifold should cover most of the floor of the lauter tun.
- The manifold should have a pipe spacing of 2–3 in. (5–7 cm).
- Use balanced spacing to get the best results with the fewest pipes.
- Choose a cooler size that will give a good grain bed depth for your typical batch. I recommend a depth of 4–12 in. (10–30 cm).

In a circular cooler, the best shape for a manifold is a circle divided into quadrants, although an inscribed square seems to work just as well (see fig. E.7, page 502). In a rectangular cooler, the best shape is rectangular with several legs to adequately cover the floor area (fig. F.13). Whether circular or rectangular, when designing your manifold, keep in mind the need to provide full coverage of the grain bed while minimizing the total distance the wort has to travel to reach the drain.

In addition, it is very important to avoid channeling of the water down the sides from placing the manifold too close to the walls. The distance of the outer manifold tubes to the cooler wall should be half (or slightly greater) of the manifold tube spacing. This results in water along the wall not finding a shorter path to the drain than wort that is dead center between the tubes.

The transverse tubes in the rectangular tun should not be slotted. The longitudinal slotted tubes adequately cover the floor area and the transverse tubes are close enough to the wall to encourage channeling. The slots should face down—any wort physically below the slots will not be collected. In a circular tun, the same guidelines apply, but if you are using an inscribed square the transverse tubes can be slotted where they are away from the wall.

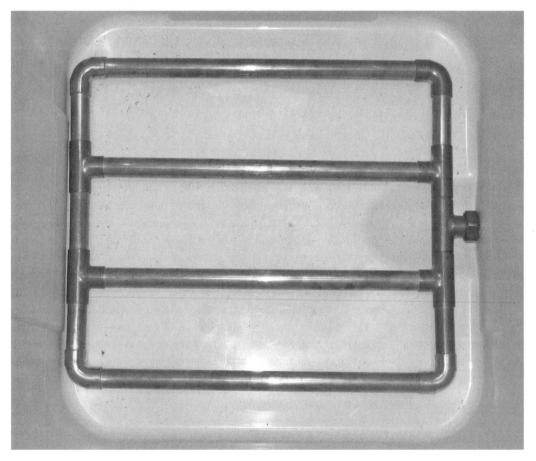

**Figure F.13.** Multi-pipe manifold in square cooler.

## Designing Ring Manifolds for Continuous Sparging

Ring-shaped manifolds or stainless steel braids are an elegant looking system for cylindrical beverage coolers, but how efficient are they? It turns out that they are pretty good if the balanced spacing concept is applied. For a single ring, balanced spacing means having an equal volume inside and outside the ring. The calculate the diameter of a ring that will divide the tun volume in half, simply multiply the tun diameter by 0.707:

Ring diameter = 0.707 × tun diameter

The chart in figure F.15 plots the efficiency quantities of rings and false bottoms in a Sankey keg as a function of diameter. The total diameter of a Sankey keg is 15 in. (38 cm), so the half volume diameter is 10.6 in. (27 cm). It is interesting to note that a single ring is more efficient than a false bottom until the false bottom diameter is at least 80% of the tun diameter. These ratios hold true for any tun diameter. If more rings were added in a balanced spacing manner, the efficiency would improve and approach that of false bottoms, just like rectangular manifolds in rectangular coolers.

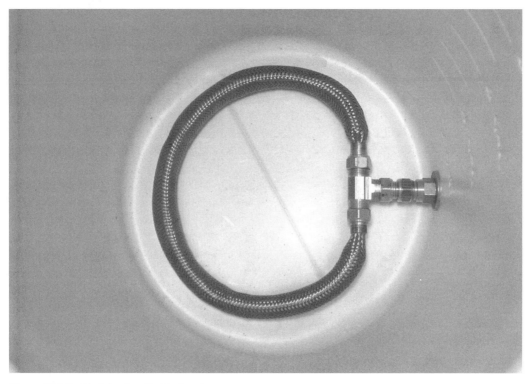

**Figure F.14**. Braided ring in cylindrical cooler.

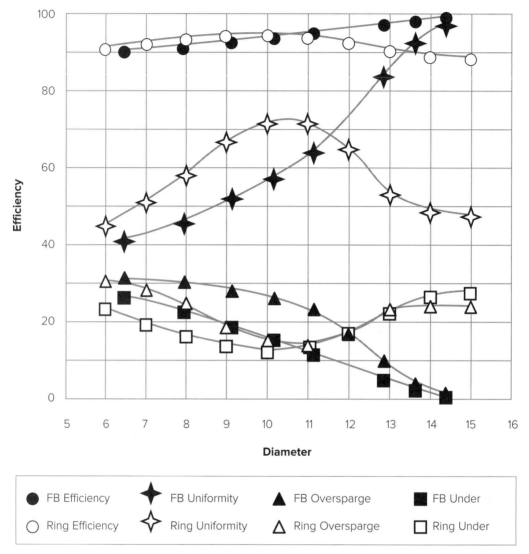

**Figure F.15.** Plotting the efficiency quantities of rings and false bottoms (FB) in a 15 in. diameter Sankey keg as a function of ring or FB diameter.

# Appendix G

## Brewing Metallurgy

This appendix may contain a lot more detail than most brewers need, but, being a metallurgist, I frequently get asked about metals used in brewing, so I thought my book would be a good place to write it all down for reference. The three main topics are cleaning, corrosion, and joining. I will discuss how to best clean the various metals and how the different metals corrode. Homebrewers make a lot of their own equipment from off-the-shelf parts at the hardware store, but you should be aware how building an item from dissimilar metals can accelerate corrosion. Likewise, the joining of metals can present a challenge, so I will provide tips for soldering, brazing, and welding.

The primary concern in brewing is flavor. We want to taste the beer, not the materials used in its production. While some metals like aluminum, iron, and carbon steel will merely taste bad, other metals can be toxic to yeast if the concentrations are high enough. Human toxicity from metals used in brewing equipment is rarely a real concern. The fourth and final section of this appendix discusses toxicity information for metals commonly found in brewing and plumbing.

## General Information and Cleaning

### Aluminum

Aluminum is a good choice for brew pots and direct-heat mash-lauter tuns. It has high heat conductivity, which helps prevent hot spots and scorching of the wort or mash, and is less expensive than stainless steel. The aluminum alloys most commonly used for cookware are alloys 3003 and 3004, which have very high corrosion resistance. Under the temperature and pH conditions normally encountered in brewing, aluminum (by itself) will not corrode and should not contribute any metallic flavor to your beer. However, when using aluminum for a brewing pot, do not clean the metal shiny bright between uses or you may get a metallic off-flavor. Like all metals, aluminum depends on a passive surface oxide layer for corrosion resistance, and scouring the metal shiny bright will remove this passive layer. Allow it to grow dull and gray with use. To encourage a passive layer in a brand new pot, wash it thoroughly, dry it thoroughly, and then put it in your oven (dry) at 350°F (180°C) for about 10 min. This will help the anhydrous oxide layer to thicken. To clean aluminum, I recommend percarbonate-based cleaners, such as Straight A and PBW, or an unscented dishwashing detergent brand, like Ivory®. Do not use bleach because it can cause pitting of the aluminum.

Aluminum will corrode if placed adjacent to another metal like copper in wort or beer, but even this most aggressive corrosion situation is usually insignificant in home brewing. I will discuss this in the galvanic corrosion section later in this appendix.

### Copper

Copper has a long history in brewing. It has high heat conductivity, is easy to form, and was traditionally used for making the brewing kettles, or "coppers." These days, professional brewers typically choose stainless steel because it is stronger, more inert, and easier to maintain. But for the homebrewer, copper and brass are still the cheapest and best choices for wort chillers and fittings. Copper is relatively inert to both wort and beer. With regular use, it will build up a stable oxide layer (evidenced by a dull copper color) that will protect it from any further interaction with the wort. Only minimal cleaning to remove surface grime, hop bits, and wort protein is necessary. There is no need to clean copper shiny bright after every use or before contact with your wort. It is better that you allow the copper to form a dull copper finish with use.

However, you need to be aware that copper can develop a toxic blue-green oxide called *verdigris*. Verdigris includes several chemical compounds, such as cupric acetate, copper sulfate, and cupric chloride. These blue-green compounds should not be allowed to contact your beer, or any other food item, because they readily dissolve in weakly acidic solutions (like beer), which can lead to copper poisoning. To clean copper that has heavy oxidation (black) and verdigris (green-blue), use vinegar, or use oxalic acid-based cleaners like Revere® Copper and Stainless Steel Cleaner.

For regular cleaning of copper and brass, unscented dish detergent or sodium percarbonate-based cleaners are preferred. Cleaning and sanitizing copper wort chillers with bleach solutions is not recommended. Oxidizers, like bleach and hydrogen peroxide, attack copper and these cleaning agents will quickly cause copper and brass to blacken as oxides form. These black oxides do not protect the surface from further corrosion, and since they are formed under alkaline conditions, they will quickly dissolve in the acidic wort. If a wort chiller is cleaned or sanitized with bleach, therefore, the yeast will be exposed to potentially harmful levels of dissolved copper, although yeast does have a high tolerance. Under normal brewing conditions, no off-flavors are associated with copper because almost all of it is removed (chelated) from solution by the yeast.

Always thoroughly rinse copper! Do not soak copper in acidic or alkaline cleaning solutions for extended periods of time (>1 hour). Do not leave water inside counterflow or plate-style chillers—drain them or blow them dry.

## Brass

Brass is a group of alloys made from copper and zinc, with some lead thrown in for machinability. The lead percentage varies, but for the alloys used in plumbing fittings it is 3% or less. In fact, most plumbing parts are required to be made from lead-free brass these days. Lead does not alloy or mix with the copper and zinc in brass, but instead exists as tiny globules, like bananas in Jell-O®. These globules act as a lubricant during machining and result in a microthin film of lead being smeared over the machined surface. It is this lead that can be dissolved off by the wort.

While this teeny, tiny amount of surface lead is not a health concern, most people would be happier if wasn't there at all. Fortunately, this surface lead is very easy to remove by soaking the parts in a solution of vinegar and hydrogen peroxide. You can get these at the grocery store or drug store. You can use either white distilled vinegar or cider vinegar, just check the label to be sure it is 5% acid by volume. The hydrogen peroxide should be 3% by volume. To make the solution, mix them at a 2:1 volume ratio of vinegar to peroxide. Simply immerse the brass parts in the solution and watch for the color of the parts to change. The process takes less than five minutes to clean and brighten the surface. The color of the brass will change to buttery yellow-gold when the lead is removed. If the solution starts to turn blue or green or the parts start darkening, it means that the parts have been soaking too long; the result is that the peroxide is used up and the copper is dissolving, which will expose more lead. Make up a fresh solution and soak the parts again. This treatment only needs to be done once before the first use of the parts.

While zinc is an important nutrient for yeast, it can be too much of a good thing. High zinc levels (>5 ppm) from the corrosion of brass can cause soapy or goaty flavors due to excessive yeast growth, plus increased acetaldehyde and fusel alcohol production. But, like copper, brass is usually very stable in wort and will turn dull with regular use as it builds up a passive oxide layer. Brass should be treated like copper for normal cleaning. Do not clean it with bleach.

## Carbon Steel

Carbon steel is predominantly iron, alloyed with carbon and other trace elements. In homebrewing carbon steel is commonly used for porcelain-enamel cookware and as rollers in grain mills. Many homebrewers get started in the hobby with an enamelware brew pot because of their low cost. The drawback with these pots is that the porcelain can become cracked or chipped with use, exposing the steel to the wort. While a little extra iron (or rust) in your diet won't hurt you, it will taste bad and promotes haze via oxidation reactions. There is no practical way to fix these flaws in the porcelain, and the steel will rust between uses. A rusty pot will cause metallic, bloodlike off-flavors in the wort.

Some brewers like to build their own roller mills for crushing grain. Carbon steel is not stainless steel and needs to be protected against rusting by oiling or plating. If the roller steel is kept clean and dry between crushes, then it usually won't rust. It can be cleaned with a nylon or brass wire brush to remove any light rusting that may occur. Cleaning with steel wool or a steel wire brush will actually promote corrosion.

You can improve the corrosion resistance of carbon steel slightly by rubbing it with vegetable oil and buffing it off like car wax. By doing this you protect the surface oxides from hydration, producing a black oxide rather than rust. The black oxide is more adherent and will eventually cover the entire surface, inhibiting further corrosion. The oil will become more waxlike too as the volatile components vaporize over time. This oxide-wax coating has limited corrosion resistance and direct contact with water will usually induce red rust. The rust can be cleaned away as described above to restore the more passive black oxide surface layer.

## Stainless Steel

Stainless steels are iron alloys containing chromium and nickel. The most common type of stainless steels used in the food and beverage industry are the 300 series—typically containing 18% chromium and 8% nickel. The specific alloys that are most often used are AISI[1] grades 304 and 316, which are very corrosion resistant and are basically inert to beer. The presence of chromium and its oxides inhibit rust and corrosion. Stainless steel is referred to as being "passivated" when the protective chromium oxide surface layer is unbroken. If this oxide layer is breached by iron (from a wire brush or drill bit), dissolved by chemical action (e.g., bleach), or compositionally altered by heat (brazing or welding), it will rust. The problem with stainless steel corrosion is usually not an off-flavor, but more often a hole in a valuable piece of equipment.

In industry, stainless steel is passivated by dipping the parts in a strong nitric acid solution. This treatment removes some of the iron from the surface of the alloy and improves the overall corrosion resistance, especially in strong chemical or saltwater environments. This kind of treatment is not necessary for brewing equipment. If the protective oxide layer is compromised, stainless steel can be repassivated by thoroughly cleaning to remove the contamination. The key to achieving a passive surface is getting the steel clean and free of contaminants. The easiest way to do this at home is to use a kitchen cleanser made for cleaning stainless steel cookware. Three examples are Bar Keeper's Friend®, Kleen King® Stainless Steel & Copper Cleaner, and Revere® Copper and Stainless Steel Cleaner. The active ingredient in these cleansers is oxalic acid, which serves the same cleaning purpose as nitric acid. Once the surface has been cleaned to bare metal, the passive oxide layer will reform immediately. Citric acid cleansers also work well for this purpose. Both these types of cleanser also work very well for cleaning copper.

What this means is that you can perform cutting, grinding, soldering, or welding on your stainless steel and with just a few minutes of work with cleaner and a green scrubby (Scotch-Brite™), it will be passive again. Be sure to rinse thoroughly with clean water afterward so you don't leave any acid behind. Do not use steel wool or even a stainless steel scrubby; these will cause rust. Be careful to not scratch or heavily roughen the surface, because deep scratches can also lead to corrosion. Smoother is better.

As you may be realizing, stainless steel is not invulnerable. Unfortunately, people tend to assume it is and then are shocked when it does corrode. Stainless steel has an Achilles' heel, and that weakness is chlorine, which is common in cleaning products. Chlorine can dissolve the protective oxides, exposing the metal surface to the environment. Let's suppose you are sanitizing a Cornelius (corny) keg with bleach. If there is a scratch, or a rubber gasket against the steel that creates a crevice, then these secluded areas can lose their passivation. Inside the crevice, on a microscopic scale, chlorides combine with oxygen from the oxide to form chlorite ions. That crevice becomes a tiny, highly

---

[1] American Iron and Steel Institute.

chemically active site compared to the more passive stainless steel around it, so it corrodes. This mechanism is known as crevice corrosion.

The same thing can happen at the water's surface if the keg is only half full. In this case, the steel above the waterline is in air and the passive oxide layer is stable. Beneath the surface, the oxide layer is less stable due to chloride ions present, but the layer is uniform. With a stable area above, and a less stable but very large area below, the waterline becomes the "crevice." Usually this type of corrosion will manifest as pitting or pinholes. The mechanism described is accelerated by localization, so a pit is most often the result, which can cause pinholes to form within a few hours in kegs half-filled with bleach solution.

Biofouling (trub deposits) and beerstone scale (calcium oxylate) can cause corrosion by a similar mechanism. The metal underneath the deposit can become oxygen depleted via biological or chemical action. When this happens, it will lose passivation and become pitted. This is why the removal of beerstone from stainless steel storage or serving tanks is important. The dairy industry has the same problem with calcium oxylate and uses phosphoric acid to dissolve the buildup. Phosphoric acid is a good choice as it does not attack the steel. Do not use swimming pool acid to dissolve beerstone or clean stainless steel. The acid used for swimming pools is actually hydrochloric (muriatic) acid, which is very corrosive to stainless steel.

A second way that chlorides can cause corrosion of stainless steel is by concentration. This mode is very similar to the crevice mode described above. Chlorides become concentrated when chlorinated solutions evaporate and dry on a steel surface. The next time the surface is wetted it will corrode at that spot, creating a shallow pit. The next time the keg is allowed to dry that pit will probably be one of the last sites to dry, causing chloride concentration again. At some point in the life of the keg, the site will become deep enough for crevice corrosion to take over and the pit will corrode through.

To prevent stainless steel from being attacked and pitted by the use of chlorinated cleaning products like bleach, follow these simple guidelines:

- Do not allow the stainless steel vessel to sit for extended periods of time (i.e., hours) filled with bleach water or another chlorinated cleaning solution.
- If you use bleach for cleaning, rinse the vessel thoroughly with water and dry it to prevent evaporation concentration.
- Some of the chemical company sales reps also recommend a final rinse with an acidic product, such as phosphoric acid rinse or sanitizer, but this is overkill. Just be sure to rinse thoroughly with water.
- Use percarbonate-based cleaning products, such as PBW, Straight A, B-Brite™, or One Step, which won't attack the protective oxide layer.
- Hydrochloric acid can be very damaging to stainless steel. It is a very effective rust remover, but rinse thoroughly after use.

## Galvanic Corrosion

All corrosion is basically galvanic, although this is an over-generalization. The electrochemical difference between two metals in an electrolytic solution causes an electric current to flow, which results in one of the metals becoming ionized. These ions combine with oxygen or other elements to create corrosion products. Cleaning off the corrosion products will not solve the problem. The cause of the corrosion is usually the environment (electrolytic solution) and the metals themselves.

Think back to your high school chemistry class and I will explain. An electrolytic solution can be defined as any liquid containing dissolved ions or salts, like tap water or sea water. Metals will corrode faster in strong electrolytic solutions (sea water) than in weak electrolytic solutions (tap water). For example, sticking a copper wire and a nail in a potato will give a different voltage (and therefore a different corrosion rate) compared to putting them in a glass of beer. And the ratio of the surface areas between the two metals directly affects the corrosion rate too.

Because the galvanic corrosion potential of two metals depends on several factors, including electrolytes and surface area, the standard electrolytic solution for comparing galvanic potential is seawater. A galvanic series lists the corrosion potential of different metals from most active to most passive (table G.1). When two metals are placed in contact with one another in the presence of the electrolytic solution, the most active metal of the pair will corrode. The metals don't actually have to be wet, although that helps. If you place two dissimilar metals together under pile of table salt, they will corrode due to the moisture in the atmosphere. The degree of separation between any two of the metals in the list gives an indication of the aggressiveness of the corrosion between them, all other factors being equal. The corrosion rate between platinum and magnesium would therefore be the highest.

The surface area factor works like this: if you have an active metal coupled to a passive metal, and the passive metal has a larger surface area than the active metal, the corrosion rate of the active metal will be increased. If the active metal area is larger than the area of the passive metal, the corrosion rate of the active metal will decrease significantly. In either case, most of the corrosion will take place at the interface of the two metals.

This means that, if you have small area of aluminum in contact with a large area of brass, the aluminum will corrode quickly. But, if you mount a small brass fitting on an aluminum pot, very little corrosion of the aluminum will take place because of the large difference in surface areas. Brass, copper, stainless steel, and silver solder are close enough together on the galvanic series that there is not much potential for corrosion between them. Copper has the additional property of polarization, where it tends to be become more galvanically similar to its neighbor, reducing the electrical potential and overall corrosion rate. I've had brass and copper fittings mounted or soldered to my stainless steel converted kegs for the past 10 years and have not seen any corrosion to speak of. Randy Mosher, noted author and DIY guy, has copper brazed to stainless steel with silver braze and copper welded with aluminum bronze in use for several years with no problems either.

This brings up the deciding factor in galvanic corrosion situations—exposure time. As homebrewers, our equipment is not operating seven days a week. The equipment is only exposed to an electrolytic solution for a few hours at a time, every couple of weeks or so. This is not much exposure when compared to the situation in a professional brewery, or other industrial application. So, even if we design and build equipment with galvanic couples, the useful life of our equipment is still pretty long.

## TABLE G.1—GALVANIC SERIES IN SEA WATER

### MOST ACTIVE

Magnesium

Zinc

Galvanized steel

Aluminum
(pure, and alloys 3003 and 3004)

Cadmium

Carbon steel and cast iron

Unpassivated stainless steels

Lead-tin solder

Lead

Tin

Brass

Copper

Bronze

Passivated stainless steels

Silver solder

Silver

Titanium

Graphite

Gold

Platinum

### MOST PASSIVE

## Soldering, Brazing, and Welding

### Soldering

Soldering is the only non-mechanical joining process you need 90% of the time when you are building homebrewing equipment. The other 10% usually consists of welding a stainless steel nipple onto a converted stainless steel keg (more on this later). Soldering with lead-free silver plumbing solder allows you to join any of the metals we've discussed. Modern lead-free silver plumbing solder (in accordance with ASTM Standard B-32) is an alloy of tin, bismuth, and silver and typically has a melting range of 420–460°F (215–238°C).

The most common difficulty encountered when trying to solder to stainless steel is lack of wetting—the solder just balls up and sits there. This is caused by not having the proper flux. The surface oxides that protect stainless steel also make it difficult for the solder to wet it. Use a water-soluble flux that contains hydrochloric acid or zinc chloride. The second most common difficulty is getting the parts hot enough. Most of the time a propane torch is sufficient, but sometimes a methylacetylene-propadiene (MPS)-type gas (e.g., MAPP® gas) is needed if the parts are very large. Although MPS gas burns hotter than propane, it does not burn as hot as acetylene and does not need special equipment. A good strategy is to "tin" one of the parts with solder beforehand to create a pre-wetted surface. Flux is then applied to the other parts and the joint is fitted together and heated. In this way, the surfaces are protected from oxidation until the solder can melt and make the joint. Once hot, more solder can be fed into the joint to finish it.

### Brazing

Brazing is exactly like soldering except the filler metals are stronger and melt at higher temperatures. Unless you are going to butt braze a nipple onto the side of a keg, there is no real reason to use brazing instead of soldering. Brazing provides for a stronger joint, but usually the strength of soldering is more than adequate. The problem with torch brazing stainless steel is that the brazing temperatures 1000–1600°F (540–870°C) are right in the temperature range where embrittling occurs, 800–1600°F (425–870°C). Significant time at these temperatures (>3 min.) allows the chromium to diffuse away from the grain boundaries to form chromium carbides, depleting that area of chromium and creating un-stainless steel. In other words, it will crack and rust. Steel that has been exposed to such temperatures is referred to as "sensitized." Embrittling soon leads to localized corrosion and rapid cracking at the grain boundaries. All exposure to these temperatures is cumulative and the resulting chromium diffusion cannot be corrected in any practical manner for homebrewers. It is much better to just avoid these temperatures and prevent it from occurring.

## Welding

If you need a really strong joint in stainless steel, the best method is welding, and the best welding method for adding nipples to converted kegs or pots is gas tungsten arc welding (GTAW), also known as TIG welding. Gas metal arc welding (GMAW), also known as MIG, is also popular. The advantages of TIG welding is that it has a small weld head, a lower heat input is required, and the use of filler metal is optional. Although MIG is probably the most common process for welding stainless steel, the large weld head must be held close to the work and this decreases its effectiveness in tight areas.

Nowadays it is pretty easy to search online and find a local stainless steel welder to do the job for you. You will most likely not exceed their one-hour minimum charge. One thing to keep in mind after welding is that the blue-ish or straw colored area around the weld joint is no longer passivated. The heat has created different oxides that can corrode, so you will need to use the stainless steel cleansers mentioned earlier to clean the discoloration away to bare metal so it can repassivate itself.

**Note:** Do NOT weld galvanized or cadmium-plated steel parts. These metals vaporize easily and can cause metal-fume fever as well as acute toxicity from inhalation. In fact, cadmium has no place in the brewery at all due to its high toxicity. Galvanized parts for stands can be welded if the zinc coating is first sanded or ground off within ½" (13 mm) of the weld area.

I hope these sections have given you the information you need to help choose your materials and processes for gadget building. Some points to remember:
- Metals depend on passive surface oxides for corrosion protection.
- Cleaning metals shiny bright may lead to off-flavors.
- Soldering will usually do the job, and if not, welding is easily hired out.

## Toxicity of Metals

While many people are aware of the general toxicity of lead and cadmium, most people don't know *how* they are toxic. In all cases of acute heavy metal poisoning by ingestion, the symptoms are nausea and vomiting. Chronic (long-term) poisoning symptoms are more varied, but often involve skin discoloration, weakness, and anemia. The following information comes from three books on industrial hygiene[2] and much of the data is from standard FDA animal testing. The notation "$LD_{50}$" means that half of the test animals (usually mice) in the test were killed by the test dosage, and the dosage is stated as milligrams ingested per kilogram of body weight.

### Aluminum

*Usage:* Aluminum is used in cookware and tubing. Galvanically active.

There was a concern several decades ago that the use of aluminum in cooking and the ingestion of aluminum contributed to Alzheimer's disease. The medical study that generated this controversy

---

[2]    See bibliography: Casarett *et al.* (1980), Owen (1981), and Patty (1981).

was later found to have be flawed due to contamination of the test samples. An independent experiment, conducted by Jeff Donaghue and reported in *Brewing Techniques* magazine, showed that in side-by-side trials of single mash worts boiled using aluminum versus stainless steel pots, there was no detectable difference in the amount of aluminum between the samples either before or after fermentation.[3] The amount of aluminum in the wort boiled in the aluminum pot was less than the detection limit for the test, which is 0.4 mg/L (0.4 ppm). If you drank 20 L (5.3 gal.) of that beer, you would only ingest 20 mg of aluminum, about the same amount as a single buffered aspirin tablet, and half of what you would get from a single antacid tablet.

*Acute toxicity:* Aluminum chloride—$LD_{50}$ of 770 mg/kg.

*Chronic toxicity:* No data.

## Cadmium

*Usage:* Cadmium is an ingredient in some solders and brazing alloys, none of which are approved for use with food. Like galvanizing, cadmium plating is also used as an industrial protective coating for steel (most common in nuts and bolts), but it has a more golden color. Galvanically active.

*Acute toxicity:* Symptoms are exhibited upon ingestion of 14.5 mg, which causes nausea and vomiting. A case where a 180 lb. man ingested 326 mg was not fatal. The presence of copper or zinc at the time of ingestion will lessen the rate of absorption of cadmium into the body and reduce its toxic effects. Cadmium is easily vaporized during welding and can cause acute toxicity from inhalation.

*Chronic toxicity:* A study with rats found a 50% reduction in their hemoglobin levels over a three-month period when the rats drank water containing 50 mg/L (50 ppm) cadmium. Other rats studied that were given water with 0.1–10 mg/L (0.1–10 ppm) cadmium for one year showed no change in hemoglobin levels.

## Chromium

*Usage:* Chromium is a secondary constituent of stainless steel, and is also used as electroplated coating for carbon steel. Galvanically passive.

The chromium 6 ion, which received so much publicity in the 1990s, is not encountered in homebrewing. Chromium 6 is electrically generated in solution during chromium electroplating and is a wastewater contaminant. Chromium 6 is not generated by water sitting in contact with electroplated chromium, nor by the galvanic corrosion of stainless steel.

*Acute toxicity:* Soluble chromates are of very low toxicity when ingested, up to 1500 mg/kg body weight before symptoms are seen. Chromium is most toxic when inhaled as fumes or dust, but chromium is not vaporized during typical welding of stainless steel.

*Chronic toxicity:* No documented evidence of long term toxicity from soluble chromates.

## Copper

*Usage:* Copper is used to make rigid and flexible tubing for plumbing and refrigeration systems. Galvanically passive.

Copper is an essential nutrient; the average daily intake is 2–5 mg, and 99% is excreted from the body in the feces.

*Acute toxicity:* 200 mg/kg body weight of copper salts is the lowest lethal dose.

---

[3]    Jeff Donaghue, "Testing Your Metal—Is Aluminum Hazardous to Your Beer?" *Brewing Techniques,* vol. 3, January/February, 1995, p.62.

*Chronic toxicity:* While dosages are not recorded, chronic poisoning symptoms include: headache, fever, nausea, sweating, and exhaustion. Sometimes hair, fingernails, skin, and bones will turn green.

## Iron

*Usage:* Iron is the primary constituent of carbon steel and stainless steel. Galvanically active.

While iron is an essential nutrient, overdoses of iron supplements are very dangerous.

*Acute toxicity:* Ferric chloride—$LD_{50}$ of 400 mg/kg. Symptoms of iron toxicity include: headache, nausea, vomiting, anorexia and weight loss, and shortness of breath. Skin might turn gray.

*Chronic toxicity:* No data on dose. The United States recommended daily allowance (USRDA) for iron is 10 mg/day for men and 12 mg/day for women.

## Lead

*Usage:* Lead is a tertiary constituent of brass; used in plumbing fixtures, fittings, and non-food grade solders.

*Acute toxicity:* The oral dose of soluble lead needed to kill a guinea pig is 1330 mg/kg body weight.

*Chronic toxicity:* Lead slowly accumulates in the body. Normal intake from environmental sources of lead averages 0.3 mg/day, with 92% being excreted. Blood tests are a good indicator of lead exposure. A normal blood lead level in an adult is 3–12 micrograms (μg) per 100 g of whole blood.[4] Adverse effects are not seen until a person has had a blood lead level of over 20 μg/100 g whole blood for several years. More serious symptoms are seen when blood levels test over 50 μg/100 g whole blood for a period of 20 years, or from a single massive dose. Symptoms of lead poisoning range from loss of appetite and a metallic taste in the mouth, to anxiety, nausea, weakness, and headache, to tremors, dizziness, and hyperactivity, to seizures, coma, and death from cardiorespiratory failure. Men will also suffer from impotence and sterility.

## Zinc

*Usage:* Zinc is a secondary constituent of brass. Zinc is an essential nutrient and the USRDA is 15 mg.

*Acute toxicity:* Mass poisonings have been repeatedly reported caused by drinking acidic beverages from galvanized containers. (e.g., wine punch in trash cans). Fever, nausea, stomach cramps, vomiting and diarrhea occurred 3–12 hours after ingestion. The lowest lethal dose for zinc in guinea pigs is 250 mg/kg body weight. Zinc is easily vaporized during welding and can cause acute toxicity from inhalation. Short-term symptoms mimic the flu and are indeed called "welder's flu" and "metal fume fever."

*Chronic toxicity:* No apparent injury observed in rats dosed with 0.5–34 mg of zinc oxide for periods ranging from 1 month to 1 year.

## Acknowledgement

My thanks to Mike Maag—Industrial Hygienist with the Virginia Department of Labor and Industry—for helping me track down this information.

---

[4] Note the use of micrograms (μg), not milligrams (mg). One microgram is one-thousandth of a milligram.

# Appendix H

## Metric Conversions

During the technical review for this edition, one of the editors brought up a good point about the use of significant digits in measurement systems. Basically, what this means is that if you have a ruler with only inch markings on it (i.e., no fractions of inch markings) then you can only measure anything to the nearest whole inch. So, with 7 in., for example, you can eyeball it and say that it looks like seven and a half inches, but the ruler only has a 1-inch resolution and therefore you are limited in reporting that data to one significant digit, which is 7. Does this mean that if something is about 11 in. long it must be reported as being 10 in.? No, you can report that to two significant digits (i.e., 11 is the nearest whole inch), unless the ruler is only marked in 10-inch increments (i.e., 10, 20, 30, etc.). In other words, the number of significant digits is limited by the resolution of the measurement device.

If I were to convert that measurement from inches to centimeters, and report it as data, then I would need to observe the typical resolution of the measuring devices. A precise conversion of 7 in. is 17.78 cm, but I would be better reporting it as 18 cm (i.e., the nearest whole centimeter).

Now then, data reporting is one thing, specifying recipe quantities is another. I could be very precise and specify 3.000 gal., but do you have the equipment to measure so precisely? No, probably not. Everyday logic says that 3 gal. means about three gallons, plus or minus a little bit. Not two and a half gallons, not three and a half gallons, but about 3 gal. as probably indicated by a 3 gal. volume marker on the side of the bucket. When I specify 3 gal. of water as part of a recipe, then the conversion of that value to liters needs to be precise enough that you have the best comprehension of the quantity that I have in mind. A very precise conversion of exactly 3 gal. is 11.3562353520 L. The concept we just discussed suggests I should round that to 11, but 11 L is only 96.8% of the intended quantity. If we break this down by percentage:

11 of 11.3562353520 = 96.8%

11.3 of 11.3562353520 = 99.5%

11.35 of 11.3562353520 = 99.95%

11.356 of 11.3562353520 = 99.998%

But, do we exceed the interest and enthusiasm of the average homebrewer when measuring a volume down to the nearest milliliter, as in this case? Yes, I think so. In my opinion, the best conversion of 3 gal. to liters is 11.4 L, because 11.4 of 11.3562353520 = 1.0039%. In other words, 11.4 L is actually a closer approximation than 11.35 L (+0.0039% vs. −0.05%), and it is also what engineers call "conservative," that is to say, we have a little bit of a buffer. Better to have a little bit more wort than not quite enough.

So in the table below, I am going to list the conversion conventions for this book; they may not be exact, they may in fact not be very precise, but they are conservative. Conversions of weights that are less than 1 kilogram will be given in grams. Volume conversions will depend on the scale, large volumes will generally be rounded to tenths of a liter, while milliliter measurements will be integers.

When I am speaking in general, such as when discussing kettle sizes, then a five-gallon kettle may be described as 20 L instead of 19 L. A cup of water (nominally 8 fl. oz.) will be converted to 250 mL. When I need to be very precise, such as when calculating salt and acid additions for water adjustment, then only liters and milliliters (or grams and milligrams) will be used.

The tables given in this section are intended to provide a quick reference for conversion between US standard units and metric units.

## Conversion Tables

The following tables list the standard conversions that will be used for recipes throughout this book. The difference to the actual equivalent is small and should be insignificant. Some recipe quantities will need to be more precise, for example, 0.25 lb., and in those cases the metric equivalent will be rounded to the nearest five grams, (e.g., 0.25 lb. will be converted to 115 g).

## TABLE H.1—RECIPE WEIGHT CONVERSIONS

| Nominal weight, US standard units | Metric equivalent (actual) | Metric equivalent w(for this book) |
|---|---|---|
| 0.25 oz. | 7.1 g | 7 g |
| 0.5 oz. | 14.2 g | 15 g |
| 0.75 oz. | 21.3 g | 23 g |
| 1.0 oz. | 28.3 g | 30 g |
| 0.25 lb. | 113.4 g | 115 g |
| 0.5 lb. | 226.8 g | 225 g |
| 0.75 lb. | 340.2 g | 340 g |
| 1.0 lb. | 453.6 g | 450 g |
| 1.25 lb. | 567.0 g | 565 g |
| 1.5 lb. | 680.4 g | 680 g |
| 1.75 lb. | 793.8 g | 790 g |
| 2.0 lb. | 907.2 g | 910 g |
| 2.5 lb. | 1.134 kg | 1.14 kg |
| 3.0 lb. | 1.361 kg | 1.36 kg |
| 3.5 lb. | 1.588 kg | 1.6 kg |
| 4.0 lb. | 1.814 kg | 1.8 kg |
| 4.5 lb. | 2.041 kg | 2.05 kg |
| 5.0 lb. | 2.268 kg | 2.3 kg |

## TABLE H.2—RECIPE VOLUME CONVERSIONS

| Nominal volume, US gallon | Equivalent volume, US quart | Metric equivalent (actual) | Metric equivalent (for this book) |
|---|---|---|---|
| 0.25 gal. | 1 qt. | 0.946 L | 0.95 L |
| 0.5 gal. | 2 qt. | 1.893 L | 1.9 L |
| 0.75 gal. | 3 qt. | 2.839 L | 2.8 L |
| 1.0 gal. | 4 qt. | 3.785 L | 3.8 L |
| 1.5 gal. | 6 qt. | 5.678 L | 5.7 L |
| 2.0 gal. | 8 qt. | 7.571 L | 7.6 L |
| 2.5 gal. | 10 qt. | 5.678 L | 5.7 L |
| 3.0 gal. | 12 qt. | 11.356 L | 11.4 L |
| 3.5 gal. | 14 qt. | 13.249 L | 13.25 L |
| 4.0 gal. | 16 qt. | 15.142 L | 15.15 L |
| 4.5 gal. | 18 qt. | 17.034 L | 17.0 L |
| 5.0 gal. | 20 qt. | 18.927 L | 19 L |
| 5.5 gal. | 22 qt. | 20.820 L | 21 L |
| 6.0 gal. | 24 qt. | 22.712 L | 23 L |
| 6.5 gal. | 26 qt. | 24.605 L | 24.6 L |
| 7.0 gal. | 28 qt. | 26.498 L | 26.5 L |
| 7.5 gal. | 30 qt. | 28.391 L | 28.4L |
| 8.0 gal. | 32 qt. | 30.283 L | 30.3 L |
| 8.5 gal. | 34 qt. | 32.176 L | 32.2 L |
| 9.0 gal. | 36 qt. | 34.069 L | 34 L |
| 9.5 gal. | 38 qt. | 35.961 L | 36 L |
| 10.0 gal. | 40 qt. | 37.854 L | 38 L |
| 11.0 gal. | 44 qt. | 41.639 L | 41.6 L |
| 12.0 gal. | 48 qt. | 45.425 L | 45.4 L |
| 13.0 gal. | 52 qt. | 49.210 L | 49.2 L |
| 14.0 gal. | 56 qt. | 52.996 L | 53 L |
| 15.0 gal. | 60 qt. | 56.781 L | 57 L |
| 20.0 gal. | 80 qt. | 75.708 L | 76 L |

## TABLE H.3—TEMPERATURE CONVERSIONS

| °F | °C | | °F | °C | | °F | °C | | °F | °C |
|----|----|---|----|----|---|----|----|---|----|----|
| 32 | 0 | | 88 | 31 | | 144 | 62 | | 200 | 93 |
| 34 | 1 | | 90 | 32 | | 146 | 63 | | 202 | 94 |
| 36 | 2 | | 92 | 33 | | 148 | 64 | | 204 | 96 |
| 38 | 3 | | 94 | 34 | | 150 | 66 | | 206 | 97 |
| 40 | 4 | | 96 | 36 | | 152 | 67 | | 208 | 98 |
| 42 | 6 | | 98 | 37 | | 154 | 68 | | 210 | 99 |
| 44 | 7 | | 100 | 38 | | 156 | 69 | | 212 | 100 |
| 46 | 8 | | 102 | 39 | | 158 | 70 | | | |
| 48 | 9 | | 104 | 40 | | 160 | 71 | | | |
| 50 | 10 | | 106 | 41 | | 162 | 72 | | | |
| 52 | 11 | | 108 | 42 | | 164 | 73 | | | |
| 54 | 12 | | 110 | 43 | | 166 | 74 | | | |
| 56 | 13 | | 112 | 44 | | 168 | 76 | | | |
| 58 | 14 | | 114 | 46 | | 170 | 77 | | | |
| 60 | 16 | | 116 | 47 | | 172 | 78 | | | |
| 62 | 17 | | 118 | 48 | | 174 | 79 | | | |
| 64 | 18 | | 120 | 49 | | 176 | 80 | | | |
| 66 | 19 | | 122 | 50 | | 178 | 81 | | | |
| 68 | 20 | | 124 | 51 | | 180 | 82 | | | |
| 70 | 21 | | 126 | 52 | | 182 | 83 | | | |
| 72 | 22 | | 128 | 53 | | 184 | 84 | | | |
| 74 | 23 | | 130 | 54 | | 186 | 86 | | | |
| 76 | 24 | | 132 | 56 | | 188 | 87 | | | |
| 78 | 26 | | 134 | 57 | | 190 | 88 | | | |
| 80 | 27 | | 136 | 58 | | 192 | 89 | | | |
| 82 | 28 | | 138 | 59 | | 194 | 90 | | | |
| 84 | 29 | | 140 | 60 | | 196 | 91 | | | |
| 86 | 30 | | 142 | 61 | | 198 | 92 | | | |

## TABLE H.4—EXTRACT YIELD CONVERSIONS

| PPG | PKL | PPG | PKL |
|---|---|---|---|
| 1 | 8 | 24 | 200 |
| 2 | 17 | 25 | 209 |
| 3 | 25 | 26 | 217 |
| 4 | 33 | 27 | 225 |
| 5 | 42 | 28 | 234 |
| 6 | 50 | 29 | 242 |
| 7 | 58 | 30 | 250 |
| 8 | 67 | 31 | 259 |
| 9 | 75 | 32 | 267 |
| 10 | 83 | 33 | 275 |
| 11 | 92 | 34 | 284 |
| 12 | 100 | 35 | 292 |
| 13 | 108 | 36 | 300 |
| 14 | 117 | 37 | 309 |
| 15 | 125 | 38 | 317 |
| 16 | 134 | 39 | 325 |
| 17 | 142 | 40 | 334 |
| 18 | 150 | 41 | 342 |
| 19 | 159 | 42 | 351 |
| 20 | 167 | 43 | 359 |
| 21 | 175 | 44 | 367 |
| 22 | 184 | 45 | 376 |
| 23 | 192 | 46 | 384 |

Notes: This is a quick reference chart for converting between gravity points per pound per gallon (PPG) and gravity points per kilogram per liter (PKL). 1 PPG = 8.3454 PKL.

# Appendix I

## The Trouble with Producing Gluten-Free Beer

Barley is a member of the grass family, subfamily Pooideae, belonging to tribe Triticeae. The same is true for wheat and rye. Also in subfamily Pooideae are oats, but oats belong to tribe Aveneae. Rice is subfamily Oryzoideae and belongs to tribe Oryzeae. Corn, or maize, is subfamily Panicoideae and belongs to tribe Andropogoneae. What all of these cereal grains have in common is that they are used in making beer.

There are at least three different conditions that can affect people who experience an adverse reaction to beer: an allergic reaction; an autoimmune response; or a response that is non-allergic and non-autoimmune, but may have similar symptoms.

An allergic reaction to barley may or may not be associated with gluten. People can be allergic to barley just as they can to wheat, horses, eggs, and peanut butter. There are two dozen different allergens in wheat, for instance, of which gluten is only one. A typical allergic reaction, whether triggered by gluten or not, can manifest as watery eyes, runny nose, and respiratory problems.

However, people can also be specifically sensitive to gluten due to an autoimmune disorder. One example is "gluten rash," which is a type of dermatitis caused by an autoimmune response. Celiac disease is a very serious autoimmune disease that damages the mucosal layer of the small intestine, thereby compromising its ability to absorb nutrients. The typical initial symptoms include gastro-intestinal "distress," including bloating, diarrhea, and abdominal pain. Sufferers of celiac disease are also at increased risk of cancer.

Gluten sensitivity (more properly called non-celiac gluten sensitivity) is a condition where sufferers are intolerant of dietary gluten via a mechanism that is not autoimmune in nature. Gluten sensitivity will cause very similar initial symptoms to celiac disease, but without the damage to the small intestine. Sufferers of celiac disease have an immunoreactive response to gliadin, which is a prolamin (a storage protein, as will be explained below) in wheat that combines with other wheat proteins (such as glutenin) to form gluten, which gives bread dough both elasticity and structure.

## Gluten in Beer

There are 20 different amino acids that serve as building blocks for construction of larger structures in all living organisms. A polypeptide is the result of several amino acids joined together via peptide bonds. Proteins are constructed of one or more long-chain polypeptides and have different properties and functionalities based on their three-dimensional physical structure. In plant seeds, several different classes of protein serve as amino acid reservoirs (what brewers would call FAN) for the embryo. These are called *storage proteins*, and include albumins, globulins, glutelins, and prolamins. Storage proteins are characterized based on their solubility in the laboratory: albumins are soluble in water, globulins are soluble in dilute salt solutions, glutelins are soluble in dilute acid or base solutions, and prolamins are soluble in alcohol solutions. The storage proteins important for cereals are prolamins, globulins, and glutelins. (Actually, glutelins are a type of globulin, but they are treated separately due to differences in solubility.)

Cereal chemists organize barley proteins into two main groups, storage and non-storage, based on their location and function within the kernel. The major storage proteins in barley are the *hordeins* (a class of prolamin) and glutelins. Non-storage proteins are the structural proteins and metabolic proteins (i.e., enzymes). During malting, the endosperm protein matrix is hydrolyzed into polypeptides, oligopeptides, and free amino acids. These endosperm proteins are primarily a mixture of hordeins and, to a lesser extent, glutelins. It is the breakdown of this endosperm matrix during germination that provides the vast majority of FAN to the wort. The non-storage proteins are the source of the enzymes that are present in barley before malting, such as beta-amylase; they are also a source of (non-storage) albumins, such as protein Z, which along with hordein is a primary foam-forming component in beer.

Prolamins are also the major storage protein for wheat and rye. In wheat, prolamins include the *gliadins*, which are structurally very similar to hordeins. The equivalent prolamin in rye is *secalin*. Gliadin is the most studied class of cereal prolamin, and a major component of wheat gluten.

The structural similarity between prolamins of the Triticeae is the reason that, although barley doesn't technically contain gliadin, beer can still be a problem for people with gluten sensitivity. Celiac disease sufferers can be sensitive to any of these prolamins, but gliadin and hordein account for 90% of immunoreactive response in T cell testing. About 10% of celiacs exhibit an immunoreactive response to oats and corn (maize). Oats contain much less prolamin (called *avenin*) than wheat, barley or rye, and this may be why many people with celiac disease do not react to it; or it may be that some celiacs are less sensitive to prolamins in general and only react to particular types. There are in fact several hundred polypeptides within each prolamin group that are immunoreactive. To frame this discussion another way, it is like saying that all fish, mammals, and reptiles are dangerous man-eaters because they have large teeth, when in fact some are and some aren't, for various reasons.

## Prolamins and Beer Haze

Beer haze, including chill haze, is a combination of haze-active proteins and haze-active polyphenols that come together via hydrogen bonding to create large visible molecules. Hydrogen bonding is strongest at colder temperatures. At warm temperatures, the component molecules are vibrating too much for the weak hydrogen bonds to hold the complex together, and that's why chill haze disappears as the beer warms. With time, these complexes can oxidize and polymerize into permanent haze, but it all starts with haze-active proteins and haze-active polyphenols.

Why do we say, "haze-active"? What does that mean? Basically, it refers to the types and size of these proteins and polyphenols—at a certain size these entities cause haze. Haze-active proteins are of the same size class as foam-active proteins, which typically includes hordeins. Prolamins are so called because they contain an unusually high proportion of the amino acids proline and glutamine. It appears that the hydrogen bonding between haze-active polyphenols and proteins occurs at the proline sites on the protein.[1] The point is that the basis for haze, at least in part but perhaps a majority, is the hydrogen bonding at proline sites in hordeins. These haze-active proteins can be broken down by protease enzymes (e.g., the enzyme papain, used in papaya-based meat tenderizer). This creates smaller peptide chains that are not haze-active. Unfortunately, those protease enzymes will also break down the foam-active proteins, which is obviously a problem.

## Enzyme Clarifiers

There are two main categories of enzyme that degrade barley endosperm proteins during malting. The first are endoproteases and endopeptidases, which act to break up the protein molecules from inside the protein structure. There are at least 40 such enzymes involved in this stage.[2] The second group are exoenzymes, such as carboxypeptidase, that produce individual amino acids from the carboxyl ends of the peptide chains (i.e., only at the outside ends of the protein structure). To clarify, peptides are short chains of amino acids joined together, whereas polypeptides are longer chains of amino acids (usually >50) or protein segments. Peptides, polypeptides, and proteins can be affected by various protease-class enzymes.

Proline-specific endoproteases, such as Brewers Clarex® (from DSM) and Clarity Ferm (White Labs), act by specifically cleaving the haze-active proteins at the proline segments in the chain, reducing them to non-haze-active sizes (i.e., cleaving polypeptides into peptides). The proline sites of the peptides will still bond with haze-active polyphenols, but the size of the complex can't grow to form a haze. Thus, most beer haze should be prevented by using these enzymes.

The proline-specific endoprotease in Brewers Clarex and Clarity Ferm has the bonus benefit of breaking up the hordein proteins that form barley gluten (it has a similar effect on gliadin and secalin). Industry studies have shown that beer treated with this enzyme measured less than 20 ppm gluten based on the R5 Mendez Competitive enzyme-linked immunosorbent assay (ELISA). Further analysis by mass spectrometry of the concomitant proteolytic residues

---

[1]   Siebert and Lynn (1997).
[2]   Bamforth (2006, 54).

also indicated that the residues would not be immunoreactive.[3] However, even though the current benchmark for considering a food to be gluten-free is <20 ppm gluten, there is enough variation among people with gluten-related allergies that marketing a beer as "gluten-free" may be impossible from a legal liability standpoint. For example, in the case of one celiac patient, the consumption of 1 mg of gluten per day from a communion wafer was sufficient to prevent mucosal recovery of the small intestine (i.e., it caused chronic inflammation). Avoiding the wafer allowed recovery within six months.

Confounding the issue is the fact that the current antibody tests (i.e., ELISA) only target a handful of the hundreds of gluten proteins that can cause immunoreactions. Furthermore, these tests are calibrated on commercial wheat gliadin, which is not representative of barley hordein that has been malted, mashed, and boiled. To quote a paragraph from the conclusion of Tanner *et al.* (2014):

> *Definitive evidence of the safety of treated beer for celiacs ideally requires a double-blind crossover dietary challenge. In this experiment, the effect on circulating T-cells and mucosal appearance of a large number of celiacs, including sensitive subjects, who have been challenged with either PEP-treated beer or untreated beer followed by a crossover to the other treatment regime, would provide convincing evidence for the efficacy of A.* niger *PEP on eliminating gluten peptides for the whole celiac population. In order to achieve this, subjects would have to drink 10 L of an average beer (at 100 ppm) per day to consume sufficient hordein (1 g) for a useful short term challenge. Sourcing volunteers for such an experiment may not be a problem, but ethics approval would be unlikely. (p.46)*

In other words, achieving FDA "Gluten-Free" status for barley, wheat, and rye based beers treated with proline specific endoproteases is probably not going to happen. Nonclinical experimental approval would require isolation and identification of hundreds and hundreds of prolamin polypeptides that could cause an immunoreactive response, and also require characterization of those polypeptides and their post-treatment residues by concomitant mass spectrometry, high performance liquid chromatography (HPLC), and other tests. Currently, beers treated by proline-specific endoproteases are being marketed as "gluten-reduced," and that is probably the best we can hope for in the foreseeable future.

### Author's Note

Portions of this work were originally published as a feature article, For Geeks Only, Gluten-free Beer, in *Zymurgy*, Volume 39, No. 2. March/April 2016.

---

[3]   Tanner *et al.* (2014).

# Bibliography

Algazzali, Victor, and Thomas Shellhammer. 2016. "Bitterness Intensity of Oxidized Hop Acids: Humulinones and Hulupones." *Journal of the American Society of Brewing Chemists* 74(1):36–43. doi:10.1094/ASBCJ-2016-1130-01.

Arendt, Elke. 2015. "The influence of lactic acid bacteria in malting and brewing." YouTube video, http://youtu.be/9a-ZpF2LDm8. Slide show PDF, http://belgianbrewingconference.org/2015/11 _Arendt.pdf. Presentation at the Belgian Brewing Conference, KU Leuven, Belgium, September 2015.

Bamforth, C.W. and W.J. Simpson. 1995. "Ionic equilibria in brewing." *Brewer's Guardian*, vol. 124, December, p.18–24.

Bamforth, Charles W. 1999. "Beer Haze." *Journal of the American Society of Brewing Chemists* 57(3):81–90. doi:10.1094/ASBCJ-57-0081.

———. 2006. *Scientific Principles of Malting and Brewing*. St. Paul, MN: American Society of Brewing Chemists.

Baril, Randy. 2015. *Hosting Cask Ale Events: Practical Advice for Preparing and Serving Real Ale for the Publican, Homebrewer, and Cask Ale Enthusiast*. Self-published by author.

Barth, Roger, and Rameez Zaman. 2015. "Influence of Strike Water Alkalinity and Hardness on Mash pH." *Journal of the American Society of Brewing Chemists* 73(3):240–2. doi:10.1094 /ASBCJ-2015-0621-01.

Beechum, Drew, and Denny Conn. 2014. *Experimental Homebrewing: Mad Science in the Pursuit of Great Beer*. Minneapolis: Voyageur Press.

———. 2016. *Homebrew All-Stars: Top Homebrewers Share Their Best Techniques and Recipes*. Minneapolis: Voyageur Press.

Boulton, C., and W. Box. 2003. "Formation and disappearance of diacetyl during lager fermentation." In *Brewing Yeast Fermentation Performance*. 2nd ed. K. Smart, editor. Oxford: Blackwell Science. 183–95.

Boulton, Chris, and David Quain. 2001. *Brewing Yeast and Fermentation*. Oxford: Blackwell Science.

Briggs, D.E., J.S. Hough, R. Stevens, T.W. Young. 1981. *Malt and Sweet Wort*. 2nd ed. Vol. 1 of *Malting and Brewing Science*. New York: Kluwer Academic / Plenum Publishers.

Brungard, Martin. 2015. *Bru'n Water*. http://sites.google.com/site/brunwater.

Casarett, Louis J., John Doull, Curtis D. Klaassen, and Mary O. Amdur. 1980. *Casarett and Doull's Toxicology: The Basic Science of Poisons*. 2nd ed. New York: MacMillan Publishing.

Colby, Chris. 2016. *Home Brew Recipe Bible*. Salem, MA: Page Street Publishing.

Crumplen, R.M., ed. 1997. *Laboratory Methods for Craft Brewers*. St. Paul, MN: American Society of Brewing Chemists.

Curtis, David. 2014. "Putting some numbers on first wort and mash hop additions." Presentation at the AHA National Homebrewers Conference, Grand Rapids, MI, June 12–14, 2014.

Daniels, Ray. 1995. "Beer Color Demystified—Part III: Controlling and Predicting Beer Color." *Brewing Techniques*, vol. 3, no. 6, 56–63.

———. 1996. *Designing Great Beers*. Boulder: Brewer's Publications.

De Rouck, Gert, Barbara Jaskula, Brecht De Causmaecker, Sofie Malfliet, Filip Van Opstaele, Jessika De Clippeleer, Jos De Brabanter, Luc De Cooman, and Guido Aerts. 2013. "The Influence of Very Thick and Fast Mashing Conditions on Wort Composition." *Journal of the American Society of Brewing Chemists* 71(1):1–14. doi:10.1094/ASBCJ-2013-0113-01.

Evans, D. Evan, Helen Collins, Jason Eglinton, and Annika Wilhelmson. 2005. "Assessing the Impact of the Level of Diastatic Power Enzymes and Their Thermostability on the Hydrolysis of Starch During Wort Production to Predict Malt Fermentability." *Journal of the American Society of Brewing Chemists* 63(4):185–98.

Evans, D. Evan, Mark Goldsmith, Robert Dambergs, and Ralph Nischwitz. 2011. "A Comprehensive Revaluation of Small-Scale Congress Mash Protocol Parameters for Determining Extract and Fermentability." *Journal of the American Society of Brewing Chemists* 69(1):13–27. doi:10.1094/ASBCJ-2011-0111-01.

Fritsch, A., and T. Shellhammer. 2007. "Alpha Acids do not Contribute Bitterness to Lager Beer." *Journal of the American Society of Brewing Chemists* 65(1):26–28. doi:10.1094/ASBCJ-2007-0111-03.

Hahn, Christina, Scott Lafontaine, and Thomas Shellhammer. 2016. "A holistic examination of beer bitterness." Abstract 65. Technical Session 19: Beer Bitterness. Paper presented at the World Brewing Congress, Denver, August 2016.

Hansen R. and J. Guerts. 2015. "Specialty malt acidity." Proceedings of the MBAA National Conference, Jacksonville, FL, October 8–10, 2015.

Hertrich, Joseph D. 2013. "Topics in Brewing: Brewing Adjuncts." *Technical Quarterly of the Master Brewers Association of Americas.* 50(2):72–81. doi:10.1094/TQ-50-2-0425-01.

Hieronymus, Stan. 2005. *Brew Like a Monk: Trappist, Abbey, and Strong Belgian Ales and How to Brew Them.* Boulder: Brewers Publications.

———. 2010. *Brewing with Wheat: The 'Wit' & 'Weizen' of World Wheat Beer Styles.* Boulder: Brewers Publications.

———. 2012. *For the Love of Hops: The Practical Guide to Aroma, Bitterness, and the Culture of Hops.* Boulder: Brewers Publications.

Humbard, Matthew. *A Ph.D in Beer* (blog). http://phdinbeer.com/.

Ishibashi, Y., Y. Terano, N. Fukui, N. Honbou, T. Kakui, S. Kawasaki, and K. Nakatani. 1996. "Development of a New Method for Determining Beer Foam and Haze Proteins by Using the Immunochemical Method ELISA." *Journal of the American Society of Brewing Chemists* 54(3):177–82. doi:10.1094/ASBCJ-54-0177.

Jones, Berne L. 2005. "Endoproteases of barley and malt." *Journal of Cereal Science* 42(2):139–56. doi:10.1016/j.jcs.2005.03.007.

Jones, Berne L., and Allen D. Budde. 2005. "How various malt endoproteinase classes affect wort soluble protein levels." *Journal of Cereal Science* 41:95–106. doi:10.1016/j.jcs.2004.09.007.

Klimovitz, Ray, and Karl Ockert, eds. 2014. *Beer Packaging.* 2nd ed. St. Paul: Master Brewers Association of Americas.

Krogerus, Kristoffer, and Brian R. Gibson. 2013. "125th Anniversary Review: Diacetyl and its control during brewery fermentation." *Journal of the Institute of Brewing* 119(3):86–97.

Kunze, Wolfgang. 2014. *Technology Brewing and Malting.* Edited by Olaf Hendel. Translated by Sue Pratt. 5th rev. ed. Berlin: VLB Berlin.

Lee, W.J. 1990. "Phytic Acid Content and Phytase Activity of Barley Malt." *Journal of the American Society of Brewing Chemists* 48(2):0062.

MacGregor, A.W. and C. Lenoir. 1987. "Studies on Alpha Glucosidase in Barley and Malt." *Journal of the Institute of Brewing* 93:334–37.

Malowicki, Mark G., and Thomas H. Shellhammer. 2005. "Isomerization and Degradation Kinetics of Hop (*Humulus lupulus*) Acids in a Model Wort-Boiling System." *Journal of Agricultural and Food Chemistry* 53:4434–39.

Maskell, Dawn L. 2016. "Brewing Fundamentals, Part 2: Fundamentals of Yeast Nutrition." *Technical Quarterly of the Master Brewers Association of Americas* 53(1):10–16.

McMurrough, L., G.P. Hennigan, and K. Cleary. 1985. "Interaction of proteases and polyphenols in worts, beers and model systems." *Journal of the Institute of Brewing* 91(2):93–100. doi:10.1002/j.2050-0416.1985.tb04312.x.

Mosher, Randy. 1994. *The Brewer's Companion*. Seattle: Alephenalia Publications.

———. 2004. *Radical Brewing: Tales and World-Altering Meditations in a Glass*. Boulder: Brewers Publications.

———. 2009. *Tasting Beer: An Insider's Guide to the World's Greatest Drink*. North Adams, MA: Storey Publishing.

Muller, Robert. 1995. "Factors Influencing the Stability of Barley Malt beta-Glucanase During Mashing." *Journal of the American Society of Brewing Chemists* 53(3):136–40.

Owen, Charles A. 1981. *Copper Deficiency and Toxicity: Acquired and Inherited, in Plants, Animals, and Man*. Park Ridge, NJ: Noyes Publications.

Palmer, John, and Colin Kaminski. 2013. *Water: A Comprehensive Guide for Brewers*. Boulder: Brewers Publications.

Palmer, John. 2016. "A Study of Mash pH and its effects on Yield and Fermentability." Presentation. Proceedings of the World Brewing Congress, Denver, August 13–17, 2016.

Papazian, Charlie. 1984. *The Complete Joy of Home Brewing*. New York: Avon Books.

Pattinson, Ronald. 2014. *The Home Brewer's Guide to Vintage Beer: Rediscovered Recipes for Classic Brews Dating from 1800 to 1965*. Beverly, MA: Quarry Books.

[Patty, Frank A.] 1981. *Toxicology*. Vol. 2A of *Patty's Industrial Hygiene and Toxicology*. 3rd rev. ed. New York: John Wiley and Sons.

Preis, F., and W. Mitter. 1995. "The re-discovery of first wort hopping." *Brauwelt International* 13:308–15.

Sammartino, M. 2015. "Fermentation and Flavor: A perspective on Sources and Influence." Proceedings of the MBAA National Conference, Jacksonville, FL, October 8–10, 2015.

Siebert, Karl J. and P.Y. Lynn. 1997. "Mechanisms of Beer Colloidal Stabilization." *Journal of the American Society of Brewing Chemists* 55(2):73–78.

Smythe, John E., and Charles W. Bamforth. 2000. "Shortcomings in Standard Instrumental Methods for Assessing Beer Color." *Journal of the American Society of Brewing Chemists* 58(4):165–66. doi:10.1094/ASBCJ-58-01650.

Speers, Alex. "Brewing Fundamentals, Part 3: Yeast Settling – Flocculation." *Technical Quarterly of the Master Brewers Association of Americas* 53(1):17–22.

Steele, Mitch. 2012. *IPA: Brewing Techniques, Recipes and the Evolution of India Pale Ale*. Boulder: Brewers Publications.

Stenholm, Katharina, and Silja Home. 1999. "A New Approach to Limit Dextrinase and its Role in Mashing." *Journal of the Institute of Brewing* 105(4):205–10.

Stewart, G.G., and I. Russell. 1998. *Brewing Science and Technology, Series III, Brewers Yeast*. London: Institute of Brewing.

Stewart, Graham G. 2016. "Brewing Fundamentals, Part 1: Yeast – An Introduction to Fermentation." *Technical Quarterly of the Master Brewers Association of Americas* 53(1):3–9. doi:10.1094/TQ-53-1-0302-02.

Strong, Gordon. 2015. *Modern Homebrew Recipes: Exploring Styles and Contemporary Techniques*. Boulder: Brewers Publications.

Sykes, Walter J., and Arthur R. Ling. 1907. *The Principles and Practice of Brewing*. 3rd ed. London: Charles Griffin.

Tanner, Gregory J., Michelle L. Colgrave, Crispin A. Howitt. 2014. "Gluten, Celiac Disease, and Gluten Intolerance and the Impact of Gluten Minimization Treatments with Prolylendopeptidase on the Measurement of Gluten in Beer." *Journal of the American Society of Brewing Chemists* 72(1):36–50.

Taylor, David G. 1990. "The Importance of pH Control during Brewing." *Technical Quarterly of the Master Brewers Association of Americas*. 27(4):131–6.

Tinseth, Glenn. 1995. "Glenn's Hop Utilization Numbers." *Real Beer*. Accessed September 6, 2016. http://www.realbeer.com/hops/research.html.

Troester, Kai. 2009. "The effect of brewing water and grist composition on the pH of the mash." *Braukaiser.com*, PDF document. October 31, 2009. http://braukaiser.com/documents/effect_of_water_and_grist_on_mash_pH.pdf.

White, Chris, and Jamil Zainasheff. 2010. *Yeast: The Practical Guide to Beer Fermentation.* Boulder: Brewer Publications.

Zainasheff, Jamil, and John Palmer. 2007. *Brewing Classic Styles: 80 Winning Recipes Anyone Can Brew.* Boulder: Brewers Publications.

# Index

Entries in **boldface** refer to tables, photos, and illustrations.